David Huntley   £75.00
Plymouth

August 1994

Mixing and Transport
in the Environment

Mixing and Transport in the Environment

A MEMORIAL VOLUME FOR CATHERINE M. ALLEN (1954–1991)

Edited by
KEITH J. BEVEN
Lancaster University, UK

PHILIP C. CHATWIN
The University of Sheffield, UK

JOHN H. MILLBANK
Salford University, UK

JOHN WILEY & SONS

Chichester · New York · Brisbane · Toronto · Singapore

Other Wiley Editorial Offices

John Wiley & Sons, Inc., 605 Third Avenue,
New York, NY 10158-0012, USA

Jacaranda Wiley Ltd, 33 Park Road, Milton,
Queensland 4064, Australia

John Wiley & Sons (Canada) Ltd, 22 Worcester Road,
Rexdale, Ontario M9W 1L1, Canada

John Wiley & Sons (SEA) Pte ltd, 37 Jalan Pemimpin #05-04,
Block B, Union Industrial Building, Singapore 2057

Library of Congress Cataloging-in-Publication Data

Mixing and transport in the environment: a memorial volume for
   Catherine M. Allen (1954–1991)/edited by Keith J. Beven, Philip C.
   Chatwin, and John H. Millbank.
       p.   cm.
   Includes bibliographical references and index.
   ISBN 0-471-94142-5
   1. Ocean circulation. 2. Lakes — Circulation. 3. Oceanic mixing.
   I. Beven, K.J. II. Chatwin, Philip C. III. Millbank, John H.
   IV. Allen, Catherine M., 1954–1991.
   GC231.M59   1994
   627′.042—dc20   93-8777
                                                                 CIP

British Library Cataloguing in Publication Data

A catalogue record for this book is available from the British Library

ISBN 0-471-94142-5

Typeset in 10/12pt Times by MHL Typesetting Ltd, Coventry
Printed and bound in Great Britain by Bookcraft (Bath) Ltd.

Contents

List of Contributors

A.A.I. Al-Hamid
King Saud University, Riyadh, Saudi Arabia

U. Arnold
IKW GmbH, Berlin, Germany

K.J. Beven
Centre for Research on Environmental Systems and Statistics, Institute of Environmental and Biological Sciences, Lancaster University, Lancaster LA1 4YQ, UK

D. Booth
Département d'Océanographie, Université de Québec à Rimouski, 310 Allé des Ursulines, Rimouski, Quebec, Canada G5L 3A1

K. Buckley
Centre for Research on Environmental Systems and Statistics, Institute of Environmental and Biological Sciences, Lancaster Univesity, Lancaster LA1 4YQ, UK

P.A. Carling
Institute of Freshwater Ecology, Ferry House, Far Sawry, Ambleside, Cumbria LA22 0LP, UK

P.C. Chatwin
School of Mathematics and Statistics, University of Sheffield, Sheffield S10 2TN, UK

M.S. Curé
School of Oceanography, Southampton University, Highfield, Southampton SO9 5NR, UK

R.A. Falconer
Department of Civil Engineering, University of Bradford, Bradford BD7 1DP, UK

D. Flatt
Proudman Oceanographic Laboratory, Bidston Observatory, Birkenhead L43 7RA, UK

D.G. George
Institute of Freshwater Ecology, Ferry House, Far Sawry, Ambleside, Cumbria LA22 0LP, UK

M.S. Glaister
Institute of Freshwater Ecology, Ferry House, Far Sawry, Ambleside, Cumbria LA22 0LP, UK

H.M. Green
Centre for Research on Environmental Systems and Statistics, Institute of Environmental and Biological Sciences, Lancaster University, Lancaster LA1 4YQ, UK

Li Guiyi
Southern Science Ltd, Premium House, Worthing, W. Sussex BN11 2EN, UK

S.E. Heslop
Department for Continuing Education, Bristol University, Bristol BS8 1HR, UK

M.J. Holland
Centre for Research on Environmental Systems and Statistics, Institute of Environmental and Biological Sciences, Lancaster University, Lancaster LA1 4YQ, UK

J. Höttges
Hydro-Ingenieure GmbH, Düsseldorf, Germany

C.F. Jago
University of Wales Bangor, School of Ocean Sciences, Menai Bridge, Gwynedd LL59 5RH, UK

S. Jones
University of Wales Bangor, School of Ocean Sciences, Menai Bridge, Gwynedd LL59 5RH, UK

M.B. Jordan
Plymouth Marine Laboratory, Prospect Place, The Hoe, Plymouth PL1 3DH, UK

A. Kelsey
Centre for Research on Environmental Systems and Statistics, Institute for Environmental and Biological Sciences, Lancaster University, Lancaster LA1 4YQ, UK

D.W. Knight
School of Civil Engineering, The University of Birmingham, Birmingham B15 2TT, UK

M.F. Lavin
Departemento de Oceanografia Fisica, Ensenada, Baja California, Mexico

P.F. Linden
Department of Applied Mathematics and Theoretical Physics, University of Cambridge, Silver Street, Cambridge CB3 9EW, UK

H.G. Orr
Institute of Freshwater Ecology, Ferry House, Far Sawry, Ambleside, Cumbria LA22 0LP, UK

D. Prandle
Proudman Oceanographic Laboratory, Bidston Observatory, Birkenhead L43 7RA, UK

D. Pugh
Institute of Oceanographic Sciences, Deacon Laboratory, Brook Road, Godalming, Surrey GU8 5UB, UK

G. Rouvé
Institute for Hydraulic Engineering and Water Resources, RWTH, Aachen, Germany

J.E. Simpson
Department of Applied Mathematics and Theoretical Physics, University of Cambridge, Silver Street, Cambridge CB3 9EW, UK

J.H. Simpson
University of Wales Bangor, School of Ocean Sciences, Menai Bridge, Gwynedd LL59 5RH, UK

L.A. Smith
Mathematics Institute, University of Oxford, Oxford OX1 3LB, UK

R. Smith
Mathematical Sciences, Loughborough University, Loughborough LE11 3TU, UK

A.J. Souza
University of Wales Bangor, School of Ocean Sciences, Menai Bridge, Gwynedd LL59 5RH, UK

P.J. Sullivan
Department of Applied Mathematics, University of Western Ontario, London, Ontario, Canada N6A 5B9

S.A. Thorpe
School of Oceanography, Southampton Univesity, Highfield, Southampton SO9 5NR, UK

R.J. Uncles
Plymouth Marine Laboratory, Prospect Place, The Hoe, Plymouth PL1 3DH, UK

G.C. van Dam
Rijkwaterstaat, Tidal Waters Division, PO Box 20907, 2500 EX The Hague, The Netherlands

J.M. Vassie
Proudman Oceanographic Laboratory, Bidston Observatory, Birkenhead L43 7RA, UK

S.G. Wallis
Department of Civil and Offshore Engineering, Heriot-Watt Univesity, Riccarton, Edinburgh, UK

J.R. West
School of Civil Engineering, The University of Birmingham, Birmingham B15 2TT, UK

L.R. Wyatt
School of Mathematics and Statistics, University of Sheffield, Sheffield S10 2TN, UK

P.C. Young
Centre for Research on Environmental Systems and Statistics, Institute of Environmental and Biological Sciences, Lancaster University, Lancaster LA1 4YQ, UK

K.W.H. Yuen
Department of Civil Engineering, University of Birmingham, Birmingham B15 2TT, UK

43

Preface

CATHERINE M. ALLEN (1954–1991)

It is perhaps a little unusual for a scientist still at the beginning of her career, with relatively few publications to her name and still at the stage of showing great potential, to be the subject of a memorial volume. Cath Allen was, however, a very special person. Her research interests were wide and interdisciplinary in nature and, as a result, her colleagues, associates and collaborators were many in number. Having proposed the idea of a memorial conference and book, we were delighted to find that many scientists from around the world were eager to participate in such a tribute in recognition of the benefit they had gained from knowing and working with Cath. It also provided a focus for some of her colleagues to complete some of the projects that she had initiated and the diversity of her work is aptly indicated by the number of papers within this book that bear her name as a co-author. The memorial conference itself, entitled 'Physical mechanisms of mixing and transport in the environment' was held at Lancaster University over a three day period in June 1992.

As an undergraduate Cath studied at the University of Liverpool, where she obtained a Bachelor of Science degree in physics and earth sciences with first class honours; in both of her last two years winning the Royal Institution (Iliffe) scholarship. It was during her final undergraduate year that she began to develop a serious interest in the 'science of the sea' and in the mathematics that describe marine processes. After graduation in 1974, she began what was to become her long association with the oceanographers of the Bidston Observatory, now called the Proudman Oceanographic Laboratory. During that summer at Bidston she undertook a short project on the modulation of tides with David Pugh and Ian Vassie (see paper 13 in this book). Cath subsequently returned to Liverpool to take a Master's degree in the Department of Applied Mathematics and Theoretical Physics, where she was taught marine dynamics by the late Norman Heaps, who became a great source of inspiration to her. Her research project in that year was on surface wave refraction under the supervision of Gordon Crapper. Cath's characteristically strong will was already manifesting itself by this stage. One of the editors (PCC) remembers being reprimanded by Cath for not knowing enough about the M_2 tide after asking a question during the discussion following a departmental seminar!

Cath clearly identified with this type of research because she then went on to pursue a PhD at the University College of North Wales Bangor on the structure and variability of shelf sea fronts under the supervision of John Simpson, now Professor of Physical Oceanography at Bangor. Her project involved not only theoretical work, but also ship surveys in the Irish Sea, which revealed complex patterns of variability in the vicinity of fronts between well mixed and stratified bodies of water. Early on in this work she had considerable problems trying to collate and disseminate the enormous volumes of data involved, a task that was not helped by the difficulties in formatting it for computer input on what would now be considered as distinctly 'user unfriendly' machines. At this point Cath almost abandoned her research in despair, but with her own tenacity and encouragement from her new husband (JHM) she persevered, made sense of it all, and produced a commendable doctoral thesis.

Her PhD research may have given Cath a taste for impossible problems because she returned to Liverpool in 1978 to work as a research fellow with Philip Chatwin, now Professor of Applied Mathematics at Sheffield, on an NERC funded project: 'The use of numerical simulation techniques in the analysis of data on the diffusion of chemical and biological species in estuaries'; an even more complex environment. This work led to the much cited papers in the *Proceedings of the Royal Society* (1982) and *Annual Reivew of Fluid Mechanics* (1985).

In 1981 Philip Chatwin was offered a further three year research grant funded by the Ministry of Defence on 'The statistical properties of distributions of concentration in free turbulent shear flows'. This was offered to Cath who, despite some reservations about such basic and environmentally important work being funded by the MoD, accepted. It was during this period in her research career that she also produced, in December 1981, her first child, Matthew, and, as a result, gained some direct experience of working to improve the conditions of employment of female research staff by persuading Liverpool University and the MoD to make an appropriate allowance for maternity leave.

At this time Cath and John were living on the Wirral peninsular in a rural setting surrounded by VW Beetles, a memory invoked by many of the scientists who enjoyed the hospitality of their homely cottage. Cath and an old Beetle always seemed to go together.

In 1985 Cath came to Lancaster to teach fluid dynamics within the Department of Environmental Sciences and to expand and continue her research into mixing processes in estuaries and coastal environments. She was soon involved in a number of projects in the department, most notably with Mike Kelly and others in work on the river Esk estuary on the Cumbrian coast, where the significance of sediment transport on the movement of radionuclides away from the Sellafield nuclear reprocessing site gave her another challenging problem to tackle.

Cath managed to balance her substantial, largely self-imposed, workload with a commitment to family life and, in October 1988, gave birth to Jessica. Her joy at the new arrival gave her a great sense of fulfilment and satisfaction that she had successfully combined the roles of caring mother and dedicated scientist.

Within a few months, however, Cath was diagnosed as having cancer. Throughout her illness she continued to develop her work on turbulence and sediment transport, with determination, commitment and enthusiasm. Collaborations with Judy Wolf and David Prandle of the Proudman Laboratory, Sally Heslop (now at Bristol University), and Paul Carling of the Institute of Freshwater Ecology (IFE) at Windermere, looked at environments ranging from the River Severn to the North Sea, the Dee Estuary and Liverpool Bay.

Cath was successful in obtaining an SERC grant with Keith Beven and Ian Guymer (now at Sheffield University) to take advantage of the experimental facilities of the SERC Flood Channel at Hydraulics Research Wallingford to study dispersion processes. She also had a long standing collaboration with John Hamilton-Taylor (Lancaster) and Glen George (IFE, Windermere) attempting to understand the physical, chemical and biological controls on the workings of Esthwaite in the English Lake District. She was keen to explore the use of remote sensing techniques in understanding the fluid dynamics of the natural environment and spent a lot of time co-ordinating NERC aircraft flights sitting in a small boat on Esthwaite or the River Esk, waiting for the right conditions to enable quality data to be recorded.

Cath's research student, Carolyn Simon, completed her PhD at Lancaster and then went on to become one of the very select band of woman scientists at the British Antarctic Survey. There were many other colleagues, both here and abroad, with whom Cath enjoyed associations of long standing, including Ron Smith of Cambridge, now Professor of Applied Mathematics at Loughborough; Keith Dyer and Reg Uncles at Polytechnic South West, Plymouth; and also Brenda Topliss (Department of Fisheries and Oceans, Dartmouth, Nova Scotia) and David Booth (Université de Québec à Rimouski), both of whom were research students during Cath's time at Bangor.

Cath thrived on these collborations, and on the work in the field or at sea with colleagues and students. She also enjoyed the gatherings and discussions with friends at meetings and conferences, and many colleagues benefited from the open welcome that she and John gave to visitors to Lancaster. She was elected by her fellow scientists as a council member of the Challenger Society for Marine Science and was an Associate Fellow of the Institute of Mathematics and its Applications.

Cath gave to her research great mathematical facility, a very bright mind that could get very quickly to the heart of a problem, and a temperament that would not stand nonsense easily. Add to this her love of the outdoors and early recognition of the value of fieldwork and she had what is a very rare combination of talents, ideally suited to addressing the problems of understanding, modelling and predicting the complex processes that occur in the environment, those impossible problems referred to earlier. The basis of much of Cath's work was a desire to try and resolve the tension between the complexity we see in the natural world and the need to use the limited mathematical tools and computers we have available to make predictions in a way that reflects that complexity. In her work with random particle tracking models she had found a pathway towards the resolution of that tension, one that the hydrology and fluid dynamics research group at Lancaster is continuing to follow.

However, Cath's contribution at Lancaster was not limited to her research. It was typical of her unselfish professionalism that she devoted a great deal of time to teaching and administration. She carried by choice a high teaching load and was a supportive and deeply caring teacher and, towards the end, she insisted on completing her teaching duties despite the tremendous toll it took on her physically. In fact, she continued to teach until the week before she died, even though she had to be driven to the University and it could take many minutes for her to walk the tens of yards to her office.

Cath also worked as a tutor for the Open University, for the Society of Women in Science and Engineering, and on the Equal Opportunities Committee in the University. She was certain that the world would only become a better place if women became more assertive, and she went about proving her right to believe that by succeeding in the male-dominated world of UK science. She did so with a warm smile, the commitment to her family, a concern for her students, and an amused tolerance of her male colleagues.

For oceanography and the environmental sciences, the loss of Cath is great indeed, and many of us will reflect on the many more problems we could have tackled with her help, stimulating discussion and enthusiasm. We should, however, remind ourselves of the magnitude of the legacy she has left us, not only in her science, but perhaps even more importantly in the inspiring way she coped with all that beset her during the two years of her illness. It proved to be another impossible task, but she tackled it with just as much care, preparation, perseverance, and satisfaction in a job well done as in her research. She set an example in both these spheres that affected many people deeply and will live on in their memories.

Keith Beven   March 1993
Philip Chatwin
John Millbank

Catherine M. Allen: Publications and Theses

Catherine M. Allen (1954−1991)

Allen, C.M. (1975) Surface wave refraction. *Unpublished MSc Thesis*. University of Liverpool.

Simpson, J.H., Allen, C.M. and Morris, N.C.G. (1978) Fronts on the continental shelf. *J. Geophys. Res.* **83**(C9), 4607−4614.

Allen, C.M. (1980) The structure and variability of shelf sea fronts. *Unpublished PhD Thesis*. University of Wales.

Allen, C.M., Simpson, J.H. and Carson, R.M. (1980) The structure and variability of shelf sea fronts as observed by an undulating CTD system. *Oceanol. Acta* **3**, 59−68.

Allen, C.M. (1982) Numerical simulation of contaminant dispersion in estuary flows. *Proc. Roy. Soc. London* **A381**, 179−194.

Allen, C.M. (1984) The statistical properties of distributions of concentration in free turbulent shear flows. *Report for MoD (Admiralty Research Establishment) under contract AT2067/046*, 112pp.

Chatwin, P.C. and Allen, C.M. (1985) A note on time averages in turbulence with reference to geophysical applications. *Tellus* **37B**, 46−49.

Chatwin, P.C. and Allen, C.M. (1985) Reply to "Letter to the editor" on "A note on time averages in turbulence". *Tellus* **37B**, 315−316.

Chatwin, P.C. and Allen, C.M. (1985) Mathematical models of dispersion in rivers and estuaries. *Ann. Rev. Fluid Mech.* **17**, 119−149.

Allen, C.M. (1986) Walking randomly through fluids. *Computat. Fluid Dynam.* 2/86.

George, D.G., Allen, C.M. and Smith, D.G. (1987) The remote sensing of phytoplankton chlorophyll in Esthwaite Water, Cumbria. *Proceedings of the NERC 1987 Airborne Campaign Workshop*. NERC, Swindon, pp. 59−74.

Heslop, S.E. and Allen, C.M. (1989) Turbulence and dispersion in larger UK rivers. *Proceedings of the XXIII IAHR Congress, Ottawa*, pp. D75−D82.

Allen, C.M. (1990) Particle tracking models for pollutant dispersion. In: *Computer Modelling in the Environmental Sciences* (Eds D.G. Farmer and M.J. Rycroft). Clarendon Press, Oxford, pp. 65−74.

Guymer, I., Brockie, N.J.W. and Allen, C.M. (1990) Towards random walk models in a large scale laboratory facility (Eds E.E. Adams and G.E. Hecker). *International Conference on Physical Modelling of Transport and Dispersion, MIT, Cambridge, Ma, USA*, Aug. 7−10, 1990, pp. 12B1−12B6.

Guymer, I., Allen, C.M. and Brockie, N.J.W. (1991) Solute mixing from river outfalls during over-bank flood conditions. In: *Environmental Hydraulics* (Eds J.H.W. Lee and Y.K. Cheung). Balkema, Rotterdam, pp. 447−452.

Brockie, N.J.W., Allen, C.M. and Guymer, I. (1991) An initial comparison between 3D random walk models and tracer studies in a large experimental facility. *Proceedings of the XXIV IAHR Congress, Madrid, Spain*, Sept. 9−13, 1991, vC, pp. 537−544.

Allen, C.M. (1993) Dispersion in rivers and estuaries. In: *Concise Encyclopaedia of Environmental Systems* (Ed. P.C. Young). Pergamon Press, Oxford, pp. 146−150.

Allen, C.M. (1993) Lakes: measurement of physical processes. In: *Concise Encylcopaedia of Environmental Systems* (Ed. P.C. Young). Pergamon Press, Oxford, pp. 332−337.

Heslop, S.E. and Allen C.M. (1993) Modelling contaminant dispersion in the River Severn using a random-walk model. *J. Hydraul. Res.,* **31**, 323−331.

George, D.G. and Allen, C.M. (1994) Turbulent mixing in a small thermally stratified lake. In: *Mixing and Transporting the Environment* (Eds K.J. Beven, P.C. Chatwin and J.H. Millbank). Wiley, Chichester (this volume).

Heslop, S.E., Holland, M.J. and Allen, C.M. (1994) Turbulence measurements in the River Severn. In: *Mixing and Transporting the Environment* (Eds K.J. Beven, P.C. Chatwin and J.H. MIllbank). Wiley, Chichester (this volume).

Kelsey, A., Beven, K.J., Allen, C.M. and Carling, P. (1994) Particle tracking models of sediment transport. In: *Mixing and Transporting the Environment* (Eds K.J. Beven, P.C. Chatwin and J.H. Millbank). Wiley, Chichester (this volume).

Part I

TRANSPORT AND DISPERSION IN FRESH WATER SYSTEMS

1 Turbulent Mixing in a Small Thermally Stratified Lake

D.G. GEORGE

Institute of Freshwater Ecology, Ambleside, Cumbria, UK

and

C.M. ALLEN

1.1 INTRODUCTION

Freshwater ecologists have become increasingly aware that variations in the intensity of wind mixing have a major effect on the seasonal succession of plankton (Reynolds, 1980; 1990). In many lakes we can now identify the mixing conditions that lead to the growth of certain algae (Reynolds, 1984; Steinberg and Hartmann, 1988) and predict the consequences of changes in the weather from year to year (George *et al.*, 1990). Seasonal shifts from one plankton assemblage to another are generally brought about by major changes in the stability of the water column. More subtle changes in the composition of the plankton can, however, be produced by localized changes in the intensity of vertical and horizontal mixing.

The physical processes responsible for vertical and horizontal mixing typically cover a broad range of spatial scales (Boyce, 1974; Imboden, 1990). Mechanistically, these processes can be visualized as a field of nested eddies where turbulent energy cascades from the largest eddy to smaller and smaller eddies until viscous scales are reached. In practice, these complex mixing processes are usually represented by a small number of empirically defined eddy diffusion coefficients. These coefficients tell us little about the basic physics, but they provide a convenient method of integrating the processes responsible for vertical and horizontal mixing. The physical processes responsible for vertical mixing have received considerable attention, but rather less is known about the processes that influence the dispersion of particles in the horizontal plane. Simple dimensional arguments suggest that there should be a functional relationship between the intensity of horizontal mixing and the size of a water body (Boyce, 1974). Some tenuous links between mixing rates and basin morphometry have been identified in the Great Lakes of North America (Csanady, 1963), but little is known about the critical scales of motion in small lakes.

This paper summarizes the results of a series of dispersion experiments designed to identify the dominant scales of turbulent motion in a small lake in Cumbria, England. The preliminary experiments were simple and used small patches of drogues tracked by an observer on the shore. Later experiments included some airborne surveys where patches of drogues were photographed at regular intervals and surface temperature variations measured with an airborne radiometer.

Mixing and Transport in the Environment. Edited by K.J. Beven, P.C. Chatwin and J.H. Millbank
© 1994 John Wiley & Sons Ltd

1.2 DESCRIPTION OF SITE

Esthwaite Water is a small, biologically productive lake in the English Lake District (latitude 54° 22′ N, longitude 0° 49′ W). The lake is approximately 2.5 km long and has a mean depth of 6.4 m and a maximum depth of 15 m. The bathymetric map in Figure 1.1 shows that the lake is topographically complex and is divided into three relatively distinct basins. The lake typically remains thermally stratified from the beginning of May until early October. The seasonal thermocline is usually less than 6 m deep and effectively divides the lake into a turbulent epilimnion and a quiescent hypolimnion. Lagrangian measurements of current speed and direction at different depths (George and Heaney, 1978) reveal complex patterns of motion that can broadly be described as distorted Ekman spirals (see Pond and Pickard, 1983). These progressive rotations are qualitatively similar to those described by George (1981) in Windermere, but occupy a much more restricted portion of the water column. Modelling studies (Falconer et al., 1990) confirm that these vertical rotations are governed by Coriolis effects and demonstrate that they are readily suppressed by strong winds.

1.3 METHODS

1.3.1 Drogue experiments

Several workers have used clusters of drogues to measure relative two-dimensional diffusion. Monahan and Monahan (1973) have reviewed the different types of drogues in common use and Sanderson (1987) provides an up to date account of their application in diffusion studies. The drogues used in this study were of a simple construction and were produced at low cost. Each drogue was made from a square of white-painted plywood measuring 60 × 60 cm fixed to an upturned plastic bucket. The plywood float offered little resistance to the wind and the bucket acted as an efficient 'drag' which was not strongly influenced by the orbital motion of surface waves.

In preliminary experiments, triangles of drogues were deployed at a number of locations to compare dispersion rates in different parts of the basin. Each triangle was tracked for an hour or so, and the positions of the drogues plotted on a large-scale map of the lake. Typical plotting intervals ranged from 10 to 15 minutes, depending on the wind speed and the final size of the patch. The position of each drogue in the patch was determined from a fixed point on shore using a Geodimeter Model 216 electronic distance meter mounted on a Zeiss 20A theodolite. The surveyor on shore was in radio contact with the tracking crew on the lake and noted the number of each drogue when a reflective target was held near the float. Tests with fixed targets suggested that the position of each drogue could be determined with an absolute accuracy of 1−2 m. As there was usually a one to two minute delay between individual measurements, a simple linear interpolation scheme was used to estimate the 'true' position of the second and third drogues in each patch. The boat used for tracking was fitted with an electrically powered outboard which did not disturb the relative disposition of the drogues. Nevertheless, preliminary trials showed that it was wise to deploy the drogues at least 10 m apart to minimize the risk of convergence and collision.

The triangle experiments were later supplemented with more detailed measurements on larger patches of dispersing drogues. As the synoptic position of individual drogues in a large patch could not be recorded by ground measurements, these patches were photographed from the air at regular intervals. A patch of 12 drogues was usually deployed in a rough circle

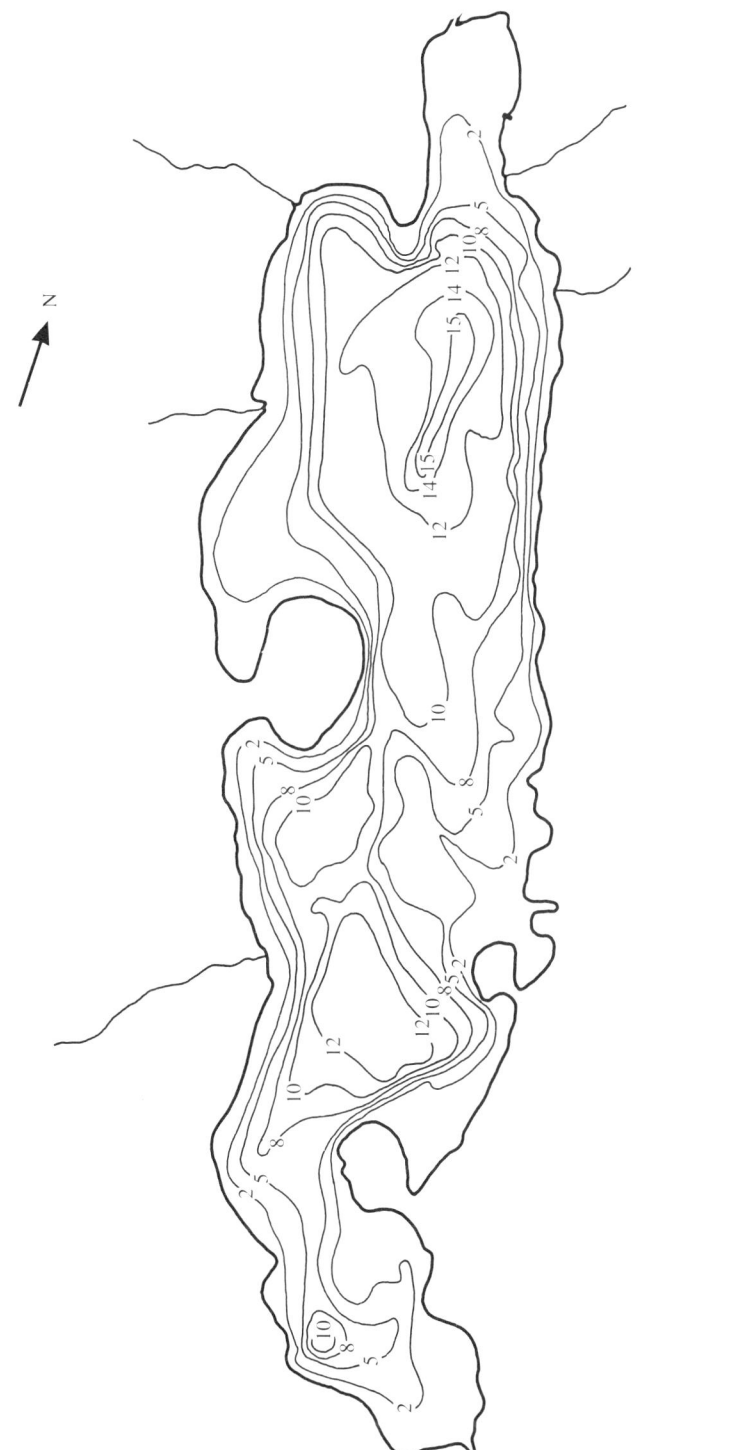

Figure 1.1 Bathymetric map of Esthwaite Water showing the position of the three deep basins

and the dimensions of the patch measured at successive intervals of time. The area of each patch was measured planimetrically and the maximum width and length estimated from the relative position of the outlying drogues.

1.3.2 Remote sensing studies

Patches of dispersing drogues were tracked from the air on two occasions in 1986 and on three occasions in 1987. In these surveys, the aircraft flew over the lake at $10-15$ minute intervals to photograph patches of drifting drogues and to record the spatial changes in surface temperature. The aircraft was a twin-engine Piper Navahoe Chieftain fitted with a Wild RC8 aerial camera and a Daedalus 1268 multispectral scanner. The Daedalus scanner is one of the most advanced scanner systems currently available. Data can be acquired in 11 spectral bands ranging from the 'blue' visible (band 1) to the thermal infra-red (band 11). The thermal band is normally calibrated internally by alternately scanning a cooled and a heated reference plate. Surface 'skin' temperatures can then be recorded with an absolute accuracy of $0.2\,°C$ and a relative accuracy of $0.1\,°C$. In the drogue tracking surveys we used the output from a near infra-red channel (band 8) to outline the surface area of the lake, which was then used as the 'region of interest' for all subsequent image processing. Most of the surveys were flown at an altitude of 600 m to produce images with a nominal ground resolution of 1.5 m. All the flights took place between 11.00 and 14.00 hours GMT to reduce the effects of sunglint. The digital data from the Daedalus scanner were processed on an International Imaging Systems (I^2S) Model 75 processor at the Institute of Terrestrial Ecology in Bangor (North Wales). Selected 512×512 images were then transferred to floppy disk for subsequent processing on an R-CHIPS system at the Institute of Freshwater Ecology, Windermere.

1.3.3 Analysis of patch growth

The turbulent dispersal of a 'patch' of floating objects is usually treated as a problem of relative diffusion. Several theoretical approaches have, over the years, been used to describe the behaviour of dispersing clouds. One of the most useful models is that suggested by Batchelor (1952) to describe the dispersal of floating objects in a uniform, unbounded flow field. According to this model, small patches of floating objects are entrained progressively into a series of larger and larger eddies. The eddies that are most effective in producing relative motion, and hence dispersal, are those that are about the same size as the patch. Very small eddies produce very little relative movement and very large eddies simply move the whole patch to another location. In the initial stages of dispersal, a small patch therefore grows at a rate that is directly proportional to the time elapsed

$$\sigma = (2Kt)^{1/2} \tag{1.1}$$

where σ is the standard deviation of the patch (a measure of patch size), K is the coefficient of eddy diffusion and t is the time elapsed. When the patch encounters larger eddies, its rate of growth increases and Equation 1.1 is replaced by

$$\sigma = \text{constant} \times t^{3/2} \tag{1.2}$$

In most lakes, this period of rapid growth is relatively short and the patch is soon entrained in larger eddies that tend not to distort the relative position of drogues in a patch. In this final stage of dispersal the diffusivity becomes constant and the rate of patch growth becomes proportional to the square root of the time elapsed

$$\sigma = \text{constant} \times t^{1/2} \qquad (1.3)$$

In a large lake, a diffusing patch of drogues may continue to grow as the square root of the time elapsed until the patch is broken by large-scale water movements. In small lakes, however, further growth is constrained by the bounding effects of the shoreline and patches of drogues may even decrease in size if they encounter converging currents.

1.4 RESULTS

1.4.1 Dispersal of large patches of drogues

Measurements on large patches of drogues deployed in open water showed that very different dispersion rates were often recorded at different locations under comparable meteorological conditions. The behaviour of these large patches usually followed one of three basic patterns: (1) an 'isotropic' pattern of dispersion where the patch increased slowly in size without deforming the original disposition of the drogues; (2) a 'non-isotropic' pattern of dispersion where there was relatively little lateral movement but the patch became stretched in the direction

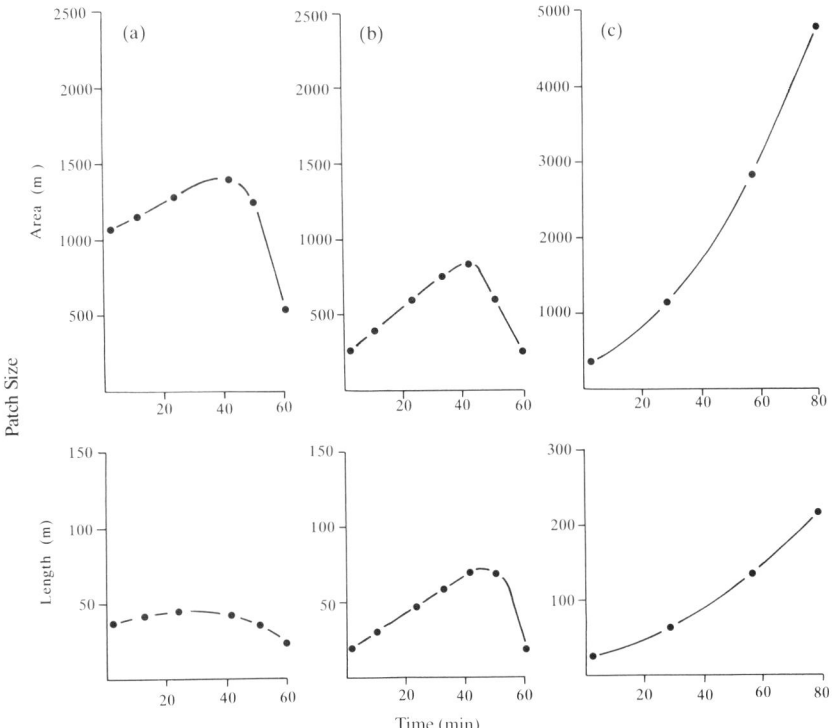

Figure 1.2 Time dependent changes in the area and length of three large patches of drogues. (a) Isotropic dispersion of a patch deployed in open water. (b) Constrained latitudinal dispersion of a patch trapped in a convection cell. (c) Accelerated dispersion of a patch along a shear boundary

of the prevailing wind; and (3) an accelerated pattern of dispersion where individual drogue spacing increased rapidly along two principal axes.

Figure 1.2 shows some typical examples of the three types of dispersion. All these experiments were performed in the open water with winds blowing from the south-west at speeds ranging from 5 to 6 m/s. Large patches of drogues deployed in the open water usually followed an isotropic pattern of dispersion (Figure 1.2a). In this example, the drogues drifted across the lake for 40 minutes and then converged in a downwelling area near the downwind shore. Smaller patches of drogues sometimes dispersed in an isotropic manner, but more commonly formed an elongated patch aligned with the prevailing wind (Figure 1.2b). In this example the drogues formed a 'slick' only 15–20 m across after becoming entrained in a series of convection cells or roll vortices. 'Accelerated' dispersion rates (Figure 1.2c) were generally recorded when the drogues encountered shearing currents. Such shearing currents sometimes appeared near headlands, but were more commonly associated with the thermal fronts that develop in the shallow bays. In the example given, the drogues were entrained in a shear zone that had formed along the boundary of a thermal front in the north-western bay.

Figure 1.3 shows the results of these three dispersion experiments plotted on a semilogarithmic scale. The square root of patch area is taken as an 'average' measure of patch size, and the two broken lines indicate the asymptotic rate of change predicted by the Batchelor model. In the first experiment (a; ▲), the relative separation of the drogues increased as the one-tenth power of time elapsed. In the second experiment (b; ●), the rate of patch growth was proportional to the one-quarter power of time elapsed. Only the patch released in the third experiment (c; ■), grew at the one-half power rate predicted by the Batchelor model. These drogues were, however, entrained in an organized flow, so only a proportion

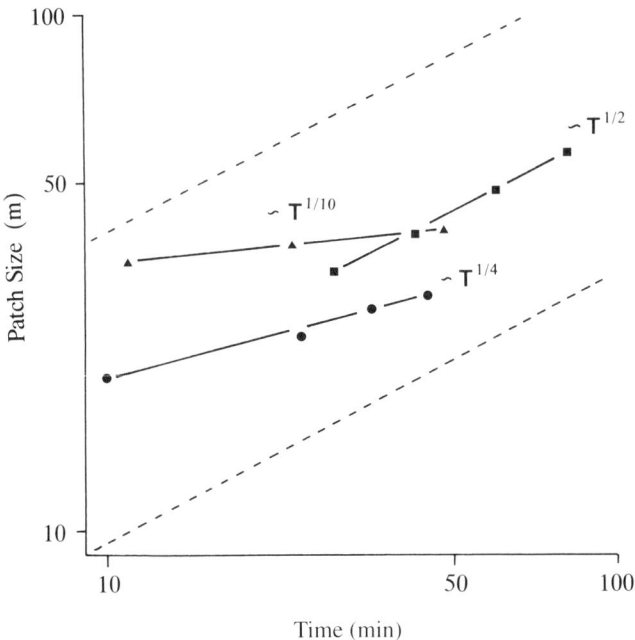

Figure 1.3 Logarithmic plot of the dispersion rates in experiments (a), (b) and (c) of Figure 1.2

of their relative motion can strictly be ascribed to the stochastic motions that are commonly regarded as turbulence.

1.4.2 Dispersal of the drogue triangles

Experiments with large patches of drogues are time consuming and are inevitably restricted to a few horizontal locations. As turbulent diffusion is a stochastic process, more information can be gained from repeated simple experiments than from a single more sophisticated one. The drogue triangle experiments were designed to provide quick estimates of the spatial variation in diffusion over short periods of time. In practice, four or five experiments could be completed in one afternoon before there was any appreciable change in the wind speed or direction.

Figure 1.4 shows the results of a series of drogue triangle experiments performed in the open water under steady wind conditions. The drogues were deployed approximately 25 m apart to minimize the risk of convergence and tracked for at least 40 minutes before being recovered. The size of a diffusing patch was taken to be the square root of the measured triangle area. A logarithmic regression fitted to the observations in Figure 1.4 showed that the rate of patch growth was proportional to the one-third and not the one-half power of elapsed time ($R = 0.76$).

1.4.3 Estimating the integral scale of turbulence in Esthwaite Water

Our observations on the dispersal of large patches of drogues in Esthwaite Water suggest

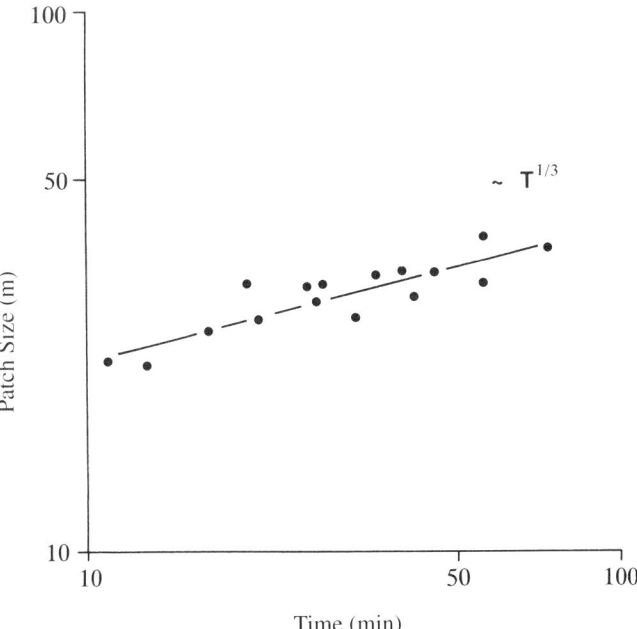

Figure 1.4 Relative growth of small patches (triangles) of drogues deployed in open water. The 'patch size' is the square root of the triangular area

that the size of the most energetic eddies was close to the initial patch size. Experiments
with smaller patches were therefore designed to estimate the length scales over which the
rate of dispersion changed from the rapid intermediate rate to the slow asymptotic rate
postulated by the Batchelor model. Okubo and Farlow (1967) suggest that this critical
dimension, the 'integral scale of turbulence' (Hinze, 1975), can be estimated from logarithmic
plots of effective diffusivity versus time. Effective diffusivities (K) can be computed from
a series of positional measurements using the formula originally suggested by Einstein (1908)

$$K = \tfrac{1}{2}(D^2/t).t \tag{1.4}$$

where D is the standard deviation of the drogue patches with respect to their initial positions.
Okubo and Farlow (1967) assume that the integral scale of turbulence is approximately equal
to $4D$, where D is the lowest value of D which is proportional to the one-half power of time
elapsed. Figure 1.5 is a logarithmic plot of D versus elapsed time for a series of experiments
in Esthwaite Water. According to Batchelor (1952), D becomes proportional to the one-half
power of time elapsed in the asymptotic phase of diffusion. A linear regression fitted to all
the asymptotic phase observations in Figure 1.5, however, had a slope of only one-third,
not one-half as suggested by the model. The lowest value of D for which this reduced rate
of dispersion was recorded was 10 m and the drogues appeared to reach this state in 10–15
minutes. This implies an integral scale of turbulence of about 40 m, a dimension consistent
with the slow rates of dispersion recorded for the largest patches of drogues. This critical
dimension is much smaller than that reported for drogues dispersing in large lakes. Okubo
and Farlow (1967) report an integral scale of turbulence of about 800 m in Lake Erie, where
drifting drogues took about one hour to reach an asymptotic rate of dispersion.

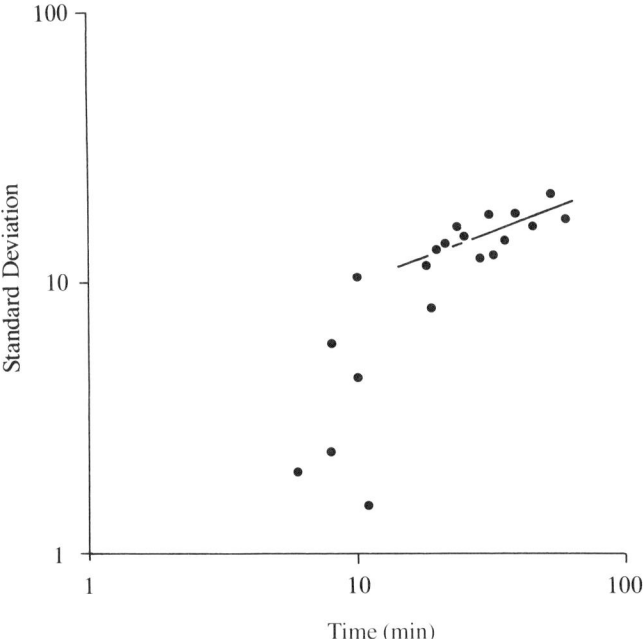

Figure 1.5 Time behaviour of the standard deviation (D) of the distribution of small patches of drogues
with respect to their original positions

1.4.4 Remote sensing evidence for small-scale organized flows

Passive remote sensing techniques cannot be used directly to record variations in the surface flow, but they can be used to track natural markers or record the resulting change in surface temperature. Patches of buoyant blue-green algae are easily deformed by water movements and can readily be monitored from the air by measuring the change in 'green' reflectance (George *et al.*, 1988).

On 24 July 1987, a series of multispectral scanner flights over Esthwaite Water were organized to coincide with sampling activities on the lake. These surveys showed that patches of the buoyant alga *Oscillatoria agardhii* were being transported into the western bay and forming windrows in the open water. Plate 1 is a constrast-stretched image of the north basin taken at 14.00 hours GMT when the wind was blowing steadily from the north. Three spectral bands have been combined in this image to show the spatial distribution of the algae and the horizontal variation in surface temperature. The most obvious feature is the strong transverse gradient that implies a westward deflection of the wind-induced flow. Coriolis deflections of this kind are commonly recorded in Cumbrian lakes (George, 1981) and reflect the low current speeds generated in these sheltered basins.

Several workers have suggested that Coriolis effects of this kind can influence the orientation of the roll vortices known as Langmuir circulations (Faller, 1964). The windrows that mark the position of these vortices in Plate 1 are, however, aligned more or less parallel to the wind. An unusual feature of these roll vortices is the distance between the individual windrows. It is now generally accepted that the distance between windrows is related to the depth to which the circulations penetrate. Early observations in the ocean implied that the distance between windrows was always less than twice the depth of the mixed layer (Assaf *et al.*, 1971; Maratos, 1971). Laboratory experiments by Faller (1969) and Buranthanitt and Cockrell (1979) have, however, shown that width to depth ratios of 2.5−3.5 are also common where the lower boundary is a controlling factor. On 24 July 1987, the thermocline in Esthwaite was only 6 m deep, but the *Oscillatoria* windrows visible in the open water were all over 18 m apart. A number of theories have periodically been advanced to explain how these circulations are generated by the wind (Scott *et al.*, 1969; Pollard, 1977; Leibovich, 1983). All the recent theories assume some form of wave interaction, so it is of interest to note that the windrows in the image were restricted to the open water where there were well developed waves.

Turbulent motions in lakes can be generated directly by the driving force of the wind and indirectly by the shearing action of established currents. In Esthwaite Water the strongest shearing currents usually develop along the boundary of the western bays. On some occasions these currents are generated by topographic effects, but are more commonly associated with the heating imbalances that develop when there is little wind (see Monismith *et al.*, 1990). On 27 April 1987 a well defined thermal 'front' developed along the boundary of the shallow north-western bay. The thermal images in Plate 2 show the progressive erosion of this front as the wind veered from the north towards the north-west. At 13.30 hours GMT (Plate 2a) the main mass of water in the bay was still separated from the open water by a steep temperature gradient. By 14.20 hours GMT the southern section of this front had developed into a counter-rotating eddy almost 200 m in diameter (Plate 2b). Whether organized structures of this size should strictly be regarded as 'eddies' is a matter for some debate, but they clearly have a major effect on the horizontal dispersion of suspended particles.

1.4.5 Relative scale of mixing processes in Esthwaite Water

It is customary to compare the results of diffusion studies in different environments by constructing two-dimensional diffusion diagrams (Stommel, 1949; Okubo, 1971). One of the most commonly used diagrams is a plot of effective diffusivity (K) versus the critical length scale (L). This critical length scale is either calculated from the standard deviation of the drogue positions or estimated from the average size of a diffusing patch. Figure 1.6 compares a number of diffusivity and length scale estimates for Esthwaite Water with similar data from other water bodies. The estimates for Esthwaite Water are all based on time-averaged measurements of large patches of drogues dispersing in the open water. The 'ocean' line is taken from a paper by Okubo (1971) and the 'large lakes' results from studies by Murthy (1970), Huang (1971) and Zivi et al. (1975). The horizontal diffusivities recorded in Esthwaite Water are relatively low but are similar to those measured by Bengtsson (1973) in a number of small lakes. The spatial disposition of turbulent energy in Esthwaite Water is also different to that reported from the Great Lakes. The flow field in these large lakes is typically dominated by large-scale eddies, whereas much of the turbulent energy in Esthwaite Water is concentrated at short 'wavelengths'.

1.5 DISCUSSION

Although a great deal of work has been carried out on the theory of turbulence [see reviews by Shermn et al. (1978) and Fischer et al. (1979)], we still have to resort to experiments

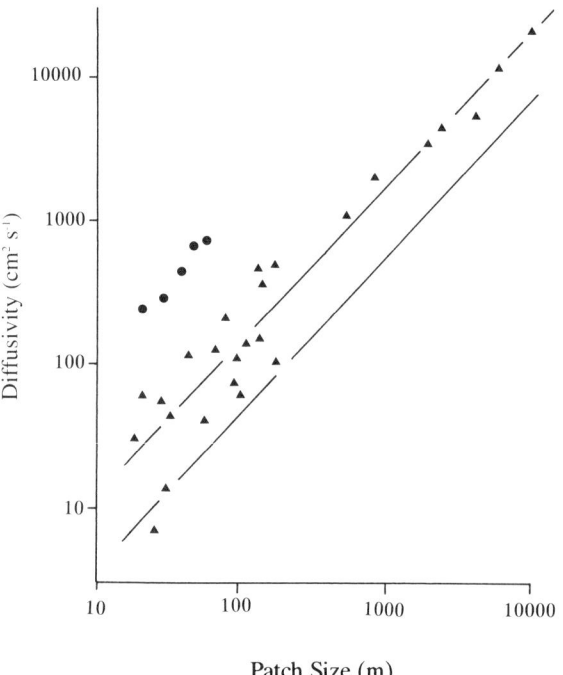

Figure 1.6 Horizontal eddy diffusivity as a function of 'patch size' in Esthwaite Water (•), the Great Lakes (-▲-) and the ocean (−). After Murthy, 1970; Huang, 1971; Okubo, 1971

when we wish to estimate dispersion rates in a particular lake. In this study we used some simple experiments to demonstrate that the critical scales of turbulent mixing in Esthwaite Water are very different from those commonly reported in large lakes. In the large lakes of North America, Okubo and Farlow (1967) found that patches of drogues dispersed at the hypothesized 'intermediate' rate for up to 40–50 minutes. In Esthwaite Water the patches of drogues outgrew this intermediate rate in 10–15 minutes and then dispersed at rates that varied with their horizontal location. In the open water, patches of drogues more than 30 m in diameter typically increased slowly, but isotropically, in size. This suggests that they were entrained in a relatively homogeneous flow field and had already outgrown the length scale of the most energetic eddies. Smaller patches of drogues deployed in the open water sometimes dispersed isotropically, but often became trapped in roll vortices that prevented any further lateral movement. The 'average' rate of drogue dispersion in the open water was, consequently, much lower than the asymptotic rate predicted by the Batchelor model. The only drogues to disperse at the rate suggested by Batchelor were those entrained in the shear flows that commonly developed in the shallower bays.

The Batchelor model of dispersion is strictly applicable to the analysis of movements in an unbounded, homogeneous field of turbulence. In Esthwaite Water the horizontal eddy size is constrained, not only by the irregular shoreline, but also by the 'diffusion floor' formed by the shallow thermocline. Relatively little is known about the effect of stratification on the Batchelor spectrum but Dillon and Caldwell (1980) have demonstrated that turbulence can be suppressed in strongly stratified environments. In Esthwaite Water a relatively shallow, stable thermocline is present throughout the summer. Average Brunt–Vaisala frequencies (N^2) typically range from $3 \times 10^{-4} \, \text{s}^{-2}$ in a windy summer to $18 \times 10^{-4} \, \text{s}^{-2}$ in a calm summer (Heaney and Butterwick, 1985). The generation of turbulence in the epilimnion is further complicated by the pronounced Coriolis rotations that appear when the wind speed is relatively low. In Esthwaite Water the Ekman spirals that develop in the wind drift current merge imperceptibly with the reverse Ekman spirals that develop in the return flow. This compression of shearing forces in the epilimnion must have a major effect on the spectrum of turbulence and may explain the concentration of energy at very small time and length scales.

The Batchelor model of turbulence also assumes a degree of random motion that is seldom encountered in the open waters of a small lake. Our remote sensing studies confirm that it is unwise to regard the epilimnion as an isotropic mass of nested eddies. Many of the 'turbulent' structures visualized by these techniques are clearly highly organized but cannot currently be treated as explicitly prescribed flows. Most numerical models of lake circulation implicitly assume that all physical processes smaller than the chosen grid size can either be neglected or parameterized by an 'average' eddy diffusion coefficient. The limitations of modelling dispersion using average values are widely recognized (Chatwin and Allen, 1985), but spatially variable coefficients cannot readily be derived from 'deterministic' mean flows. A practical distinction can, however, be drawn between the turbulence generated directly by the wind and that generated indirectly by the action of shear currents. The turbulence generated by the wind is systematically distributed across the surface of a lake, but the turbulence generated by shear flows is typically concentrated at a few locations. These local shear structures are, however, inherently unstable and may release a great deal of mixing energy when they eventually start to disintegrate.

ACKNOWLEDGEMENTS

Thanks are due to Matthew McCartney for carrying out most of the drogue experiments while he was an undergraduate at the University of Lancaster. Ms D.P. Hewitt and Mr L. Rosser helped with the fieldwork and Mr W. Slade provided invaluable advice on image processing. The aerial surveys were co-ordinated by Mr J. Cook and formed part of the Natural Environment Research Council's Airborne Remote Sensing Campaign.

REFERENCES

Assaf, G., Gerard, R. and Gordon, A.L. (1971) Some mechanisms of oceanic mixing revealed in aerial photography. *J. Geophys. Res.* **76**, 6550−6572.

Batchelor, G.K. (1952) Diffusion in a field of homogeneous turbulence. II. The relative motion of particles. *Proc. Cambridge Philos. Soc.* **48**, 345−362.

Bengtsson, L. (1973) Conclusions about turbulent exchange coefficients from model studies. *Proceedings of the Helsinki Symposium, July 1973. IAHS Publ. No. 109*, pp. 306−312.

Boyce, F.M. (1974) Some aspects of Great Lakes physics of importance to biological and chemical processes. *J. Fish. Res. Board Can.* **31**, 689−730.

Burnathanitt, T. and Cockrel, D.J. (1979) Some effects of Langmuir circulations on suspended particles in lakes and reservoirs. *Rep. 79-6.* Department of Engineering, University of Leicester.

Chatwin, P.C. and Allen, C.M. (1985) Mathematical models of dispersion in rivers and estuaries. *Ann. Rev. Fluid Mech.* **17**, 119−149.

Csanady, G.T. (1963) Turbulent diffusion in Lake Huron. *J. Fluid Mech.* **17**, 360−384.

Dillon, T.M. and Caldwell, D.R. (1980) The Batchelor Spectrum and dissipation in the upper ocean. *J. Geophys. Res.* **85**, 1910−1916.

Einstein, A. (1908) The elementary theory of Brownian Motion. Reprinted in Furth, R. (Ed.) (1956) *Investigations on the Theory of the Brownian Movement by Albert Einstein*. Dover, New York, pp. 68−85.

Falconer, R.A., George, D.G. and Hall, P. (1991) Three-dimensional numerical modelling of wind-driven circulation in a shallow homogeneous lake. *J. Hydrol.* **124**, 59−79.

Faller, A.J. (1964) The angle of windrows in the ocean. *Tellus* **16**, 363−370.

Faller, A.J. (1969) The generation of Langmuir circulations by the eddy pressure of surface waves. *Limnol. Oceanogr.* **14**, 504−513.

Fischer, H.B., List, E.J., Koh, R.C.Y., Imberger, J. and Brooks, N.H. (1979) *Mixing in Inland and Coastal Waters*. Academic Press, New York, 483pp.

George, D.G. (1981) Wind-induced water movements in the South Basin of Windermere. *Freshwater Biol.* **11**, 37−60.

George, D.G. and Heaney, S.I. (1978) Factors influencing the spatial distribution of phytoplankton in a small productive lake. *J. Ecol.* **66**, 133−155.

George, D.G., Allen, C.M. and Smith, D.G. (1988) The remote sensing of phytoplankton chlorophyll in Esthwaite Water, Cumbria. *Proceedings of the NERC 1987 Airborne Campaign Workshop*. NERC, Swindon, pp. 59−74.

George, D.G., Hewitt, D.P., Lund, J.W.G. and Smyly, W.J.P. (1990) The relative effects of enrichment and climate change on the long-term dynamics of Daphnia in Esthwaite Water, Cumbria. *Freshwater Biol.* **23**, 55−70.

Heaney, S.I. and Butterwick, C. (1985) Comparative mechanisms of algal movement in relation to phytoplankton production. In: *Migration: Mechanisms and Adaptive Significance* (Ed. M.A. Rankin), *Contrib. Mar. Sci. Suppl. 27*, pp. 114−134.

Hinze, J.O. (1975) *Turbulence*. McGraw-Hill, New York, 790pp.

Huang, J.C.K. (1971) Eddy diffusivity in Lake Michigan. *J. Geophys. Res.* **76**, 8147−8152.

Imboden, D.M. (1990) Mixing and transport in lakes: mechanisms and ecological relevance. In: *Large Lakes: Ecological Structure and Function* (Ed. M.M. Tilzer and C. Serruya). Springer, Berlin and Heidelberg, pp. 47−80.

Leibovich, S. (1983) The form and dynamics of Langmuir circulations. *Ann. Rev. Fluid Mech.* **15**, 391−427.

Maratos, A. (1971) Study of the near shore surface characteristics of windrows and Langmuir circulation in Monterey Bay. *MS Thesis*. US Navy Postgraduate School, Monterey, California.

Monahan, E.C. and Monahan, E.A. (1973) Trends in drogue design. *Limnol. Oceanogr.* **18**, 981−985.

Monismith, S.G., Imberger, J. and Morison, M.L. (1990) Convective motions in the sidearm of a small reservoir. *Limnol. Oceanogr.* **35**, 1676−1702.

Murthy, C.R. (1970) An experimental study of horizontal diffusion in Lake Ontario. *Proceedings of the 13th Conference on Great Lakes Research*, Internat. Assoc. Gt. Lakes Res. Ann Arbor, Michigan, pp. 477−489.

Okubo, A. (1971) Oceanic diffusion diagrams. *Deep-Seal Res.* **18**, 789−802.

Okubo, A. and Farlow, J.S. (1967) Analysis of some Great Lakes drogue studies. *Proceedings of the 10th Conference on Great Lakes Research*, Internat. Assoc. Gt. Lakes Res. Ann Arbor, Michigan, pp. 299−308.

Pollard, R.T. (1977) Observations and theories of Langmuir circulations and their role in near surface mixing. In: *A Voyage of Discovery: George Deacon 70th Anniversary Volume* (Ed. M. Angel). Pergamon Press, Oxford, pp. 235−251.

Pond, S. and Pickard, G.L. (1983) *Introductory Dynamical Oceanography*. Pergamon Press, Oxford, 329pp.

Reynolds, C.S. (1980) Phytoplankton assemblages and their periodicity in stratifying lake systems. *Holarctic Ecol.* **3**, 141−159.

Reynolds, C.S. (1984) *The Ecology of Freshwater Plankton*. Cambridge University Press, Cambridge, 384pp.

Reynolds, C.S. (1990) Temporal scales of variability in pelagic environments and the response of phytoplankton. *Freshwater Biol.* **23**, 25−53.

Sanderson, B. (1987) An analysis of Lagrangian kinematics in Lake Erie. *J. Great Lakes Res.* **13**, 559−567.

Scott, J.T., Myer, G.E., Stewart, R. and Walther, E.G. (1969) On the mechanisms of Langmuir circulations and their role in epilimnion mixing. *Limnol. Oceanogr.* **14**, 493−503.

Sherman, F.S., Imberger, J. and Corcos, G.M. (1978) Turbulence and mixing in stably stratified waters. *Ann. Rev. Fluid Mech.* **10**, 267−288.

Steinberg, E.W. and Hartmann, H.M. (1988) Planktonic bloom-forming Cyanobacteria and the eutrophication of lakes and reservoirs. *Freshwater Biol.* **20**, 279−287.

Stommel, H. (1949) Horizontal diffusion due to oceanic turbulence. *J. Mar. Res.* **8**, 199−225.

Zivi, S.M., Frye, D.E., Buell, R.E. and Van Loon, S. (1975) Measurements in eddy diffusivities in nearshore regions of Lake Michigan. In: *Measurements of Physical Phenomena Related to Power Plant Waste Discharges* (Eds J.V. Toker, S.M. Zivi, A.A. Frigo, L.S. Van Loon, D.E. Frye and C. Tome), *ANL/WR-75-1*. Argonne National Laboratory, Argonne, Illinois, pp. 164−210.

2 One-dimensional Dispersion in a Lake Inferred from Sonar Observations

S.A. THORPE and M.S. CURÉ

School of Oceanography, Southampton University, Southampton, UK

2.1 INTRODUCTION

Cath was interested in dispersion. We believe that she would have been amused by the simplistic attempts to infer dispersion in a lake which are described here.

Imagine that two vertical parallel plane meshes made of chicken wire are erected at about 90° to the wind direction across a lake to form a narrow cage through which the water can flow unimpeded, but which prevents particles in the cage from moving up- or downwind. We suppose that the particles, which float, and which we shall call parsnips in tribute to the exploratory study of dispersion made using pieces of these vegetables by Richardson and Stommel (1948), are otherwise constrained by the cage but are advected by cross-wind currents and are dispersed within the cage across the lake.

Our observations will allow us, with some simplifying assumptions, to make estimates of the dispersion and of its variation with the prevailing conditions. The cage is envisaged as being along our sonar beam. Although we might infer something of the downwind motion and so make estimates of the dispersion of free floating bodies, or even of neutrally buoyant particles, these estimates require additional assumptions (Thorpe, Curé, Graham and Hall, 1994). We have chosen instead to use the information which is available (examples of which are presented in Plates 3−5) with the fewest assumptions, hoping therefore to obtain the most reliable, if not the most interesting or useful, estimates.

The key elements are these. Clouds of subsurface bubbles, generously and randomly formed by breaking wind waves, are 'visible' to high frequency sonar (Thorpe, 1982). The mean bubble population decays rapidly in depth; the acoustic scattering cross-section has an e-folding scale of 1−2 m. Individual 'clouds' typically persist for five minutes (Thorpe and Hall, 1983), during which time they are carried by Langmuir circulation (LC; Leibovich, 1983) towards or (if they persist long enough and start close enough) into the regions of convergence below the floating bands of flotsam or foam known as windrows. The rates of advection can be measured by monitoring the changing range of clouds on the sonar. The clouds accumulate and persist near the centre of the convergence zones (Thorpe, 1984), subsequent lateral dispersion occurring when the LC pattern breaks down (Csanady, 1973; Faller and Auer, 1987; Thorpe, 1992a). The sonar beam can also monitor this breakdown and hence infer where in the beam and how often these dispersive events occur. We use the real data to simulate the motion of floating parsnips within the beam — that is, constrained within the chicken wire cage.

Mixing and Transport in the Environment. Edited by K.J. Beven, P.C. Chatwin and J.H. Millbank

This paper describes the instrumentation used and the lake site, some of the observations, the method used to infer dispersion, and some results. These are as yet preliminary and will form a major component of the second author's PhD dissertation; he contributed most to the collection of the data, together with others whose help is acknowledged at the end, and will eventually have to make sense of it.

2.2   INSTRUMENTATION AND PROCESSING OF SONAR DATA

We carried out experiments in Loch Ness, Scotland. This 180 m deep body of fresh water with a fetch at our site of 20 km to the prevailing south-westerly wind, provides an ideal test tank. We used a simple 90 kHz sidescan sonar with a slant range of about 250 m. The bubbles generated by breaking waves provide a high near-surface sonar target strength at this frequency. The sonar transducer was kept horizontal by attaching it to the bottom of a pendulum mounted in a supporting frame which was placed at a depth of 40 m on a steep slope, 50 m from the shore with the vertical fan-shaped beam pointing out across the loch (Figure 2.1a). The axis of the main lobe was inclined upwards at an angle of 20° to the horizontal. The energy in the beam, although widely distributed in the vertical plane, was narrowly confined in the horizontal plane to an arc of less than 5° (Figure 2.1b). Bubble targets are detected in a thin strip of water near the surface, the 'cage' effect referred to in Section 2.1, with an axis perpendicular to the shoreline. The movement of bubble clouds along this thin strip can be seen in the sonar images.

The sonar pulse repetition frequency (PRF) was varied between 1.5 and 4 Hz, allowing different ranges to be sampled. The signal was sampled at 20 kHz and the slant range was converted to a surface range, with zero surface range corresponding to a point 100 m from the loch edge where the water depth was approximately 50 m. Time averaging of the data depended on the PRF, with a number of successive pulses of sound being averaged to increase the signal to noise ratio: 32 pulses at a PRF of 4 Hz, 16 at 2 Hz, and 12 at 1.5 Hz. The sonar transmission pulse lasted 0.1 ms, giving a range resolution of 0.15 m for a sound speed of 1500 ms^{-1}. The sonars have not been calibrated to establish the strength of targets, but the effect of range attenuation has been removed simply by dividing the signal by the range average. A digital Gaussian edge enhancement and lowpass filter with a cutoff at 2.0 m was applied in range, and a lowpass filter with a cutoff of 60 s applied in time. All calculations to produce sonar images were performed on a personal computer. Observations of wind, temperature and precipitation were made using a Didcott Met. Station at a site 40 m above the level of the loch on the hillside. This was not an ideal position, particularly for winds, and frequent observations were made of wind direction and of speed using a hand-held anemometer from an inflatable dinghy on the loch for intercalibration.

2.3   OBSERVATIONS

Plates 3–5 are sonograph images of three periods of data obtained over a range of wind speeds chosen from a total of 420 hours of observations in late winter and autumn at Loch Ness. The patterns seen are typical of times when either there was sufficient wind to cause breaking waves or when heavy rain produced bubbles. The most obvious features, such as at A in Plate 4, are congregations of bubbles, or bubble bands, which typically persist in the sonar beam for five minutes. These mark the convergent regions of LC. They have a variable cross-wind velocity, an observation important when we later consider lateral diffusion

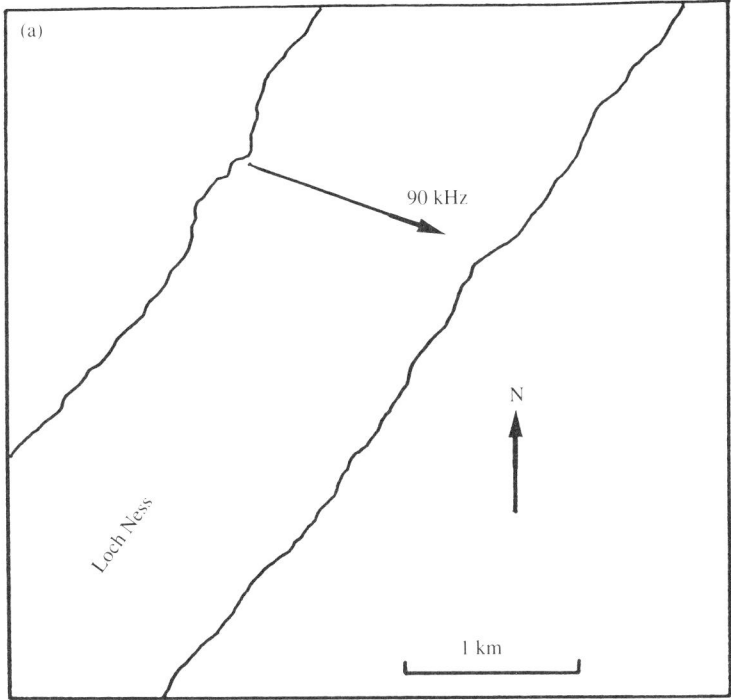

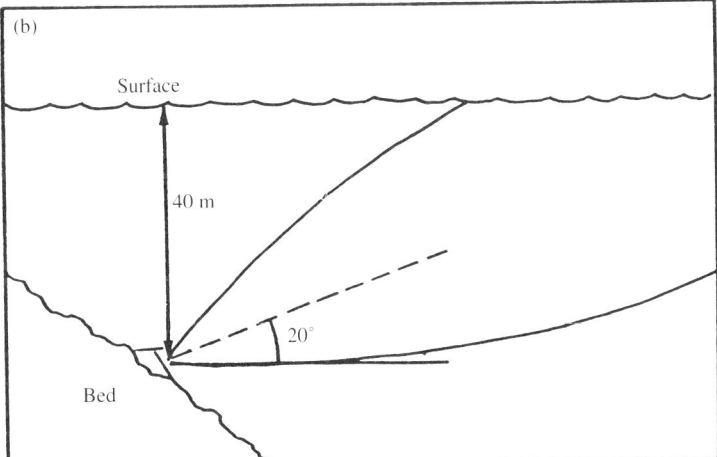

Figure 2.1 The sonar was mounted on the bottom of the loch and directed across the loch almost perpendicular to the loch axis. (a) Plan view of the site. The signal cables were taken back to the shoreside recording hut. The meteorological station was on the hillside behind the hut. (b) Beam pattern of the sonar. The bottom shelved at roughly 45°. During all periods of data presented here, the wind and waves were from the south-west

from a fixed point. High backscatter regions (e.g. at A) often line up to form a larger scale structure (e.g. BB, Plate 4) which can persist for periods of up to one hour. Video time-lapse sequences of the loch surface from the nearby hill have shown that windrows composed of floating foam are often situated close to, or directly above, these bubble patches. Small-scale convergence regions such as at C and D are drawn into the larger LC from the left and right of the downwind direction, respectively. A hierarchy of scales is especially evident in Plate 5 during high wind speeds. In the late winter period when no well-developed thermocline was present, the larger scale LC usually moves to the right of the downwind direction, the bubble bands being advected by the cross-wind component of current at a few centimetres per second (see also Thorpe and Hall, 1982). The speeds of smaller scale regions (e.g. C, D, Plate 4) towards the larger features increase with increasing wind speed. Plates 3 and 5 show two extreme wind speeds at which the patterns have been detected. Plate 4, at an intermediate wind speed, shows some data from an autumn deployment when the effects of stratification may have some influence.

The reader will be struck by the persistence of patterns of high sonar backscatter on the loch surface on typically 1−20 minute time-scales and by their variability in scale and over longer periods of time. As the targets are accumulations of bubbles either from waves which break in a random pattern over the loch, or from rain drops which inject bubbles fairly uniformly over the loch, the region of high sonar returns can be regarded as providing a map of convergence (or at least a time integral of convergence over a typical bubble lifetime, say, two minutes) on the surface of the loch (Thorpe, 1984). Given the known structure of LC, the targets will also mark regions of downward motion between neighbouring cells, although not necessarily the magnitude of the motion; bubbles may remain in regions of low vertical current (Stommel, 1949), but will be carried by the circulation and dispersed if the speed is too high.

To estimate our cage diffusion we will consider what happens to a floating particle (i.e. a parsnip) when it is placed onto the loch surface in the beam. The important measurements to be made are firstly the approach velocity of the parsnip to the nearest LC convergence marked by bubbles, the way that this convergent region moves across-wind, and how long it lasts before breaking up and thus freeing the parsnip, although still confined within the cage defined by the sonar beam, to head towards another convergence. The next section shows how the sonar data undergoes further processing to achieve these results.

2.4 DISPERSION

Every 160 seconds parsnips are released every five metres in range along the beam into a set of data, a 'skeleton', resembling and derived from the sonographs. Each parsnip is made to approach the nearest convergent region as represented by the skeleton extracted from sonar images such as those in Plates 3−5. They move with a constant velocity until they intersect the skeleton, after which they follow it until the convergent line they have entered ceases to exist. The process repeats itself until either the particle leaves the top or bottom of the image or leaves the end. In each instance it is re-introduced from the other side at the same range or time. The ratio of the number of times a particle leaves a bubble band to a particle in a windrow leaving the image is 19 or more, and as such the effect of a finite image size is not expected to be very important. The path of each parsnip is recorded and dispersion calculated from these paths.

The first stage in the processing is the isolation of the 'skeleton', which essentially represents the convergence pattern of the sonar image. It has proved satisfactory to select the 15% of the pixels in the sonograph with the highest intensity sonar backscatter. These pixels tend to congregate together in large blobs or features which do not merge with one another at this value of threshold, nor are small features missed. The set of blobs constitute the reduced image. Each resulting high intensity blob has any holes within it filled and then has its perimeter followed recursively. Each pixel on the perimeter is considered in turn and is deleted if, in so doing, the connectivity of the blob as a whole is not destroyed. In this way the blob becomes smaller on each circuit until finally a connected set of arcs results and no further pixels can be deleted. When considering each pixel for deletion a set of rules is used (see Appendix). An example of the resulting skeleton is shown in Figure 2.2.

The inflow velocities are measured manually from the passage of the smallest scale convergences (e.g. Plate 3, C, D) into the larger scale LC. Usually a clear distinction can be made between the high intensity blobs which line up to form the large LC and are skeletized and the very low intensity but nevertheless distinct small features which are drawn into them. Typically 150 velocities will be averaged from each two hour period to give values for the inflow velocities from the left and from the right of the downwind direction. The determination of the inflow velocities is subjective but, as yet, no efficient algorithm for their calculation has been devised.

Figure 2.2 shows an example of the multiple release of the parsnips and their subsequent tracks. These calculations have been performed on an IBM 3090 mainframe computer. Parsnips can be seen to congregate in the convergent regions. There are three sources of relative dispersion between particles: (1) that caused by the movement of parsnips onto different skeletal lines (i.e. onto different windrows); (2) the relative motion of different lines of convergence or windrows, skeleton lines, on which parsnips are trapped; and (3) the relative motion of particles 'released' when a skeleton line, on which they had previously been trapped, breaks up.

2.5 RESULTS

Individual tracks of surface-following particles (parsnips) can be represented as a distance y measured along the sonar beam from their initial release point, at a time t from their release, the diffusion time. If we make the assumption that the particles are moving in a stationary turbulent field and mean flow which are independent of range, then the distribution of all particle positions for each value of t can be used to infer one-dimensional dispersion or, more specifically, the spreading of the particles released continuously into the 'cage' from a fixed point. Figure 2.3 shows six plots of position y, the offset from the release point, against t, three from late winter and three from autumn data for various wind speeds. The mean cross-wind flow and dispersion is apparent for all of the plots. The variance σ_y^2 of the cross-wind position may be calculated as a statistic for a series of diffusion times. Following Csanady (1973), we define a cross-wind dispersion coefficient as

$$K_y = \frac{1}{2}\left(\frac{\partial}{\partial t}\sigma_y^2\right) \tag{2.1}$$

Following Bowden and Lewis (1973), the variance is represented as

$$\sigma_y^2 = at^m \tag{2.2}$$

and so

$$K_y = \frac{am}{2}t^{m-1}. \tag{2.3}$$

22

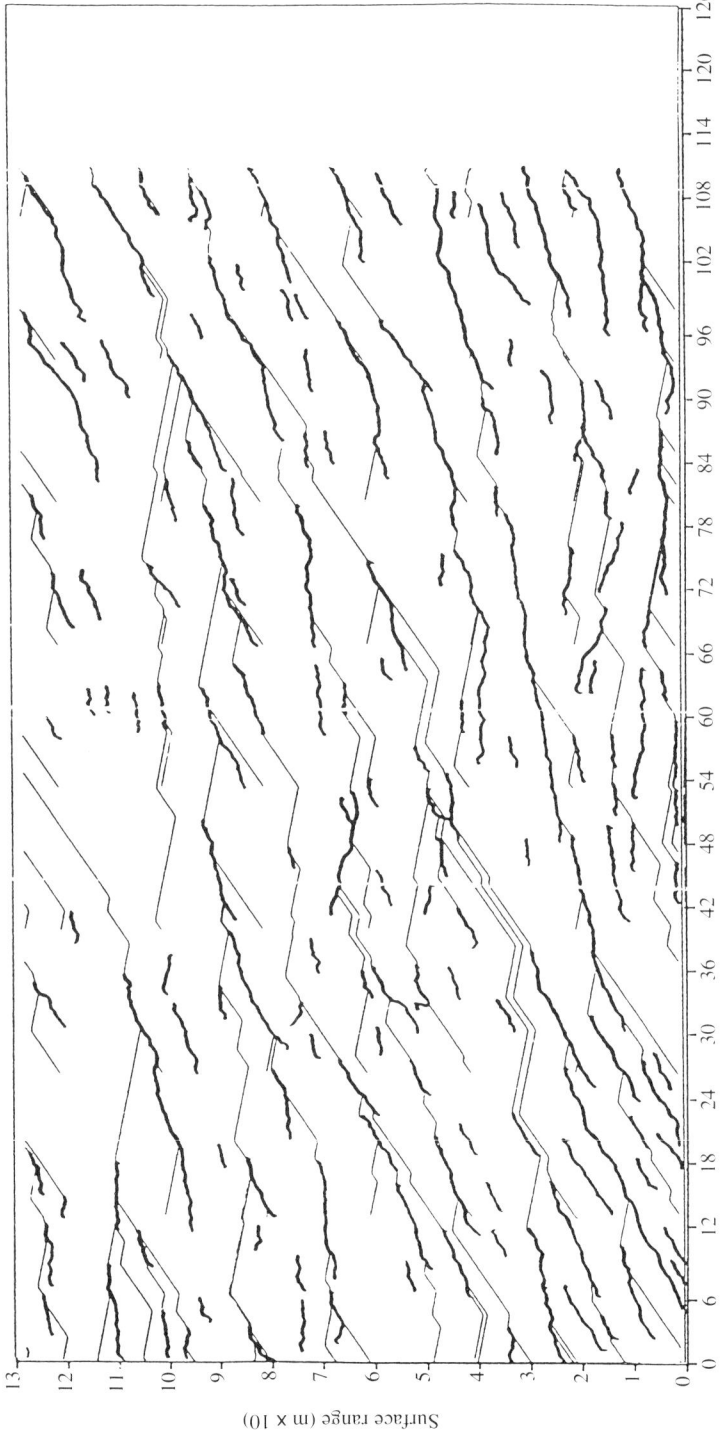

Figure 2.2 Tracks of particles in a 'skeleton' plot. The thick lines are derived from sonographs and represent the regions of surface convergence: the skeleton. Particles are released at regular intervals and at regular times. Their paths are shown by the thin lines (for clarity, only a few are shown here). The time history of these tracks is used to calculate the lateral dispersion

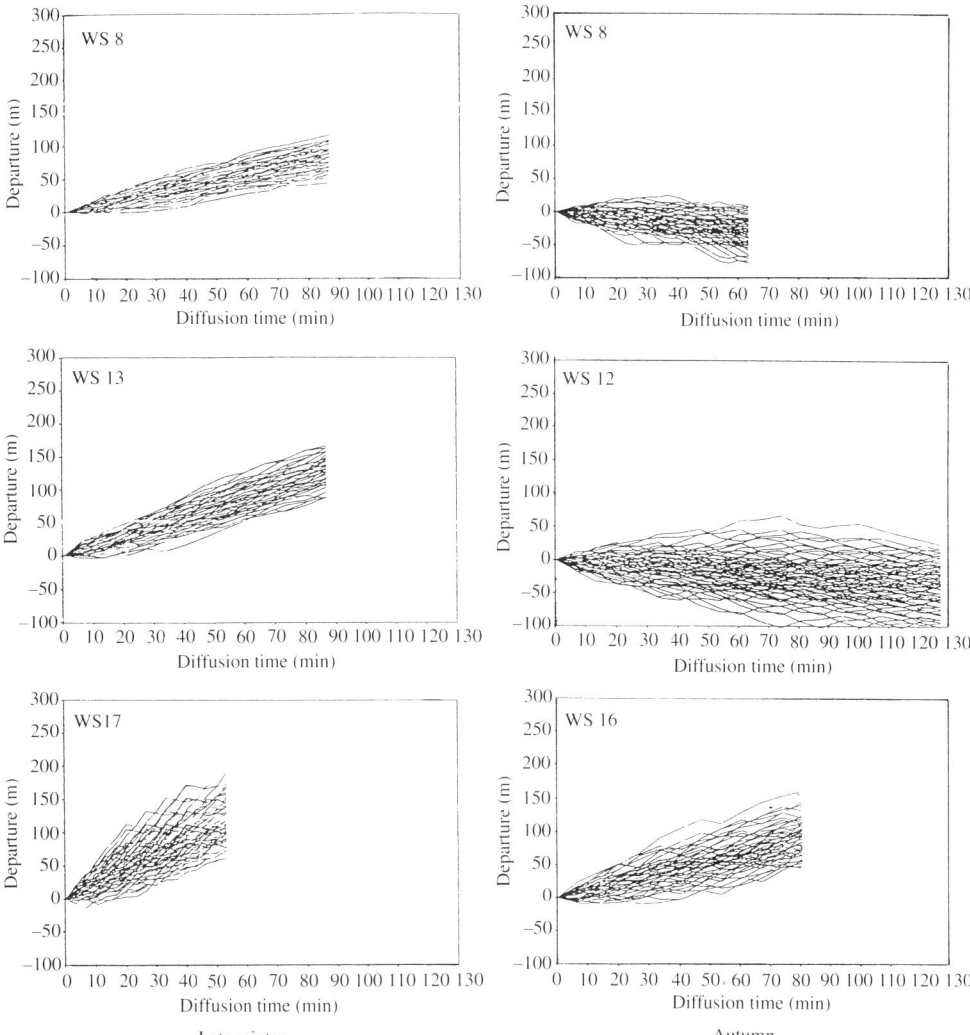

Figure 2.3 Particle paths such as in Figure 2.2 are presented here as plumes (although, for clarity, only a few of the paths are shown). A comparison can be drawn between different wind speeds (WS in m s^{-1}) and between late winter data and autumn data when a seasonal thermocline was present. It can be seen that the plumes initially grew rapidly and that the rate then slowed. Plumes grew more rapidly with diffusion time as wind speed increased. A steady cross-loch advection to the right of the downwind direction presumably caused by an Ekman effect was observed during the late winter experiments (though not always in the autumn)

It is traditional in dye diffusion experiments to take many transects of a dye plume at some distance downstream from the release point and to average the concentration profiles about their respective mean concentrations, from which the variance may be evaluated. This statistic results from 'relative diffusion' (see Csanady, 1973) and sometimes reflects an observational difficulty in accurately specifying the sampling positions. In our observations, the position in the sonar beam is accurately measured. Similarly, in the numerical simulation, the position,

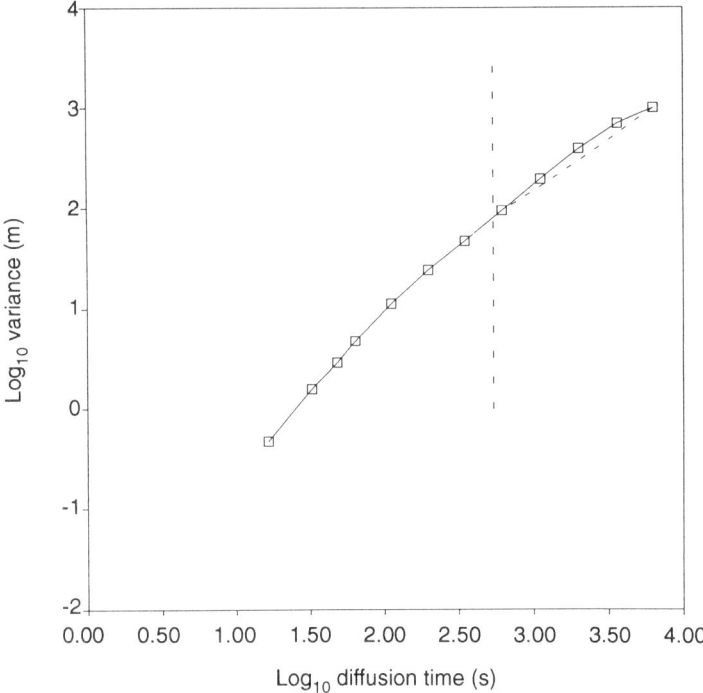

Figure 2.4 Diffusion results for an autumn period with a wind speed of 12 m s^{-1}. The rate of increase of variance of cross-loch position decreases with diffusion time. The vertical broken line is the mean time for a particle to enter and then leave its first convergence line after having been placed on the water surface. At subsequent times diffusion should result from movements of the convergent lines themselves. The diffusion curve for later times is simplified by a dotted line in the manner described in the text. The gradient m of this line and the intercept $\log_{10} a$ are measured and used to plot the data in Figure 2.5. Convergence from left, 3.2 cm s^{-1}; convergence from right, -5.0 cm s^{-1}

y, of the particles is known and is averaged over an ensemble of releases. The results presented here are of 'absolute diffusion', which in general will be larger than relative diffusion, but are inherently more interesting.

The value of m is not constant with t. Figure 2.4 shows how σ_y^2 varies with t, plotted on log scales, in one particular, but typical, case. It can be seen that the gradient decreases with diffusion time. At small values of t, K_y depends on the mean inflow speed and mean spacing of the LC. We are mainly concerned with longer diffusion times. The vertical broken line represents the mean time, t_1, required for a particle to find its first convergence and then exit from it. At times greater than this diffusion appears to result mainly from the relative movement and break-up of the convergent regions themselves rather than from a difference in inflow speeds. From diffusion curves such as in Figure 2.4, a line is drawn between the intersection of the curve with the vertical broken line and the variance at 3600 seconds. The gradients m and the proportional term $am/2$ in Equation 2.3 for this line are then calculated (Figure 2.5a and 2.5b).

The values of m show considerable scatter but are near unity in accordance with Fickian diffusion (Csanady, 1973). We have chosen to estimate K_y for a diffusion time of one hour, much greater than t_1 but sufficiently short for the mean flow to be regarded as 'steady'.

Figure 2.5c shows how these K_y estimates vary with wind speed. Figure 2.5c also shows estimates from dye diffusion experiments by some other workers (references are given in the figure caption). A direct comparison is not valid as we are dealing with floating tracers in a cage and absolute, as opposed to relative, dispersion. We note nevertheless a broad order of magnitude agreement and that in general, the restriction of dispersion both to surface floating parnsips and to the 'cage' leads to estimates of K_y which are less than those for dyes which, in addition to being carried and dispersed by differential motions in the wind direction, may be more freely exchanged between neighbouring Langmuir cells and hence dispersed more rapidly than parsnips.

2.6 CONCLUSIONS

(1) Plates 3−5, showing the sonograph records obtained at different wind speeds in Loch Ness, vividly emphasize the highly anisotropic and variable patterns of flow at scales of 1−200 m on the lake surface, and the variety of scales on which convergence may occur.

(2) The observations support the idea advanced by Csanady (1973) and Faller and Auer (1987) that cross-wind dispersion at scales of 1 m to 1 km may be dominated by the presence of LC and, in particular, that the rate of dispersion will be determined by the frequency of cell instability, transience, amalgamation or bifurcation, all of which may be observed using sonar.

(3) The rates of dispersion estimated using the sonar data have a level of confidence which is yet to be firmly established. The diffusivity, however, appears to increase with wind speed. We emphasize the special and rather particular geometry, parsnips confined to a chicken wire cage, to which our estimates apply. Not only are the tracers confined to the surface of the lake, but they are also constrained so that they cannot follow the 'eddies' or turbulent structures as they move and evolve in space and time. The estimated dispersion may depend on the ratios of several time-scales — for example, the lifetime of windrows, the time for their advection through the sonar beam and the time for floating particles to enter a windrow after release. Windrows have a finite length, and the greater the advection through the chicken wire cage, the more often are particles freed to move into neighbouring convergence regions when a windrow leaves the sonar beam, and the shorter is the time that a particle resides in a windrow relative to the time spent in travelling between them. Some results indicate that K_y increases with advection speed (Thorpe *et al.*, 1994). Although advection is generally small in Loch Ness compared with shelf seas where, for example, tides may produce relatively large currents, it may not be negligible and further study of its effects is merited.

(4) Sonar offers a remote sensing tool suitable for continuous survey in lakes, even in high winds. The simultaneous use of frequencies higher than those reported here allows the study of currents and breaking waves (Thorpe and Hall, 1983; Thorpe, 1992b) and, for example, their relation to LC which may provide an insight into how it is generated.

ACKNOWLEDGEMENTS

The sonar belongs to the Institute of Oceanographic Sciences Deacon Laboratory and our thanks are due to Mr Alan Hall for preparing and installing it and in helping with the collection of data, a task in which Andrew Carrington, Alan Forster, Angus Graham and Dave Thorpe

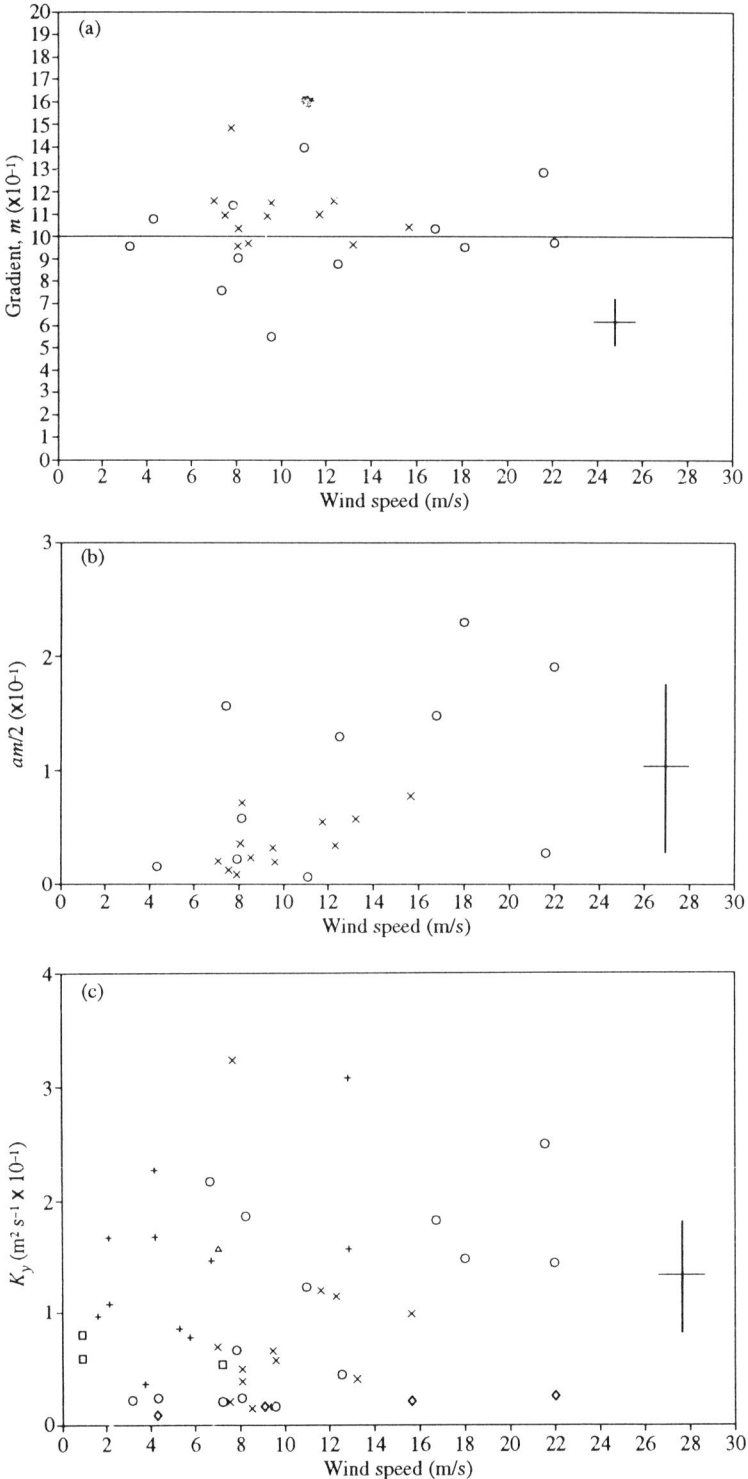

were also involved. The project was funded by the NERC and ONR, and we are most grateful for this support which made the project possible.

REFERENCES

Bowden, K.F. and Lewis, R.E. (1973) Dispersion in flow from a continuous source at sea. *Wat. Res.* **7**, 1705−1722.

Csanady, G.T. (1973) *Turbulent Diffusion in the Environment*. Reidel, Dordrecht.

Elliott, A.J. and Wallace, D.C. (1989) Dispersion of surface plumes in the southern North Sea. *Dtsch. Hydrogro. A.* **42**, 1−16.

Faller, A.J. and Auer, S.J. (1987) The role of Langmuir circulation in the dispsersion of surface tracers. *J. Phys. Oceanogr.* **18**, 1108−1123.

Leibovich, S. (1983) The form and dynamics of Langmuir circulations. *Ann. Rev. Fluid Mech.* **15**, 391−427.

Meerburg, A.J. (1972) An experimental study of the turbulent diffusion in the upper few metres of the sea. *Neth. J. Sea Res.* **5**, 492−509.

Richardson, L.F. and Stommel, H. (1948) Note on eddy diffusion in the sea. *J. Meteorol.* **5**, 238−240.

Schott, F., Ehlers, M., Hubrich, L. and Quadfasel, D. (1978) Small-scale diffusion experiments in the Baltic surface-mixed layer under different weather conditions. *Dtsch. Hydrogr. Z.* **31**, 195−215.

Stommel, H. (1949) Trajectories of small bodies sinking slowly through convection cells. *J. Mar. Res.* **12**, 157−172.

Thorpe, S.A. (1982) On the clouds of bubbles formed by breaking wind-waves in deep water, and their role in air-sea gas transfer. *Phil. Trans. R. Soc. London A* **304**, 155−210.

Thorpe, S.A. (1984) The effect of Langmuir circulation on the distribution of submerged bubbles caused by breaking waves. *J. Fluid Mech.* **142**, 151−170.

Thorpe, S.A. (1992a) The break-up of Langmuir circulation and the instability of an array of vortices. *J. Phys. Oceanogr.* **22**, 350−360.

Thorpe, S.A. (1992b) Bubble clouds and the dynamics of the upper ocean. *Q. J. R. Meteorol. Soc.* **118**, 1−22.

Thorpe, S.A. and Hall, A.J. (1982) Observations of the thermal structure of Langmuir circulation. *J. Fluid Mech.* **114**, 237−250.

Thorpe, S.A. and Hall, A.J. (1983) The characteristics of breaking waves, bubble clouds, and near-surface currents observed using side-scan sonar. *Cont. Shelf Res.* **1**, 353−384.

Thorpe, S.A., Curé, M.S., Graham, A. and Hall, A.J. (1994) Sonar observations of Langmuir circulation, and estimation of dispersion of floating particles. Submitted to *J. Atmos. Ocean. Technol.*

APPENDIX: GENERATION OF SKELETONS FROM THE SONOGRAPHS

The blob is defined as a set of pixels of value 1, other pixels have a value of 0. The origin of the blob is the pixel with the earliest time and least range. All pixels are eligible for deletion except for this one. The next clockwise perimeter pixel of the blob is identified and called X. The values of the boundary pixels 0−7 are examined (see Figure 2.A1a). In this set, the number of non-zero pixels $= nb$ and the number of zero to one transitions $nzero$. If pixels 0, 2 and 4 all equal 1, $p024 = 1$. If pixels 2, 0, 6

Figure 2.5 Gradients of the linear portions of diffusion curves for a variety of data from both late winter (o) and autumn (x), as referred to in the caption to Figure 2.4. They show a scatter around 1.0 which is equivalent to Fickian diffusion. (b) The constant part of K_y, given by $am/2$, while having a large scatter, shows a tendency to increase with increasing wind speed. (c) K_y is derived from the results in (a) and (b) for a diffusion time of 3600 seconds. Dye diffusion results from Schott *et al.* (1978; □), Meerburg (1972; +) and Elliott and Wallace (1989; △) are shown for comparison. The results from the Monte Carlo simulation of Faller and Auer (1987) for floating tracers is also shown (◇). For this last comparison, typical mean inflow velocities to convergent lines and mean spacings of those lines as derived from the sonar data have been used in their non-dimensional relationship for K_y

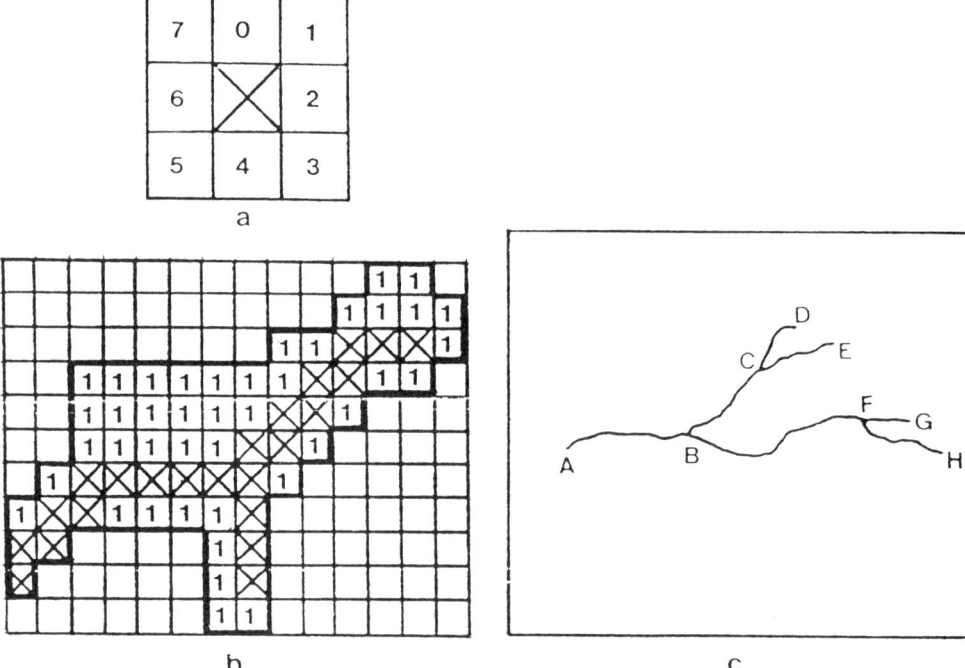

Figure 2.A1 (a) Pixel X lies on the perimeter of a blob. When considering whether such a pixel should remain or be deleted from the skeleton, the boundary pixels 0–7 are examined. (b) The solid boundary is the original blob. Pixels marked X are the skeleton representation of the blob, the 'convergent line', after redundant pixels marked 1 have been deleted. (c) In this skeleton, ABFH is isolated as the first main 'trunk'. Other branches are detached. Subsequently, BCE is also kept, but branches CD and FG are of insufficient duration and are discarded

all equal 1, $p206 = 1$. The algorithm used to decide whether a boundary pixel is set to zero and deleted or kept is presented here in the 'C'-like pseudo code:

```
if (nb > =2 and nb < =6) {
    if (nzero = =1) {
        if (p024 = =0  or  (nzero for pixel 0 !=1) {
            if (p206 = =0  or  (nzero for pixel 2 !=1) {
                set the pixel X to 0;
            }
        }
    }
}
```

The blob is circuited in a clockwise fashion until no more pixels are deleted. Figure 2.A1b shows how a blob is reduced to a connected set of pixels by this algorithm. The final structures resemble trees with many branches, some of them small. Branches which have a duration of less than 60 seconds (features lasting less than this time are considered inconsequential) are pruned as follows. For each structure the pixels with the minimum time and the maximum time are identified. If the time difference between this is less than 60 seconds, the whole is deleted; otherwise the intermediate pixels are saved and all of the branches from this main 'trunk' are detached. The detached branches each become eligible for consideration as structures in their own right. The process continues until every single line of pixels has been processed (see Figure 2.A1c).

3 Turbulence Measurements in the River Severn

S.E. HESLOP
Department for Continuing Education, Bristol University, Bristol, UK

M.J. HOLLAND
Centre for Research on Environmental Systems and Statistics, Institute of Environmental and Biological Sciences, Lancaster University, Lancaster, UK

and

C.M. ALLEN

3.1 INTRODUCTION

The prediction of water quality — for example, the spatial and temporal variation of toxic concentrations resulting from pollution incidents in rivers — using existing physically-based computer models requires a better understanding of the fundamental processes involved in pollutant dispersion and sediment transport. In particular, the turbulent characteristics of the fluvial environment need further investigation. This includes the basic length and time-scales of turbulent motions which affect the dispersion, the forces acting within the fluid, and the level of organized motion to be expected in a particular environment. This paper describes work carried out to collect and analyse turbulence data from the upper reaches of the River Severn in Shropshire. The Severn is England's longest river, and measurements were made there because data on the bulk flow structure of several reaches were already available (Beven and Carling, 1992). The work described here was part of a collaborative project between Lancaster University and the Institute of Freshwater Ecology (IFE, formerly the Freshwater Biological Association) at Windermere.

Three periods of field work in 1988 and 1989 yielded large amounts of two- and three-component turbulence data, which were used initially to provide parameters for the further development of a particle tracking (random walk) model (Heslop and Allen, 1989; 1993). A more detailed analysis of these data sets was carried out in 1990–1, followed by more fieldwork in April 1991, to investigate further the structural features of turbulence in the River Severn. This later work concentrated on spectral analysis, bursting phenomena and aspects of the detailed differences between straight and curved reaches on the river. The dynamic behaviour of the data has been investigated by L.A. Smith (this volume).

Cath was very much involved with this work until a few days before her death in March 1991. She was particularly concerned to see the equipment modified and the 1991 fieldwork completed.

Mixing and Transport in the Environment. Edited by K.J. Beven, P.C. Chatwin and J.H. Millbank
©1994 John Wiley & Sons Ltd

3.2   DATA COLLECTION

Data were collected from five cross-sections on two reaches of the upper River Severn near Shrewsbury (see Figure 3.1). A detailed description of four of these sites, at Montford and Leighton, is given in Heslop and Allen (1989) and a summary of their main features is given in Table 3.1. The field sites were selected to give a variety of channel form, and measurements were made over a fairly wide range of discharges. Most of the turbulence work in 1988 and 1989 was carried out in conjunction with standard current metering (by IFE staff) and also with dye tracing work (by other workers from Lancaster). The propeller current meters provided mean velocity profiles for the selected cross-sections, whereas the dye-tracing measurements gave contaminant dispersal profiles at the same sections, plus estimates of discharge through the whole reach.

In 1988 most of the data were collected from Montford upstream and downstream riffles (MURIF and MDRIF), with some from the Leighton new upstream pool and downstream sites (LNUP and LDSB). In 1989 attention was focused on the two Leighton sites. The 1991 fieldwork was also carried out at Leighton, this time at the 'launch site' (LLS) just upstream of LNUP. The 1991 work was again carried out in collaboration with IFE staff and included the investigation of a large area of recirculation in the lee of an alluvial fan and its associated shear zone where the main flow is deflected. This so-called 'dead zone' is a larger scale version of those structures found throughout the body of a natural channel, which are thought to contribute to the skewness of tracer/pollutant distributions to give a long tail (Wallis, Young and Beven, 1989).

The turbulence measurements were made using Colnbrook (series 3) electromagnetic current meters (ECMs), which allow the simultaneous measurement of two orthogonal components of velocity on each sensor. In 1988 only one ECM was mounted, giving two-component data sets; from 1989 two sensors were deployed and fully three-dimensional velocity records were

Table 3.1 Site descriptions, summary of conditions, and components investigated. Typically, low stage represents a discharge of $6-10\,\mathrm{m^3\,s^{-1}}$, and very high stage one of $120-260\,\mathrm{m^3\,s^{-1}}$. (Note: the components indicated are those with respect to the rig, which are not guaranteed to correspond exactly to the fluvial downstream, transverse and vertical components, unless physical or numerical manipulation is used.)

Site name and code	Description	Dates	Components	Stage
Montford Upstream Riffle (MURIF)	Site on bend 30 m wide, pool on outer bank	26 April 1988 28 July 1988	x, z & y, z x, z & y, z	Low Low
Montford Downstream Riffle (MDRIF)	Site on bend near Montford Bridge, 27 m wide	30 April 1988	y, z	Low
Leighton New Upstream Pool (LNUP)	Straight stretch, reasonably deep, 45 m wide	1 May 1988 27 July 1988 4 April 1989	x, z & y, z y, z $x, y\,z$	Low Low Very high
Leighton Downstream Site B (LDSB)	Bend, mainly riffle but with a deep pool on the outside, 60 m wide	28 April 1988 6 April 1989	y, z x, y & x, z	Low Very high
Leighton Launch Site (LLS)	Area of recirculation and shear formed by deflection of flow around alluvial bar	25 April 1991	x, y, z	Low

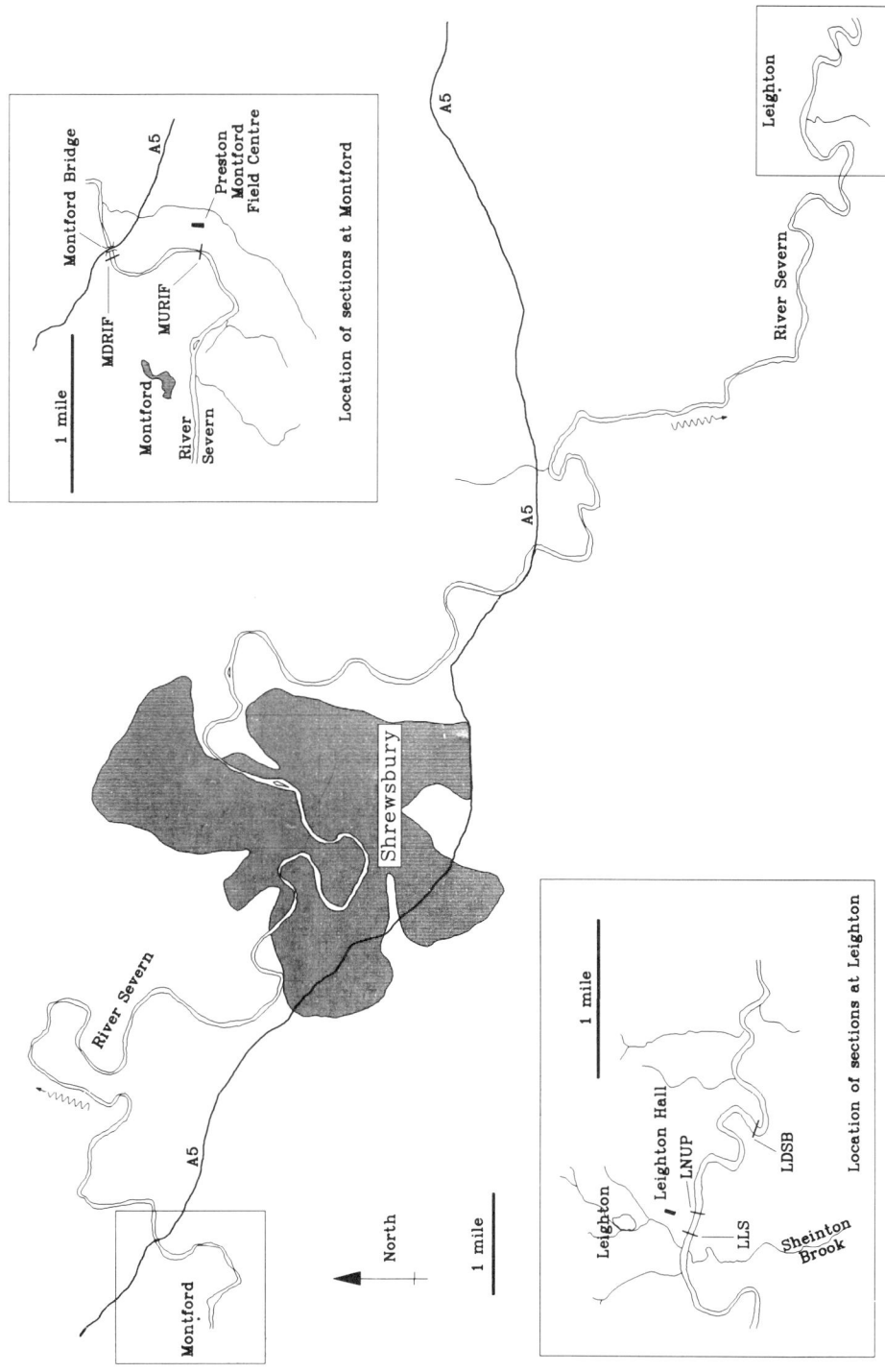

Figure 3.1 Map of River Severn from Montford to Leighton showing location of sections studied

obtained. The turbulence levels were measured at several heights above the river bed so that the relationship between the velocity fluctuations and depth could be investigated.

If an ECM was used independently to generate a one- or two-component velocity record, then the instrument's performance was satisfactory. However, the investigation of a three-dimensional velocity field, involving the combination of records from separate meters to yield the full U, V, W component set (i.e. the mean plus the fluctuating parts of the resultant field as resolved along the *x, y, z* axes of the rig), demanded a greater uniformity of response characteristics between instruments than was available at the time. This non-uniformity resulted in the combination of signals from instruments with various noise levels, frequency responses and drift resistances; the characteristics of the resultant three-component record was therefore representative of the particular instruments deployed in a particular configuration, as well as the velocity field being measured. Upgrading of all instruments to a (higher) common standard has now been effected.

The ECMs were mounted on a heavy frame which was deployed from the back of a boat and rested on the river bed. This equipment underwent several modifications as experience was gained, and the latest version is shown in Figure 3.2: extra weights may be incorporated in the base, and numerous heights (up to 3.5 m from the bed) and attitudes of sensor operation are attainable. Sensor separation is an important variable to be considered, as this, rather than individual sensor size, determines the frequency-response maxima when using two instruments as shown in Figure 3.2. The rig allows flexibility in this respect: a centre to centre distance of 12 cm was used in 1989, and one of 16.5 cm in 1991.

A problem with the early method of data collection was the orientation of the sensors with respect to the main flow of the river. Misalignment of the sensor heads can lead to a 'shadowing' effect, resulting in a false velocity record, so it is important to ensure that the plane of the sensor head disc is aligned with the resultant velocity, *V*, at the point of measurement, i.e. no net flow parallel to the shaft. When two meters are used simultaneously, the best results are obviously achieved when *V* is coincident with the line of intersection of

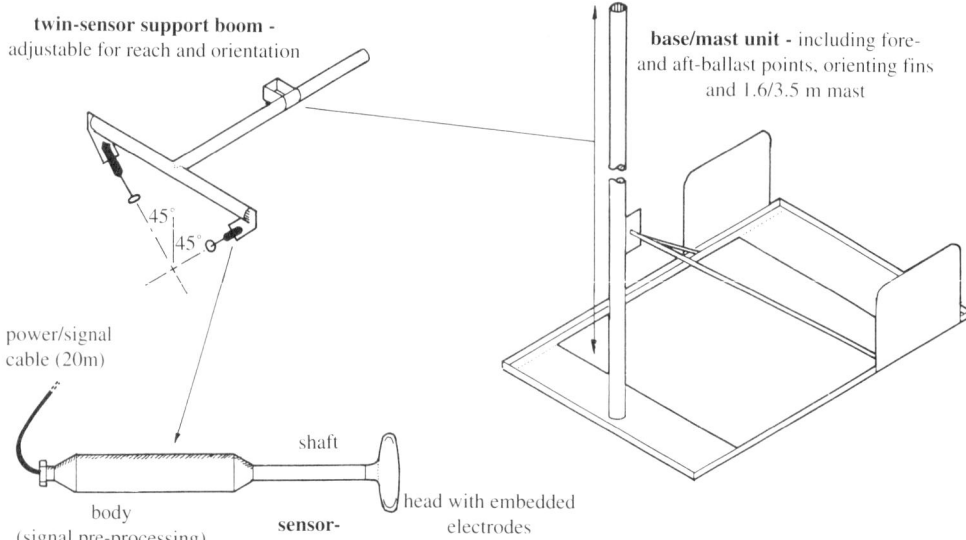

twin-sensor support boom -
adjustable for reach and orientation

base/mast unit - including fore-
and aft-ballast points, orienting fins
and 1.6/3.5 m mast

45°

45°

power/signal
cable (20m)

shaft

body
(signal pre-processing)

sensor-

head with embedded
electrodes

Figure 3.2 Sketch of electromagnetic current meter (ECM) deployment rig

the two planes defined by the sensor heads (see Figure 3.3a). To ensure correct alignment of the rig (and therefore the sensor heads), two large wooden fins are attached to the rig. These fins are acted upon by any residual transverse flow to twist the rig around its vertical axis as it is lowered to the bed.

In this way a single turbulence measurement at a point results in four time series (two from each meter), and a combination across three sets of two (shown in Figure 3.3c) then provides a three(orthogonal)-component record (see Soulsby, 1981). The effect of helicoidal flow, which is expected in the region of river bends, requires assessment of the attitude of the rig during measurement, so that the mean velocity components from points over one cross-section may be better related. This is achieved with a simple compass and levelling bubble arrangement.

Data logging was carried out initially with an Epson microcomputer, which used a 12-bit analogue to digital converter to digitize data in the field at 10 Hz. This system, developed at Lancaster, worked very well but the memory available for data storage was limited. In 1989 a Kyowa multi-channel analogue tape recorder was borrowed from Heriot-Watt University and the tapes digitized at a later date, again at 10 Hz. In 1991 an Epson portable computer with a large storage capacity allowed a return to on-site digitization and also allowed real-time visualization of the velocity time series.

To make a better assessment of possible low frequency components, the data-collection time was increased from five or six minutes in 1988 to 10 minutes in 1989. Currently, a total of 2^{13} points (819.2 seconds at 10 Hz, approximately 14 minutes) is the normal length of data set.

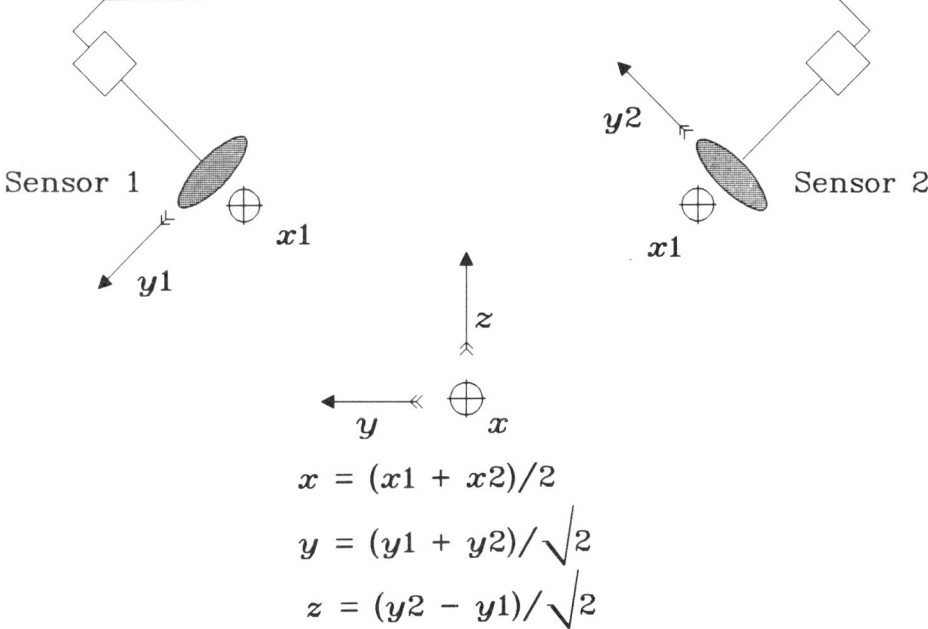

$$x = (x1 + x2)/2$$

$$y = (y1 + y2)/\sqrt{2}$$

$$z = (y2 - y1)/\sqrt{2}$$

Figure 3.3 Method of combining electromagnetic current meter (ECM) outputs to provide three-dimensional records. (a) Recording situation, looking downstream in the plan of both ECM sensors. (b) Desired components, ($\oplus$) into page and ($\leftarrow$) parallel to it. (c) Method of combination using available components

3.3   ANALYSIS

3.3.1   Two-component data

The time series from Montford and Leighton were subject to a de-spiking algorithm, necessary because of interference on one of the ECM circuits. The mean and fluctuating components (U, V, W, u, v and w) were obtained and the turbulent intensities calculated. In the first period of fieldwork in 1988 there were some problems with the alignment of the frame in the downstream direction. However, when it was possible to check the ECM velocity data against the mean flow data collected by the IFE (Figure 3.4), the agreement was good. This suggested that misalignment of the frame was not critical in calculating the mean and RMS turbulent velocities, unlike the evaluation of Reynolds stresses. In all later fieldwork, the fins fitted to the frame provided much better alignment with the flow. Figure 3.4 also shows that a typical velocity profile from the sections on the River Severn is reasonably log-linear.

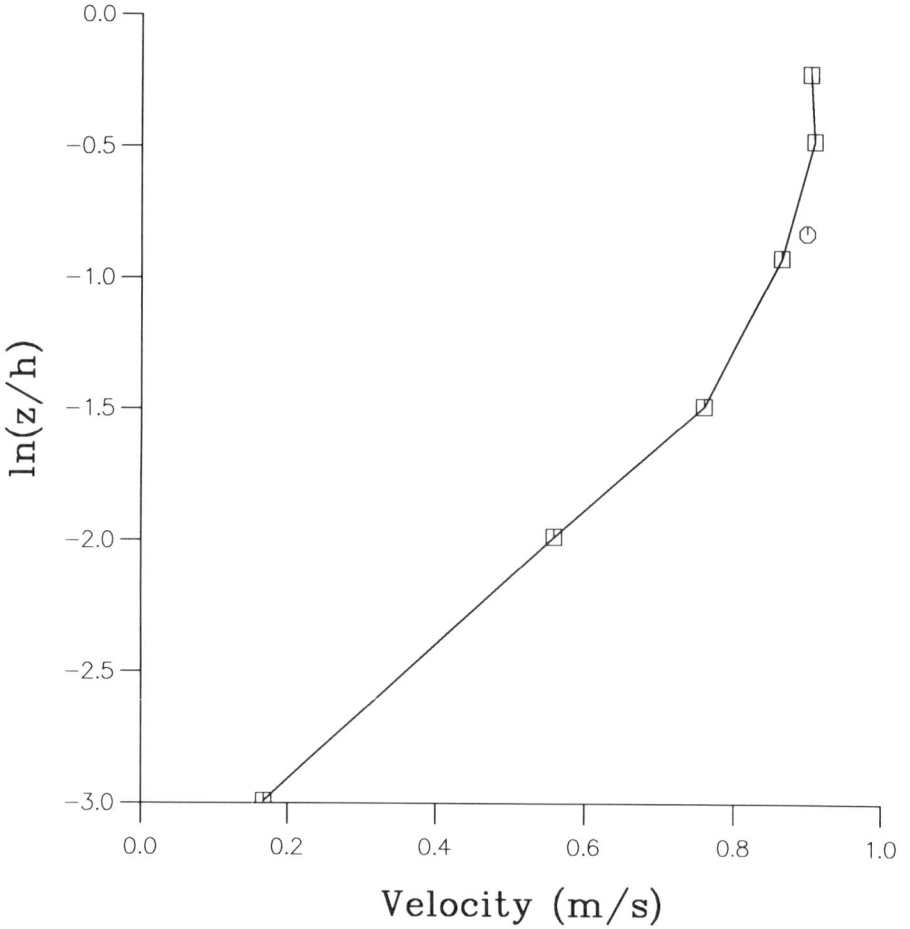

Figure 3.4 Velocity profile for Montford Upstream Riffle. Recorded on 29 April 1988, 18 m from left bank. (o) Electromagnetic current meter output; (□) mean velocity profile at 0.5 m

3.3.2   Three-component data

For all available 1989 data records, the large amplitude—short duration spikes due to electrical noise from the ECM circuits were removed. However, another source of noise had been incorporated into the records at or before digitization, giving time series, dominated by a high frequency component, which bore little resemblance to those recorded in 1988. The analogue tapes were re-digitized, using a 5 Hz anti-alias filter, to give a data set which was of comparable quality (i.e. signal to noise ratio) to the 1988 set. The following analyses were then carried out on all the 1989 and 1991 data sets.

(1) For each time series, the mean velocities and magnitudes of the (normalized) turbulent intensities were determined, and then the turbulent length scales in downstream, lateral and vertical directions were estimated. This was followed by spectral analysis and comparison with results from other turbulent environments.
(2) An analysis of two time series (either uv, uw, or vw) from a single point was then carried out to evaluate the components of Reynolds stress and to give an insight into the 'bursting' phenomena. These phenomena are thought to be linked to intermittent small-scale coherent turbulent structure and to be a cause of sediment migration and suspension in natural channels. They are also possibly linked to the intermittency of small-scale turbulent features such as hairpin vortices or low speed streaks. Any three-component records were subject to a process of re-orientation to estimate the variability of turbulent characteristics with the frame of reference, and so to indicate the sensitivity of the turbulent characteristics to correct alignment. As the only suitable records from 1989 were taken from a straight reach (LNUP), the sensors appeared well aligned with the mean flow direction, and so the results of re-orientation proved to be inconclusive.
(3) Time series from different points were compared to show the variation of turbulence levels over both single verticals and across single sections. This comparison was repeated for series from different reaches to illuminate the variety of processes occurring on straight and meander reaches and their possible interrelations through the identification of secondary flow circulation.

3.4   RESULTS

3.4.1   Two-component data

Plots of mean turbulence against cross-sectional position for LNUP, LDSB and MURIF are shown in Figure 3.5, transverse and vertical components only, as very few measurements in the downstream direction were made at this time. Figure 3.5a shows that the variation in turbulence across the river is very small on the straight stretch LNUP, as might be expected. On the more curved sections (Figure 3.5b and 3.5c) turbulence levels vary across the section, the exact behaviour depending on the height above the bed and the discharge. Figure 3.5c clearly shows the increase in turbulence with proximity to the river bed.

Table 3.2 compares the turbulence intensity results for the River Severn with values obtained by other workers. Owing to the lack of results for non-tidal river environments, the comparison is with estuarine and marine measurements (Soulsby, 1981; West, Knight and Shiono, 1986; Shiono and West, 1987). This comparison shows generally good agreement between all the sets of measurements.

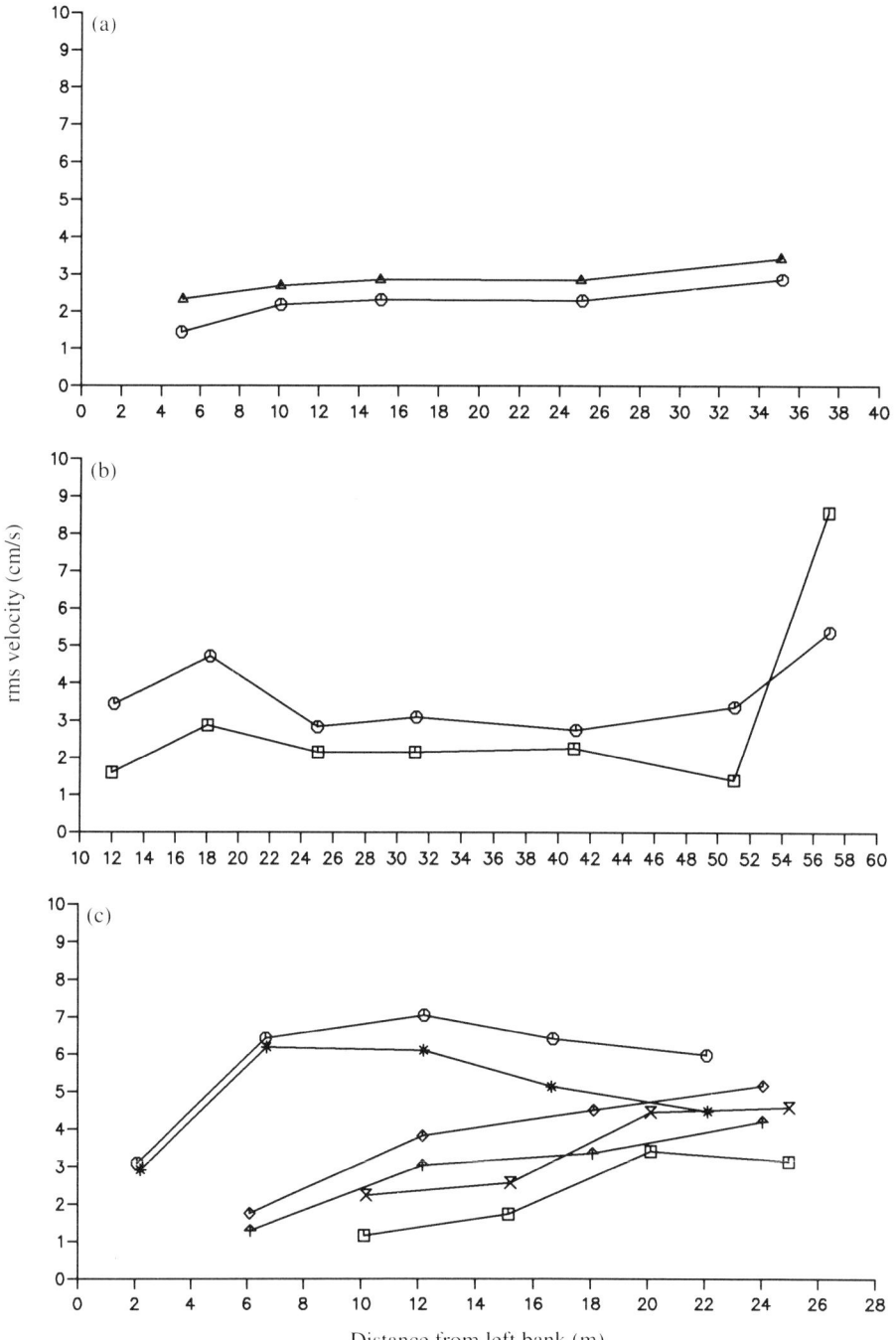

Figure 3.5 Variation of turbulent intensity: measured component and height above bed. (a) Leighton New Upstream Pool, 50 cm (May 1988). (▵) v′; (○) w′. (b) Leighton Downstream Site B, 50 cm (April 1988). (○) v′; (□) w′. (c) Montford Upstream Riffle. July 1988: (○) v′; (*) w′, both 30 cm above bed. April 1988: (◇) v′; (↑) w′, both 50 cm above bed. (▽) v′; (□) w′, 65 cm above bed

Table 3.2 Comparison of turbulence intensities in fluvial, estuarine and marine environments. River Severn data from 1988 fieldwork

Site	Conwy	Great Ouse	Start Bay	Severn
Situation	Estuary	Estuary	Marine sandbank	Non-tidal river
Reference	Shiono and West (1987)	West, Knight and Shiono (1986)	Soulsby (1981)	This work
u'/U	0.1−0.2	0.04−0.16	0.12	0.14−0.15
$w'U$	0.05−0.1	0.02−0.08	0.06	0.02−0.07

These results were used to provide length and time-scale parameters for the random walk model developed by Allen (1982) for the prediction of passive contaminant dispersion. The model was then used to simulate the dye dispersion experiments carried out on the Montford and Leighton sections of the River Severn. Reasonable agreement between observed and predicted concentrations was obtained. A full account of this work is given in Heslop and Allen (1993).

3.4.2 Three-component data

Table 3.3 shows turbulent intensities normalized by mainflow velocity, together with sample variance; the results for three different sites are shown as ratios of the normalized turbulent intensity. The results for straight sections (e.g. LNUP) indicate that conditions there are more uniform (i.e. the sample variance is lower), and show better agreement with Soulsby (1981) than do the data collected at bends (e.g. LDSB). Soulsby's data were collected at a single point over a marine sandbank, within a period of 24 minutes, so we should expect a narrower range of turbulent intensity ratios than in the fluvial example, where the results represent an ensemble of data from different heights and verticals over the same section. In this light the close agreement between accepted data for the marine example and the straight-reach fluvial case is very marked, compared with the curved-reach fluvial situation.

Figure 3.6 shows the variation of turbulent intensity with sampled height (normalized against downstream velocity and flow depth, respectively), confirming, if the data points are reliable, the trend towards higher turbulent energies with proximity to the bed. Note also the *generally* lower turbulence levels of the vertical components compared with the lateral and longitudinal components. This figure represents an ensemble of data from many different flow depths and environments. Detailed data from a single vertical would be expected to show a lower degree of scatter.

Table 3.3 Variation of normalized turbulent intensity. LDSB is classified as a meander reach, and LNUP as a straight reach. (Note: Soulsby's data originally quoted as u'/u_*)

	LDSB (Severn)	LNUP (Severn)	Soulsby (1981; Start Bay)
u'/U	0.116 ± 0.047	0.107 ± 0.009	n/a
v'/U	0.139 ± 0.095	0.099 ± 0.018	n/a
w'/U	0.024 ± 0.035	0.062 ± 0.012	n/a
u'/v'	0.835 ± 0.663	1.080 ± 0.216	1.273 ± 0.003
v'/v'	1.000 ± 0.696	1.000 ± 0.257	1.000 ± 0.032
w'/v	0.173 ± 0.267	0.626 ± 0.166	0.676 ± 0.019

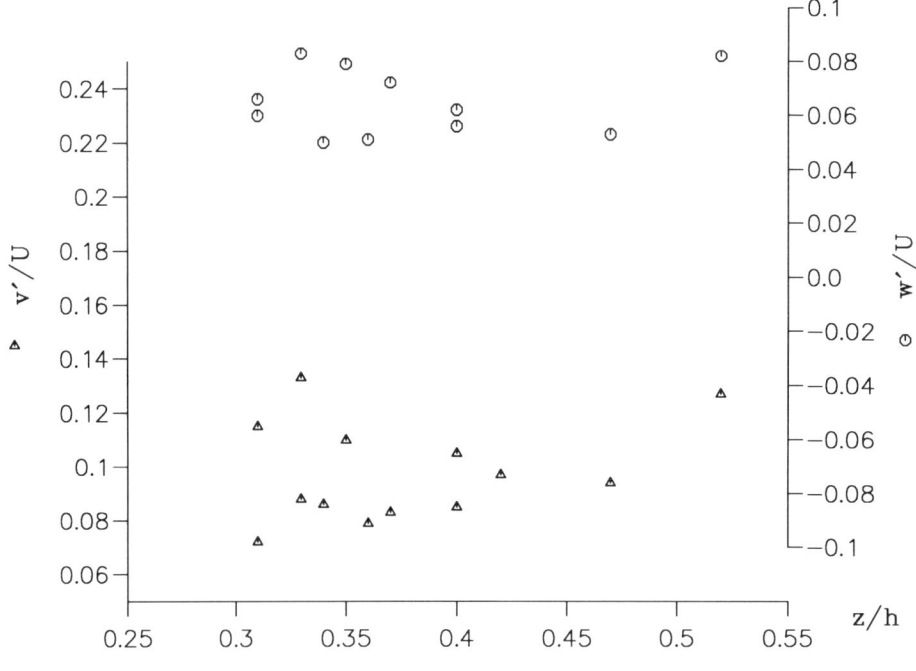

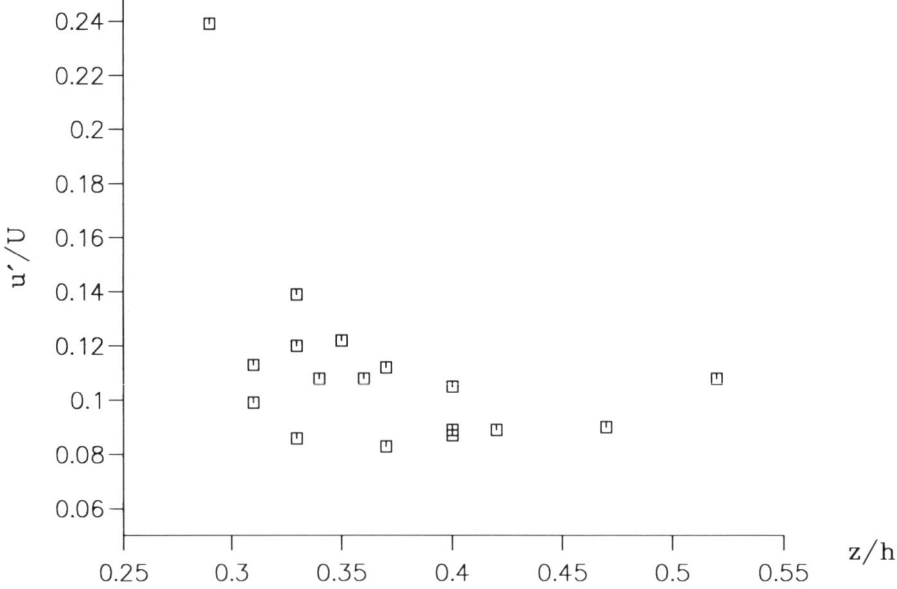

Figure 3.6 Variation of turbulent intensity with relative height. Data from 1989, straight and meander reaches

The average, over time, of the product of any two (zero-mean) velocity fluctuation samples gives an estimate of the relevant component of kinematic Reynolds stress, significant in the initiation of bedload movement and the maintenance of suspended sediment. Although uncertain rig-alignment gives reason to treat the result of this analysis with caution, qualitative differences between different components are not expected to be badly masked. Over a single vertical, laboratory channel measurements of the uw component of Reynolds stress (τ_{uw}) show its linear decay with increasing height above the bed (e.g. McQuivey and Richardson, 1969). Although the Severn results apply across a range of sites, this relationship is reasonably consistent when τ_{uw} (for all 1989 data) is plotted against relative height (see Figure 3.7). The general decay of τ_{uw} with z/h is well demonstrated, unlike that of τ_{uv} and τ_{vw}.

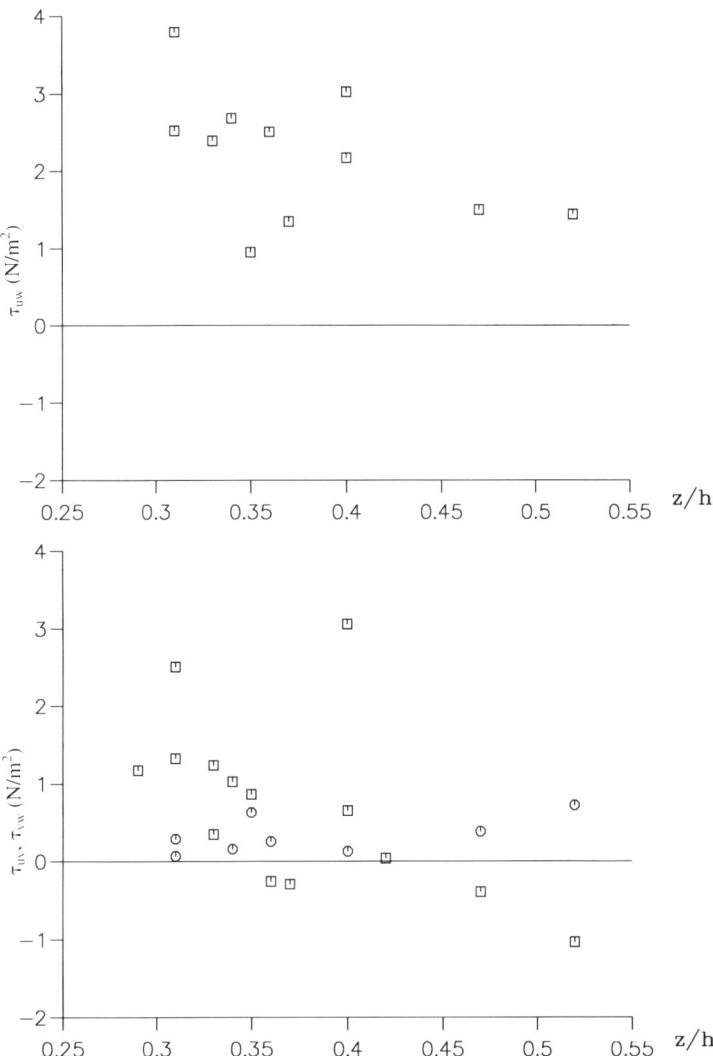

Figure 3.7 Variation of Reynolds stress components with relative height (all 1989 data). (o) vw and (□) uv component

The variation of Reynolds stress with downstream velocity, as investigated by Bowden and Ferguson (1980) in the Irish Sea, was shown to follow

$$\tau_{uw} = cU^2$$

where $2.0 < c < 6.0$.

A comparison of Reynolds stresses from both straight and curved reaches is presented in Figure 3.8. The relation between the uw component and U^2 is well defined, especially when only data from the straight reach are considered (in which case c was found to lie between 2.5 and 4.0). The other two components show no clear relationship with U^2.

Related to the kinematic Reynolds stress, and a cause of the skewed nature of τ_{uw}, is the phenomenon of bursting, or more correctly the burst/sweep cycle. Bursting is the (upward) ejection of low velocity fluid from close to the bed, whereas sweeping is the (downward) inrush of high velocity fluid from higher in the water column. Quadrant analysis (Lu and Willmarth, 1973) of two turbulent components, together with the specification of a threshold stress, H, allows the classification of turbulent events into one of five regions (Figure 3.9a). This then allows analysis by either time or event magnitude, as demonstrated in Figure 3.9b, where the second quadrant (burst) events are identified by shading. Each quadrant may be seen to contribute to the total Reynolds stress in a characteristic manner. Comparison of contributions to τ_{uw} over sections from curved (LDSB) and straight (LNUP) reaches of the river are shown in Table 3.4.

Table 3.5 shows similar results published by Gordon (1974) for a marine example. Note

Table 3.4 Reynolds stresses by quadrant for the River Severn

	% contribution to total τ_{uw} Reynolds stress, by quadrant, H = 0.0			
	1	2	3	4
89LNUP1	−8.23	55.09	−8.26	61.38
89LNUP2	−6.97	61.85	−6.40	51.52
89LNUP3	−6.44	59.16	−5.02	52.30
89LNUP4	−11.89	72.79	−15.29	54.39
89LNUP5	−6.44	64.63	−5.57	47.36
89LNUP6	−5.59	60.60	−3.84	48.83
89LNUP7	−4.70	61.43	−5.63	48.90
89LNUP8	−22.65	84.87	−29.10	66.88
Mean ± σ	−9.11 ± 5.51	65.05 ± 8.87	−9.89 ± 7.99	53.95 ± 6.39
89LDSB8	−107.24	180.34	−115.02	141.92
89LDSB9	−59.82	134.59	−68.35	93.58
89LDSB10	−28.90	97.68	−27.47	58.69
Mean ± σ	−65.32 ± 32.22	137.54 ± 33.81	−70.28 ± 35.77	98.06 ± 34.13

Table 3.5 Percentage contribution to total Reynolds stress, by quadrant, after Gordon (1974)

	Quadrant			
	1	2	3	4
z = 1.0 m	−15 ± 4	65 ± 3	−15 ± 4	67 ± 6
z = 2.5 m	−17 ± 6	73 ± 8	−20 ± 8	65 ± 8

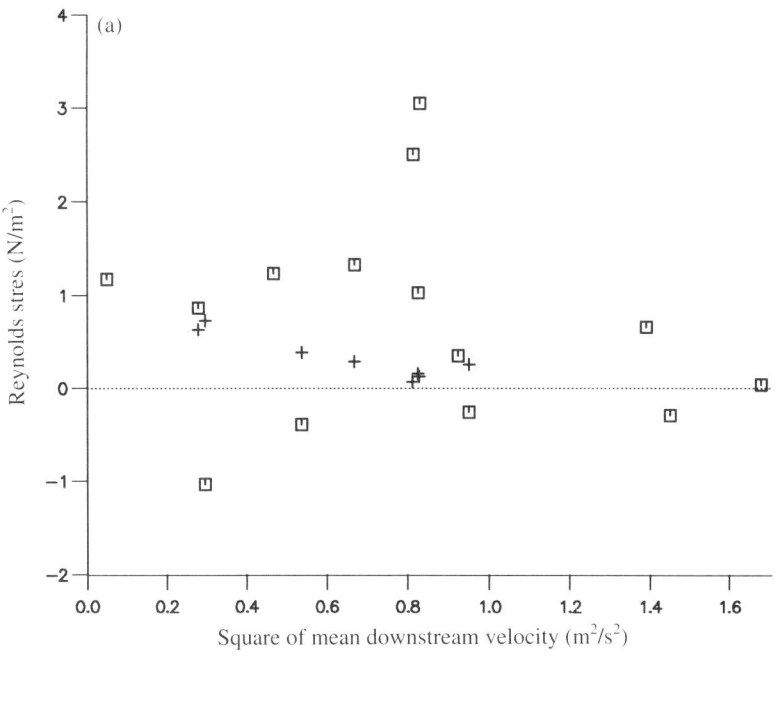

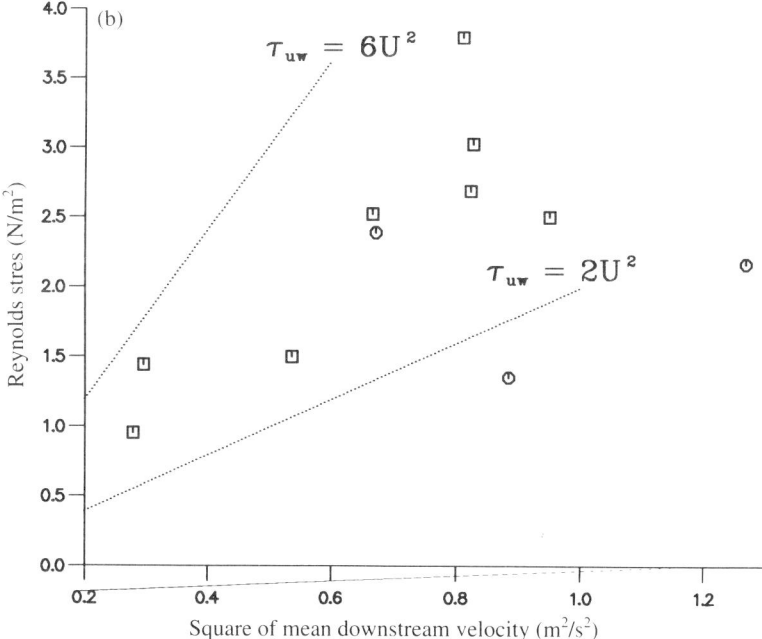

Figure 3.8 Variation of Reynolds stress components with velocity. (a) Comparison of uv (□) and vw (+) components of Reynolds stress from the River Severn. (b) Comparison of straight and meander uw Reynolds stress with observed limits from Irish Sea data. Component (□) uw straight; (○) uw meander

the similarity with the 89LNUP results in all quadrants; the 89LDSB results exhibit a far wider variation in both mean and standard deviation than is shown in either the data of Gordon (1974) or the straight-section data.

If a burst is considered as one or more ejections resulting from the same streak instability, probabilistic analysis (Luchik and Tiederman, 1987) of the time interval between ejections (i.e. quadrant 2 events) can divide the intervals into two classes, namely those separating ejections arising from a common burst and those which separate unrelated ejections. By increasing H (from 0.0 to the point where all events are included in the 'hole' region), and therefore excluding more and more events from quadrants 1 to 4, a narrow range of H may sometimes be found where a kink in the probability against interval plot indicates a likely value for τ, the maximum interval between related events. Although this technique is more commonly applied to smooth-walled laboratory channels than natural channels, where measurement very close to the rough bed is difficult, the 1989 data were analysed in this way. The results showed at least one time series product (89LNUP4uw) where this kind of grouping was indicated. The maximum interval between causally related events in this instance is of the order of 10 seconds, for a value of H of 0.5 to 0.7 (Figure 3.10). Whether this phenomenon is more widely observable in the fluvial environment is obviously important, as it may indicate a temporal aspect of independent transport phenomena.

The main aim of the original fieldwork was the determination of length scales for inclusion in the random walk model to enable different size jumps to be made in appropriate directions

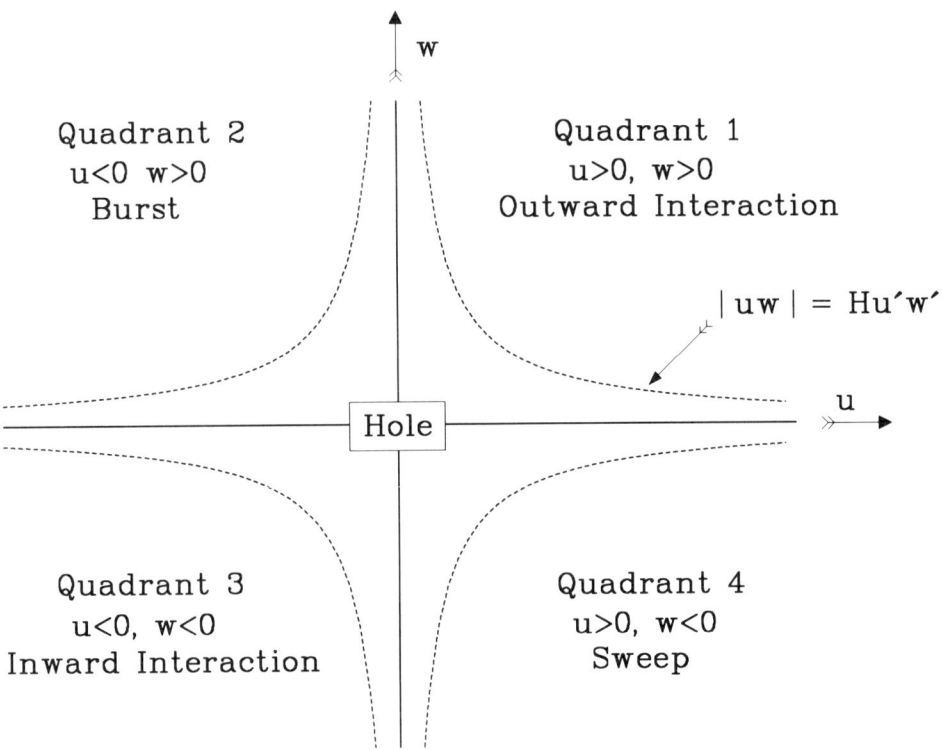

Figure 3.9 (a) Event nomenclature.

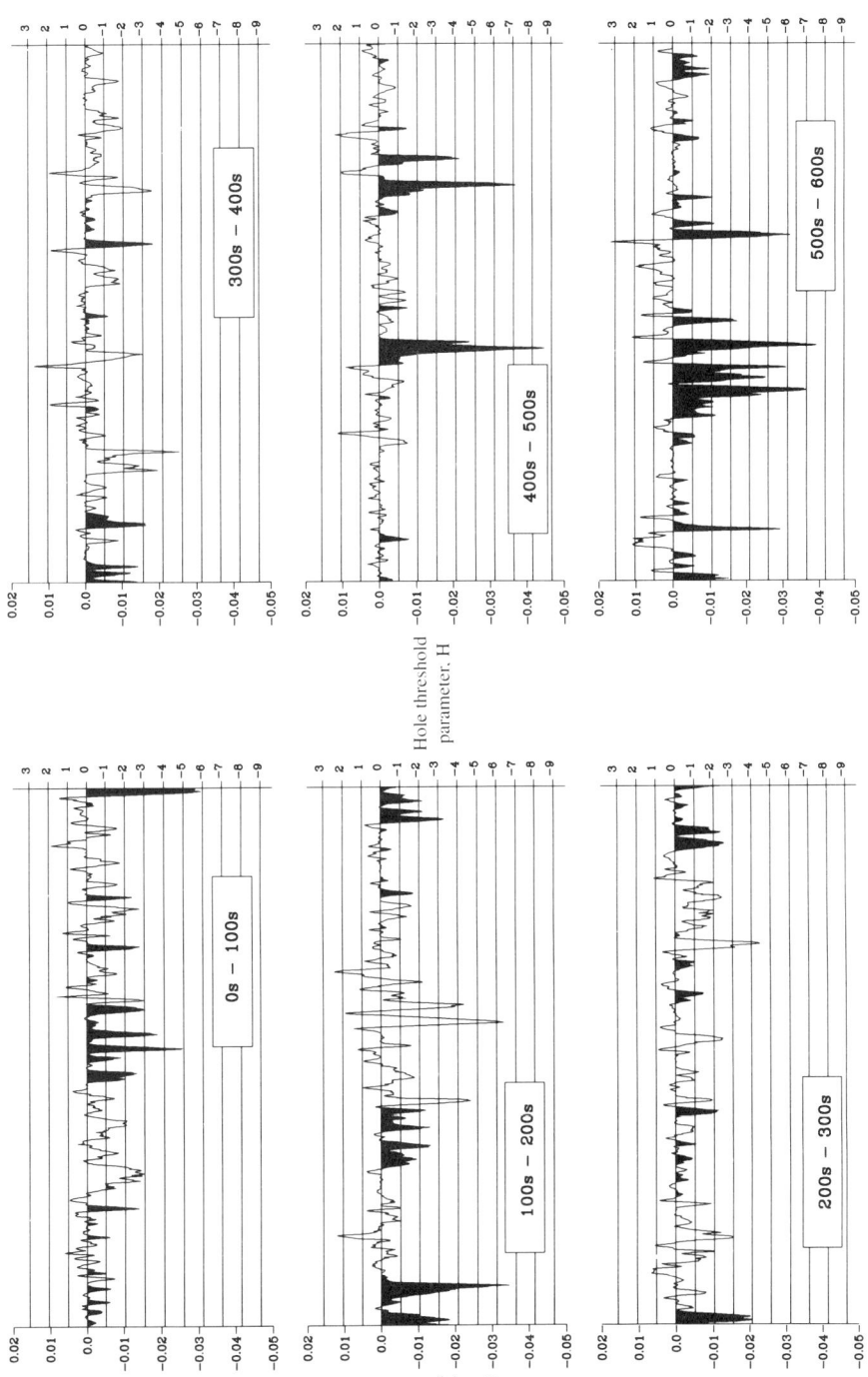

Figure 3.9 (*continued*) (b) Identification of quadrant 2 (burst) events in velocity fluctuation record product (LNUPC4uw)

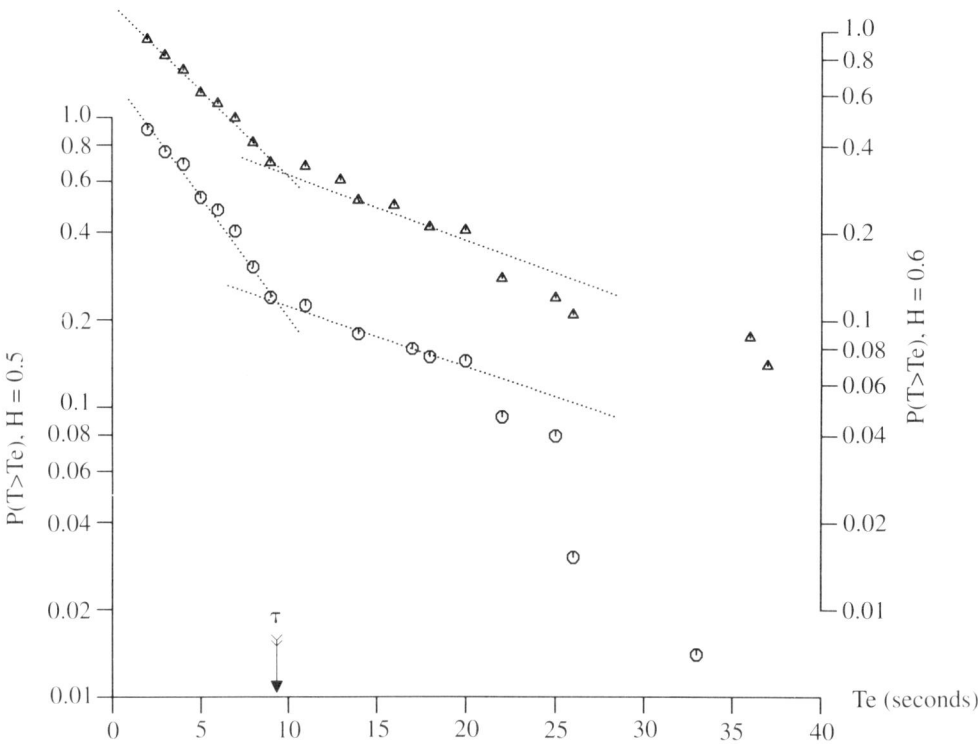

Figure 3.10 Determination of the inter-burst interval, where $P(T > Te)$ indicates the probability of a given interval (T) exceeding the measured inter-event interval (Te), for a specified 'hole factor' H

and at different positions in the modelled channel. For the 1989 data, the method of length scale determination was the construction of a correlogram relating the autocorrelation of a fluctuating velocity component, i.e. the u, v or w turbulence record, to imposed time lags (West, Knight and Shiono, 1986). Integration under the correlogram from zero lag to where the autocorrelation decays to zero, and then multiplying by U generates an estimate of the Eulerian length scale. The results plotted in Figure 3.11 show the much smaller scales of motion which operate in the vertical, and also (more generally) at the extremities of the water column (especially the top).

Length scales have been investigated by West, Knight and Shiono (1986) in a canalized tidal section of the Great Ouse, and by McQuivey and Richardson (1969) in their laboratory channel. Comparisons between these and the fluvial situation are difficult, but, taking the neutrally buoyant data set from the tidal case, broad agreement in behaviour is shown, with maxima being reached for z/h of 0.35–0.4.

The application of a fast Fourier transform technique (combined with a moving average smoothing process) to u, v and w time series shows in which frequencies lie the greatest power contributions. As higher wavenumbers contribute a higher energy for the same amplitude than lower wavenumbers, multiplying the power spectral component by the relevant wavenumber generates a so-called 'equal-area, equal-energy spectrum' (Soulsby, 1983) when working in terms of 'log wavenumber'. Peaks on these spectra are normally observed to shift to higher frequencies for u, v and w components, respectively, behaviour which is observed

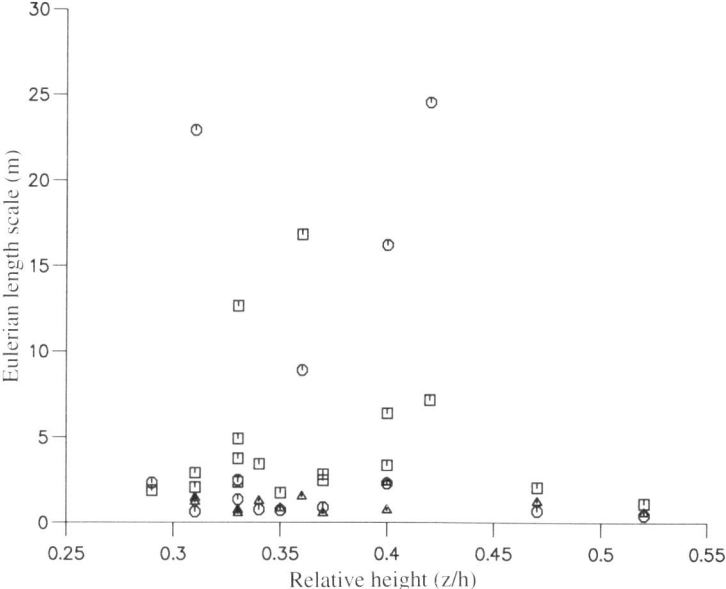

Figure 3.11 Integral length scales by component. (□) u; (○) v; (△) w

to some degree with the 89LNUP data sets (see Figure 3.12). The $k^{-5/3}$ fall off at high frequencies of a power spectrum is characteristic of the inertial subrange, where turbulence is considered to be isotropic. A linear range is indeed observed in log−log plots of the fluvial spectra, but is well within the frequency range of the meter's output filter roll-off curve, and consequently may be much modified, making firm conclusions difficult to draw, although slopes of approximately k^{-4} are observed, implying severe instrumental influence.

A non-stationary spectral analysis of the 1991 data shows some detailed low and mid range variations in amplitude, indicating that changes in character are undergone by the velocity field, even within the time frame of a single measurement.

A determination of turbulent kinetic energies at the sample points was made using the 1991 data only. Turbulent kinetic energies relate all three components of turbulent intensity in the equation

$$E = \frac{\rho}{2}(\overline{u}^2 + \overline{v}^2 + \overline{w}^2)$$

Results from 91LLS illustrate the variation of E from inside the dead zone to the shear zone and then the main flow (Figure 3.13). These data were of use to IFE researchers in verifying a model of sediment settlement in the dead zone. Note the wide range of energies found across the single section, and the relative level of internal noise within the metering/logging system (recorded as a control), and also that, outside the dead zone, the highest value of E is found closest to the bed.

3.5 CONCLUSIONS

This work has examined the turbulent characteristics of flows from selected points on straight and meandering reaches of the River Severn, including a cross-section spanning a shear zone.

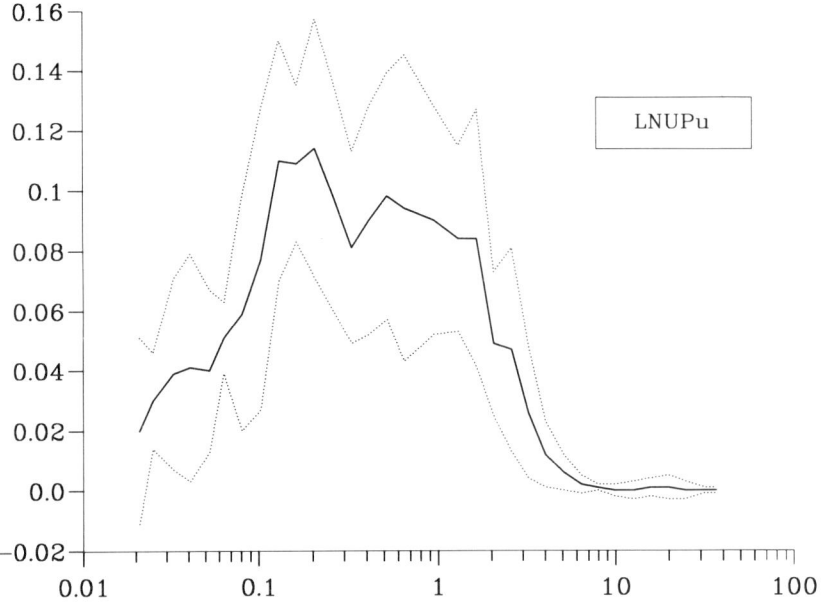

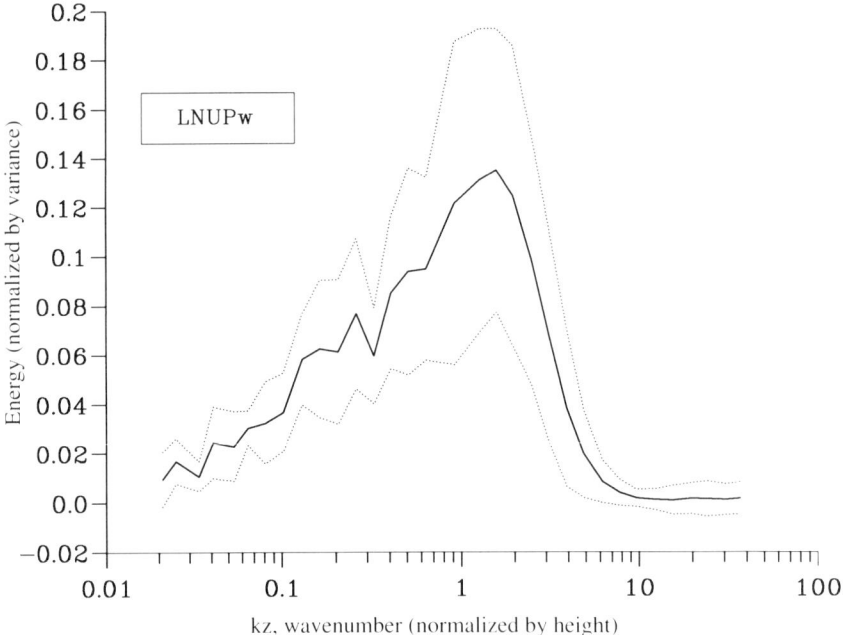

Figure 3.12 Energy spectra for u and w components of turbulent velocity, ensembled average plus standard deviation

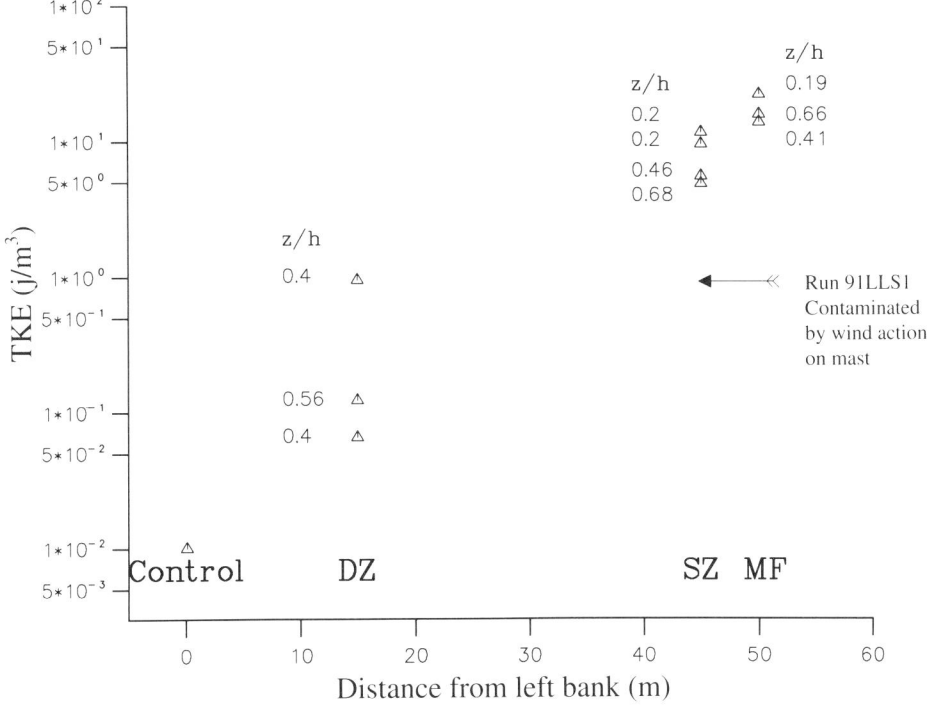

Figure 3.13 Variation of turbulent kinetic energy (TKE), dead zone to shear zone to main flow, illustrating variation with environment and relative height

For the straight-reach data the turbulent characteristics of the flows are fairly similar, considering the relatively low depth and lateral restrictions of the river, to those found in 'two-dimensional' estuarine and marine studies in terms of both the magnitude of scaled velocity fluctuations, Reynolds stresses and 'burst and sweep' characteristics. For the meander reaches, however, much less consistent patterns have been found. This probably arises from the interaction of the bend-induced secondary flows and the structure of the downstream velocity component, but the limited data available have not yet allowed a proper assessment of this interaction.

One major aim of the work was to evaluate the assumptions used in random walk particle tracking models of dispersive transport in rivers. The work has shown that even for straight reaches, it may be necessary to take account of three-dimensional mean and fluctuating velocity fields, especially in the region close to the bank. For certain problems in straight reaches, the two-dimensional velocity profile originally used at Lancaster may be acceptable.

The data analyses that have been carried out suggest that our understanding of the relationship between flow structures and turbulent characteristics in natural rivers is limited at present. The next step might be to mount a small number of intensive field measurements on specific reaches, combining current meter, ECM and tracer studies to supplement laboratory work which has been undertaken at the SERC Flood Channel Facility at Hydraulics Research Ltd, Wallingford. Such field campaigns would also allow more complete validation of the random walk particle tracking models and three-dimensional flow models, as well as the relationship between detailed flow structures and reach scale dispersion models such as the Aggregated Dead Zone (ADZ) model.

There are some points related to the effect of curved channels on turbulent characteristics which require further study. Firstly, when allowances have been made for the helicoidal skewing of the resultant mean velocity vector, does the turbulence structure become normalized and recognizable as a simple variant of the two-dimensional case, or does it take on a form governed by a new set of relationships? Secondly, are the components of Reynolds stress disorganized or reorganized by the helicoidal flow, and will they interact with the gravitational forces on suspended sediment to give implications for deposition processes?

Work will continue on the application of new analysis techniques for the characterization of turbulence data with a view to developing more realistic models of velocity fluctuations. These will be of use in the random walk model of dispersion, plus other physicaly based models which require accurate turbulence information.

ACKNOWLEDGEMENTS

NERC gave financial support to SEH (grant GR3/6432) and to MJH (grant GR9/288). Our thanks go to various members of staff at IEBS, especially to Ian Edmonson and Brian Robinson for electronics and computing support and to Harry Vasey in the mechanical workshop. Adrian Kelsey of IEBS contributed many useful suggestions. Many thanks also to Mark Glaister and Paul Carling of IFE for assistance with fieldwork, and to Paul Roberts and John Millbank for helping with hardware modifications.

REFERENCES

Allen, C.M. (1982) Numerical simulation of contaminant dispersion in estuary flows. *Proc. Roy. Soc. A*, **381**, 179–194.

Beven, K.J. and Carling P. (1992) Velocities, roughness and dispersion in the lowland River Severn. In: *Lowland Rivers: Geomorphological Perspectives* (Eds G.F. Petts and P. Carling. Wiley, Chichester, 71–93.

Bowden, K.F. and Ferguson, S.R. (1980) Variations with height of the turbulence in a tidally induced bottom boundary layer. In: *Marine Turbulence* (Ed. J.C.J. Nihous. Elsevier, Amsterdam, 259–286.

Gordon, C.M. (1974) Intermittent momentum transport in a geophysical boundary layer. *Nature* **248**, 329–394.

Heslop, S.E. and Allen, C.M. (1989) Turbulence and dispersion in larger UK rivers. *Proceedings of the XXIII Congress of the International Association for Hydraulic Research, Ottawa, Canada*, IAHR Publication, Volume D, pp. 75–82.

Heslop, S.E. and Allen, C.M. (1993) Modelling contaminant dispersion in the River Severn using a random walk model. *J. Hydraul. Res.*, **31**–3, 323–331.

Lu, S.S. and Willmarth, W.W. (1973) Measurements of structure of Reynolds stress in a turbulent boundary layer. *J. Fluid Mech.* **60**, 481–512.

Luchik, T.S. and Tiederman, W.G. (1987) Timescale and structure of ejections and bursts in turbulent channel flows. *J. Fluid Mech.* **174**, 529–553.

McQuivey, R.S. and Richardson, E.V. (1969) Some turbulence measurements in open channel flow. *J. Hydraul. Div., ASCE* **95**, 209–223.

Shiono, K. and West, J.R. (1987) Turbulent perturbations of velocity in the Conwy estuary. *Est. Coast. Shelf Sci.* **25**, 533–553.

Smith L.A. (1994) Turbulence in the River Severn: a dynamical systems analysis. In: *Mixing and Transport in the Environment* (Eds K.J. Beven, P.C. Chatwin and J.H. Millbank). Wiley, Chichester, Ch. 20.

Soulsby, R.L. (1981) Measurements of the Reynolds stress components close to a marine sand bank. *Mar. Geol.* **42**, 35–47.

Soulsby, R.L. (1983) The bottom boundary layer of shelf seas. In: *Physical Oceanography of Coastal and Shelf Seas* (Ed. B. Johns). Elsevier, Amsterdam, 189–266.

Wallis, S.G., Young, P.C. and Beven, K.J. (1989) Experimental investigation of the aggregated dead zone model for longitudinal solute transport in stream channels. *Proc. Inst. Civ. Eng., Part 2* **87**, 1–22.

West, J.R., Knight, D.W. and Shiono, K. (1986) Turbulence measurements in the Great Ouse estuary. *J. Hyd. Eng.* **112**, 167–179.

NOMENCLATURE

E	Turbulent kinetic energy
h	Flow depth
H	Threshold factor for burst/sweep intermittency analysis
k	Wavenumber
u_*	Friction velocity
u, v, w	Fluctuating time series of zero-mean, parallel to x, y, z
u′, v′, w′	Standard deviation of u, v, w
U, V, W	Mean velocity in x, y, z directions
U	Sum of u and U, etc.
V	Resultant of U, V, W
x	Downstream axis of the rig, or reoriented* data set
y	Transverse axis (right bank to left, looking downstream) of the rig, or reoriented* data set
z	Vertical axis (bed to surface) of the rig, or reoriented* data set
Z	Height above the bed
ρ	Fluid density
τ	Time displacement of bursts
τ_{ij}	*ij* component of Reynolds stress, one of six, determined by: $\tau_{uw} = -\rho\overline{uw}$

---

*Note that, ideally, the data collection techniques used ensure that any non-zero vertical and transverse mean flows are negligibly small; however, in the case of strong secondary flows, a trigonometric recombination may be necessary to produce this desired state.

4 Boundary Shear Stress Distributions in Open Channel Flow

D.W. KNIGHT and K.W.H. YUEN
School of Civil Engineering, The University of Birmingham, Birmingham, UK

and

A.A.I. AL-HAMID
King Saud University, Riyadh, Saudi Arabia

4.1 INTRODUCTION

The distribution of boundary shear stress around the wetted perimeter of a channel is influenced by many factors, notably the shape of the cross-section, the longitudinal variation in planform geometry, the sediment concentration and the lateral and longitudinal distribution of boundary roughness. All of these factors combine to make the prediction of the local shear stress at any point on a channel boundary a particularly difficult task. This is unfortunate given the importance of the boundary shear stress, or the corresponding shear velocity, in many problems in river engineering. Such problems might include the resistance of the channel, the sediment-carrying capacity of the channel, bank erosion and dispersion phenomena. The related topics of tractive force design for irrigation canals and sidewall correction procedures for laboratory flume studies also serve to highlight the importance of this subject.

This review draws on some recent studies of the distribution of boundary shear stress in prismatic and non-prismatic open channels under sediment-free conditions. Particular emphasis is given to simple or compound trapezoidal channels as these are often typical of river cross-sections. Following a description of the three-dimensional velocity field in such channels, typical boundary shear stress distributions are given for different aspect ratios and Froude numbers. Equations are presented for the percentage of the total shear force which acts on the walls of a channel, together with ancillary equations which give the mean bed and wall shear stresses. The influence of heterogeneous roughness distributions and planform geometry are also briefly considered. One of the aims of the review is also to highlight the relationship between local boundary shear stress and lateral eddy diffusivity, a topic on which Cath Allen was working before her untimely death.

4.2 FLOW STRUCTURES IN OPEN CHANNELS

It is important to appreciate the general three-dimensional flow structures that exist in straight or curved prismatic channels to understand the lateral distribution of boundary shear stress. These are best understood by reference to the governing equations and illustrative sketches provided by various workers. The Navier−Stokes equations for turbulent flow (Schlichting,

Mixing and Transport in the Environment. Edited by K.J. Beven, P.C. Chatwin and J.H. Millbank
© 1994 John Wiley & Sons Ltd

1979), which express the mean and fluctuating velocity components ($\bar{U}$, $\bar{V}$, $\bar{W}$ and u, v, w) in terms of space and time (xyz and t), may be rewritten in terms of vorticity, Ω, the longitudinal component of which, ξ ($d\bar{W}/\partial y - \partial \bar{V}/\partial z$), is especially important in boundary shear stress work. The pattern and number of secondary flow cells will influence the lateral distribution of boundary shear stress τ_b, and are described elsewhere by Perkins (1970), Knight *et al.* (1982), and Knight and Patel (1985b).

Rather than present a detailed mathematical analysis, we have attempted to provide a visual picture of the flow mechanisms and extensive plots of actual boundary shear stress data from which some general features emerge. The number of equations has therefore been purposely restricted. By way of illustration, Figure 4.1 shows the interaction between the primary velocity $\bar{U}$, presented in the form of isovels, the secondary flow velocities, $\bar{V}$ and $\bar{W}$, indicated by the presence of two pairs of contra-rotating secondary flow cells, and the resultant distribution of boundary shear stress, τ_b, around the wetted perimeter. This pattern arises because of the anisotropic nature of the turbulence, particularly in corner regions, which gives rise to transverse gradients of Reynolds stress components which act wholly in a plane normal to the primary flow direction. These gradients are responsible for the production of longitudinal or axial vorticity and the presence of the secondary flow cells; see for example, Einstein and Li (1958), Brundrett and Baines (1964), Tracy (1965), Perkins (1970), Gessner (1973) and Melling and Whitelaw (1976) for a fuller description.

4.3 VELOCITY DISTRIBUTIONS

Despite the insights provided by the governing equations, some people are often still surprised by the three-dimensional nature of the velocity field, even in straight prismatic channels with a simple cross-section. Figure 4.1 illustrates some isovels in a trapezoidal channel, together with the corresponding distributions of boundary shear stress measured independently from the velocity. The flow is supercritical with a Froude number, Fr, equal to 3.24 and the channel aspect ratio, B/H, is 1.52. The influence of four dominant secondary flow cells is self-evident. Figure 4.2 illustrates some subcritical velocity distributions for a wide range of aspect ratios but constant Froude number. Not surprisingly, as the aspect ratio increases the flow tends towards a more two-dimensional structure except in the corner and wall regions. Figure 4.3 illustrates these velocity distributions isometrically, together with data at aspect ratios of 1.5 and 2.0 in subcritical and supercritical flow.

The preceding figures have all illustrated the complex nature of velocity fields in simple trapezoidal channels, and anyone associated with river gauging in natural river channels will be aware of similar three-dimensional features. As it is customary to obtain the lateral distribution of depth-averaged velocity, U_d, across a channel in stream gauging, Figure 4.4 illustrates certain features of this, again for simple trapezoidal channels. The velocity data are presented in a dimensionless form by dividing U_d by the section-mean velocity, U_0. Over the central portion of the channel, U_d/U_0 is approximately constant at around 1.05 for the aspect ratios shown. However, there are clearly other features present, as the maximum depth-averaged velocity, U_d, is not at the centre of the channel in Figure 4.4a, but displaced laterally some considerable distance towards the corner. This is due to secondary flow effects and is not dissimilar from the effect on boundary shear stress, τ_b, shown in Figure 4.1. For a wider channel, Figure 4.4b shows that the maxima are at the channel centreline for supercritical flow, but not for subcritical flow. These diagrams thus serve to illustrate the

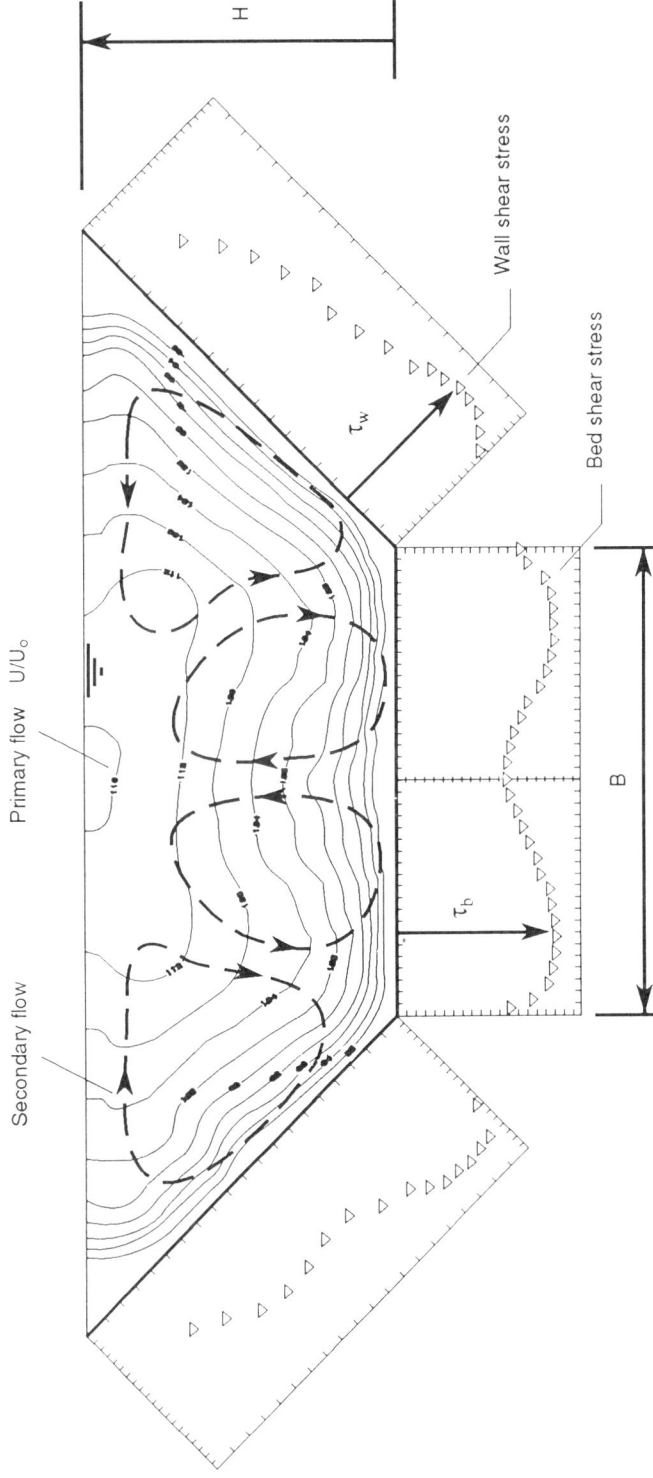

H

Wall shear stress

τ_w

Bed shear stress

Primary flow U/U_o

B

Secondary flow

τ_b

Figure 4.1 Typical relationship between boundary shear stress distribution, secondary flows and primary flow in a trapezoidal channel. $Fr = 3.24$, Asp $= B/H = 1.52$

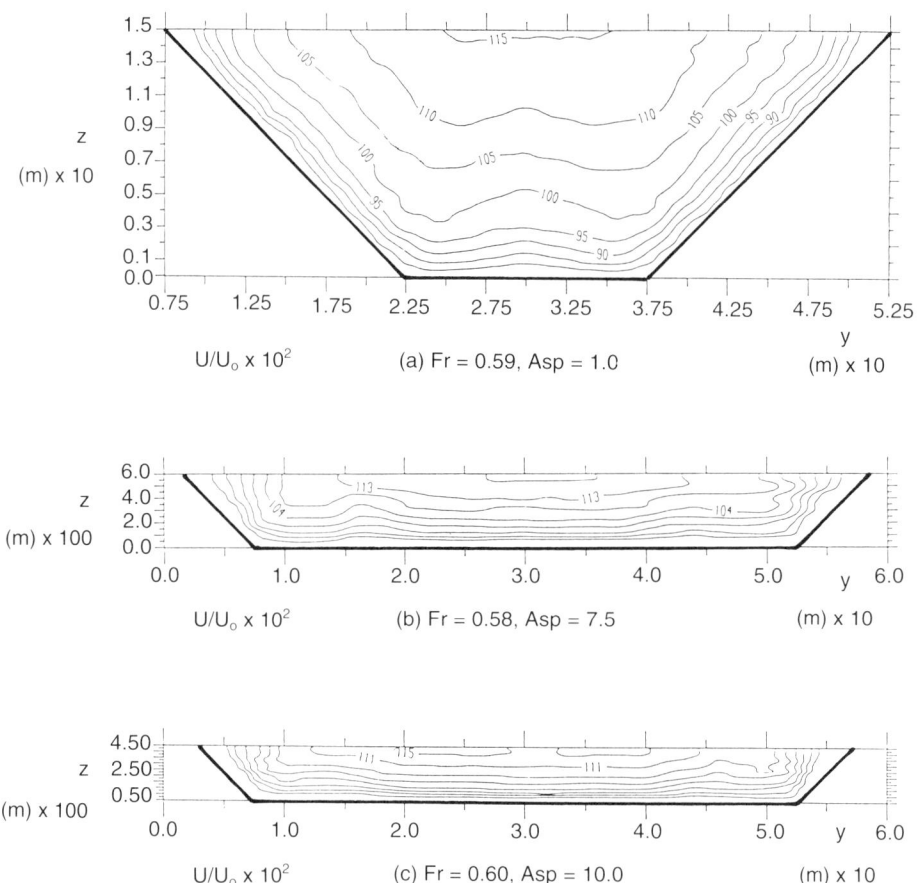

Figure 4.2 Isovel patterns in trapezoidal channels for various aspect ratios at $Fr = 0.6$

three-dimensional nature of the flow, its influence on depth-averaged parameters and the importance of longitudinal vorticity, even in simple trapezoidal channels.

Because of this it is not surprising that the determination of the distribution of boundary shear stress around the wetted perimeter of a channel of any shape is such an intractable problem. It is often assumed that a detailed knowledge of the velocity field is usually sufficient to determine the local boundary shear stresses accurately. Indeed, in field studies the logarithmic velocity distribution law is commonly adopted in the absence of any better approach. Although this may give approximate values for τ_b, it is unlikely to be sufficiently accurate for detailed analysis and for many computational models. Figure 4.5 illustrates this problem using the high quality velocity data already presented in Figures 4.1–4.4. The data in Figure 4.5 deviate from the logarithmic relationship, and as a result the boundary shear stresses derived from them would not be as accurate as those shown in Figure 4.1.

The problem is further exacerbated as the shape of the channel cross-section becomes more complex. Figures 4.6 and 4.7 show how the velocity field varies in a compound or two-stage channel over a range of overbank flow depths, $0.05 < Dr < 0.50$, where Dr equals the relative depth $[Dr = (H-h)/H]$, where H is the depth of flow in the main channel and $H-h$

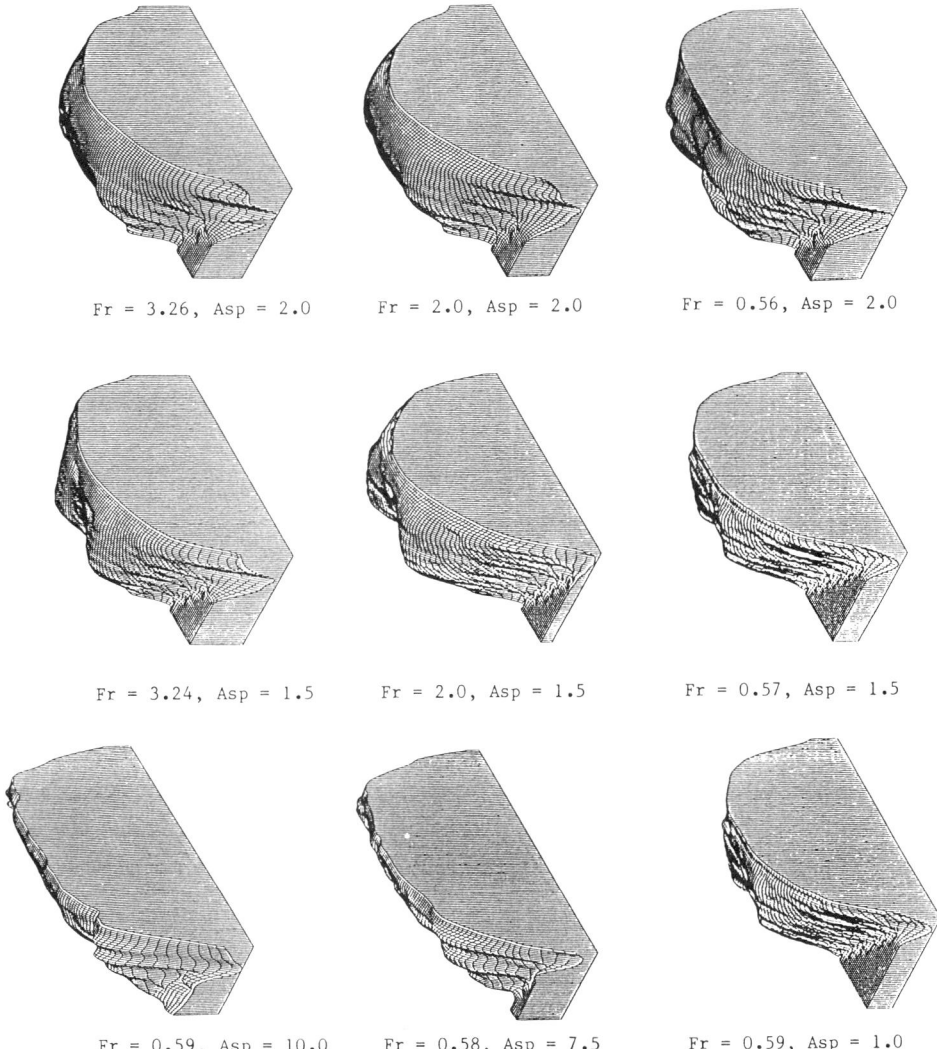

Fr = 3.26, Asp = 2.0   Fr = 2.0, Asp = 2.0   Fr = 0.56, Asp = 2.0

Fr = 3.24, Asp = 1.5   Fr = 2.0, Asp = 1.5   Fr = 0.57, Asp = 1.5

Fr = 0.59, Asp = 10.0   Fr = 0.58, Asp = 7.5   Fr = 0.59, Asp = 1.0

Figure 4.3 Three-dimensional velocity fields in trapezoidal channels for various aspect ratios and Froude numbers. Flow from right to left

is the depth of flow on the floodplain. This channel has particularly narrow floodplains as the flume geometry was selected to enable critical flow studies to be undertaken. Although such a channel might well represent certain drainage channels, most rivers will have wider floodplains and larger aspect ratios for the main channel. Experimental studies in these channels are described later. In this particular case the flow field is manifestly very three-dimensional in nature, giving rise to non-logarithmic velocity distributions normal to any boundary element. Some lateral variations of U_d/U_0 for this particular compound channel are shown in Figure 4.8 and again serve to illustrate the complex nature of the flow field. A detailed discussion of these data may be found in Yuen (1989) and a theoretical interpretation by Shiono and Knight (1989), Knight, Samuels and Shiono (1990), Knight and Shiono (1990) and Shiono

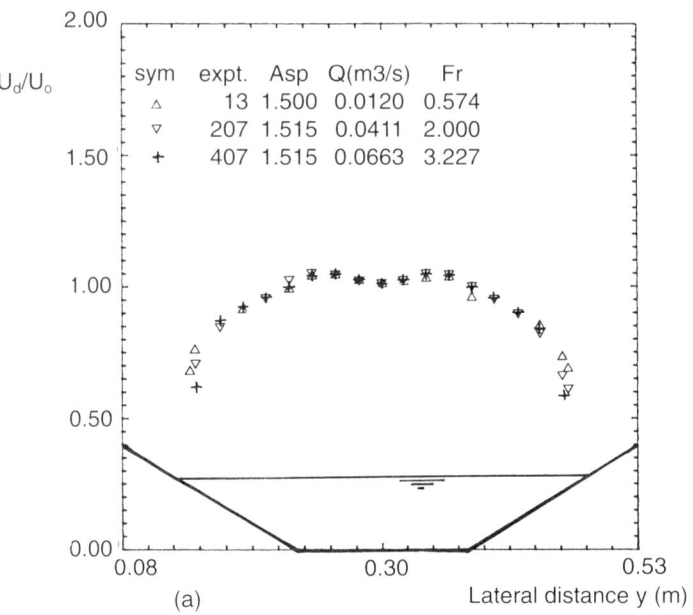

(a)

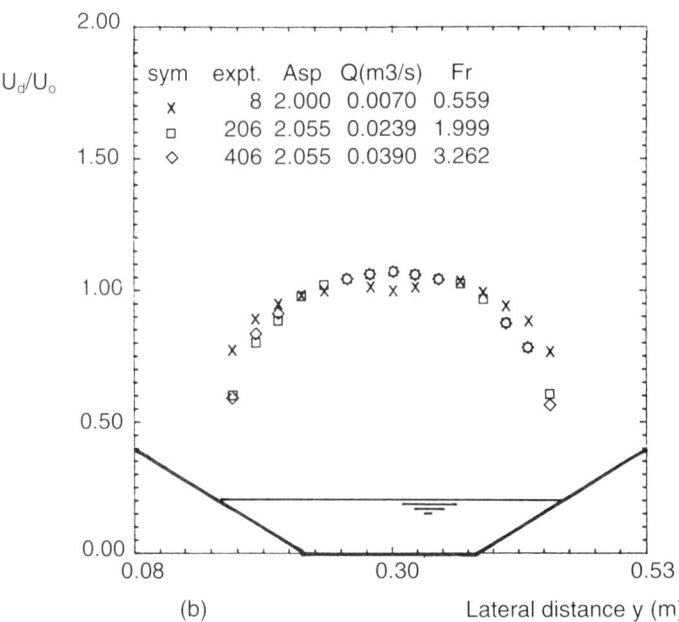

(b)

Figure 4.4 Variation of depth-averaged velocity across trapezoidal channels for constant aspect ratios and various Froude numbers

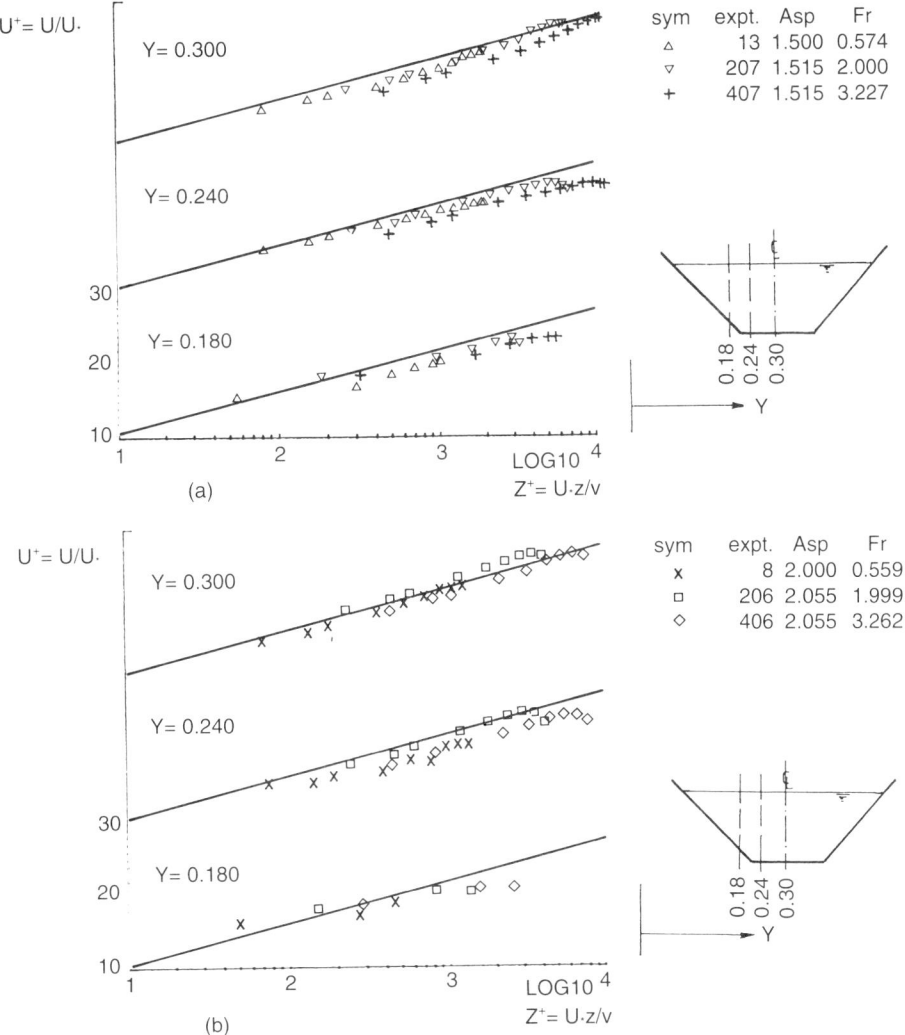

Figure 4.5 Velocity profiles at selected lateral distances in trapezoidal channels for constant aspect ratios and various Froude numbers

and Knight (1991). The velocity field at critical flow, i.e. $Fr = 1.0$, is discussed by Yuen and Knight (1990). The Froude number, Fr, is defined as $Fr = U_0/(gA/T)^{1/2}$ where A is the cross-sectional area and T is the surface width.

As a channel meanders and develops into a geomorphologically stable system, with the channel cross-sectional geometry varying in a longitudinal direction, the flow field becomes more three-dimensional in nature and the corresponding velocity distributions and boundary shear stresses yet more difficult to compute accurately; see Chang (1988), Ikeda and Parker (1989) and data from the Science and Engineering Research Council Flood Channel Facility (SERC-FCF) available through HR Wallingford, Wallingford for more details. This is illustrated in the boundary shear stress data of Knight, Yuan and Fares (1992).

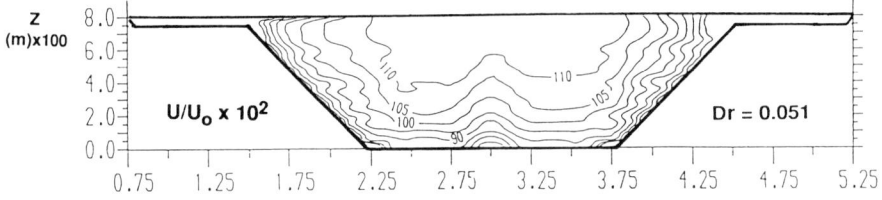

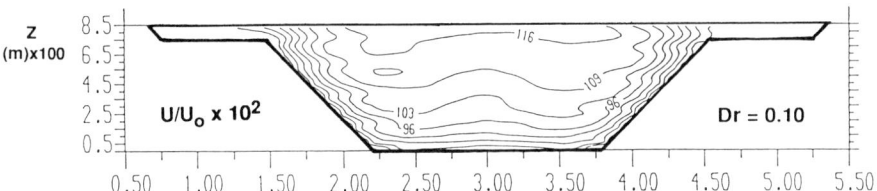

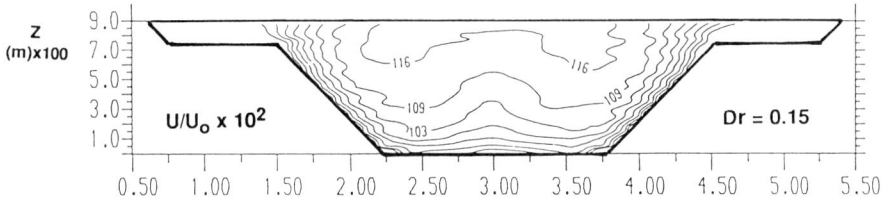

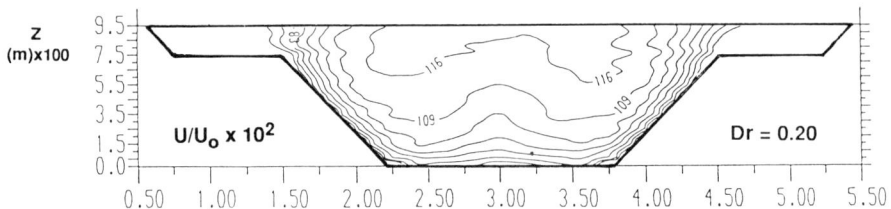

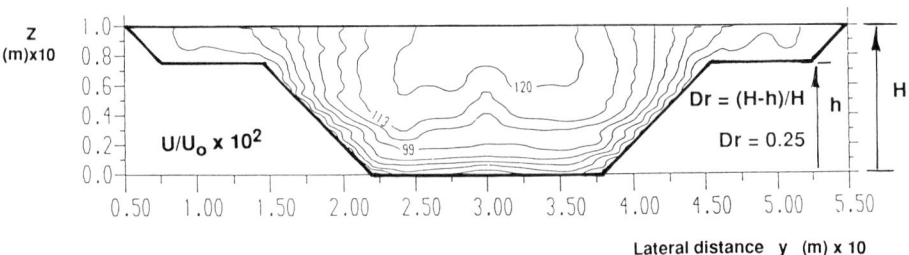

Lateral distance   y   (m) x 10

Figure 4.6 (a) Subcritical isovel patterns in compound trapezoidal channels for various relative depths, Dr ($0.051 < Dr < 0.25$)

4.4 BOUNDARY SHEAR STRESS DISTRIBUTIONS

Work at the University of Birmingham by Knight and co-workers (1978–92), Patel (1984), Lai (1986), Yuen (1989), Alhamid (1991) and Rhodes (1991) has led to an improved understanding of the lateral distributions of boundary shear stress in prismatic channels and ducts. Given the complex velocity distributions described earlier, the boundary shear stresses were determined indirectly by the tube technique of Preston (1954), using the calibration of Patel (1965). Particular attention was paid to accuracy, and the integrated mean value was always checked with the mean energy slope or pressure gradient value. This procedure is described in detail elsewhere by Knight, Demetriou and Hamed (1984). Where the two values differed by more than a small percentage, experiments were repeated to ensure a high degree of control on the quality of the data. The same procedure, that of setting several M1 and M2 profiles to obtain the normal depth precisely and checking integrated values, was also adopted for work on the SERC Flood Channel Facility. The longitudinal water surface slope

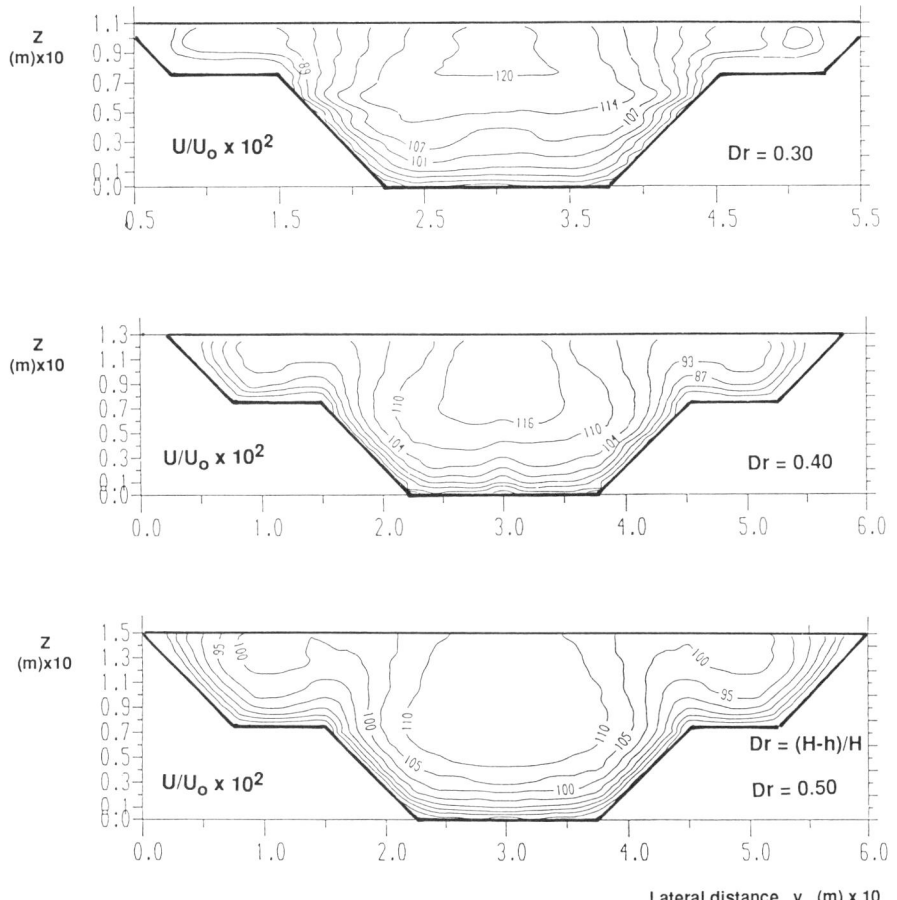

Figure 4.6 (*continued*) (b) Subcritical isovel patterns in compound trapezoidal channels for various relative depths, *Dr* (0.3 < *Dr* < 0.5)

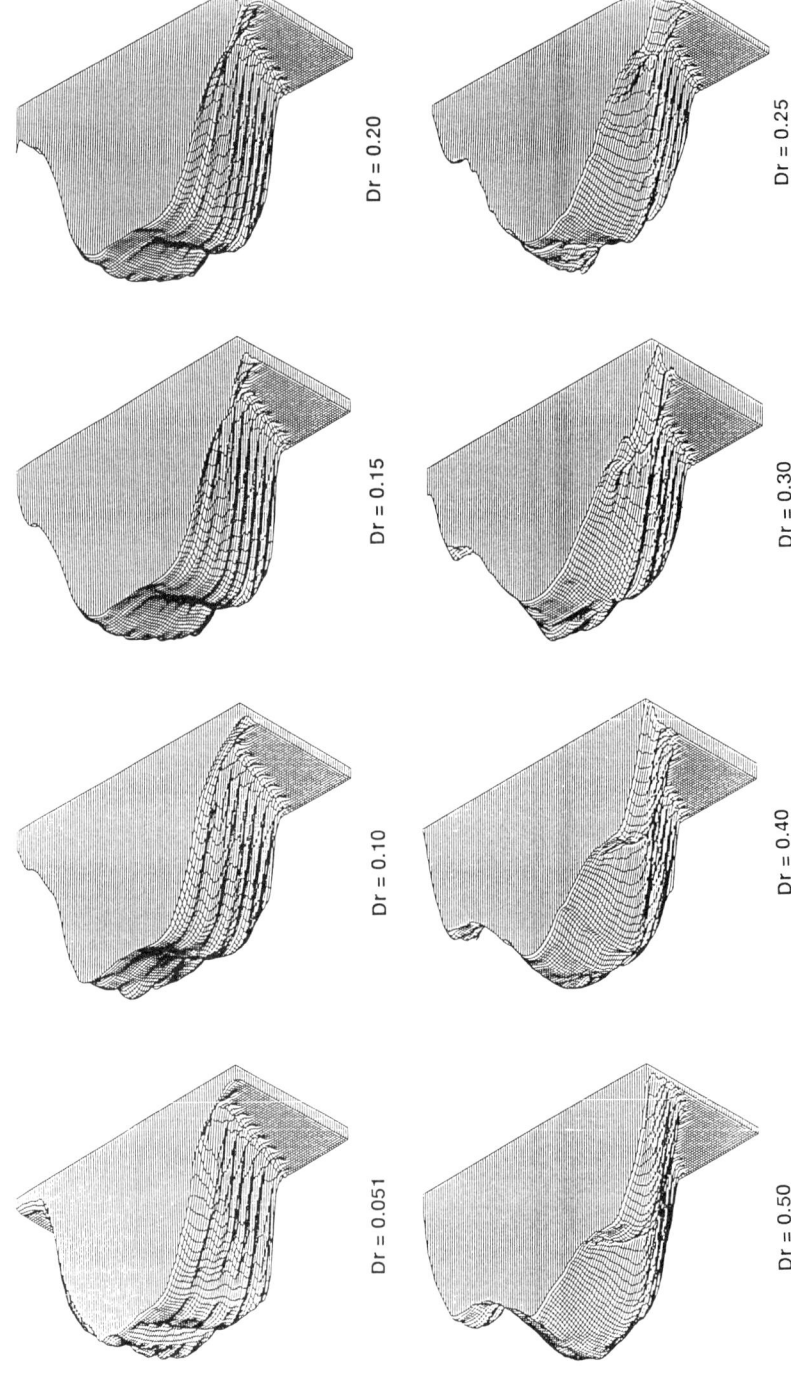

Dr = 0.20

Dr = 0.25

Dr = 0.15

Dr = 0.30

Dr = 0.10

Dr = 0.40

Dr = 0.051

Dr = 0.50

Figure 4.7 Three-dimensional velocity fields in compound trapezoidal channels for various depths. Flow from right to left

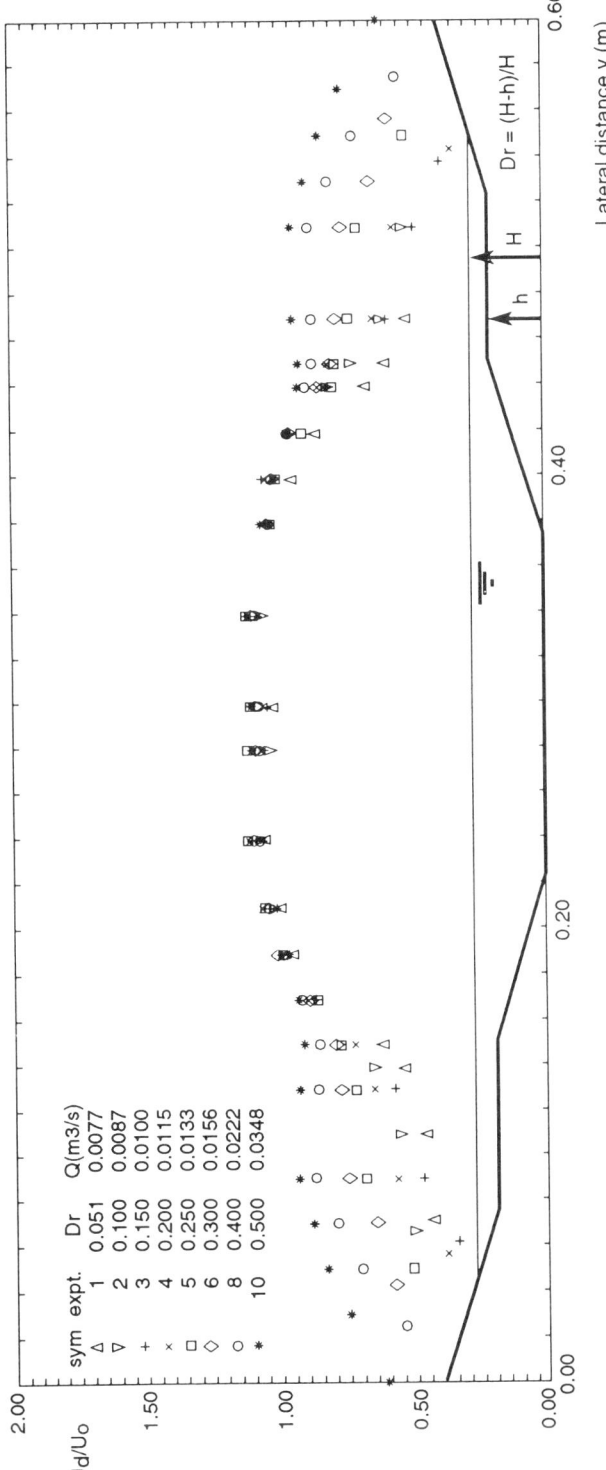

Figure 4.8 Variation of depth-averaged velocity across compound trapezoidal channels for various relative depths, *Dr*, in subcritical flow

was set precisely to within at least ±2% of the bed slope with the result that the integrated boundary shear stresses agreed generally to within about ±3% of their mean normal depth values. The Preston tube technique, combined with this quality control, meant that local values of boundary shear stress could be obtained even in those regions where strong lateral shear and secondary flows made the velocity and Reynolds stress profiles normal to a boundary highly non-linear; see, for example, Figures 6, 8 and 10 of Knight and Shiono (1990), Figure 5 of Knight, Samuels and Shiono (1990) and Figures 6, 7, 12 and 20 of Shiono and Knight (1991). The purpose of emphasizing this technique and the associated quality control is to draw attention to the care that needs to be taken in measuring local boundary shear stresses. The Preston tube technique, which depends only on a small proportion of the inner velocity law (viscous, buffer and logarithmic), is likely to be much more accurate than the method of fitting logarithmic distributions alone to the velocity fields described in the previous section. For a fuller discussion of boundary shear stress measurement, see Winter (1977).

Some typical boundary shear stress distributions are shown for straight prismatic and compound channels in Figures 4.9–4.11, taken from the work of Yuen (1989). Figure 4.9 illustrates the variation of τ_b around the wetted perimeter of a trapezoidal channel in subcritical flow, and Figure 4.10 illustrates the corresponding distributions in supercritical flow. To aid comparison between different data sets for different aspect ratios and Froude numbers, the local boundary shear stresses on either the bed or the wall, τ_{bw}, have been non-dimensionalized by the section mean value, $\tau_0 (= \rho g R S_f)$, and the length scale has been non-dimensionalized by the base width. These data thus complement some of the velocity data already presented in Figures 4.2–4.4.

The most obvious feature in Figure 4.9 is the number and magnitude of the perturbations in τ_b. The boundary shear stress is clearly a particularly sensitive parameter and the lateral

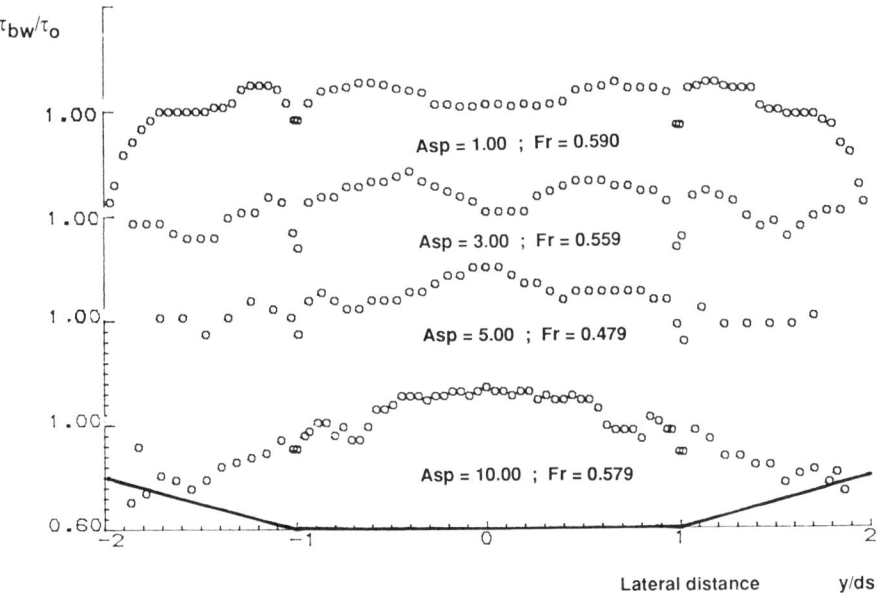

Figure 4.9 Comparison of the spanwise distributions of boundary shear stress in trapezoidal channels at different aspect ratios. $Fr = 0.48–0.59$

variations of τ_{bw}/τ_0 in Figures 4.9 and 4.10 are more pronounced than the variations in depth mean velocity, U_d/U_0, shown in Figure 4.4. Broadly speaking, this would be expected from a square law relationship. The number of perturbations is obviously linked with the number of secondary flow cells present, and this is described in more detail elsewhere by Knight *et al.* (1982), Naot and Rodi (1982) and Knight and Patel (1985a). The influence of these secondary flow cells is important, changing the centreline, τ_{bw}/τ_0 values ($y/ds = 0.0$) from 1.02 to 1.20 as the aspect ratio changes from 1.0 to 5.0 in subcritical flow. The prediction

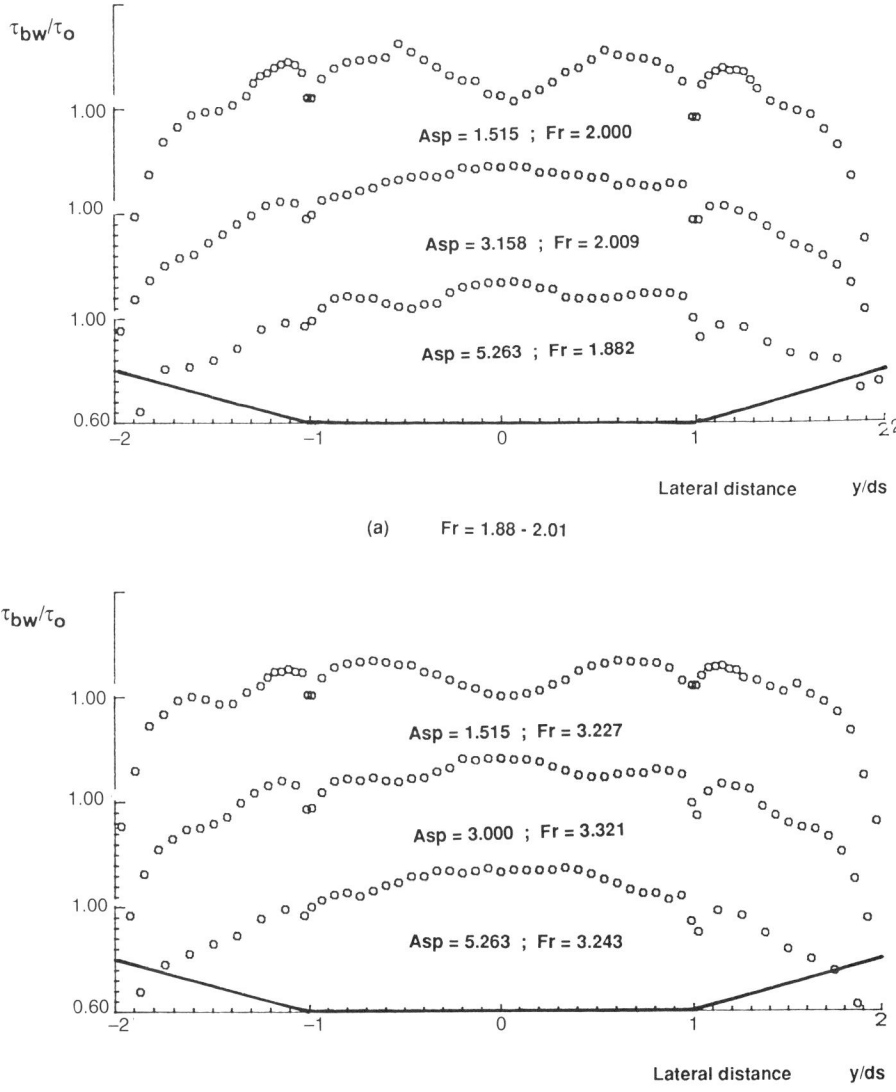

Figure 4.10 Comparison of the spanwise distributions of boundary shear stress in trapezoidal channels at different aspect ratios. $Fr = 2.0$ and 3.2

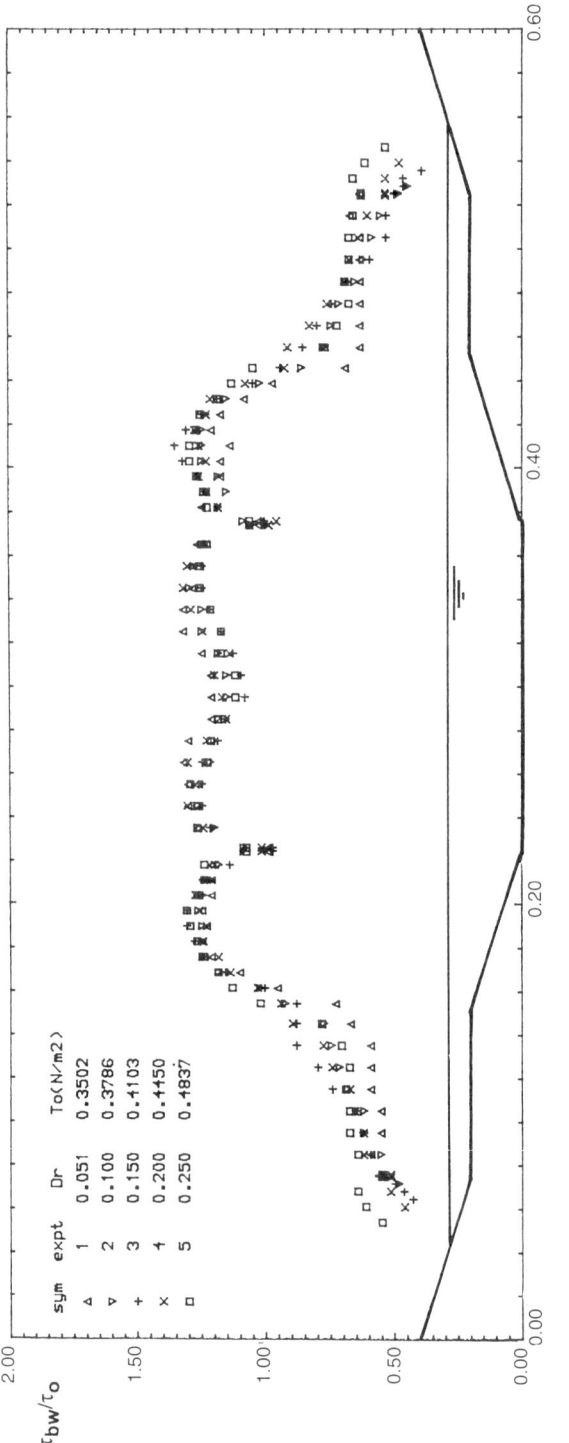

(a)

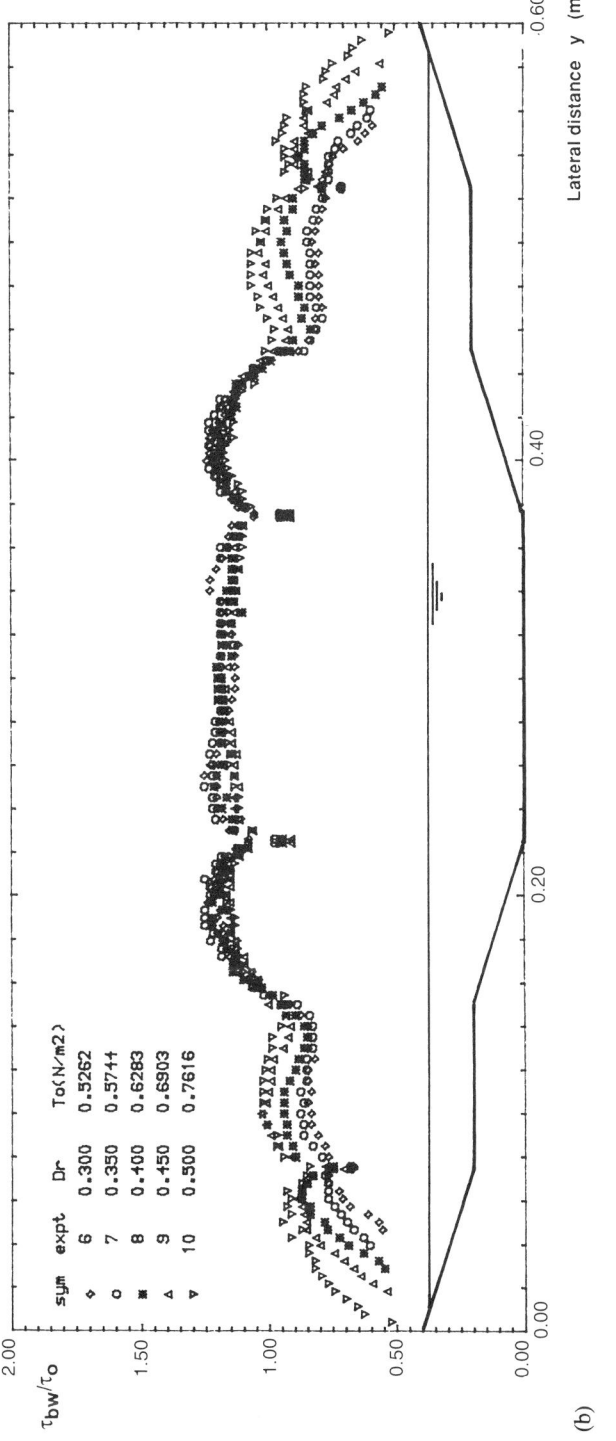

(b)

Figure 4.11 (a) Comparison of boundary shear stress distributions in compound trapezoidal channels for various relative depths, Dr (0.051 < Dr < 0.250). (b) Comparisons of boundary shear stress distributions in compound trapezoidal channels for various relative depths, Dr (0.3 < Dr < 0.5).

of the local boundary shear stress at any point on the wetted perimeter is thus inexorably linked to a complete understanding of the three-dimensional effects produced by longitudinal vorticity. Similar results may be seen in the duct flow data of Knight and Patel (1985a; 1985b). Figure 4.10 shows that similar perturbations exist in supercritical flow. For a Froude number of 3.2, the τ_{bw}/τ_0 values at the channel centreline change from 1.045, 1.175 and 1.135 as the aspect ratio changes from 1.52, 3.0 to 5.26. For a Froude number of around 2.0, the τ_{bw}/τ_0 values change from 1.0, 1.167 and 1.13 for similar changes in aspect ratio. The figures also show how widely the τ_{bw}/τ_0 values vary across the section.

Figure 4.11 illustrates the variation of τ_{bw} around the wetted perimeter of a more complex channel shape, namely a compound section identical to that illustrated in Figures 4.6–4.8. In this instance the values of τ_{bw}/τ_0 are strongly influenced by the relative depth parameter, Dr. The τ_{bw}/τ_0 values decrease on the main channel bed and increase on the shallow floodplains as the relative depth increases from 0.05 to 0.25. The redistribution of boundary shear stresses that occurs with such a change in depth is evidence of significant lateral exchanges of momentum between the deep and shallow regions of flow, together with a yet more complex distribution of secondary flow cells. Figure 4.11b shows that for $0.25 < Dr < 0.50$ the τ_{bw}/τ_0 values in the main channel remain within a relatively narrow bandwidth, whereas the values on the floodplain increase rapidly when Dr exceeds 0.3. The mechanisms behind these changes are described in more detail by Shiono and Knight (1991), who show that for a straight prismatic channel the local bed shear stress, τ_b, may be given by

$$\tau_b \left(1 + \frac{1}{s^2} \right)^{1/2} = \rho g H S_0 - \frac{\partial}{\partial y} \{ H[(\rho \bar{U} \bar{V})_d - \bar{\tau}_{yx}] \} \tag{4.1}$$

where H is the local depth, $(\rho \bar{U} \bar{V})_d$ is the depth-averaged secondary flow term, $\bar{\tau}_{yx}$ is the depth-averaged Reynolds stress and s is the channel sideslope (1:s, vertical:horizontal). Lateral variations in longitudinal vorticity, turbulence structure and depth are therefore bound to influence the lateral distribution of boundary shear stress, τ_b. The more complex the shape of the channel cross-section, the more complicated will be the distribution of τ_b. Meandering and non-straight channels will obviously exhibit other flow mechanisms and the distributions of τ_b will be governed by a more general equation than that for straight channels; see Bathurst, Thorne and Hey (1979), Chang (1988), Dietrich and Whiting (1989), Nelson and Dungan Smith (1989) and Knight, Yuan and Fares (1992) for more details.

4.5 BOUNDARY SHEAR FORCES

Given the complex three-dimensional flow structures that lead to such complicated patterns of boundary shear stress distribution, we might well wonder whether it will ever be possible to predict the distribution of τ_b around the wetted perimeter of a given channel for a certain flow-rate. One line of approach, from a theoretical viewpoint, is to use a fully three-dimensional turbulence model, with all that this implies, concerning a detailed knowledge of the spatial distribution of numerous turbulence coefficients. Alternatively, we could adopt an empirical approach, fitting equations to the sort of data illustrated earlier. As a first step in this alternative direction it would be sensible to obtain equations for the mean boundary shear stress or force on particular boundary elements. Once these had been obtained it would then be possible to fit equations for the local variations around the mean on a particular boundary element. The first step in this second approach has been attempted by many workers; see, for example,

Reploge and Chow (1966), Rajaratnam and Muralidhar (1969), Kartha and Leutheusser (1970), Ghosh and Roy (1970), Knight (1981), Knight and Hamed (1984), Knight, Demetriou and Hamed (1984), Knight and Patel (1985a) and Flintham and Carling (1988).

It now seems appropriate to review progress on this experimental approach, and the example of boundary shear force distributions in trapezoidal channels will be examined in detail. Such channel cross-sections are not only one of the basic shapes, but are also representative of the geometries that are used in schematizing natural rivers in numerical models. Non-polygonal shallow lens-shaped channels may be dealt with by other means; see, for example, Lundgren and Jonsson (1964).

In trapezoidal channels the total shear force per unit length along the channel, SF_T, is the sum of the individual shear forces carried by the walls and the bed, SF_w and SF_b, respectively. These individual shear forces were obtained by integration of the experimentally determined local boundary shear stresses, and then by multiplying the mean values, $\bar{\tau}_w$ and $\bar{\tau}_b$, by the appropriate subperimeter lengths, P_w and P_b, respectively. Thus for a trapezoidal channel of base width B, depth H, and sideslope s (1:s, vertical:horizontal or angle θ where $s = \cot\theta$) the equations are

$$
\begin{aligned}
SF_T &= SF_w + SF_b \\
SF_w &= P_w\bar{\tau}_w \\
SF_b &= P_b\bar{\tau}_b \\
\%SF_w &= 100SF_w/SF_T \\
\%SF_b &= 100SF_b/SF_T \\
P_w &= 2H(1+s^2)^{1/2} \\
P_b &= B
\end{aligned}
\tag{4.2}
$$

The percentage of the total shear force which is carried by the walls, $\%SF_w$, is a useful parameter through which to present different sets of experimental data. The mean boundary shear stresses, $\bar{\tau}_w$ and $\bar{\tau}_b$, may also be readily determined or visualized from this parameter via Equation 4.2.

Some experimental data are presented in Figure 4.12 for uniformly roughened and smooth trapezoidal channels. The $\%SF_w$ values show a systematic reduction with increase in aspect ratio, B/H, typical of an exponential function first proposed by Knight (1981). The data in Figure 4.12 are based on the work of Cruff (1965), Ghosh and Roy (1970), Kartha and Leuthheusser (1970), Myers (1978) and Noutsopoulos and Hadjipanos (1982) in smooth rectangular channels, Patel (1984) and Rhodes (1991) in smooth rectangular ducts, Yuen (1989) in smooth trapezoidal channels with 45° side walls slope, Flintham and Carling (1988) with 6 mm gravel in uniformly roughened trapezoidal channels with 68° side walls slope and Alhamid (1991) with 18.0 mm gravel (R1) and 9.3 mm gravel (R2) in uniformly roughened trapezoidal channels with 45° side walls slope. The use of duct and open channel flow data together, although strictly not compatible, will be referred to later. Most of the data, however, relate to open channel flow and Figure 4.12 immediately shows that there is a clear distinction between rectangular and trapezoidal data sets. However, using the parameter P_b/P_w, rather than B/H accounts for the side walls slope effect moderately well and Figure 4.13 shows the same data reduced to a single curve. Figure 4.14 shows the entire data set for $0 < P_b/P_w < 50$, together with the original equation of Knight et al. (1984), modified by Flintham and Carling (1988) to account for the P_b/P_w ratio, but still using the same experimental coefficients as originally proposed by Knight based on rectangular channel data alone.

At first sight this original equation appears to correlate well with the data, although on

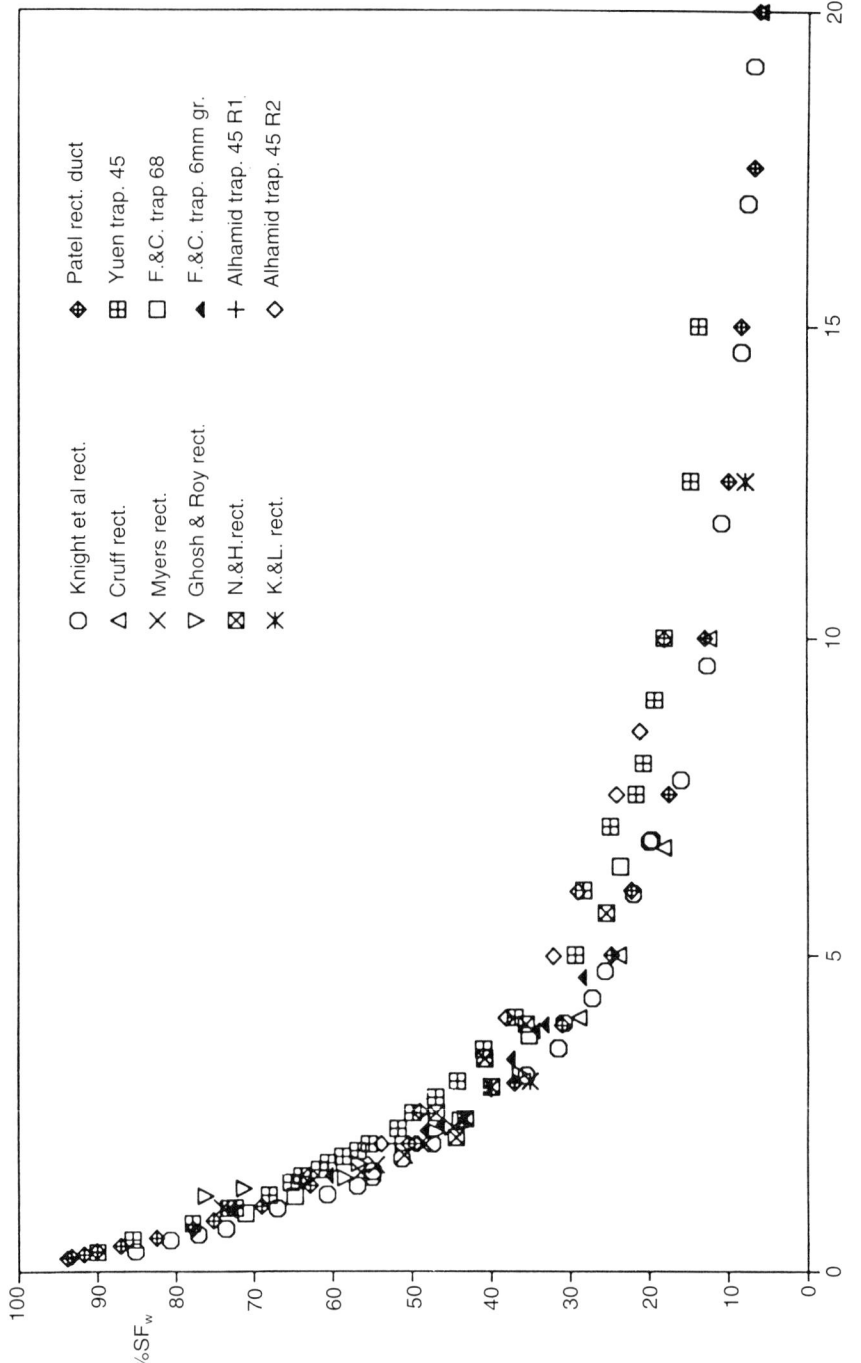

Figure 4.12 Percentage shear force carried by the walls versus *B/H* ratios for smooth and rough channels (0 < *B/H* < 20)

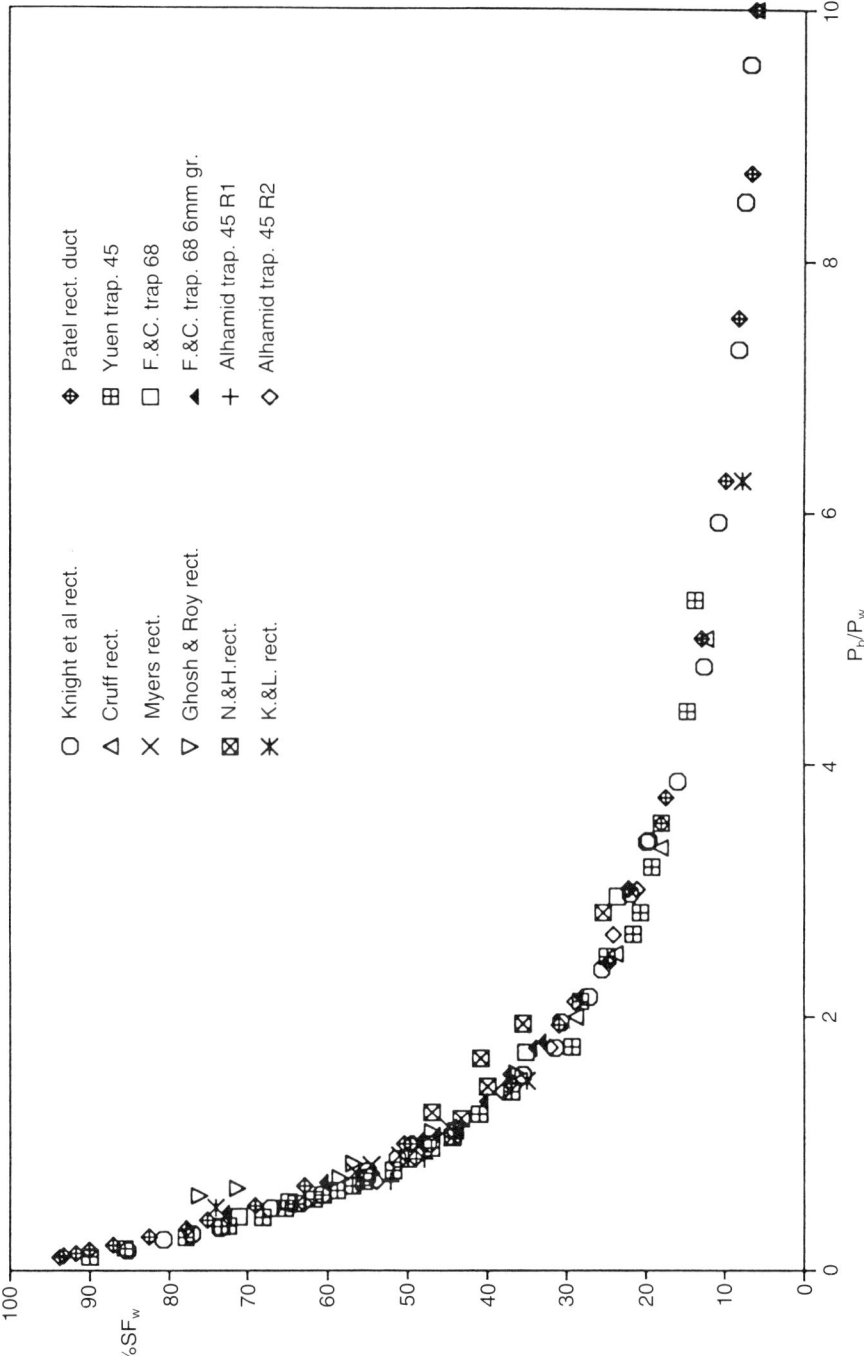

Figure 4.13 Percentage shear force carried by the walls versus P_b/P_w ratios for smooth and rough channels ($0 < P_b/P_w < 10$)

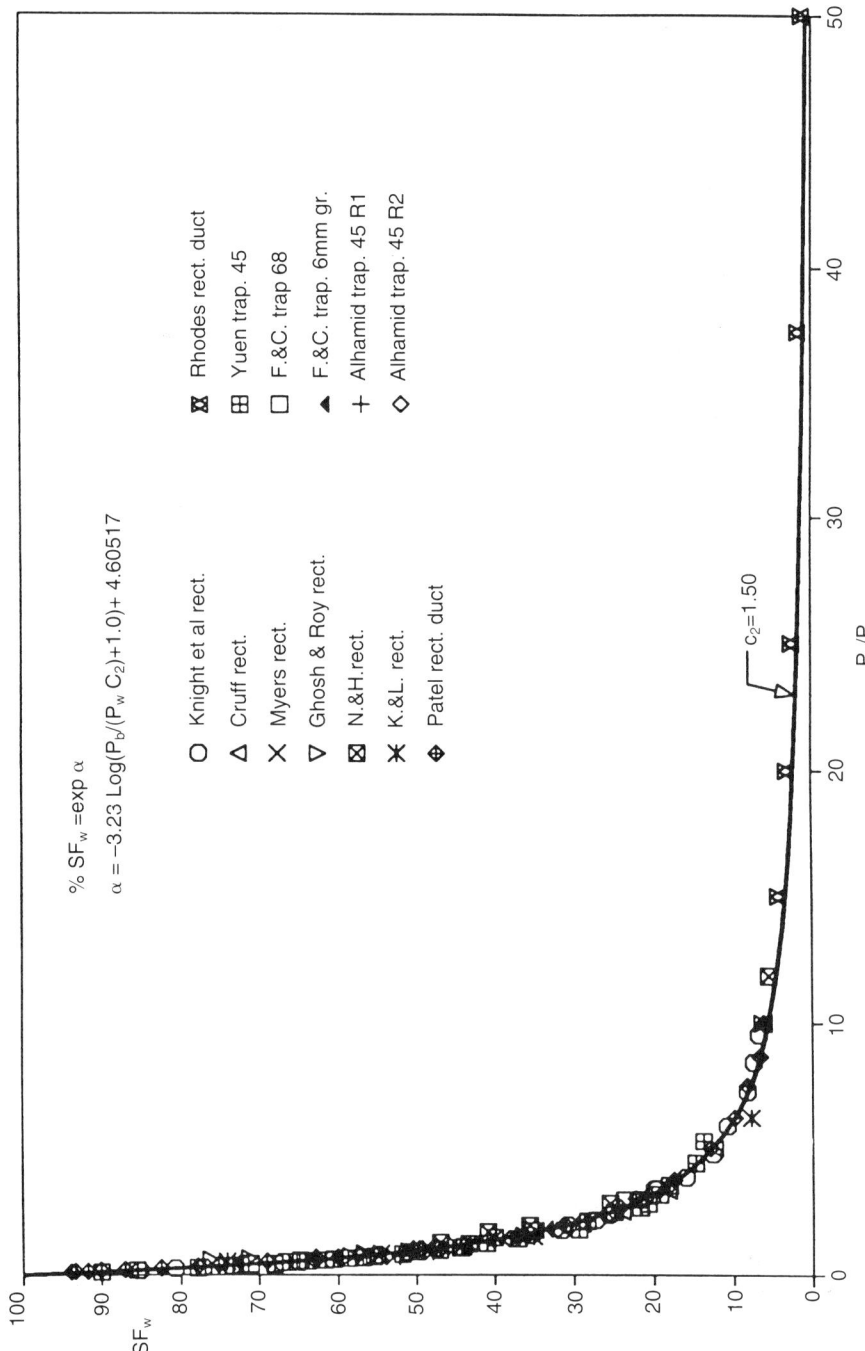

Figure 4.14 Percentage shear force values, $\%SF_w$, for different channels versus P_b/P_w, together with Equation 4.3 and $C_2 = 1.5$ $(0 < P_b/P_w < 50)$

closer inspection some subtle differences may be detected. These differences arise because of Froude and Reynolds number effects, aspect ratio effects (particularly for very wide channels, i.e. $10 < P_b/P_w < 50$) and the incompatibility of duct and open channel flow data sets.

The original equation proposed by Knight et al. (1981; 1984) assumed that $\%SF_w$ varied exponentially with B/H in the form

$$\%SF_w = e^\alpha \tag{4.3}$$

where α was a function of the aspect ratio, B/H. By plotting to log-log scales and assuming a simple linear relationship between $\%SF_w$ and B/H of the form

$$\log_{10}(\%SF_w) = A_1 \log_{10}\left(\frac{B}{H} + A_2\right) + A_3 \tag{4.4}$$

Knight obtained the coefficients $A_1 = -1.4026$, $A_2 = 3.00$ and $A_3 = 2.6692$. Relating Equations 4.3 and 4.4 by $\ln(10)$ thus gives

$$\alpha = 2.30259\left[A_1 \log_{10}\left(\frac{B}{H} + A_2\right) + A_3\right] \tag{4.5}$$

or

$$\alpha = B_1 \log_{10}\left(\frac{B}{H} + A_2\right) + B_2 \tag{4.6}$$

where $B_1 = -3.230$ and $B_2 = 6.146$.

Flintham and Carling (1988) replaced Equation 4.4 by the more general expression

$$\log_{10}(\%SF_w) = C_1 \log_{10}\left(\frac{P_b}{P_w} + C_2\right) + C_3 \tag{4.7}$$

where C_1, C_2 and C_3 are coefficients, and fixed the limiting case by defining $\%SF_w = 100\%$ for $P_b/P_w = 0$, thus eliminating one constant. Hence

$$C_3 = 2 - C_1 \log_{10}(C_2) \tag{4.8}$$

A similar procedure applies to duct flow as, by symmetry, $\%SF_w = 50\%$ when $B/H = 2.0$. Equation 4.7 thus reduces to

$$\log_{10}(\%SF_w) = C_1 \log_{10}\left(\frac{P_b}{P_w C_2} + 1.0\right) + 2.0 \tag{4.9}$$

or

$$\alpha = 2.30259\left[C_1 \log_{10}\left(\frac{P_b}{P_w C_2} + 1.0\right) + 2.0\right] \tag{4.10}$$

This is of a similar format to Equation 4.5 and may be expressed alternatively as Equation 4.6 in the form

$$\alpha = D_1 \log_{10}\left(\frac{P_b}{P_w C_2} + 1.0\right) + D_2 \tag{4.11}$$

where $D_2 = 2\ln(10) = 4.6052$. Using the same value for C_1 as Knight ($C_1 = A_1 =$

-1.4026) gives $D_1 = -3.230$ leaving one constant C_2 to be determined. This is linked to Knight's original coefficient $A_2 (C_2 = A_2/2)$ giving $C_2 = 1.50$. The transformed version of Equation 4.6 is therefore

$$\alpha = -3.23\log_{10}\left(\frac{P_b}{P_w C_2} + 1.0\right) + 4.6052 \qquad (4.12)$$

Using the second original coefficient, $C_2 = 1.50$ (or $A_2 = 3.0$), Figure 4.14 shows that Equation 4.12 appears to fit the data fairly well, although for the reasons already given further development of the basic equation is still required.

It is evident from Figure 4.14, and more so when plotted logarithmically, that Equation 4.12 underestimates $\% SF_w$ by a small amount for large P_b/P_w values. This may be accounted for by introducing a shape factor, C_{sf}

$$\% SF_w = C_{sf}\, e^{\alpha} \qquad (4.13)$$

where $C_{sf} = 1.0$ for $P_b/P_w < 6.546$

else

$$C_{sf} = 0.5857\, (P_b/P_w)^{0.28471} \qquad (4.14)$$

Figure 4.15 presents the data and Equations 4.13 and 4.14 with $C_2 = 1.50$ and the agreement is again fairly good, except perhaps for some of the supercritical trapezoidal data of Yuen. This may be further accounted for by altering C_2 to 1.38, which has the effect of reducing the $\% SF_w$ values around $0.5 < P_b/P_w < 4$ by 1 or 2% as required. The corresponding equations for C_{sf} are then

$$C_{sf} = 1.0 \text{ for } P_b/P_w < 4.374$$

else   (4.15)

$$C_{sf} = 0.6603\, (P_b/P_w)^{0.28125}$$

Figure 4.16 presents the data and Equations 4.13 and 4.15 with $C_2 = 1.38$.

To understand the influence of Froude and Reynolds numbers on boundary shear stress distributions, Yuen (1989) conducted a series of experiments in smooth trapezoidal channels at fixed aspect ratios but variable bed slopes. The results are shown in Figure 4.17 where, for a given aspect ratio, a slight downward trend in $\% SF_w$ may be observed with increasing flow-rate. Although the trend is not very pronounced, it might account for some of the scatter in Figures 4.15 and 4.16. On the other hand, the fact that the $\% SF_w$ values vary so little with the Froude and Reynolds numbers indicates that these laboratory results might be scaled up to natural phenomena with relatively little adjustment. Further data on boundary shear stresses in large-scale channels will no doubt eventually resolve this issue.

A striking feature of Figures 4.15 and 4.16 is that the data for smooth and homogeneously roughened channels for all sideslope angles and aspect ratios fall onto a single curve described by Equations 4.12–4.15. Considering that the Nikuradse roughness values for the channels vary from 0.05 to 32.98 mm (roughness R1), this is a surprising result and indicates the general robustness of the equations.

There are, however, many ways of fitting equations to the data and by altering C_1 in Equation 4.10, and maintaining $C_{sf} = 1.0$ for all P_b/P_w, the results may be fitted by $C_1 = -1.11609$ and $C_2 = 1.0$. The correlation coefficient is not as high as previously and the equation deviates a little from the data for $0.5 < P_b/P_w < 1.0$ and $5.0 < P_b/P_w < 10$.

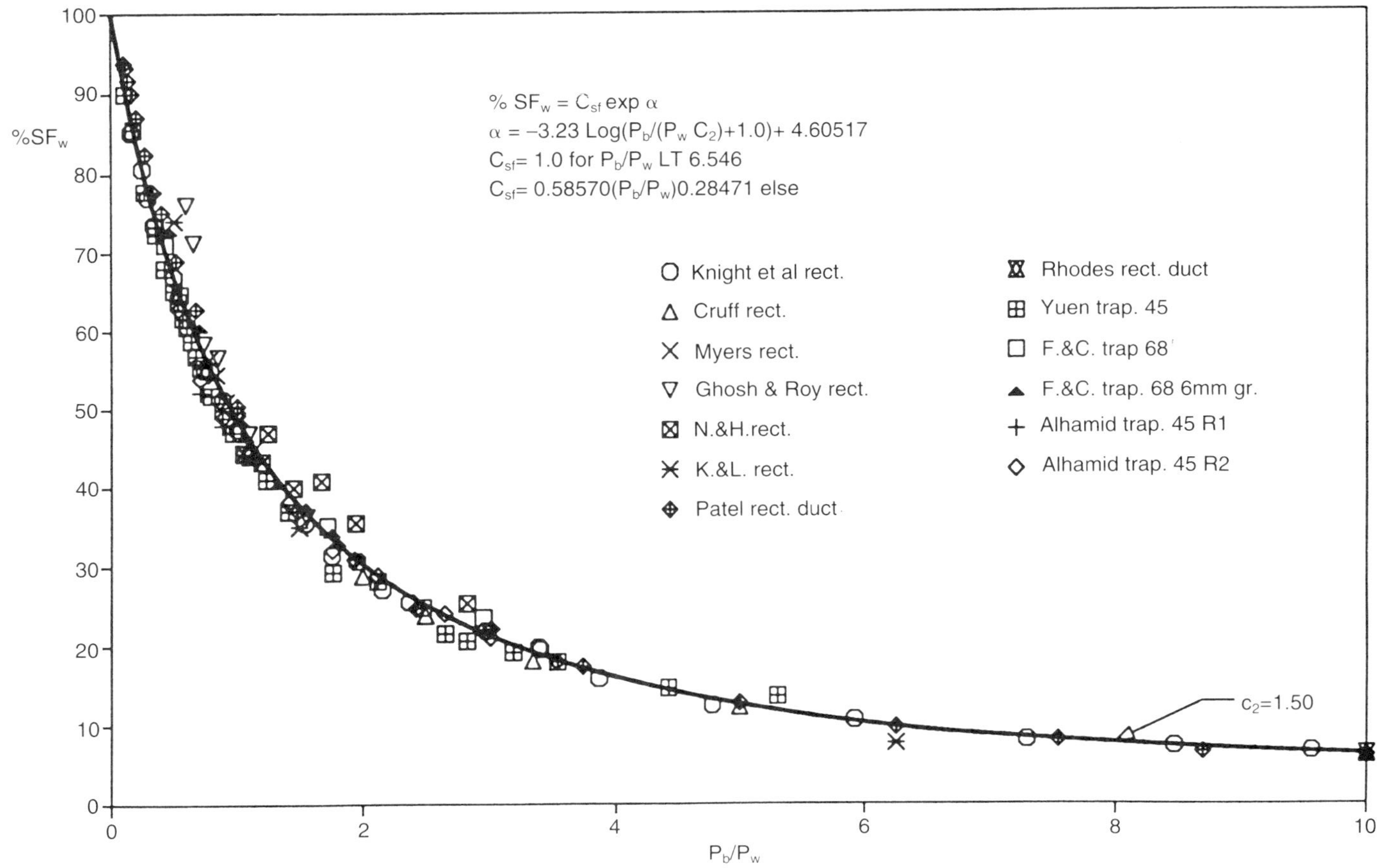

$$\% \, SF_w = C_{sf} \exp \alpha$$

$$\alpha = -3.23 \, Log(P_b/(P_w \, C_2)+1.0)+4.60517$$

$$C_{sf} = 1.0 \text{ for } P_b/P_w \text{ LT } 6.546$$

$$C_{sf} = 0.58570(P_b/P_w)0.28471 \text{ else}$$

○ Knight et al rect.
△ Cruff rect.
✕ Myers rect.
▽ Ghosh & Roy rect.
⊠ N.&H.rect.
✳ K.&L. rect.
⊕ Patel rect. duct

⊠ Rhodes rect. duct
⊞ Yuen trap. 45
□ F.&C. trap 68
◀ F.&C. trap. 68 6mm gr.
+ Alhamid trap. 45 R1
◇ Alhamid trap. 45 R2

$c_2 = 1.50$

P_b/P_w

%SF_w

Figure 4.15 Percentage shear force values, $\%SF_w$, for different channels versus P_b/P_w together with Equation 4.13 and $C_2 = 1.5$ ($0 < P_b/P_w < 10$)

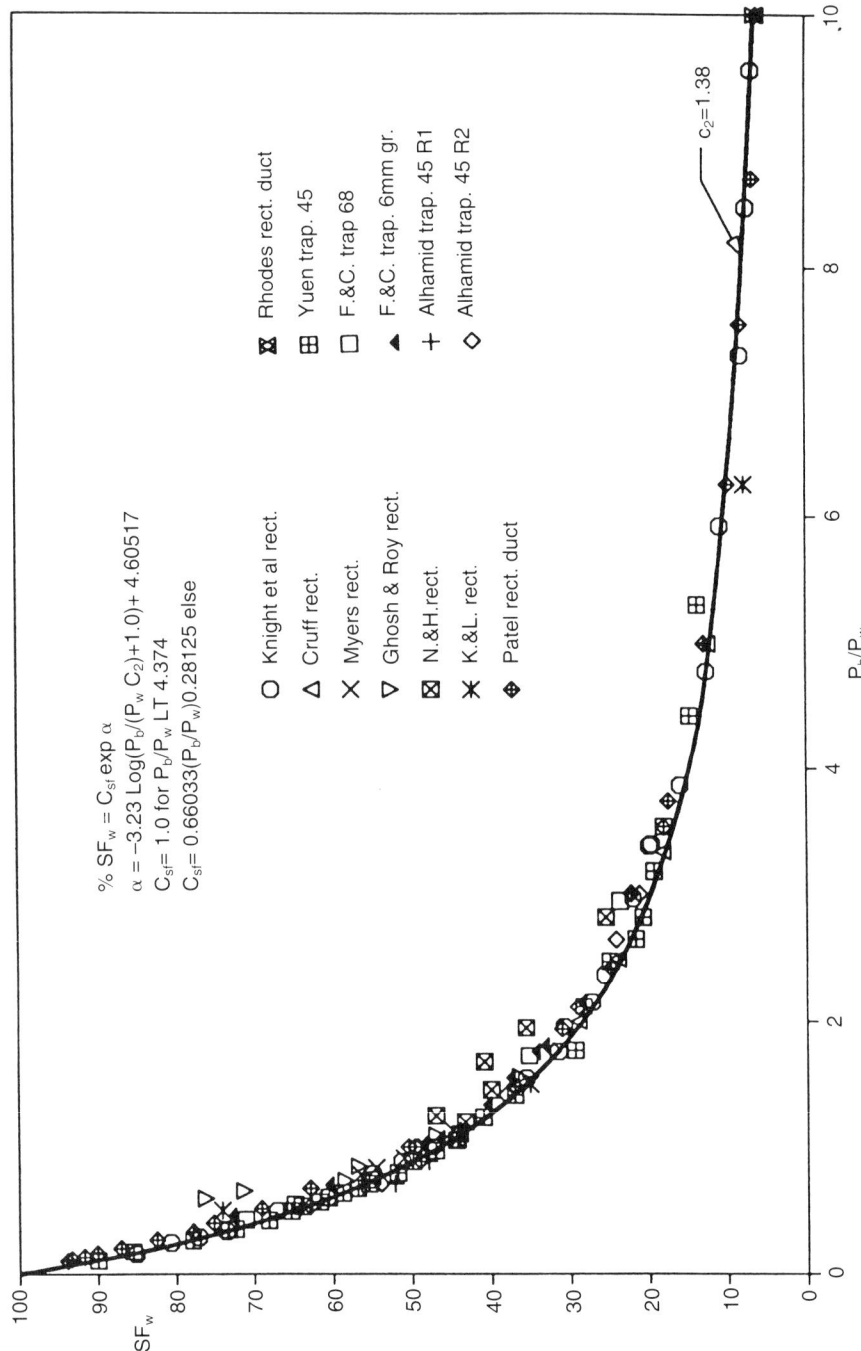

Figure 4.16 Percentage shear force values, $\%SF_w$ for different channels versus P_b/P_w, together with Equation 4.13 and $C_2 = 1.38$ ($0 < P_b/P_2 < 10$)

An alternative approach would be to use the geometrical relationship

$$\%SF_w = \frac{100}{1 + P_b/P_w} \tag{4.16}$$

as a basis for the trend line with deviations from it determined by regression using a power law; see Rhodes (1991) for an application of this approach to duct flow. It should be noted here that open channel flow data does not give $\%SF_w = 50\%$ for $P_b/P_w = 1.0$ (i.e. $B/H = 2.0$) due to the influence of the free surface. Equation 4.12 gives $\%SF_w = 48.8\%$ for $C_2 = 1.50$ and 46.6% for $C_2 = 1.38$. This feature was first noticed and commented on by Knight and Patel (1985a).

4.6 MEAN AND MAXIMUM BOUNDARY SHEAR STRESSES

The mean wall and bed shear stresses may be determined from Equation 4.2 using the correlations between $\%SF_w$ and P_b/P_w obtained in the previous section. Equation 4.2 gives

$$\frac{\bar{\tau}_w}{\rho g H S_f} = 0.01\,\%SF_w\left[\frac{s}{2(1+s^2)^{1/2}} + \frac{P_b}{P_w} \right] \tag{4.17}$$

$$\frac{\bar{\tau}_w}{\rho g R S_f} = 0.01\,\%SF_w\left[1.0 + \frac{P_b}{P_w} \right] \tag{4.18}$$

$$\frac{\bar{\tau}_b}{\rho g H S_f} = (1.0 - 0.01\,\%SF_w)\left[1.0 + \frac{s}{2(1+s^2)^{1/2}}\frac{1}{(P_b/P_w)} \right] \tag{4.19}$$

$$\frac{\bar{\tau}_b}{\rho g R S_f} = (1.0 - 0.01\,\%SF_w)\left[1.0 + \frac{1}{(P_b/P_w)} \right] \tag{4.20}$$

Figures 4.18 and 4.19 present the available open channel data for trapezoidal channels with three sideslope angles of 45°, 68° and 90°. Equations 4.17 and 4.18, together with Equations 4.12, 4.13 and 4.15, i.e. $C_2 = 1.38$, are shown superimposed on the data. The correlation for $\bar{\tau}_b$ is fairly good except that the data of Flintham and Carling (1988) fall below the line for $P_b/P_w < 2.0$.

The curves in Figure 4.19 provide the first substantial evidence for the variation of $\bar{\tau}_w$ and $\bar{\tau}_b$ with channel geometry and replace the corresponding curves for laminar flow given in Chow (1959). Such curves form the basis of the tractive force approach to channel design in which the maximum allowed boundary shear stresses on the walls and bed are given by

$$\tau_{wm} = k_{wm}\rho g H S_0$$
$$\tau_{bm} = k_{bm}\rho g H S_0 \tag{4.21}$$

where k_{wm} and k_{bm} are coefficients. The USBR suggest values of 0.75 and 1.0, respectively, which are clearly too conservative for the design of narrow channels; see also Figure 2.4 of Hemphill and Bramley (1989) where guidance on bank erosion and practical design are discussed in full. The data in Figures 4.18 and 4.19 could equally well be defined by Equations 4.12, 4.13 and 4.15 with $C_2 = 1.50$. It should be remembered, however, that in tractive force design it is the maximum boundary shear stresses that are important. For homogeneously roughened channels Alhamid (1991) gives the following equations

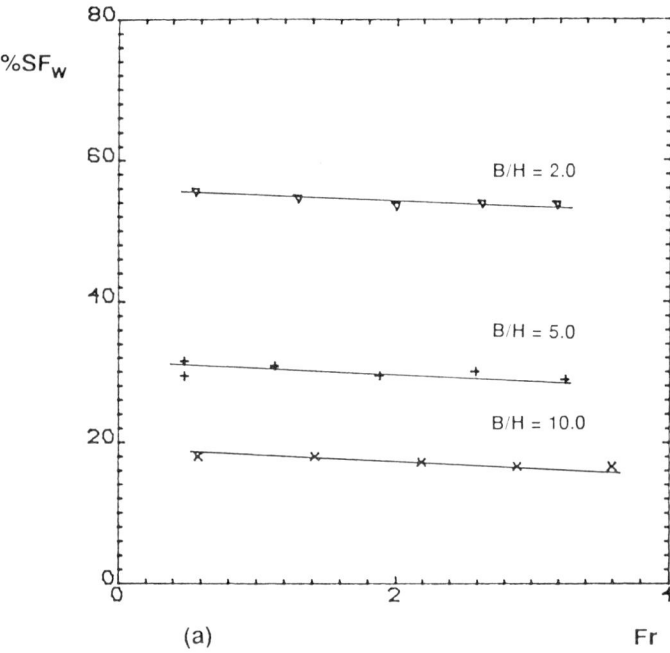

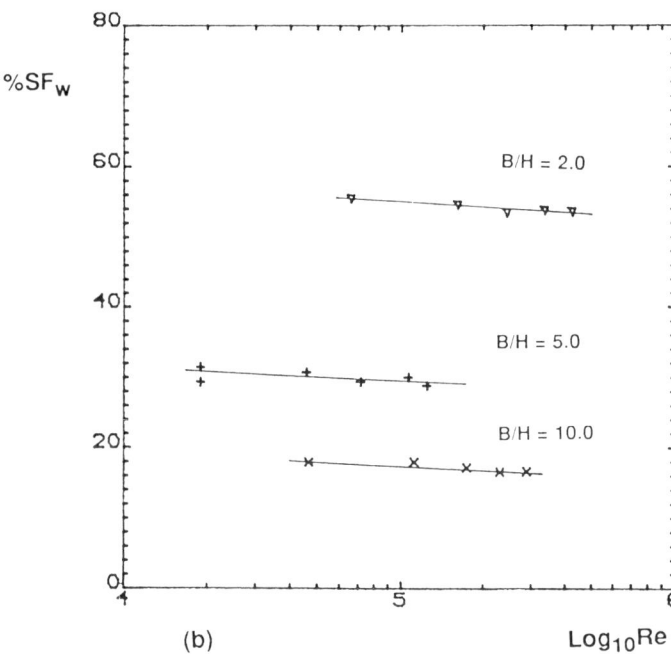

Figure 4.17 Variation of $\%SF_w$ with (a) Froude number and (b) Reynolds number for different aspect ratios

$$\frac{\tau_{wm}}{\rho g H S_f} = 0.01 \% SF_w \left[2.37 \left(\frac{P_b}{P_w} \right)^{0.85} \right] \qquad (4.22)$$

$$\frac{\tau_{bm}}{\rho g H S_f} = (1.0 - 0.01 \% SF_w) \left[0.8 + \left(\frac{P_b}{P_w} \right)^{-0.35} \right] \qquad (4.23)$$

These are shown with some rough channel data (R_w/R_b with R1 and R2) in Figures 4.20 and 4.21. The smooth channel data of Yuen (1989) lie below the rough channel data of Alhamid (1991) and at this stage it is not possible to fit a single equation for both smooth and rough channels as was the case for the mean boundary shear stresses. From a practical standpoint the important thing to note is that the k_w and k_b factors in the equivalent Equation 4.21 for $\bar{\tau}_w$ and $\bar{\tau}_b$ would be lower than the maximum shear stress values.

4.7 DESIGN EQUATIONS

Figures 4.15−4.17 and Equations 4.12−4.15 offer general guidance on open channel flow design. Because of the reduction in $\% SF_w$ with Froude number it is suggested that the following criteria and equations be adopted

(i) *Subcritical flow* $Fr < 1.0$
                            Equations 4.12−4.14 with $C_2 = 1.50$

(ii) *Supercritical flow* $Fr > 1.0$
                            Equations 4.12−4.14 with C_2 defined by
                            $$C_2 = -0.04 Fr + 1.54 \qquad (4.24)$$

This then reduces C_2 from 1.50 at $Fr = 1.0$ to 1.38 at $Fr = 4.0$, and has the effect of systematically reducing $\% SF_w$ values as required. The corresponding equation for C_{sf} for $P_b/P_w > 6.546$ might later need very slight adjustment when more wide open channel data become available. The equations are, however, probably accurate enough for design purposes. It should be noted that for supercritical flow in wide channels ($P_b/P_w > 6.546$) the equation for C_{sf} reverts back to the subcritical equation, despite C_2 being variable.

(iii) *Mean and maximum bed shear stresses*   Having obtained the appropriate $\% SF_w$ based on the P_b/P_w value, Equations 4.17−4.20 give the mean wall and bed stresses and Equations 4.22 and 4.23 the maximum wall and bed shear stresses. Differentially roughened channels may be designed using the same basic approach but relating C_2 to k_{sw}/k_{sb} where k_{sw} and k_{sb} are the Nikuradse roughness sizes for the walls and bed, respectively. An appropriate equation for subcritical flow is

$$C_2 = 1.50 \, (k_{sw}/k_{sb})^{0.2115} \qquad (4.25)$$

Further details on how the homogeneous roughness data presented in this review might be extended to heterogeneous roughness problems are given in Alhamid (1991) and Knight, Alhamid and Yuen (1992).

4.8 DEPTH-AVERAGED EDDY VISCOSITIES

Another use to which boundary shear stress data may be put is the determination of depth-averaged eddy viscosities. Integrating Equation 4.1 gives the depth-mean apparent shear stress on a vertical interface, $\bar{\tau}_a$

$$\overline{\tau}_a = -\frac{1}{H}\int_0^y\left[\rho g H S_0 - \tau_b\left(1 + \frac{1}{s^2}\right)^{1/2}\right]dy \qquad (4.26)$$

where

$$\overline{\tau}_a = [\overline{\tau}_{yx} - (\rho\overline{U}\overline{V})_d] \qquad (4.27)$$

In open channel flow analysis based on depth-averaged parameters, the apparent shear stress on a vertical interface is often related to the lateral gradient of depth-averaged velocity by the depth-averaged eddy viscosity, $\overline{\epsilon}_{yx}$, where

$$\overline{\tau}_a = \rho\overline{\epsilon}_{yx}\frac{\partial U_d}{\partial y} \qquad (4.28)$$

Expressing this in dimensionless form, using the local boundary shear velocity, $U_*[=(\tau_b/\rho)^{1/2}]$, gives

$$\overline{\epsilon}_{yx} = \lambda U_* H \qquad (4.29)$$

where λ is a dimensionless constant.

The boundary shear stress and velocity data described earlier may be processed through Equations 4.26–4.29 to yield experimentally determined values for λ. Examples of this analysis are given in Yuen (1989), Knight and Shiono (1990), Shiono and Knight (1990; 1991) Alhamid (1991) and Rhodes (1991). One example based on the data described earlier in this review is shown in Figures 4.22 and 4.23 for the case of differentially roughened trapezoidal channels with roughened side walls and a smooth bed. This data is taken from Alhamid (1991) with roughness R2 on the walls (9.3 mm gravel, giving $k_{sw}/k_{sb} = 419$). The λ data in Figure 4.23 generally lies between 0.1 and 1.0, with a mean value of 0.45. This is close to the value of 0.5 shown in compound trapezoidal channels and indicates the importance of the secondary flow term in Equation 4.27. For the case of $(\rho\overline{U}\overline{V})_d = 0$, the λ values based on turbulence stresses alone are much lower and closer to the theoretical value of 0.067 for wide open channels. For further details on this topic see Shiono and Knight (1991).

4.9   CONCLUDING REMARKS

The design equations presented earlier represent one approach to quantifying the mean and maximum boundary shear stresses which act on the bed or walls of straight prismatic trapezoidal channels. The parameter P_b/P_w has been used to account for different wall sideslopes and the parameter C_2 for subcritical and supercritical flow. The slight reduction in $\%SF_w$ values with increasing Froude number, and the corresponding reduction in C_2 values, is put forward tentatively at this stage and needs proper validation in channels of general shape. The parameter C_2 has also been used to account for differential roughness effects by linking it to the parameter k_{sw}/k_{sb}. This also may need some modification as further data become available. It has been noted that open channel flow and rectangular duct flow data are not strictly compatible. However, at very wide aspects ratios, or large P_b/P_w values, the differences are likely to be small and some duct flow data have been used to extend the $\%SF_w$ equation in this region. The general open channel flow design equations should, however, not be applied to rectangular duct flow as alternative equations exist for this particular type of flow satisfying the special condition of $\%SF_w = 50\%$ for $P_b/P_w = 1.0$.

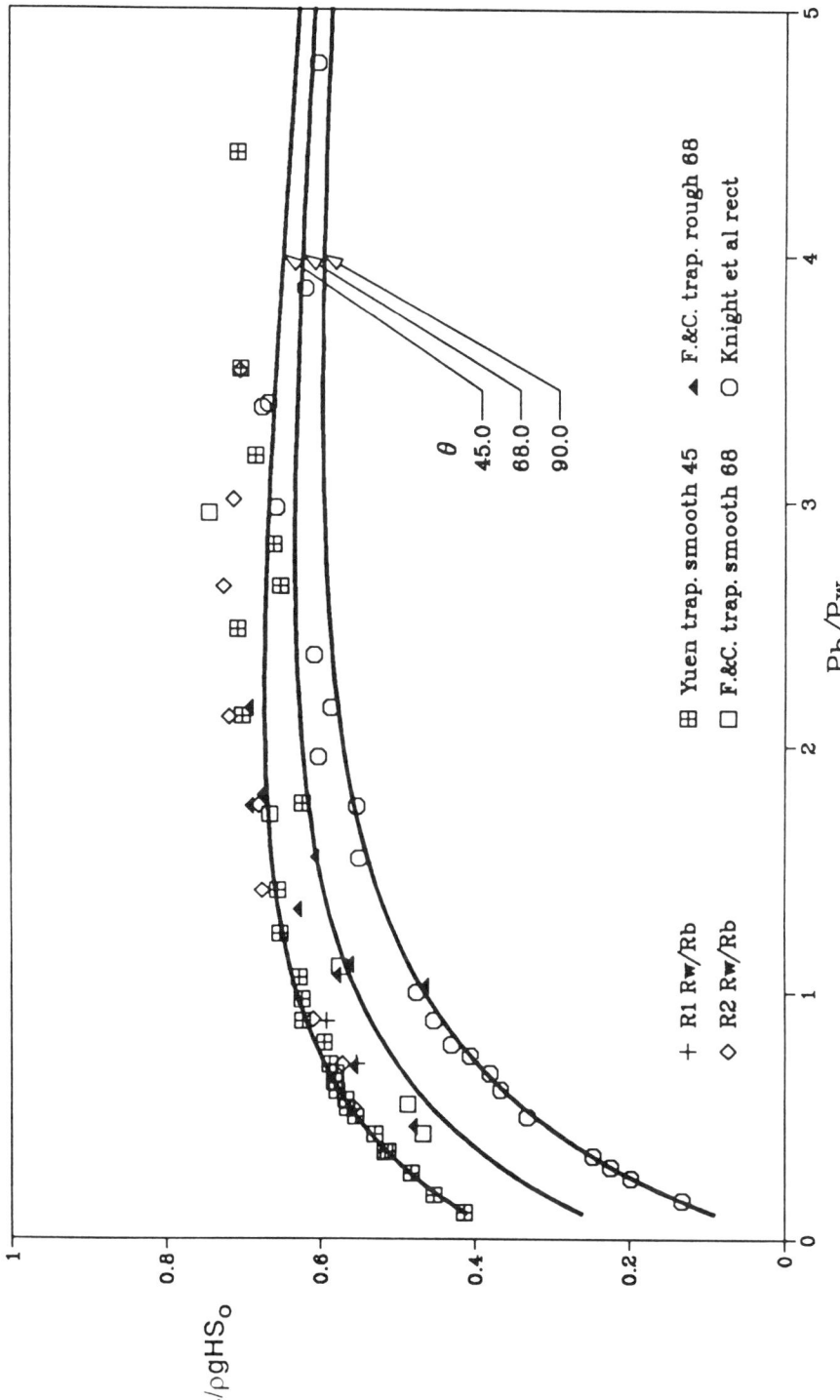

Figure 4.18 Average wall shear stress, $\bar{\tau}_w/\rho gHS_0$, for smooth and rough trapezoidal channels with different sideslope angles

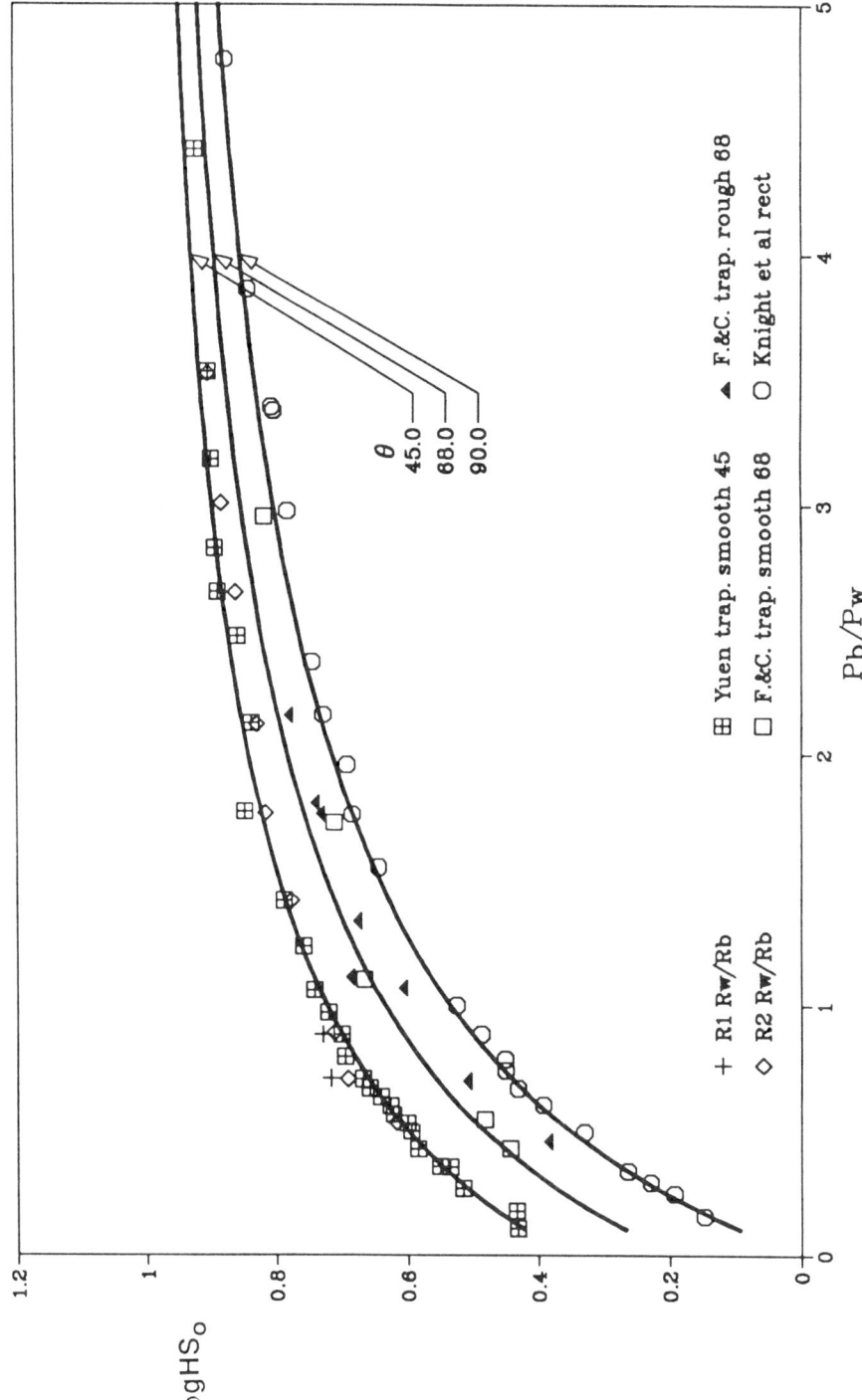

Figure 4.19 Average bed shear stress, $\bar{\tau}_b/\rho g H S_0$, for smooth and rough trapezoidal channels with different sideslope angles

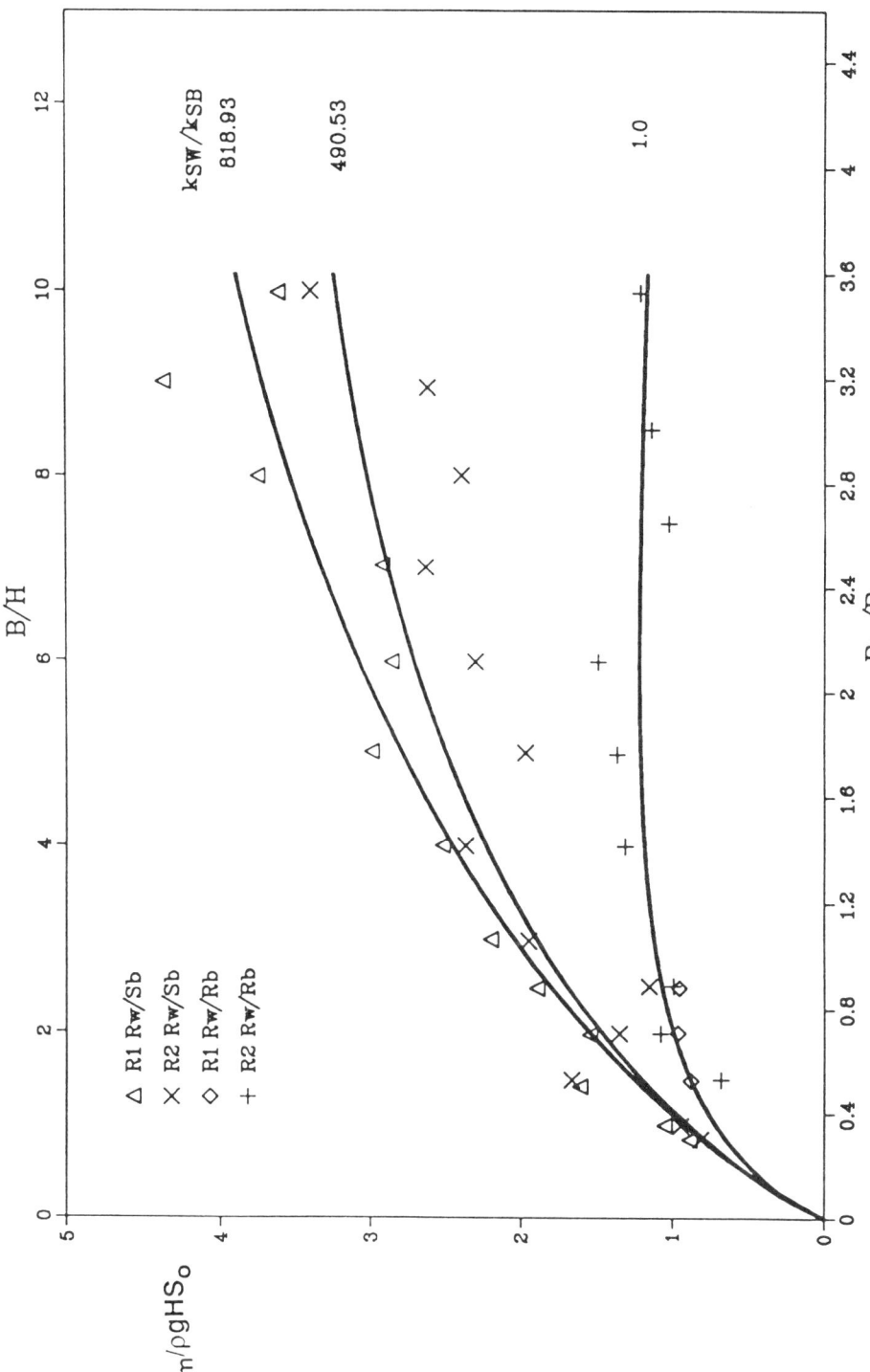

Figure 4.20 Maximum wall shear stress, $\tau_{wm}/\rho g H S_0$, for uniformly and differentially roughened trapezoidal channels

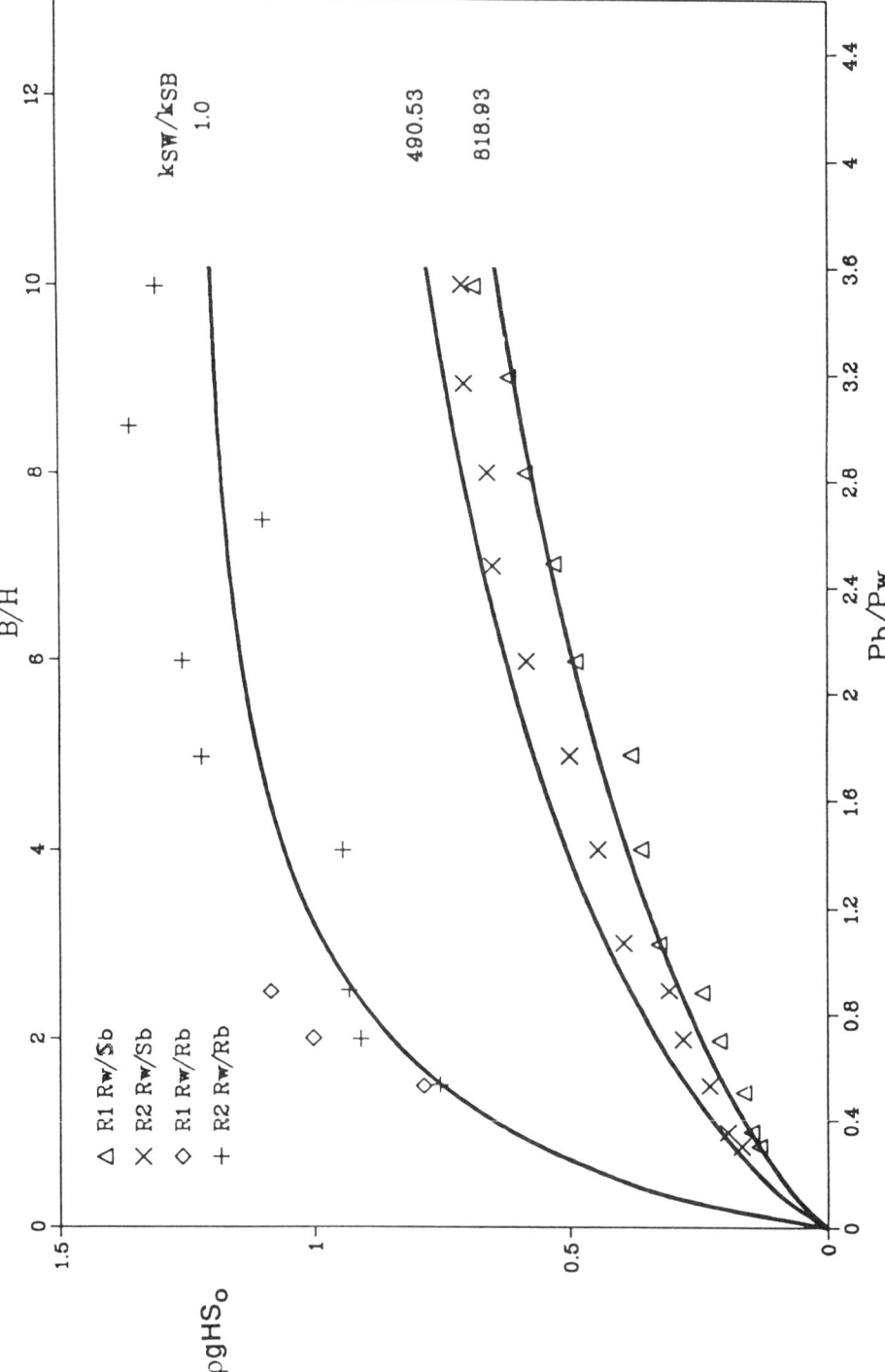

Figure 4.21 Maximum bed shear stress, $\tau_{bm}/\rho g H S_0$, for uniformly and differentially roughened trapezoidal channels

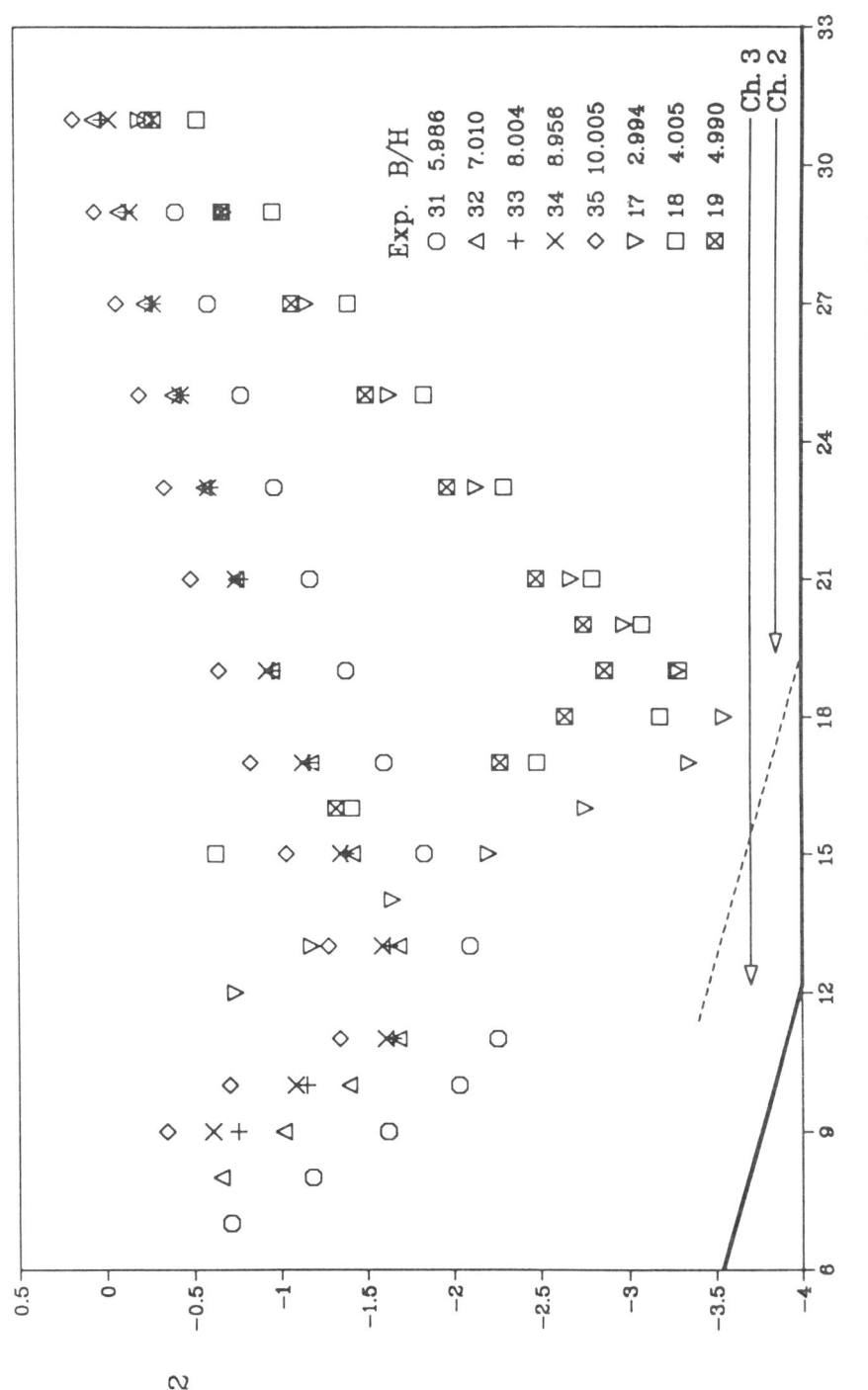

Figure 4.22 Lateral variation of depth-averaged apparent shear stress, $\bar{\tau}_a$, for trapezoidal channels with roughened walls and a smooth bed. Roughness R2

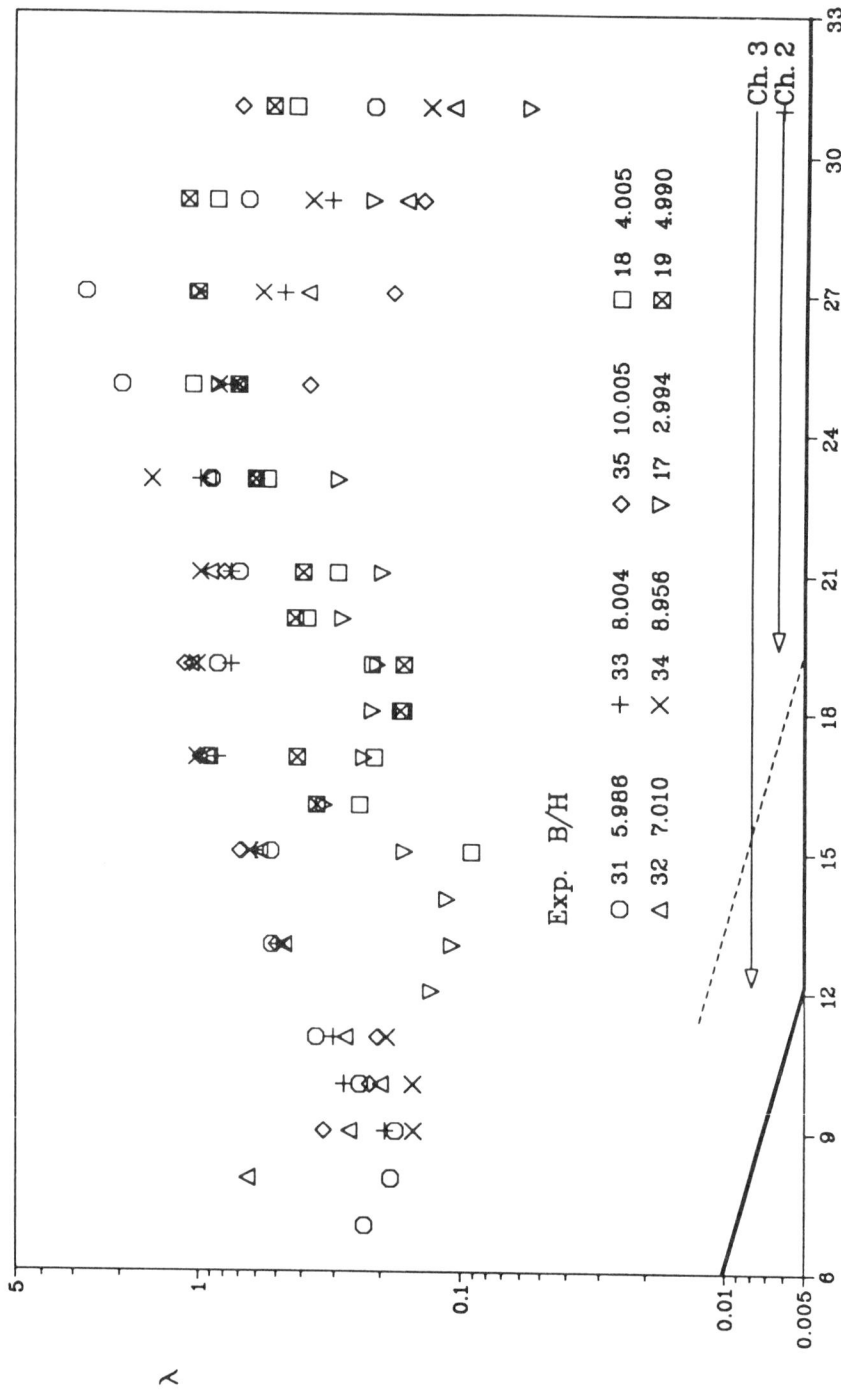

Figure 4.23 Lateral variation of the dimensionless eddy viscosity coefficient, λ, for trapezoidal channels with roughened walls and a smooth bed. Roughness R2

The experimental data presented in this review indicate that the mean boundary shear stresses, and to a lesser extent the maximum boundary shear stresses, may be quantified reasonably well by exponential or power law equations. No attempt has yet been made to specify the lateral distributions of τ_b about the mean values due to the complex influence of secondary flow cells on τ_b. Furthermore, no attempt has yet been made to extend the analysis to either straight prismatic channels of arbitrary cross-section or non-straight channels of variable planform and section geometry. Given that Cath Allen was well noted for tackling the more difficult problems in fluid mechanics, perhaps this would be a worthy challenge to throw out to other experimentalists and numerical modellers in her memory. She would have liked that.

REFERENCES

Alhamid, A.A. I. (1991) Boundary shear stress and velocity distributions in differentially roughened trapezoidal open channels. *PhD Thesis*. The University of Birmingham.

Bathurst, J.C., Thorne, C.R. and Hey, R.D. (1979) Secondary flow and shear stress at river bends. *J. Hydr. Div. ASCE* **105** (HY10), 1277−1295.

Brundrett, E. and Baines, W.D. (1964) The production and diffusion of vorticity in duct flow. *J. Fluid Mech.* **19**, 375−394.

Chang, H.H. (1988) *Fluvial Processes in River Engineering*. Wiley, Chichester.

Chow, V.T. (1959) *Open Channel Hydraulics*. McGraw-Hill, New York.

Cruff, R.W. (1965) Cross-channel transfer of linear momentum in smooth rectangular channels. *Water-Supply Pap. 1592-B*, United States Geological Survey, Boulder, B1−B26.

Deitrich, W.E. and Whiting, P. (1989) Boundary shear stress and sediment transport in river meanders of sand and gravel. In: *River Meandering* (Eds S. Ikeda and G. Parker), *Am. Geophys. Union Wat. Resour. Monogr. No. 12*, pp. 1−50.

Einstein, H.A. and Li, H. (1958) Secondary currents in straight channels. *Trans. Am. Geophys. Union* **39**, 1085−1088.

Flintham, T.P. and Carling, P.A. (1988) The prediction of mean bed and wall boundary shear in uniform and compositely rough channels. In: *Proceedings of the International Conference on River Regime* (Ed. W.R. White), Wiley, Chichester, 267−287.

Gessner, F.B. (1973) The origin of secondary flow in turbulent flow along a corner. *J. Fluid Mech.* **58**, 1−25.

Ghosh, S.N. and Roy, N. (1970) Boundary shear distribution in open channel flow. *J. Hydr. Div. ASCE* **96** (HY4), 967−994.

Hemphill, R.W. and Bramley, M.E. (1989) Protection of river and canal banks. *CIRIA Wat. Engin. Rep.*, Butterworths, London.

Ikeda, S. and Parker, G. (1989) River meandering. *Am. Geophys. Union Wat. Resour. Monogr.* No. 12.

Kartha, V.C. and Leutheusser, H.J. (1970) Distribution of tractive force in open channels. *J. Hydr. Div. ASCE* **96** (HY7), 1469−1483.

Knight, D.W. (1981) Boundary shear in smooth and rough channels. *J. Hydr. Div. ASCE* **107** (HY7), 839−851.

Knight, D.W. (1989) Hydraulics of flood channels, in *Floods: Hydrological, Sedimentological and Geomorphological Implications* (Ed. K. Beven), Wiley, Chichester, Ch. 6, pp. 83−105.

Knight, D.W. and Hamed, M.E. (1984) Boundary shear in symmetrical compound channels. *J. Hydr. Engin. ASCE* **110** 1412−1430.

Knight, D.W. and Lai, C.J. (1985) Turbulent flow in compound channels and ducts. In: *Proceedings of the 2nd International Symposium on Refined Flow Modelling and Turbulence Measurements*. Hemisphere, Washington, I2-1−I2-10.

Knight, D.W. and Patel, H.S. (1985a) Boundary shear in smooth rectangular ducts. *J. Hydr. Engin. ASCE* **111**, 29−47.

Knight, D.W. and Patel, H.S. (1985b) Boundary shear stress distributions in rectangular duct flow. In: *Proceedings of the 2nd International Symposium on Refined Flow Modelling and Turbulence Measurements*. Hemisphere, Washington, I22-1−I22-10.

Knight, D.W. and Shiono, K. (1990) Turbulence measurements in a shear layer region of a compound channel. *J. Hydr. Res.* **28**, 175–196 (Discussion in IAHR Journal, 29(2), 259–276).

Knight, D.W., Alhamid, A.A. I. and Yuen, K.W.H. (1992) Boundary shear in differentially roughened trapezoidal channels. In: *Proceedings of the 2nd International Conference on Hydraulic and Environmental Modelling of Coastal, Estuarine and River Waters* (Eds R.A. Falconer, K. Shiono and R.G.S. Matthew). Ashgate, Aldershot.

Knight, D.W., Demetriou, J.D. and Hamed, M.E. (1984) Boundary shear in smooth rectangular channels. *J. Hydr. Engin. ASCE* **110**, 405–422.

Knight, D.W., Samuels, P.G. and Shiono, K. (1990) River flow simulation: research and developments. *J. Inst. Wat. Environ. Manag.* **4**, 163–175.

Knight, D.W., Shiono, K. and Pirt, J. (1989) Prediction of depth mean velocity and discharge in natural rivers with overbank flow. In. *Proceedings of the International Conference on Hydraulic and Environmental Modelling of Coastal, Estuarine and River Waters* (Eds. R.A. Falconer, P. Goodwin and R.G.S. Matthew). Gower Technical, Aldershot, 419–428.

Knight, D.W., Yuan, Y.M. and Fares, Y.R. (1992) Boundary shear in meandering channels. In: *Proceedings of the International Symposium on Hydraulic Research in Nature and Laboratory*. (Ed. C. Jisheng) Yangtze River Scientific Research Institute, Wuhan, published by the Yangtze River Scientific Research Institute, Wuhan, China, Vol. 2, pp. 102–107.

Knight, D.W., Patel, H.S., Demetriou, J.D. and Hamed, M.E. (1982) Boundary shear stress distributions in open channel and closed conduit flows. In: *Proceedings of Euromech 156 — Mechanics of Sediment Transport* (Eds B. Mutlu Sumer and A. Muller), Balkema, Rotterdam, 33–40.

Lai, C.J. (1986) Flow resistance, discharge capacity and momentum transfer in smooth compound closed ducts. *PhD Thesis*. The University of Birmingham.

Lai, C.J. and Knight, D.W. (1988) Distributions of streamwise velocity and boundary shear stress in compound ducts. In: *Proceedings of the 3rd International Symposium on Refined Flow Modelling and Turbulence Measurements* (Eds Y. Iwasa and N. Tamai), Universal Academcy Press, Tokyo, 527–536.

Lundgren, H. and Jonsson, I.G. (1964) Shear and velocity distribution in shallow channels. *J. Hydr. Div. ASCE* **90** (HY1), 1–21.

Melling, A. and whitelaw, J.H. (1976) Turbulent flow in a rectangular duct. *J. Fluid Mech.* **78**, 289–315.

Myers, W.R.C. (1978) Momentum transfer in a compound channel. *J. Hydr. Res.* **16**(2), 139–150.

Myers, W.R.C. (1982) Flow resistance in wide rectangular channels. *J. Hydr. Div. ASCE* **108** (HY4), 471–482.

Naot, N. and Rodi, W. (1982) Calculation of secondary currents in channel flow. *J. Hydr. Div. ASCE* **108** (HY8), 948–968.

Nelson, J.M. and Dungan Smith, J. (1989) Flow in meandering channels with natural topography. In: *River Meandering* (Eds. S. Ikeda and G. Parker), *Am. Geophys. Union Wat. Resour. Monogr. No. 12*, pp. 69–102.

Noutsopoulos, G.C. and Hadjipanos, P.A. (1982) Discussion of 'Boundary shear in smooth and rough channels', by D.W. Knight. *J. Hydr. Div. ASCE* **108** (HY6), 809–812.

Patel, H.S. (1984) Boundary shear in rectangular and compound ducts. *PhD Thesis*. The University of Birmingham.

Patel, V.C. (1965) Calibration of the Preston tube and limitations on its use in pressure gradients. *J. Fluid Mech.* **23**, 185–208.

Perkins, H.J. (1970) The formation of streamwise vorticity in turbulent flow. *J. Fluid Mech.* **44**, 721–740.

Preston, J.H. (1954) The determination of turbulent skin friction by means of Pitot tubes. *J. Roy. Aeronaut. Soc.* **58**, 109–121.

Rajaratnam, N. and Muralidhar, D. (1969) Boundary shear stress distribution in rectangular open channels. *Houille Blance*, No. 5, 603–609.

Replogle, J.A. and Chow, V.T. (1966) Tractive-force distribution in open channels. *J. Hydr. Div. ASCE* **92**, 169–191.

Rhodes, D.G. (1991) An experimental investigation of the mean flow structure in wide ducts of simple rectangular and trapezoidal compound cross section, examining in particular zones of high lateral shear. *PhD thesis*. The University of Birmingham.

Schlichting, H. (1979) *Boundary Layer Theory*. McGraw-Hill, New York.

Shiono, K. and Knight, D.W. (1989) Transverse and vertical measurements of Reynolds stress in a shear layer region of a compound channel. In: *7th International Symposium on Turbulent Shear Flows, Stanford, USA, August 1989*, 28.1.1–28.1.6, Stanford University Press.

Shiono, K. and Knight, D.W. (1990) Mathematical models of flow in two or multi stage straight channels. In: *Proceedings of the International Conference on River Flood Hydraulics* (Ed. W.R. White). Wiley, Chichester, 229–238.

Shiono, K. and Knight, D.W. (1991) Turbulent open channel flows with variable depth across the channel. *J. Fluid Mech.* **222**, 617–646 (and **231**, pp. 693).

Tracy, H.J. (1965) Turbulent flow in a three-dimensional channel. *J. Hydr. Div. ASCE* **91** (HY6), 9–35.

Wagner, H. (1969) Laminar flow in open rectangular channels. In: *Proceedings of the International Association for Hydraulic Research, Kyoto, Japan*, Vol. 1, Paper A6, 43–52.

Wallis, S.G. and Knight, D.W. (1984) Calibration studies concerning a one-dimensional numerical tidal model with particular reference to resistance coefficients. *Est. Coast. Shelf Sci.* **19**, 541–562.

Whiting, P.J. and Dietrich, W.E. (1991) Convective accelerations and boundary shear stress over a channel bar. *Water Resources Res* **27**(5), pp. 783–796.

Winter, K.G. (1977) An outline of the techniques available for the measurement of skin friction in turbulent boundary layers. *Prog. Aerospace Sci* **18**, 1–57.

Yuen, K.W.H. (1989) A study of boundary shear stress, flow resistance and momentum transfer in open channels with simple and compound trapezoidal cross sections. *PhD Thesis*. The University of Birmingham.

Yuen, K.W.H. and Knight, D.W. (1990) Critical flow in a two stage channel. In: *Proceedings of the International Conference on River Flood Hydraulics* (Ed. W.R. White). Wiley, Chichester, 267–276.

5 Simulation of Solute Transport in Open Channel Flow

S.G. WALLIS

Department of Civil and Offshore Engineering, Heriot-Watt University, Edinburgh, UK

5.1 INTRODUCTION

This chapter considers the mathematical simulation of longitudinal advection and the dispersion of neutrally buoyant solutes in open channel flow. The output from such simulation models usually consists of solute concentration distributions along a river system at one or more times, or, alternatively, temporal distributions at one or more fixed locations, following the discharge of a potentially polluting substance to the river. Most interest is usually directed towards the peak values of concentration, the travel times of solute clouds and the time for which concentrations are greater than some threshold value.

A number of different approaches are currrently being used by the water industry to simulate solute transport in open channel flows. In some instances the physical processes are simulated directly, whereas in others it is the overall effect of the physical processes which is simulated. In either event, the modeller will be required not only to solve the model equations, but also to provide appropriate values for the coefficients in those equations. In the ideal world such coefficients would be measurable physical quantities such as flow-rates, but in practice they are often quantities which are either poorly defined in physical terms or are difficult to measure, e.g. mixing coefficients or depth-averaged velocities. In addition, because natural channels are longitudinally non-uniform, the modeller does not just require a single value for each coefficient, but requires their spatial distribution. In unsteady flow applications, the temporal variation of the coefficients is also needed.

There are many different types of model (deterministic; stochastic; those used for design; those used for discharge consent setting), which have different data requirements and use different mathematical procedures (such as numerical algorithms; parameter optimization). In this chapter we consider only those models operated in a deterministic simulation mode.

The aim of the work reported here is to compare three models of different origins to demonstrate to what extent they are consistent with each other, using a simple case study of solute transport in a two-dimensional turbulent shear flow.

In the following section the physical processes governing solute transport in open channels are summarized. This is followed by a description of the three models, namely the advection−dispersion equation, the aggregated dead zone model and a particle tracking model. Model results for the case study are presented and conclusions are drawn.

Mixing and Transport in the Environment. Edited by K.J. Beven, P.C. Chatwin and J.H. Millbank
© 1994 John Wiley & Sons Ltd

5.2 PHYSICAL PROCESSES OF SOLUTE TRANSPORT

The effects of the most important physical solute transport mechanisms in a typical open channel flow can easily be observed by following events after the release of a small patch of neutrally buoyant marked material on the water surface. Initially the patch spreads longitudinally, vertically and transversely due to turbulent diffusion. The patch very quickly extends over a significant part of the depth and it becomes sheared by the vertical profile of longitudinal velocity. Once the patch reaches the bed it soon becomes well mixed vertically. Transverse spreading continues, enabling the patch to sample a greater proportion of the transverse profile of longitudinal velocity and yet further shear takes place. Different parts of the patch are now being carried or advected downstream at markedly different rates, creating a rapid longitudinal dispersion of the material. Concentrations of the material rapidly decrease in these early stages as the material is diluted in an increasing volume of water.

Secondary currents and large-scale turbulent eddies enhance the cross-sectional mixing once the patch extends over a significant proportion of the cross-section, and eventually a stage is reached where the material is approximately evenly distributed within the cross-section. This represents an equilibrium condition under which the tendency for differential longitudinal advection to create solute concentration gradients is balanced by the tendency for cross-sectional mixing to break the gradients down. From now on the centre of mass of the evolving solute cloud moves downstream at the cross-sectional average flow velocity and simultaneously the cloud spreads longitudinally. Concentrations of the material now decrease much more slowly than before. Under certain ideal conditions the longitudinal dispersion occurs at a constant rate, and the transport of the solute follows an advection–diffusion law.

If the material is not neutrally buoyant, the initially rapid vertical mixing will be inhibited, but as the concentrations reduce, so the induced density differences become smaller and eventually disappear. Longitudinal variations in channel geometry and alignment can inhibit the development of the equilibrium condition referred to above. In particular, horizontal circulations and gyres caused by meanders and other geomorphological structures tend to trap material in areas of the flow which appear to be separate from that in the main stream. Smaller scale circulatory flow structures are associated with channel irregularities around the periphery of the flow and promote a similar transient trapping mechanism. Traditionally, and perhaps unfortunately, such areas are referred to as 'dead zones.'

Numerous mathematical models have been constructed to describe the effects of these physical advection and mixing mechanisms on the passage of a solute cloud along an open channel. Some only consider the effects of the turbulent shear flow, whereas others include the effects of the dead zones. All require certain assumptions to be satisfied, and this usually restricts the application of the models to certain scenarios, although this does not deter modellers from using them more widely.

In the following section we look at three models in relation to the dominant transport mechanisms described above and the assumptions under which they are theoretically valid.

5.3 ADVECTION–DISPERSION EQUATION MODEL

5.3.1 Background

The advection–dispersion equation (ADE) is the basis of most one-dimensional models of solute transport in rivers. It was initially derived by Taylor (1953; 1954) for the cases of laminar and turbulent pipe flow, and extended to open channels by Fischer (1967). Detailed derivations may also be found in Fischer et al. (1979), Holly (1985) and Young and Wallis (1993). Rather than reiterate such detailed analyses, a simpler derivation will be presented here.

Figure 5.1 shows a longitudinal section of a river which is flowing from left to right. Let us consider a solute mass balance for the incremental length of channel, δx, namely that the rate of change of mass of solute within the increment is equal to the difference between the rate at which solute enters the increment and the rate at which it leaves. Assuming that the solute is conservative, the mathematical statement of this mass balance is

$$\frac{\delta x \partial (AC)}{\partial t} = M - \left[M + \frac{\delta x \partial M}{\partial x} + O(\delta x^2) \right] \tag{5.1}$$

where A is the cross-sectional area of the flow, C is the cross-sectional average solute concentration, M is the mass flow-rate of solute, x is the longitudinal coordinate and t is time. The fourth term on the right-hand side of Euqation 5.1 indicates that there are higher order terms in the Taylor series expansion of M. Assuming that δx is small we can ignore these higher order terms. Clearly, in the limit as $\delta x \rightarrow 0$, Equation 5.1 simplifies to

$$\frac{\partial (AC)}{\partial t} + \frac{\partial (AF)}{\partial x} = 0 \tag{5.2}$$

where the mass flow-rate has been replaced by the product of cross-sectional area and cross-sectional average solute flux, F. The instantaneous solute flux at any point in the flow cross-section is given by the product of the local velocity and the local concentration. Both of these can be represented as the sum of three terms

$$u = U + U' + U'' \tag{5.3}$$

$$c = C + C' + C'' \tag{5.4}$$

where u and c are the instantaneous local velocity and concentration, respectively; U and C are cross-sectionally averaged values; U' and C' are spatial deviations; and U'' and C'' are temporal deviations. The spatial deviations account for vertical and transverse velocity gradients and solute concentration variations within the flow cross-section; the temporal deviations account for the turbulent nature of the flow. The turbulent mean cross-sectionally averaged solute flux is obtained by averaging the point flux over a suitable time interval,

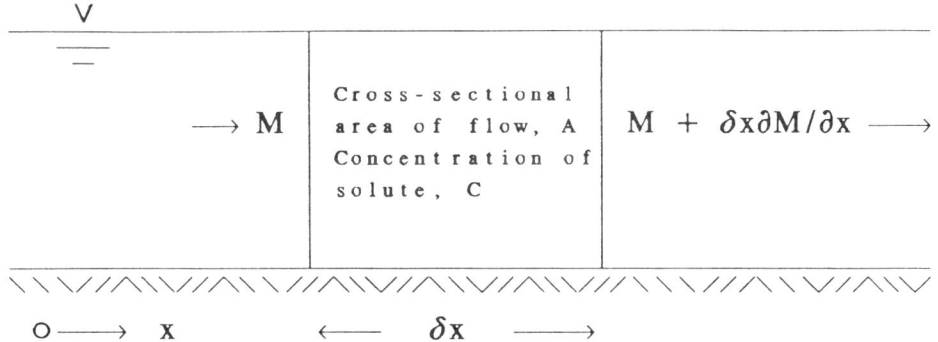

Figure 5.1 Sketch showing definitions used in advection–dispersion solute mass balance equation

followed by averaging over the flow cross-section to give

$$F = \frac{1}{A} \int_A \left[\frac{1}{\Delta t} \int uc \, dt \right] dA \tag{5.5}$$

As the average of a term containing only one appropriate deviation term is zero, Equation 5.5 simplifies to

$$F = UC + \frac{1}{A} \int_A U'C' \, dA + \frac{1}{A} \int_A \langle U''C'' \rangle \, dA \tag{5.6}$$

where the angled brackets indicate a time-averaged quantity. Clearly, by the decomposition introduced in Equations 5.3 and 5.4, U, C, U' and C' are time-averaged quantities. The first term on the right-hand side of Equation 5.6 represents solute transport by advection according to the cross-sectionally average values of solute concentration and velocity; the second term represents solute transport by longitudinal dispersion; and the third term represents solute transport by longitudinal turbulent diffusion. In most practical cases, the longitudinal turbulent diffusive flux is very much smaller than the longitudinal advective and longitudinal dispersive fluxes, and hence it is usually neglected.

Substitution of this truncated version of Equation 5.6 into Equation 5.2 gives

$$\frac{\partial(AC)}{\partial t} + \frac{\partial(AUC)}{\partial x} + \frac{\partial}{\partial x} \left[\int_A U'C' \, dA \right] = 0 \tag{5.7}$$

If we now assume that the theoretical result obtained by Taylor applies, namely that the longitudinal dispersive flux is proportional to the longitudinal concentration gradient and that it transports solute from areas of high concentration to areas of lower concentration, we can replace the integral in the third term in Equation 5.7 by $-DA\partial C/\partial x$ in which D is the longitudinal dispersion coefficient. Hence we obtain the familiar ADE

$$\frac{\partial(AC)}{\partial t} + \frac{\partial(AUC)}{\partial x} = \frac{\partial}{\partial x} \left[DA\frac{\partial C}{\partial x} \right] \tag{5.8}$$

Clearly, the assumption just made is of fundamental importance to whether or not Equation 5.8 is an appropriate mathematical description of longitudinal solute transport. Taylor's analysis considers a shear flow with no dead zones, and the assumptions under which longitudinal dispersive transport in such a flow may be described by means of a gradient diffusion term are described in Chatwin (1980), Chatwin and Allen (1985) and Young and Wallis (1993). Of particular importance is the requirement that the flow cross-section is constant. This implies that the coefficients A, U and D in Equation 5.8 should be constants, so that the following simpler equation, rather than Equation 5.8, applies

$$\frac{\partial C}{\partial t} + \frac{U\partial C}{\partial x} = \frac{D\partial^2 C}{\partial x^2} \tag{5.9}$$

Unfortunately, Equation 5.8 is of much more practical use than Equation 5.9 because natural river channels do not have longitudinally uniform cross-sections. Indeed, the reverse is the norm, so that the flow is naturally non-uniform with variations in channel shape, slope, sinuosity and roughness playing an important part in the flow hydraulics and solute transport.

Another important requirement of the gradient diffusion assumption is that the solute should have been dispersing for a sufficiently long time — long enough for full cross-sectional mixing to have occurred so that the cross-sectional deviations of solute concentration from the cross-sectional average values are small. Fischer *et al.* (1979) report that in a uniform channel the time taken for such conditions to occur is of the order of W^2/ϵ_t where W is the channel width and ϵ_t is the transverse mixing coefficient. For typical river channels this is of the order of 10^5 seconds. In a non-uniform channel this time is likely to be longer than in a uniform channel. Crucially, however, the longitudinal variation of velocity gradients and mixing rates within the cross-section may prevent the occurrence of the conditions necessary for the gradient diffusion assumption to hold (Valentine and Wood, 1979b).

Bearing in mind that most ADE based models in use are applied to non-uniform channels and rely on Equation 5.8, together with the requirement to investigate solute concentrations nearer to pollution sources rather than further away (because of the higher and potentially more dangerous concentrations), it is doubtful whether many of the applications of the ADE are truly consistent with the assumptions which underlie its validity.

5.3.2 Practical implementation

Despite the doubts aired earlier, Equations 5.8 and 5.9 have met with some success in practical problems and are used routinely by the water industry. Solutions are usually found by numerical methods to convert them into a discrete time form, enabling concentrations to be obtained by successive incremental integration over suitably chosen space and time intervals. Unfortunately, the form of Equations 5.8 and 5.9 is not well suited to the common low order numerical algorithms used in computational hydraulics. Difficulties are created by the advection term and these can be devastating when transport is advection dominated and/or when solute concentration gradients are large (Sobey, 1984; Sauvaget, 1985; Abbott and Basco, 1989).

Although numerical analysts have developed higher order numerical schemes which are capable of reducing the undesirable features encountered with low order algorithms (Gaskell and Lau, 1988; Leonard and Niknafs, 1991), there is little quantitative information available about the accuracy of these methods when applied to Equations 5.8 and 5.9.

A further difficulty is that of estimating the dispersion coefficient; this is notoriously difficult to do (Young and Wallis, 1993). If measurements of solute concentrations are available, then the dispersion coefficient can be found from the change in the variance of the concentration distributions. In fact, it is equal to twice the rate of change of the variance of the longitudinal concentration distribution, and this holds true from a much earlier time than W^2/ϵ_t (Fischer *et al.* 1979). As concentration data are usually collected in the time domain, however, the more appropriate equivalent relationship, using temporal concentration distributions, is (Valentine and Wood, 1979a; Nordin and Troutman, 1980)

$$D = \frac{U^3}{2} \frac{d(\sigma^2)}{dx} \qquad (5.10)$$

where σ^2 is the variance of temporal concentration distributions.

Although apparently easy to use with solute concentration data measured at two or more locations, the estimation of the variances of the distributions is sensitive to the information contained in the tails of the distributions. Typically, these tails are very long, and despite the concentrations being low, they are far removed from the centroid of the distribution. Therefore they have a great bearing on the variance. Unfortunately, it is often the case that the measurements do not accurately define the tail, either because of a lack of sensitivity

of the equipment or, and probably more commonly, because measurements are not taken
for long enough to ensure that all of the tail is recorded.

An alternative method for estimating the dispersion coefficient stems from Taylor's analysis.
Although of limited use in practical situations, as will be seen, it is included here because
it will be used in a later section. For the conditions under which Equation 5.9 is valid

$$D = \frac{1}{A} \int_A U'f \, dA \tag{5.11}$$

where f is a function involving the cross-sectional distribution of U' and the vertical and
transverse mixing coefficients (Young and Wallis, 1993). Clearly, such information is rarely
available. However, Fischer (1967), arguing that dispersion in rivers is dominated by the
transverse profile of longitudinal velocity, proposed a simpler version, namely

$$D = -\frac{1}{A} \int_0^w U_d d \int_0^y \frac{1}{\epsilon_t d} \int_0^y U_d d \, dy \, dy \, dy \tag{5.12}$$

where U_d is the deviation of the local depth-averaged velocity from the cross-sectional mean,
d is the local flow depth, ϵ_t is the local depth-averaged transverse mixing coefficient, y is
the transverse coordinate and w is the channel width. Again, this is not of great practical
use unless measurements or estimates are available of the transverse depth-averaged velocity
profile, the transverse variation of depth and the transverse variation of the depth-averaged
transverse mixing coefficient. However, it is of potentially more use than Equation 5.11 and
examples of its use may be found in Fischer (1968), Fischer et al. (1979) and in the present
work. Its use is also discussed in Young and Wallis (1993).

In practice, neither of these methods may be particularly useful, and recourse is often made
to empirical relationships between the dispersion coefficient and bulk flow parameters. None
of these, however, is very reliable (Young and Wallis, 1993).

Finally, it is worth noting that a potential dilemma for the modeller using ADE based models
is that if the model is found to be unable to reproduce observations of solute concentrations,
the modeller may not be able to decide whether this is because: (a) the equation on which
the model is based does not apply; (b) the coefficients being used are incorrect; (c) the
mathematical technique used to solve the equation is introducing errors or (d) there are errors
in the observations. Clearly these are not mutually exclusive.

5.4   AGGREGATED DEAD ZONE MODEL

5.4.1   Background

The aggregated dead zone (ADZ) model is a relatively recent innovation (Beer and Young,
1983) in solute transport modelling. Originally it stemmed from the belief that longitudinal
dispersion is dominated by the effects of dead zones in the flow, and it was found that by
representing a river reach in terms of one effective dead zone (which accounted for the
aggregated effect of all the individual dead zones), observations of solute transport in rivers
could be simulated very accurately.

The ADZ model equation may be derived in a similar way to that used above for the ADE,
although there are a number of crucial differences. The starting point is the same solute mass
balance as used previously, except that the solute concentration is not considered as a

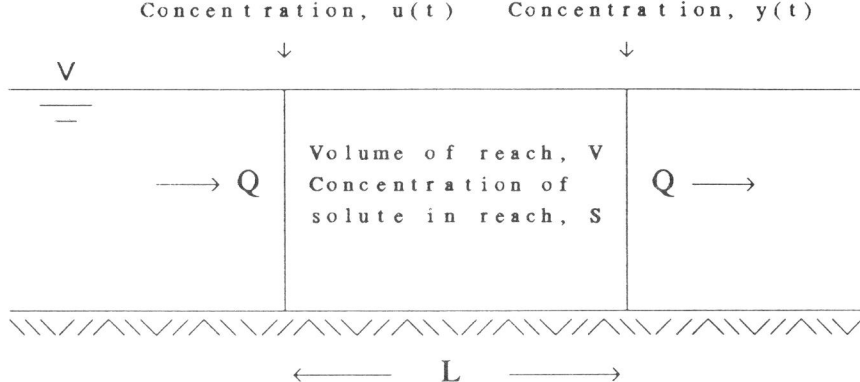

Figure 5.2 Sketch showing definitions used in aggregated dead zone model for solute mass balance

continuous variable along the river. Instead an ordinary differential storage routing equation is developed which links concentrations at two locations along a river. Referring to Figure 5.2 (which is essentially the same as Figure 5.1 apart from the notation), and applying the same solute mass balance for a conservative solute as before, but this time to a finite reach length, L, we obtain

$$\frac{d[VS(t)]}{dt} = Qu(t) - Qy(t) \tag{5.13}$$

where V is the volume of water in the reach, $S(t)$ is the average solute concentration in the reach, Q is the volumetric flow-rate of water through the reach, $u(t)$ is the cross-sectional average concentration of solute at the upstream boundary of the reach and $y(t)$ is the cross-sectional average concentration of solute at the downstream boundary of the reach. Assuming that the flow is steady, then the volume of water in the reach is a constant, and Equation 5.13 may be rewritten as

$$\frac{dS(t)}{dt} = \frac{Q}{V}\left[u(t) - y(t)\right] \tag{5.14}$$

Next, an assumption is required about how the average solute concentration in the reach depends on either or both of the solute concentrations at the boundaries of the reach. This is exactly the same problem that is encountered in hydrological flood routing in river or sewer systems (Young and Wallis, 1985). The simplest assumption to make is that $S(t)$ is proportional to $y(t)$

$$S(t) = \gamma y(t) \tag{5.15}$$

where γ is a constant. Recognizing also that the ratio V/Q is simply the travel time for the reach, $\bar{t}$, Equation 5.14 becomes

$$\frac{dy(t)}{dt} = \frac{1}{\gamma\bar{t}}\left[u(t) - y(t)\right] \tag{5.16}$$

It is instructive now to consider the characteristics of Equation 5.16. It is simply an equation which describes the dynamic behaviour of a first-order linear system (Schwarzenbach and

Gill, 1984), and it has different physical interpretations according to the value of γ. For example, if $\gamma = 1$, Equation 5.16 describes theoretical solute transport through a continuously stirred tank reactor (see, for example, Tchobanoglous and Schroeder, 1987). In this instance solute entering the reator is instantaneously and completely mixed throughout the reactor volume and has an instantaneous effect on the solute concentration leaving the reactor. This is illustrated in Figure 5.3a.

This behaviour is not characteristic of that found in rivers because solute entering a reach is neither instantaneously, nor completely, mixed within the reach volume. Instead, the response illustrated in Figure 5.3b is observed. This demonstrates the importance of the physical time lag or time delay, τ, which indicates that a change in the concentration entering the reach does not have an instantaneous effect on the concentration downstream. Equation 5.16 need only be modified in one regard for it to be able to simulate the response shown in Figure 5.3b, namely by the introduction of the time delay into the upstream concentration term. It should also be understood that as solute entering the reach is not mixed within the entire reach volume, γ will be less than unity. Hence we have

$$\frac{dy(t)}{dt} = \frac{1}{\gamma \bar{t}}\left[-y(t) + u(t-\tau)\right] \tag{5.17}$$

This is the fundamental mass balance equation underlying the ADZ approach, and it describes solute transport in terms of two physical processes, namely an advective translation represented by the time delay and a dispersive mixing represented by a time constant, which is some fraction of the travel time, i.e. $\gamma \bar{t}$. It would not be unreasonable to expect that the time delay would also be some fraction of the travel time.

5.4.2 Moment analysis

The relationship between the travel time and the time delay can be established by undertaking an analysis of the temporal moments of Equation 5.17. The nth temporal moment of a solute concentration distribution, $C(t)$, is defined as

$$M_c^n = \int_{-\infty}^{+\infty} t^n C(t) \, dt \tag{5.18}$$

The first three moments ($n = 0, 1, 2$) are of particular use because they allow three useful quantities to be evaluated, namely

$$\text{Area under distribution, } A_c = M_c^0 \tag{5.19a}$$

$$\text{Location of centroid of distribution, } \bar{t}_c = M_c^1/M_c^0 \tag{5.19b}$$

$$\text{Variance of distribution, } \sigma_c^2 = M_c^2/M_c^0 - (M_c^1/M_c^0)^2 \tag{5.19c}$$

Multiplying Equation 5.17 by t and integrating between $+\infty$ and $-\infty$ we obtain

$$\int_{-\infty}^{+\infty} \frac{t \, dy(t)}{dt} \, dt = \int_{-\infty}^{+\infty} \frac{-ty(t)}{\gamma \bar{t}} \, dt + \int_{-\infty}^{+\infty} \frac{tu(t-\tau)}{\gamma \bar{t}} \, dt \tag{5.20}$$

Integrating the first term by parts, we recognize the occurrence of the zeroth moment of the $y(t)$ distribution. Similarly, the second term contains the first moment of the $y(t)$ distribution. The third term is not as straight forward, but use of the following transformation

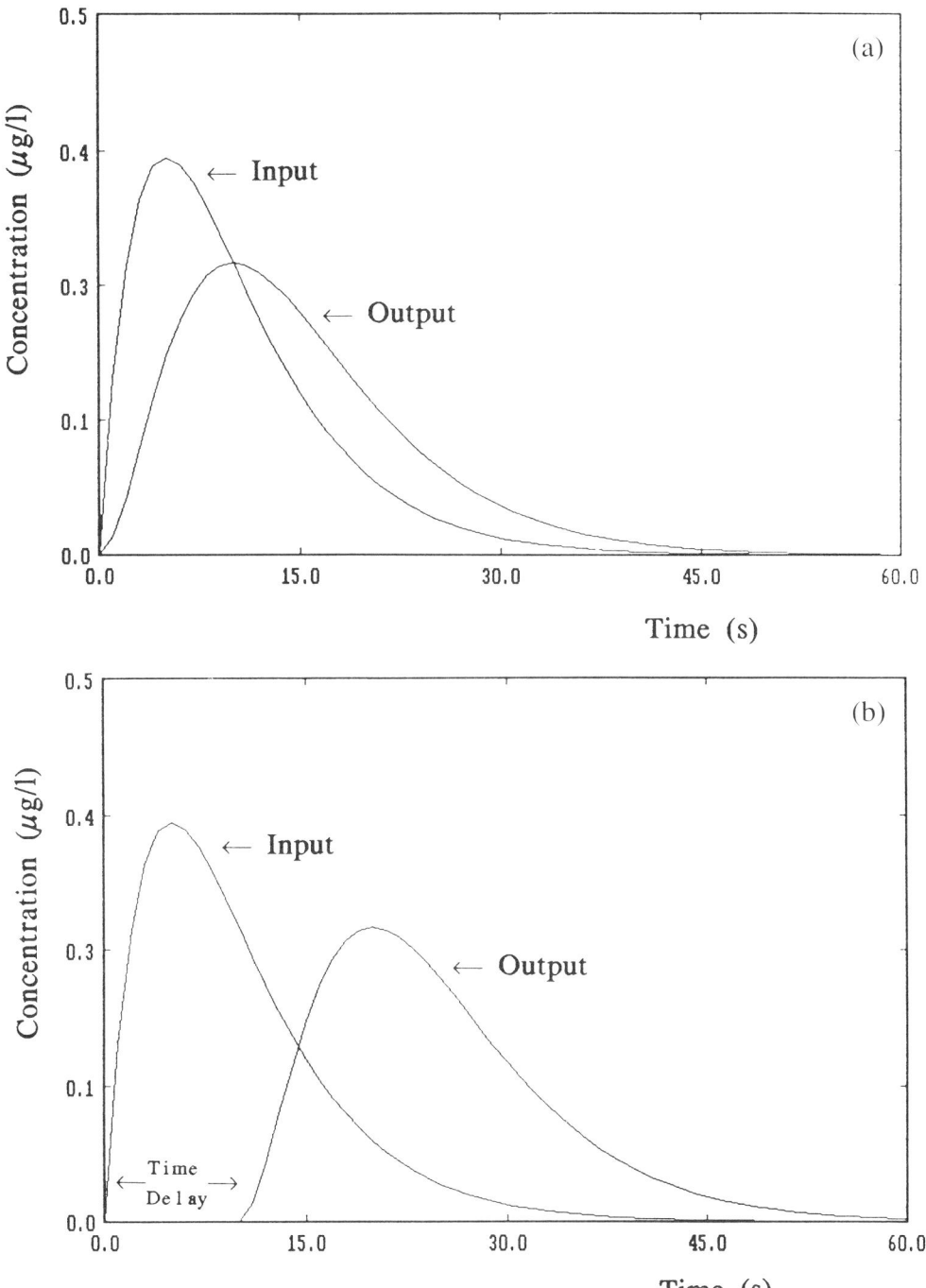

Figure 5.3 (a) Input and output functions for a continuously stirred tank reactor. (b) Input and output functions for a river

$$\int_{-\infty}^{+\infty} t^n C(\lambda) \, dt = \int_{-\infty}^{+\infty} (\lambda+\tau)^n C(\lambda) \, d\lambda \quad \text{with } \lambda = t - \tau \tag{5.21}$$

and noting that

$$\int_{-\infty}^{+\infty} t^n C(t) \, dt = \int_{-\infty}^{+\infty} \lambda^n C(\lambda) \, d\lambda \tag{5.22}$$

reveals the presence of the first moment of the $u(t-\tau)$ distribution. The result is

$$-M_y^0 = [-M_y^1 + M_u^1 + \tau M_u^0]/(\gamma \bar{t}) \tag{5.23}$$

Using Equations 5.19a–5.19c, noting that $M_y^0 = M_u^0$ for a conservative solute and rearranging we find

$$\bar{t}_y - \bar{t}_u = \gamma \bar{t} = \tau \tag{5.24}$$

The left-hand side of Equation 5.24 is simply a measure of the average time the solute spends in the reach, i.e. the travel time, $\bar{t}$. Hence we find that the travel time and the time delay are related by

$$\bar{t} = T + \tau \tag{5.25}$$

in which $\gamma \bar{t}$ is replaced by the time constant for dispersive mixing, or residence time, T.

5.4.3 Physical interpretation of the ADZ concept

It is now clear that γ has some physical significance, as it is simply the ratio $T/\bar{t}$, and it is termed the dispersive fraction. A useful alternative interpretation of the dispersive fraction is available by realizing that the temporal parameters are equivalent to volumes (by multiplication by the flow-rate): the travel time is related to the total reach volume; the residence time is related to the volume responsible for dispersion; and the time delay is related to the volume responsible for pure advection. It follows, therefore, that the dispersive fraction also represents a measure of the volume of water which is active in causing dispersion, i.e. the volume of the ADZ.

As already noted, dead zones are cells of flow which do not mix efficiently with the main flow. Instead of being characteristic of the shear flow typically assumed to describe open channel flows, they are eddy-like flow structures which provide a delay-dilution mechanism. Solute which enters a dead zone is trapped in a circulating mixing zone and is slowly released back into the main flow at reduced concentrations. The effect of dead zones is to delay the passge of solute through a river reach by temporarily storing it in local mixing zones. Traditionally (Thackston and Schnelle, 1970; Valentine and Wood 1979a; 1979b; Nordin and Troutman, 1980), dead zones were associated with physically identifiable irregularities in the bed and banks of rivers, but with regard to the ADZ model, dead zones have a much wider interpretation and also include a host of other features which impose the same sort of dynamic storage on the passage of solutes through river reaches, e.g. horizontal circulations caused by meanders, horizontal and vertical circulations caused by pool–riffle structures and transient turbulent eddies. Clearly all these exist on a number of length and time scales.

An ADZ, however, is a purely conceptual device which accounts for the overall dispersive effect of all mixing mechanisms by lumping them together into one effective dynamic storage zone. This means that even a channel with a smooth boundary, which has no apparent dead zones, can still be represented by an effective dead zone. Importantly, however, the ADZ model parameters still retain their physical meaning with the residence time, T, describing the average time spent by solute undergoing dispersion, and the time delay describing the minimum time it takes for solute to pass through a reach.

5.4.4 Practical implementation of the ADZ Model

There are two practical difficulties with Equation 5.17, namely what values of the coefficients γ, $\bar{t}$ and τ should be used, and over what reach lengths should it be applied.

If measured solute concentration distributions are available, then the model coefficients can be estimated from these data. In the absence of any solute concentration data, then two steps are necessary. Firstly, $\bar{t}$ can be estimated either from the relation V/Q introduced earlier, or from the flow hydraulics of the reach, as it is equivalent to the reach length divided by the reach mean cross-sectional average flow velocity. Secondly, the dispersive fraction can be estimated from values reported previously (Wallis, Young and Beven, 1989; Wallis, Guymer and Bilgi, 1989). To date, values of the dispersive fraction estimated from field data lie in the range 0.1–0.5, with the higher values corresponding to natural streams and rivers, and lower values corresponding to smoother man-made channels. Of particular note, however, is that for any reach the dispersive fraction appears to be approximately independent of flow-rate. This is an attractive feature from a practical point of view.

An alternative way of estimating the dispersive fraction is also available by extending the moment analysis introduced in Section 5.4.2. As we know that under certain circumstances the dispersion coefficient is simply related to the change of variance of a solute cloud, we can derive an equation linking the dispersion coefficient to the dispersive fraction by analysing the behaviour of the variance of solute concentration distributions according to the ADZ model. This is done by multiplying Equation 5.17 by t^2 and integrating between $+\infty$ and $-\infty$. After some manipulation, this results in

$$\sigma_y^2 - \sigma_u^2 = (\gamma \bar{t})^2 \tag{5.26}$$

As Equation 5.10 can be rewritten as

$$D = \frac{U^2(\sigma_y^2 - \sigma_u^2)}{2(\bar{t}_y - \bar{t}_u)} \tag{5.27}$$

then we find that

$$D = \frac{U^2 \gamma^2 \bar{t}}{2} \tag{5.28}$$

or

$$\gamma = [2D/(U^2\bar{t})]^{1/2} \tag{5.29}$$

The picture is less clear with respect to suitable reach lengths over which Equation 5.17 should be used. Intuitively, bearing in mind the assumptions made in its derivation, we would expect there to be a maximum reach length beyond which Equation 5.17 will no longer be valid. At present, it is unclear what this is or how it might be related to the hydraulic and mixing parameters of the flow. In practice, however, this need not be a serious problem, as a longer reach can be represented by two shorter reaches in series, each of which can be represented by an equation of the form of Equation 5.17.

For simulation purposes, Equation 5.17 is usually solved in a discrete time mode using a numerical algorithm. There are a number of standard numerical methods which could be used and, in general, they are far less prone to numerical errors than those used with the ADE, e.g. Runge−Kutta methods or methods for linear dynamic systems.

5.5 PARTICLE TRACKING MODELS

5.5.1 Background

Although particle tracking (PT) models have been used for about 20 years (Sullivan, 1971; Allen, 1982), they have been less often used in practice than either the ADE or ADZ models. Indeed, they are a different type of model altogether; one in which, instead of following the evolution of a solute concentration field using a differential equation description of mass balance, the motion of individual solute particles is tracked according to a specified velocity field and a specified distribution of mixing coefficients. Particles are advected longitudinally by the cross-sectional profile of longitudinal velocity and move within the flow cross-section in a manner consistent with the prescribed cross-sectional mixing mechanisms. As a result, it is the evolution of a solute cloud which is modelled, and the concentration field is estimated from the positions of all the particles tracked.

Usually, cross-sectional mixing in open channel flow is assumed to be dominated by vertical and transverse turbulent diffusion, and in a PT model this is simulated by some form of random motion of the particles. Indeed, the models are often termed random walk models because solute particles follow a random path within the flow cross-section. In principle, other cross-sectional mixing mechanisms can also be included, e.g. horizontal and vertical circulations established either by non-uniform distributions of shear stress (traditionally termed secondary currents) or by channel geomorphology such as meanders and pool−riffle structures. Longitudinal turbulent diffusive transport can also be included if necessary.

Here, however, we only consider the simplest PT model that is consistent with the main features of solute transport in open channel flow considered in previous sections. Hence longitudinal turbulent diffusive transport will be neglected, and solute transport will be assumed to be governed by the interaction of the transverse profile of longitudinal velocity with transverse mixing due to transverse turbulent diffusion. Hence a two-dimensional depth-averaged simulation will be considered. Each constituent particle of the solute cloud is tracked over a number of independent time steps, moving longitudinally and transversely in discrete jumps according to the product of the time step and an appropriate velocity coefficient.

A definition sketch is shown in Figure 5.4, in which the motion of one particle over a single time step is shown. This is represented mathematically by

$$x_i = x_{i-1} + u(y)\Delta t \tag{5.30}$$

$$y_i = y_{i-1} + vR\Delta t \tag{5.31}$$

where x_i and y_i are the coordinates of the position of the particle after the ith time step, $u(y)$ is the local turbulent mean longitudinal velocity, Δt is the time step, v is the magnitude of the transverse velocity coefficient and R is a random number. The cross-sectional motion of a particle can be simulated in a number of ways, see e.g. Dickinson (1976), Allen (1982) and Koutitas (1988). Here we follow the third of these, in which the velocity coefficient, v, is related to the local transverse mixing coefficient, ϵ_t by

$$v = (6\epsilon_t/\Delta t)^{1/2} \tag{5.32}$$

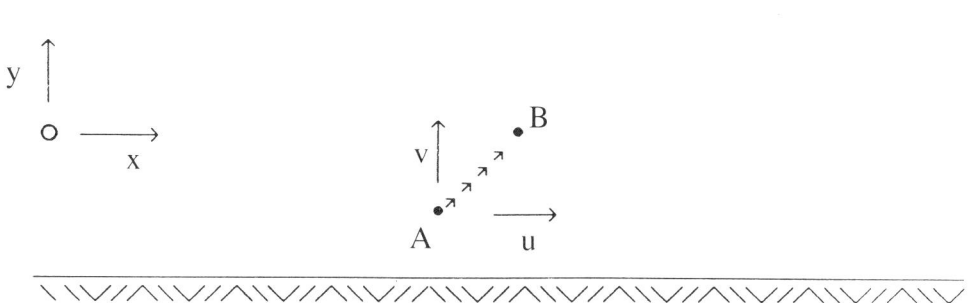

Figure 5.4 Movement of a particle during one time step of a particle tracking model. The particle at A moves to B under the combined effect of velocities u and v

and R lies in the range -1 to $+1$. It is also necessary to impose a certain pattern of behaviour on a particle when it encounters the transverse boundary of the channel. The simplest behaviour to impose is that of ideal reflection, i.e. a particle rebounds into the channel a distance equal to that it would have reached beyond the boundary, had it been able to pass through the boundary. Finally, experience suggests that a very large number of particles (of the order of 10^4) needs to be used to obtain representative results.

5.5.2 Physical interpretation

The PT model described here applies to an ideal scenario in which the transverse mixing caused by turbulent diffusion is simulated using a random walk. Each step in the random walk of each particle is assumed to be independent, and the structure of the turbulence is assumed to be homogeneous. Under these conditions, the transverse spreading of a solute cloud follows a Fickian diffusion law with the diffusion coefficient being related to the distance step and the time step of the random walk (Fischer *et al.*, 1979).

Such a description of the transverse mixing satisfies the assumptions of the analysis of Taylor (1954) of longitudinal dispersion in turbulent pipe flow and the analysis of Fischer (1967) of longitudinal dispersion in open channel flow. Hence, for longitudinally uniform flow we expect the simulations of a PT model to be consistent with those of a model based on the ADE, provided the solute has been dispersing for a time long enough for the ADE to apply.

5.5.3 Practical considerations

An important characteristic of PT models is that they seek to simulate the physical transport mechanisms themselves rather than their effects. To do this the modeller needs to furnish a detailed description of these mechanisms. In the two-dimensional model being considered here, the modeller needs to specify the transverse profile of the turbulent mean longitudinal velocity at a sufficiently frequent spacing along the channel to accurately represent the advective features of the flow. Similarly, the modeller needs to specify the magnitudes of, and the distribution of, the transverse mixing coefficients at a suitably frequent longitudinal spacing. For a longitudinally uniform flow this is relatively simple because conditions are the same at all locations along the channel, i.e. this information only needs to be specified once. For natural channels, however, which are inherently non-uniform, the amount of data required and how it is obtained is a likely impediment to the application of the method.

It is unlikely that a simple model such as that described above will accurately simulate mixing in real channels, where the turbulence structure is more complex, the effects of boundary layers are important and the flow is three dimensional. However, some work shows that more realistic implementations of such models are practically possible and are capable of simulating observed patterns of mixing and dispersion in real channels (Heslop and Allen, 1989; Brockie *et al.*, 1991).

5.6 COMPARISON OF MODELS

5.6.1 Introduction

In this section the results are presented of a case study of solute transport. The aim is to compare the ability of the three models introduced in this chapter to simulate a very simple solute transport event. The event chosen is one which is enitrely consistent with the assumptions on which the models are based and for which the solution is known. Clearly, it will be instructive to discover how closely each model is able to simulate the known solution and, if any of them perform badly, this will be indicative of a serious flaw in the model.

The example considered is that of a long uniform channel of rectangular cross-section. The width, W, is 1 m, the cross-sectional mean longitudinal velocity, U, is 0.2 m/s and the flow depth, d, is 0.2 m. The transverse profile of the depth-averaged longitudinal velocity is parabolic in shape and is given by

$$u(y) = \frac{3U}{2}(1 - 4y^2/W^2) \tag{5.33}$$

where y is the transverse coordinate direction with the origin on the channel centre line. The transverse mixing is due solely to turbulent diffusion and is described by a transverse mixing coefficient given by $0.2du_*$, where u_* is the shear velocity, specified as 0.012 m/s. Clearly, this is a very simple example of a two-dimensional shear flow. The initial condition is an instantaneous release of a solute at the upstream end of the channel in the form of a transverse line source. After an initial period of advection-dominated dispersion, the solute will eventually be transported in a manner consistent with the analysis of Taylor (1954) of turbulent pipe flow. Hence the longitudinal distribution of the solute will become a Gaussian curve and its centre of mass will move at the cross-sectional mean velocity of flow and its variance will increase linearly with time at a rate equal to twice the dispersion coefficient. Based on the time-scale introduced in Section 5.3.1, the time taken for the longitudinal concentration distribution to become Gaussian will be 2083 seconds.

5.6.2 Application of the PT Model

Particle tracking models can be implemented in a number of ways. Often they are run once with a very large number (typically 10 000) particles. A different approach has been used here. Instead of one run with 10 000 particles, 10 runs each with 1000 particles and a different sequence of random numbers were carried out. Although not used here, this form of implementation is ideally suited to a transputer based simulation because the runs could be run simultaneously, i.e. in parallel. In each run the particles were initially uniformly distributed across the upstream end of the channel and they were tracked at a time step of five seconds until they had all passed a section 750 m downstream of the release section. The results from all the runs were then combined to produce longitudinal and temporal solute concentration distributions.

Longitudinal concentration distributions were evaluated after 500 and 600 time steps of the simulation. The location of the centre of mass ($\bar{x}$) of these two longitudinal distributions, together with their variances (σ_x^2) were calculated directly from the positions of all the particles using

$$\bar{x} = \frac{1}{N} \sum_{i=1}^{N} x_i \qquad\qquad (5.34)$$

$$\sigma_x^2 = \frac{1}{N} \sum_{i=1}^{N} (x_i - \bar{x})^2 \qquad\qquad (5.35)$$

where $N = 10\,000$.

Temporal distributions were evaluated at two sections situated at 500 and 750 m downstream of the release section. In each of the 10 runs, temporal distributions were evaluated for a 2 m wide section centred at 500 and 750 m. These were then averaged across all the runs. Moments of the averaged distributions were calculated using numerical integration, allowing the travel times and temporal variances to be calculated using Equations 5.19b and 5.19c.

5.6.3 Application of the ADE model

The ADE model was used to simulate the solute transport in the case study, starting with an initial condition given by the longitudinal solute distribution evaluted after 500 steps of the PT model. The ADE was implemented using a finite element solution to Equation 5.8, which is described in Wallis *et al.* (1989). The simulation used a distance step of 8 m and a time step of 25 s. The following results were extracted: the longitudinal solute distribution after 20 time steps (corresponding to 600 time steps of the PT simulation) and the temporal solute distribution at 750 m downstream of the start of the channel.

The model coefficients were calculated from the specifications of the case study. Clearly, the flow area and average longitudinal velocity were easily obtained, whereas the dispersion coefficient was evaluated from Equation 5.12, using numerical integration, giving a value of 0.389 m^2/s.

Although this finite element solution uses only linear interpolation functions, like most of the low order numerical schemes it is capable of accurate and reliable simulations under certain conditions. These conditions are met here because concentration gradients are relatively low, there is a high spatial resolution of the solute distribution and the Courant and Peclet numbers are relatively low (5/8 and 4, respectively). Under more severe conditions, however, it breaks down and higher order schemes would be required.

5.6.4 Application of the ADZ Model

The ADZ model was used to simulate the solute transport between the sections at 500 and 750 m from the upstream end of the channel. This required the numerical solution of Equation 5.17, and following previous work with the ADZ model this was achieved by interpreting

Equation 5.17 in terms of the theory of linear dynamic systems (Schwarzenbach and Gill, 1984). Assuming that the input to the system (the upstream solute concentration) is well represented by a staircase function, i.e. it is approximately constant over small intervals of time, a numerical algorithm for evaluating the output of the system (the downstream solute concentration) is

$$y_k = -ay_{k-1} + bu_{k-\delta} \tag{5.36}$$

where $a = -\exp(-\Delta t/T)$, $b = 1 + a$, y_k is the output value at time $k\Delta t$, $u_{k-\delta}$ is the input value at time $(k-\delta)\Delta t$, Δt is the time step, T is the time constant of the system (residence time) and δ is the nearest integer representation of the time delay in terms of the time step, i.e. δ = integer value of $\tau/\Delta t$.

This algorithm usually gives accurate simulations of Equation 5.17 so long as the ratio of the time constant to the time step is greater than five, and the input is well described at the sampling frequency determined by the time step. If necessary, it is straightforward to obtain a more accurate simulation algorithm, simply by assuming a more realistic variation of the input over a time step, e.g. a linear change. Tests showed, however, that in the case being considered here there was little difference in the results between these two algorithms. Probably of more importance with regard to this type of algorithm is how closely the time delay is represented by an integer number of time steps. Again, this is related to the size of the time step.

The model coefficients were obtained from the specifications of the case study and the relationship between the dispersive fraction and the dispersion coefficient derived earlier (Equation 5.29). Hence, with a cross-sectional mean longitudinal velocity of 0.2 m/s, a reach length of 250 m and a dispersion coefficient of 0.389 m²/s, the travel time is 1250 s and the dispersive fraction is 0.125. The simulation was carried out with a time step of 25 s, so that the time delay was 44 sampling intervals and the residence time was six sampling intervals.

5.6.5 Discussion of results

Longitudinal solute concentration distributions from the PT model are shown in Figure 5.5. The elapsed time since the release of the particles is at least 2500 s, which is greater than the time-scale required for a Gaussian distribution to develop. Figure 5.6 shows Gaussian distributions plotted with these results, calculated using the locations of the centres of mass and the variances of the distributions, calculated in the manner described in Section 5.6.2. Evidently, these Gaussian distributions are well reproduced by the PT model.

The variances of these PT model distributions are 2271.32 and 1882.19 m², and the time interval is 500 s. This gives a dispersion coefficient of 0.389 m²/s, which is the same as that calculated from the flow structure in Section 5.6.3.

Figure 5.7 shows results from the ADE model plotted with the corresponding results from the PT model, i.e. at a time 3000 s after the release of the solute. Clearly, the agreement is excellent.

Figure 5.8 shows temporal concentration distributions from the PT model at two locations, namely at 500 and 750 m downstream of the release section. It is clear that these are not as smooth as the longitudinal distributions. This is probably due to the length scale (2 m) over which the distributions were evaluated, although at the time of writing this has not been investigated.

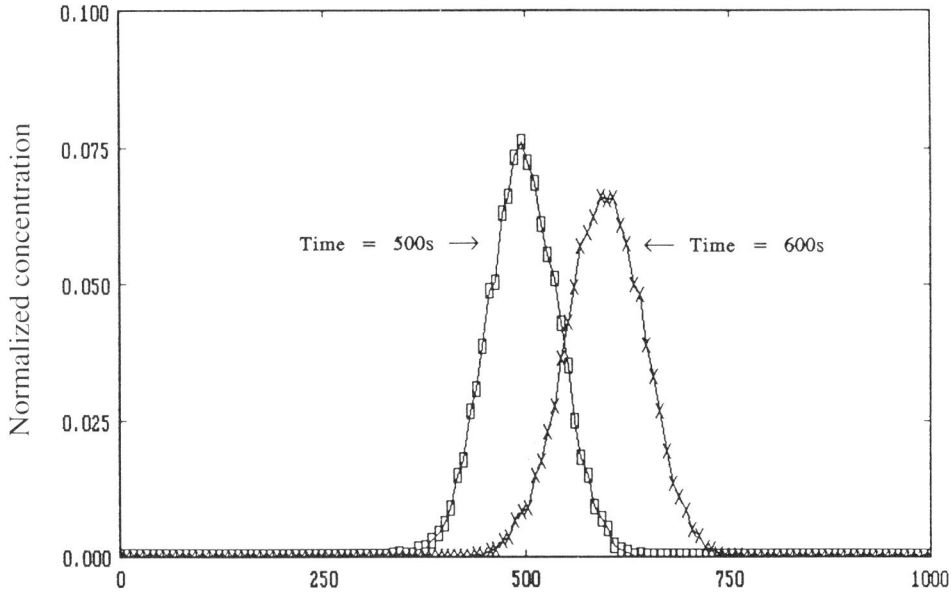

Figure 5.5 Longitudinal concentration distributions from the PT model

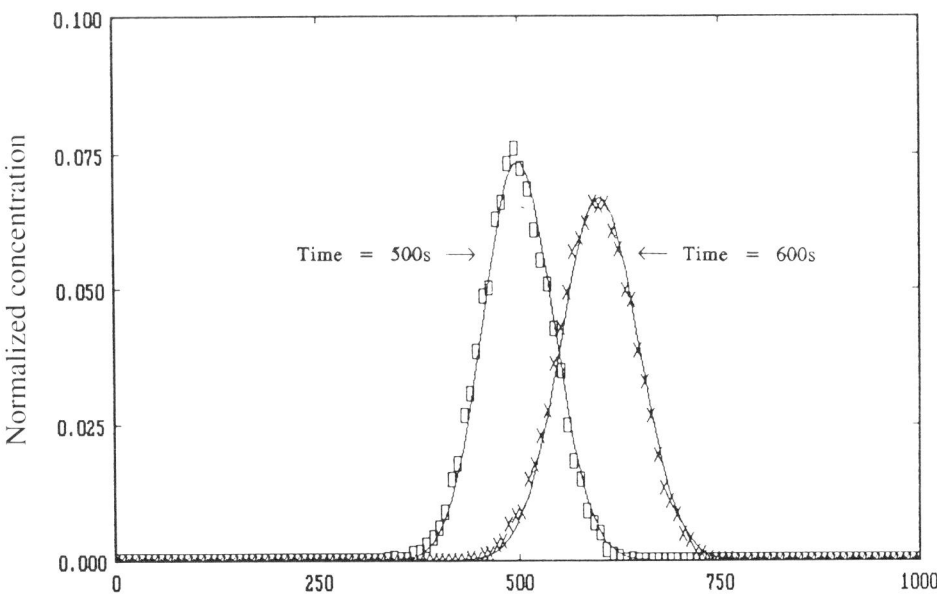

Figure 5.6 Comparison of longitudinal concentration distributions from the PT model (symbols) with Gaussian distributions (lines)

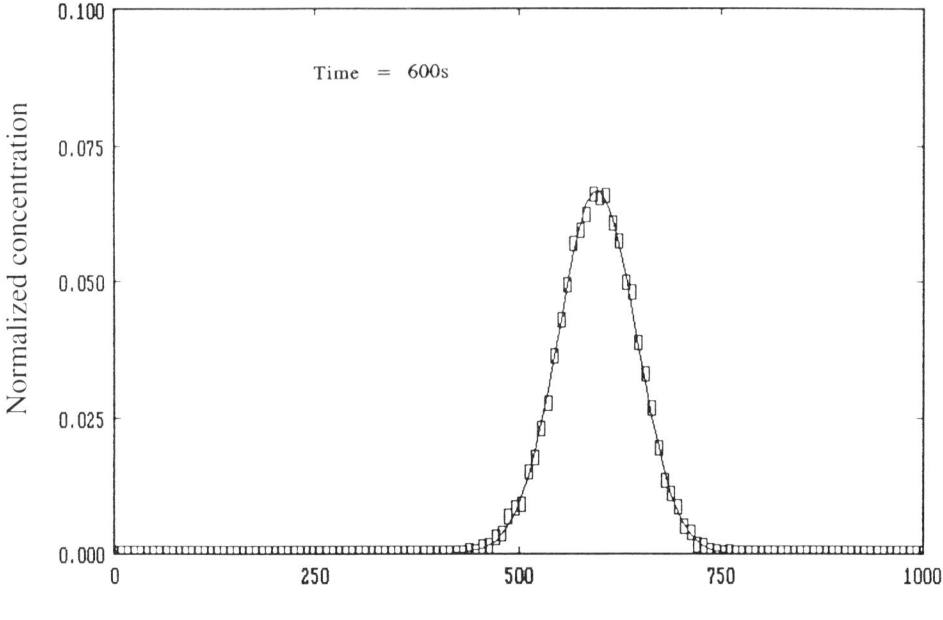

Figure 5.7 Comparison of longitudinal concentration distributions from the PT model (symbols) with distributions from the ADE model (line)

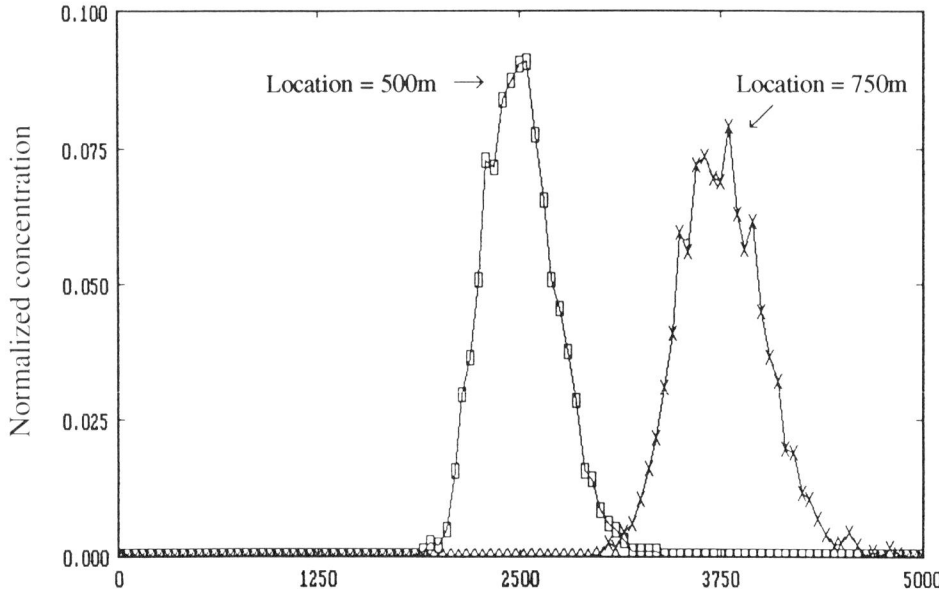

Figure 5.8 Temporal concentration distributions from the PT model

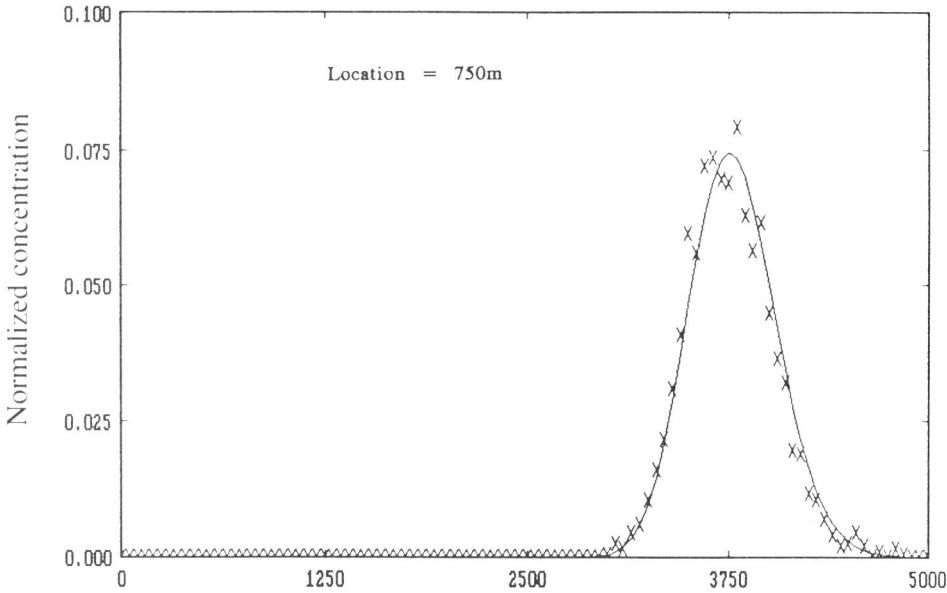

Figure 5.9 Comparison of temporal concentration distributions from the PT model (symbols) with distributions from the ADE model (line)

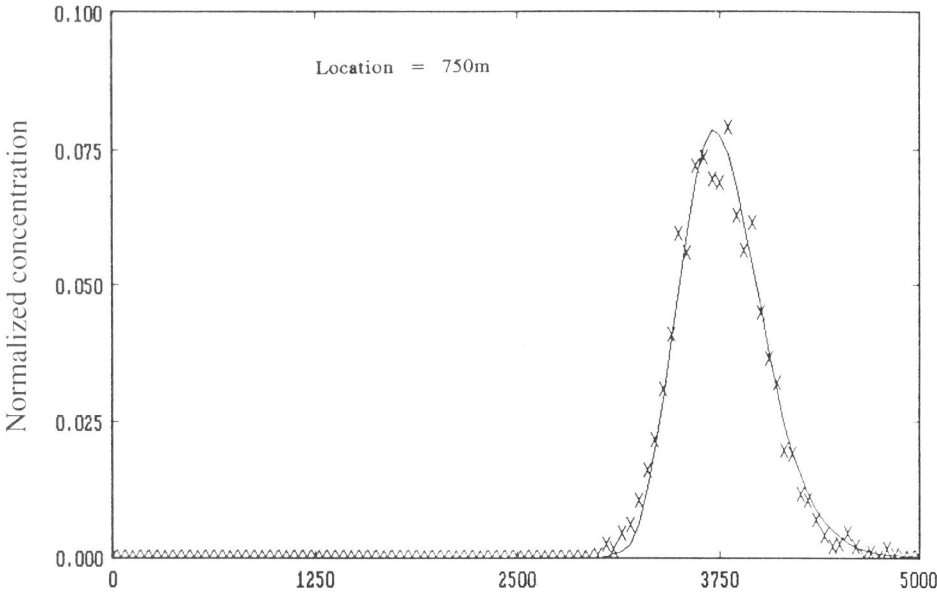

Figure 5.10 Comparison of temporal concentration distributions from the PT model (symbols) with distributions from the ADZ model (line)

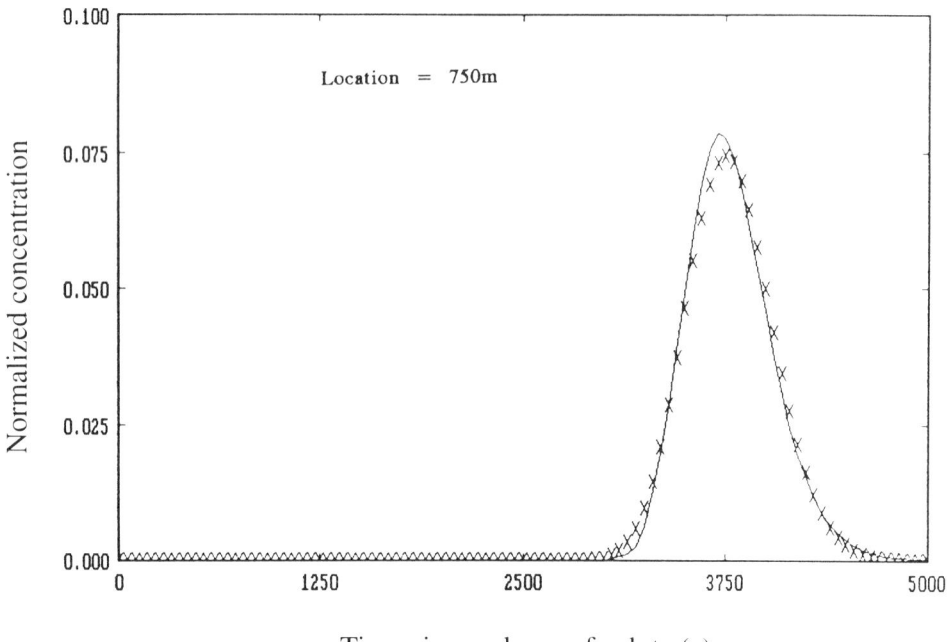

Figure 5.11 Comparison of temporal concentration distributions from the ADE model (symbols) with distributions from the ADZ model (line)

Figure 5.9 shows results from the ADE model at 750 m from the release section plotted with the corresponding results from the PT model, and Figure 5.10 shows results from the ADZ model at the same location similarly plotted with those from the PT model. Finally, Figure 5.11 compares the results from the ADE and ADZ models at this location.

It is apparent that all three models are producing consistent results in the time domain, although there are some minor differences. It is also instructive that the time delay between the starts of the PT model distributions in Figure 5.8 is close to 44 sampling intervals (the value obtained by estimating it from the flow structure via the dispersion coefficient). Clearly, the starts of these distributions are not well defined, so there is a degree of uncertainty involved in estimating the time delay in this way.

5.7   CONCLUDING REMARKS

In this chapter three different models of longitudinal solute transport in open channel flow have been described. An idealized example of a two-dimensional shear flow has been used to compare the results of the models, and they have been found to be in good agreement. Some useful links between the physical flow and mixing mechanisms and the model coefficients have been emphasized. In particular, a relationship between the dispersion coefficient of the ADE model and the dispersive fraction of the ADZ model has been derived.

The fact that these models of very different backgrounds and mathematical technique agree so closely is extremely encouraging and lends credence to all three of them. It should be understood, however, that the case study used here is only one of an infinite number of scenarios that could have been considered, and it may be simply good fortune that one has been chosen which appears to yield such a significant conclusion. Evidently, this case study is rather simplistic because the real difficulty in simulating solute transport in open channel flows lies in reproducing the behaviour observed in natural channels, which by their very nature are non-uniform. However, there would be little point in addressing this problem before establishing that the models being used were capable of accurately reproducing the simpler situation.

Turning our attention to non-uniform channels, there are two separate problems to be resolved. Firstly, how do we estimate model coefficients in such situations, and secondly how do we estimate the accuracy of the model results? The first of these is difficult to resolve, and, indeed, there are many answers to it depending on the modelling philosophy being used. For example, if the modeller takes a purely deterministic view he may wish to estimate the coefficients from measurements of the physical characteristics of the channel, the flow hydraulics and the mixing mechanisms. Alternatively, the modeller may take a more pragmatic view, and consider that as any set of observations of solute concentrations contains information embodying the effects of all of the physical processes influencing solute transport, a model can be based purely on such data. And, of course, there are many stances between these two extremes.

On a more encouraging note, however, this work suggests an intriguing possibility for the second problem. By configuring the PT model to simulate solute transport in non-uniform channels we can obtain concentration distributions and model coefficients for a wide range of non-uniform flow and mixing conditions. Although the simulations may not exactly reproduce the physical processes existing in real channels, they will produce solutions against which the ADE and ADZ models can be compared over the range of coefficient values likely to be needed in practice. Furthermore, as the PT model introduces no numerical error into its solutions, we have no reason to doubt the accuracy of the simulations. Hence, the accuracy of the ADE, ADZ and, indeed, other models can be assessed, as once the modeller has assumed that a particular model is appropriate, the only source of error in its solutions lies in the mathematical solution technique employed. It makes no difference whether the model coefficient values reflect real mixing processes or not, so long as, in this exercise, the model coefficients are consistent with the PT model simulations. As we have seen, this is easily ensured because all of the model coefficients can be calculated from the specifications of the PT simulation.

Finally, the relationship between the dispersion coefficient and the dispersive fraction has much wider significance than noted earlier. Equation 5.29 indicates that the dispersion fraction is linked to the dispersion coefficient through the longitudinal velocity and the reach mean travel time. However, as this travel time is a function of reach length, it is perhaps through this relationship that an understanding can be obtained regarding the range of reach lengths for which different order ADZ models are valid. Indeed, a cursory examination reveals that the selection of a dispersive fraction may be part of the same problem of selecting model order for a given reach length.

REFERENCES

Abbott, M.B. and Basco, D.R. (1989) *Computational Fluid Dynamics: An Introduction for Engineers.* Longman, Harlow, 425pp.
Allen, C.M. (1982) Numerical simulation of contaminant dispersion in oscillatory flows. *Proc. Roy. Soc. London, Ser. A* **281**, 179–194.

Beer, T. and Young, P.C. (1983) Longitudinal dispersion in natural streams. *J. Environ. Engin. Am. Soc. Civ. Engin.* **109**, 1049–1067.

Brockie, N.J.W., Allen, C.M. and Guymer, I. (1991) An initial comparison between 3D random walk simulations and tracer studies in a large experimental facility. *Proceedings of the 24th IAHR Congress, Madrid, Spain.*

Chatwin, P.C. (1980) Presentation of longitudinal dispersion data. *J. Hydr. Div. Am. Soc. Civ. Engin.* **106**, 71–83.

Chatwin, P.C. and Allen, C.M. (1985) Mathematical models of dispersion in rivers and estuaries. *Ann. Rev. Fluid Mech.* **17**, 119–149.

Dickinson, A. (1976) Approximate techniques for investigating contaminant dispersion in fluid flow. *PhD Thesis.* University of Liverpool.

Fischer, H.B. (1967) The mechanics of dispersion in natural streams. *J. Hydr. Div. Am. Soc. Civ. Engin.* **93**, 187–216.

Fischer, H.B. (1968) Dispersion predictions in natural streams. *J. Sanit. Engin. Div. Am. Soc. Civ. Engin.* **94**, 927–943.

Fischer, H.B., List, E.J., Koh, R.C.Y., Imberger, J. and Brooks, N.H. (1979) *Mixing in Inland and Coastal Waters.* Academic Press, New York, 483pp.

Gaskell, P.H. and Lau, A.K.C. (1988) Curvature-compensated convective transport: SMART, a new boundedness-preserving transport algorithm. *Int. J. Numer. Methods Fluids* **8**, 617–641.

Heslop, S.E. and Allen, C.M. (1990) Turbulence and dispersion in larger UK rivers. *Proceedings of the 23rd IAHR Congress, Ottawa, Canada.*

Holly, F.M. (1985) Dispersion in rivers and coastal waters — 1. Physcial principles and dispersion equations. In: *Developments in Hydraulic Engineering — 3* (Ed. P. Novak). Elsevier, London, 1–37.

Koutitas, C.G. (1988) *Mathematical Models in Coastal Engineering*, Pentech Press, London, 156pp.

Leonard, B.P. and Niknafs, H.S. (1991) Sharp monotonic resolution of discontinuities without clipping of narrow extrema. *Computers Fluids* **19**, 141–154.

Nordin, C.F. and Troutman, B.M. (1980) Longitudinal dispersion in rivers: the persistence of skewness in observed data. *Wat. Resourc. Res.* **16**, 123–128.

Sauvaget, P. (1985) Dispersion in rivers and coastal waters — 2. Numerical computation of dispersion. In: *Developments in Hydraulic Engineering — 3* (Ed. P. Novak). Elsevier, London, 39–78.

Schwarzenbach, J. and Gill, K.F. (1984) *System Modelling and Control.* 2nd edn. Edward Arnold, London, 322pp.

Sobey, R.J. (1984) Numerical alternatives in transient stream response. *J. Hydr. Engin. Am. Soc. Civ. Engin.* **110**, 749–772.

Sullivan, P.J. (1971) Longitudinal dispersion within a two-dimensional turbulent shear flow. *J. fluid Mech.* **49**, 551–576.

Taylor, G.I. (1953) Disperson of soluble matter in solvent flowing slowly through a tube. *Proc. Roy. Soc. London, Ser. A,* **219**, 186–203.

Taylor, G.I. (1954) The dispersion of matter in turbulent flow through a pipe. *Proc. Roy. Soc. London, Ser. A* **223**, 446–468.

Tchobanoglous, G. and Schroeder, E.D. (1987) *Water Quality: Characteristics, Modelling and Modification.* Addison-Wesley, Reading, 768pp.

Thackston, E.L. and Schnelle, K.B. (1970) Predicting effects of dead zones on stream mixing. *J. Sanit. Engin. Div. Am. Soc. Civ. Engin.* **96**, 319–331.

Valentine, E.M. and Wood, I.R. (1979a) Experiments in longitudinal dispersion with dead zones. *J. Hydr. Div. Am. Soc. Civ. Engin.* **105**, 999–1016.

Valentine, E.M. and wood, I.R. (1979b) Dispersion in rough rectangular channels. *J. Hydr. Div. Am. Soc. Civ. Engin.* **105**, 1537–1553.

Wallis, S.G., Guymer, I. and Bilgi, A. (1989) A practical engineering approach to modelling longitudinal dispersion. *Proceedings of the International Conference on Hydraulic and Environmental Modelling of Coastal, Estuarine and River Waters* (Eds R.A. Falconer, P. Goodwin and R.G. Matthew), Gower, Aldershot, pp. 291–300.

Wallis, S.G., Crowther, J.M., Curran, J.C., Milne, D.P. and Findlay, J.S. (1989) Consideration of a one dimensional transport model of the upper Clyde estuary. In: *Advances in Water Modelling and Measurement* (Ed. M.H. Palmer). BHRA, Cranfield, 23–41.

Wallis, S.G., Young, P.C. and Beven, K.J. (1989) Experimental investigation of the aggregated dead zone model for longitudinal solute transport in stream channels. *Proc. Inst. Civ. Engin. Part 2* **87**, 1−22.

Young, P.C. and Wallis, S.G. (1985) Recursive estimation: a unified approach to identification, estimation and forecasting for hydrological systems. *J. Appl. Math. Comput.* **17**, 299−334.

Young, P.C. and Wallis, S.G. (1993) Solute transport and dispersion in stream channels. In: *Channel Network Hydrology* (Eds K.J. Beven and M.J. Kirkby). Wiley, Chichester, pp. 128−173.

Plate 1 Contrast-stretched image of Esthwaite Water (Daedalus bands 11, 3 and 8) showing the accumulation of buoyant algae in the western bay and in a series of windrows. In this image the coldest, clearest water appears blue and the warmest, most turbid water orange or yellow. The lines of algae in the open water are about 20 m apart

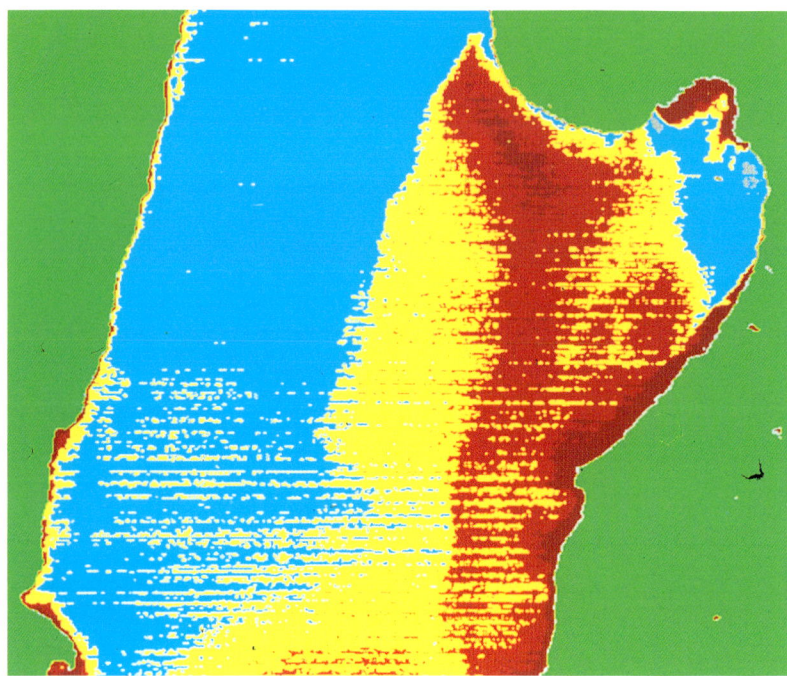

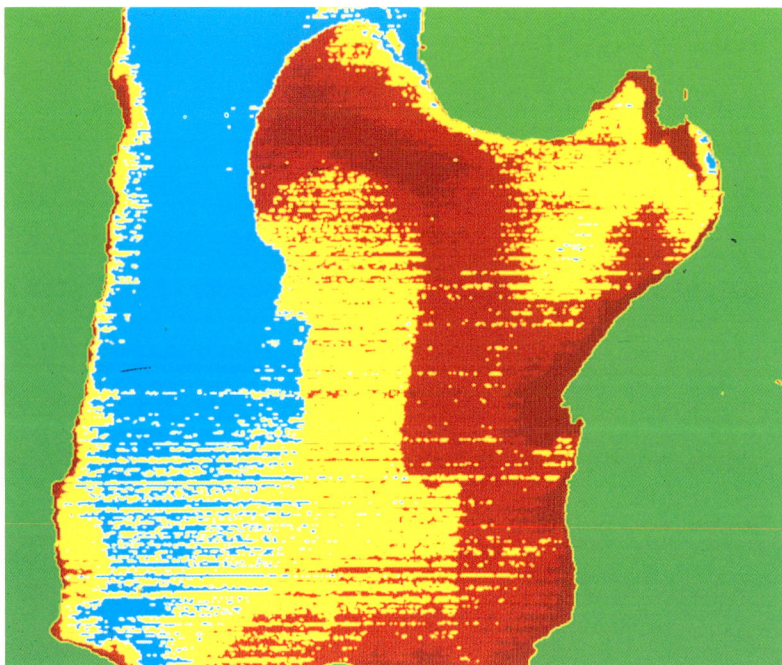

Plate 2 (a) Density-sliced thermal image of Esthwaite Water (Daedalus band 11) recorded at 13.30 hours GMT on 27 April 1987. The contours in the image represent 0.5 °C increments in the 'skin' temperature (the coldest water is blue and the warmest water is red). A thermal front has formed along the boundary of the western bay. (b) Density-sliced thermal image of the same area at 14.20 hours GMT. The front has now collapsed to form a counter-rotating eddy almost 200 m in diameter

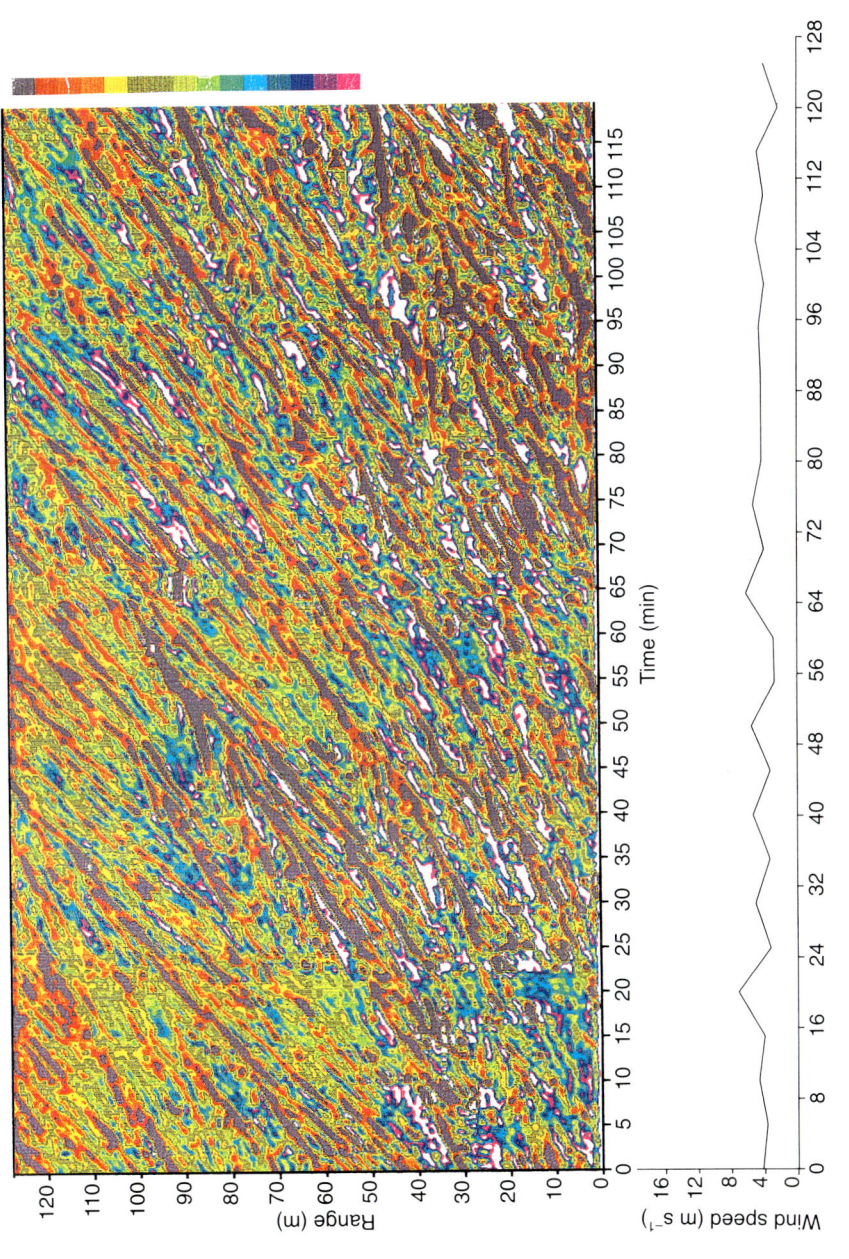

Plate 3 Sonograph collected on 23 February 1991. Purple regions represent low backscatter and brown regions high backscatter, where bubble clouds reflect sound. Intermediate target strengths are shown in the colour scale. The wind speed is shown beneath the sonograph. The wave period was 2.1 s and it was raining steadily. The pulse repetition frequency was 2 Hz. This period shows how Langmuir circulation is present even in very light winds. The sonar targets are the bubbles formed by the rain, not by the infrequent breaking waves

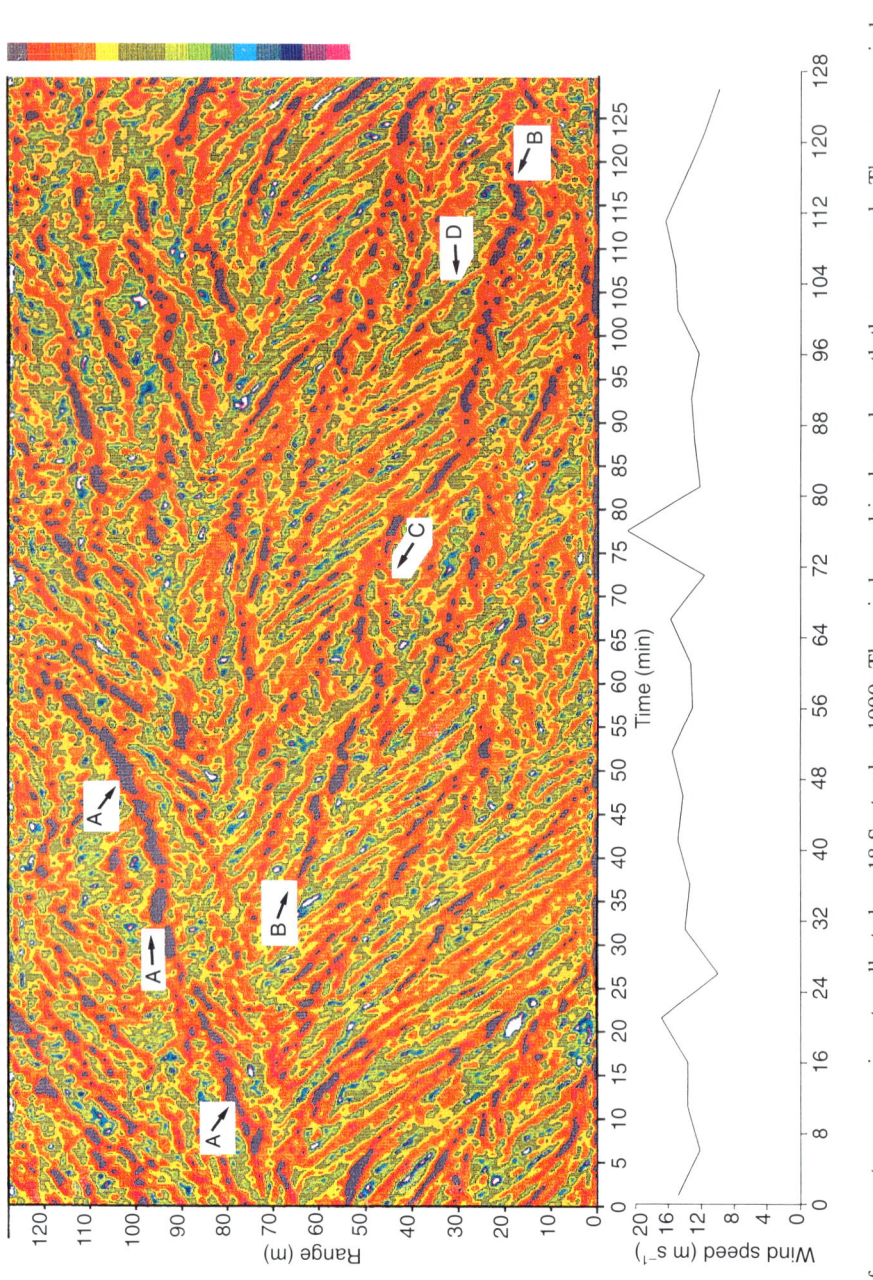

Plate 4 Sonograph from an autumn experiment, collected on 18 September 1990. The wind speed is shown beneath the sonograph. The wave period was 2.5 s and the pulse repetition frequency 1.5 Hz. By lowering a thermistor, the depth to the first stable density layer was found to be about 20 m and the main seasonal thermocline was at about 35 m. The features marked A, B, C and D are referred to in the text

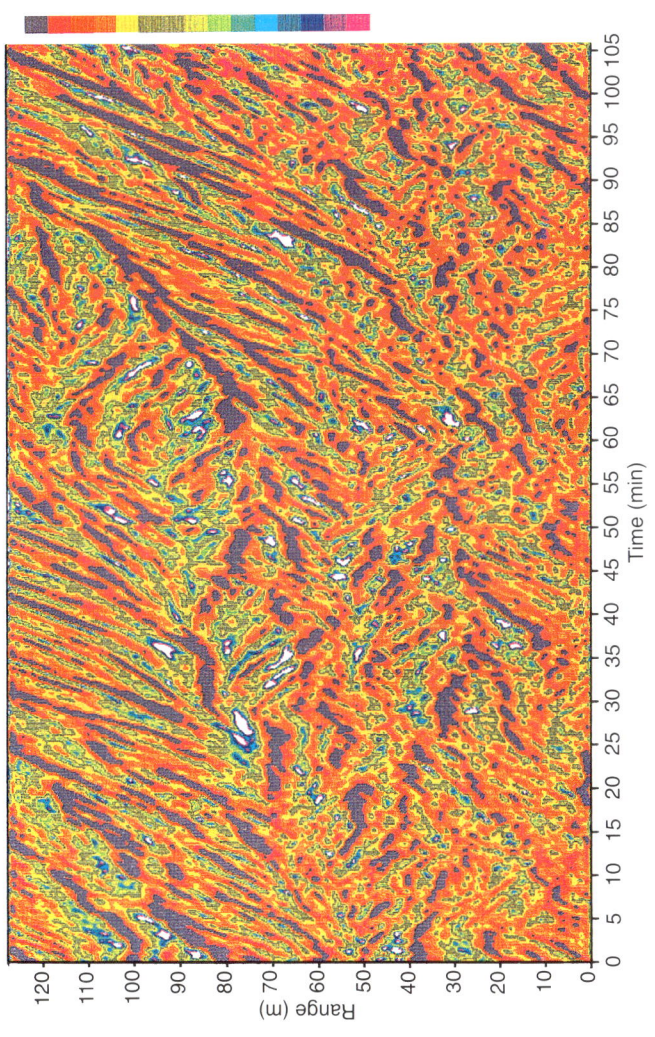

Plate 5 Sonograph collected on 14 March 1990 during a period of high winds on the loch. The Langmuir circulation convergences can be seen to move erratically. There is evidence of a larger scale of motion than in Plates 3 and 4. The wind speed over this period decreased from 20 to 16 m s^{-1}. The wave period decreased from 3.4 to 3.0 s and the pulse repetition frequency was 4 Hz

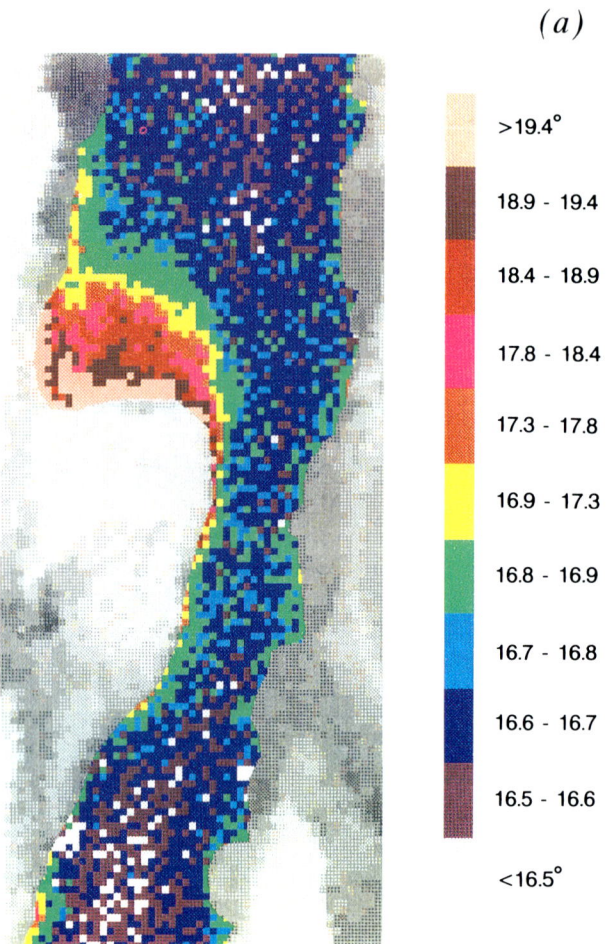

(a)

Temperature (°C)
>19.4°
18.9 - 19.4
18.4 - 18.9
17.8 - 18.4
17.3 - 17.8
16.9 - 17.3
16.8 - 16.9
16.7 - 16.8
16.6 - 16.7
16.5 - 16.6
<16.5°

Plate 6 (a) Thermal line scane of Leighton dead zone with selected temperature bands false colour coded. Pixel dimension ~ 1.25 m. (b) Simplified interpretation of selected isotherms (°C) sketched from plan (a). Calibrated thermal line scanner cross-channel temperature distributions are shown on the right with the synoptic ground-truth showed as a broken line. Note small cold water plume from stream on left, warm water in dead zone re-entrant (section B−B′) and along banks and cool waters in the pools (sections A−A′ and C−C′)

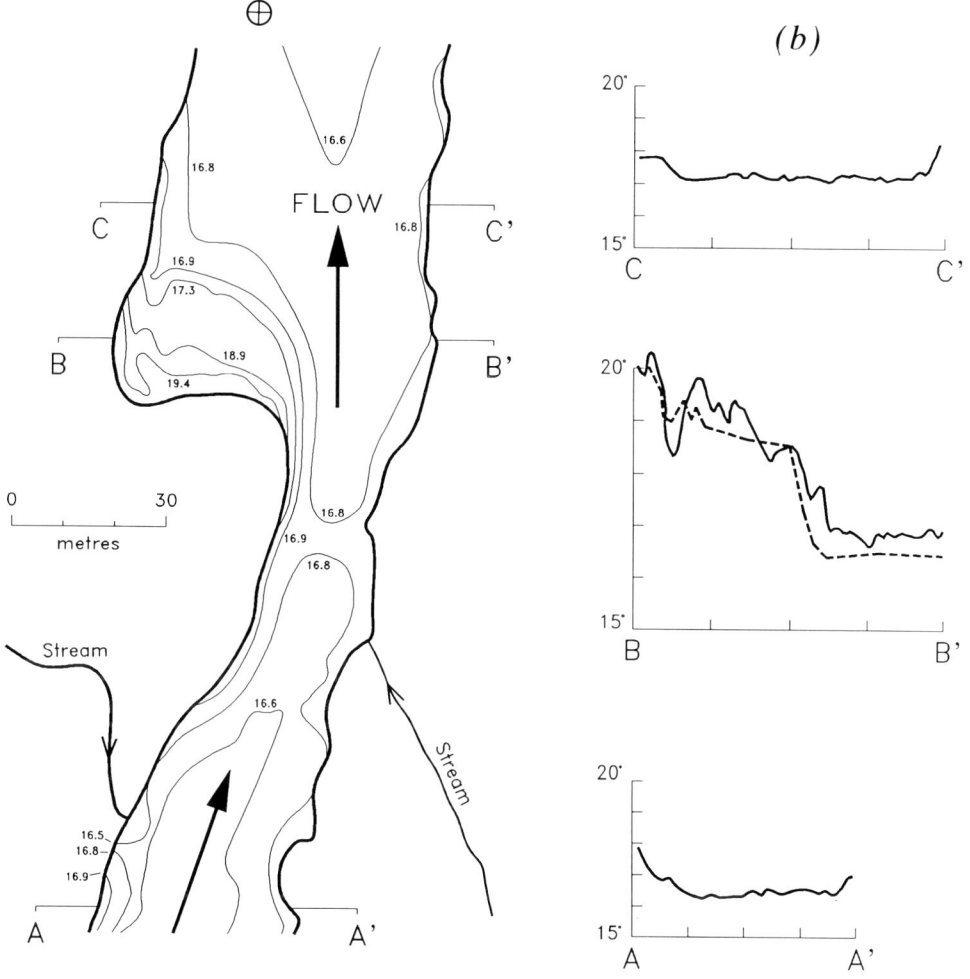

(b)

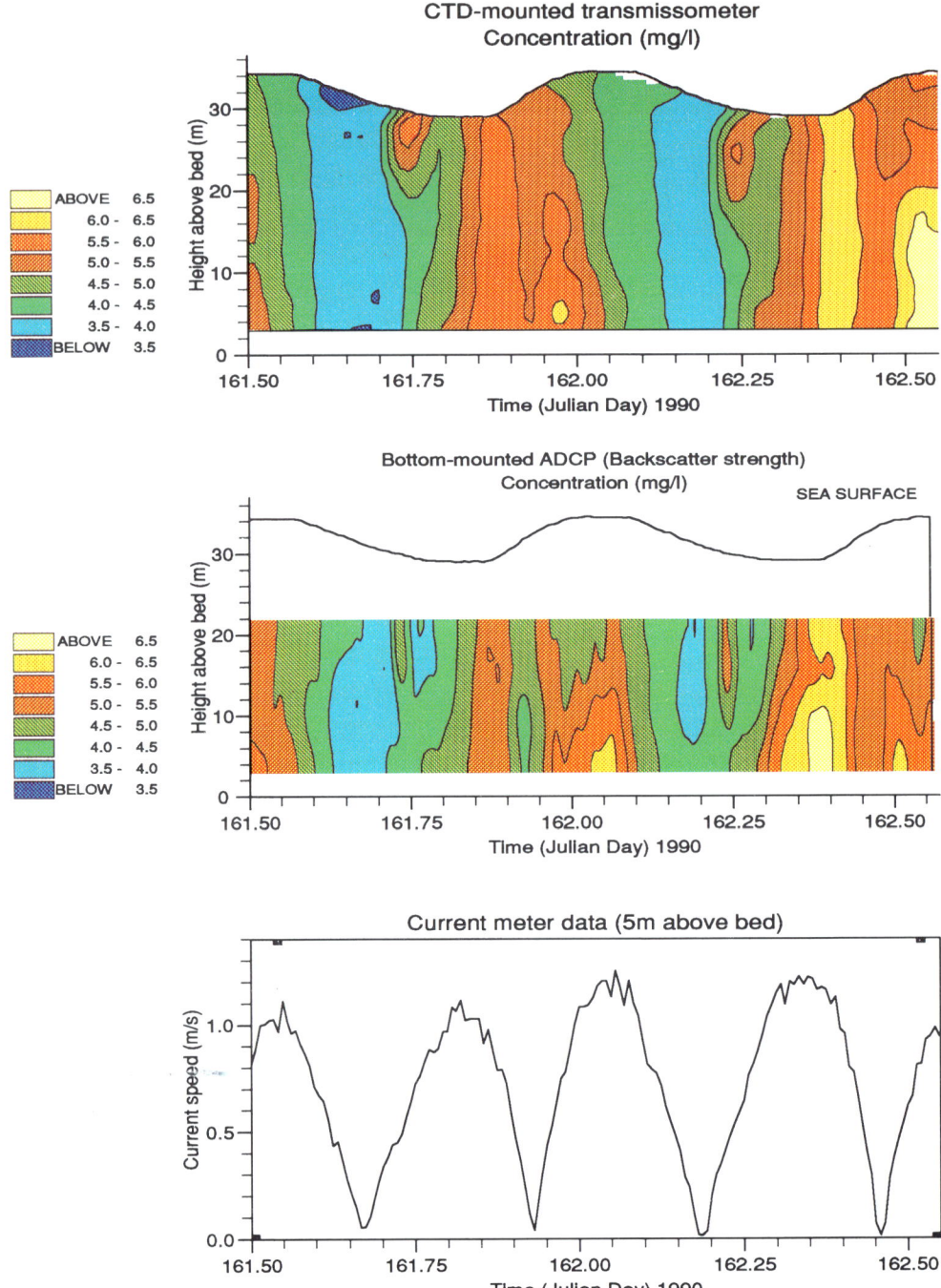

Plate 7 Contoured SPM concentration profiles from CTD-mounted transmissometer (half-hourly observations, ~0.1 m vertical resolution) and ADCP (10 minute observations, 1.5−4.5 m vertical resolution). Sea surface level taken from RCM pressure sensor

6 Pollution Incident Prediction with Uncertainty

H.M. GREEN, K.J. BEVEN, K. BUCKLEY and P.C. YOUNG
Centre for Research on Environmental Systems and Statistics, Institute of Environmental and Biological Sciences, Lancaster University, UK

6.1 INTRODUCTION

The problem of describing the dispersion of solutes in rivers has been the subject of continuous work since the mid-1920s. The study of the transport processes and their attempted description by mathematical equations helps both in understanding the relative importance of the dispersing mechanisms and in suggesting new ones; it also has practical importance in providing tools to help in predicting the effects of pollution incidents. The model described in this chapter has a completely different mathematical formulation from most other models of dispersion. It has proved successful in describing the shapes of observed concentration curves (Wallis, Young and Beven, 1989) and its structure makes it a potentially valuable practical tool for prediction purposes.

Longitudinal dispersion can be defined as the net longitudinal spreading of solutes resulting from differential advective velocities and transverse mixing. Taylor (1954), in his work on flow in uniform pipes, was the first to apply a one-dimensional Fickian diffusion equation to describe the random diffusion of particles once enough time had elapsed to allow full cross-sectional mixing. This has become embodied in the generally termed advection dispersion equation (ADE). In its most easily applied one-dimensional form the equation describes cross-sectionally averaged concentrations and predicts that the downstream concentration profiles will eventually become Gaussian in shape. This theory worked well for the pipe flow situation and for laboratory experiments in open channels. However, it has consistently been shown to be inadequate in reproducing concentration profiles in natural channels, which are characteristically long-tailed, even after correction for the fact that they are temporal rather than spatial profiles (Elder, 1959; Fischer, 1967; Nordin and Sabol, 1974; Day, 1975). This skewness is considered to be a result of the fact that the initial period of mixing has not been exceeded (Fischer, 1966), the length of which for natural channels was much investigated by Fischer (1967; 1973). Attempts to represent the profiles within the initial period have been made by considering the transverse variations in concentration using two- or three-dimensional ADE models (Valentine and Wood, 1979a and b). These have a disadvantage in that they need significantly more hydraulic data as inputs. Another explanation for the skewness was the effect of storage zones which trap solutes, only slowly releasing them back to the mainstream flow causing elevated tail portions of the curves. This is generally considered as the mechanism which causes persistence in the skewness with no ultimate conformity to Gaussian curves, as observed by Nordin and Troutman (1980). Comprehensive reviews of dispersion modelling can be found in Chatwin and Allen (1985) and Fischer *et al.* (1979).

Mixing and Transport in the Environment. Edited by K.J. Beven, P.C. Chatwin and J.H. Millbank
©1994 John Wiley & Sons Ltd

Although many of the modifications and enhancements of the ADE have proved reasonably successful in describing concentration curves, little published work exists on practical applications of the equations in a predictive rather than data-fitting role. There is a practical need for models to make predictions of times of travel, persistence and peak concentrations of pollutants to protect water quality. For instance, during pollution incidents, water intakes should be shut down to prevent contamination and here models could be used to predict the duration of closure. The main problems with ADE approaches is that the disperison coefficients used vary widely with discharge (Fischer, 1973; Whitehead *et al.*, 1986) making it inherently difficult to predict accurately values over a range of flow conditions. Relating the dispersion coefficients to the physical characteristics of depth and shear velocity has yielded differing results (Elder, 1959; Fischer, 1968; McQuivey and Keefer, 1974; Liu, 1977). The work in this paper represents first attempts to predict downstream concentration curves using the aggregated dead zone (ADZ) model, which lends itself well to the prediction of its parameters over a range of discharges.

6.1.2 Background to the ADZ model

The ADZ model was developed in the mid-1980s (Beer and Young, 1983) as an alternative to the ADE to describe dispersion in natural channels, with continued work reported in Wallis, Young and Beven (1989), Beven, Buckley and Young (1991) and Beven and Carling (1992). It took a completely new approach to the problem both in the processes modelled and in its calibration approach. The underlying concept in its development was that dispersion could be modelled by considering the storage or 'dead zones' as dominating the turbulent shear flow dispersion of the ADE to such an extent that the ADZ processes alone can be used to describe the dispersion. The model is a lumped parameter ordinary differential equation which is inherently stable and solved in discrete time. This has the advantage over the ADE, which has a distributed parameter partial differential form, in that it has a greater ease of computation. The lumped parameter form of the model at the reach scale means that the calibration does not require any *a priori* definition of where or what form the individual dead zones take and as such considers a broad spectrum of these regions as important in the process. Most commonly, dead zones are considered around the periphery of the channel associated with irregularities in the bed and bank. However, they can also be thought of in a wider sense as present within the bulk of the flow associated with turbulent eddies, wakes around roughness elements and possible reverse flow regions associated with pools and bends. Hence some dead zones will be non-stationary in time and space. Adaptations of the ADE to incorporate dead zone behaviour have usually only concentrated on peripheral dead zones (Valentine and Wood, 1977; 1979a; 1979b; Sabol and Nordin, 1978; Denton, 1990; Reichert and Wanner, 1991), dividing the flow accordingly into two regions. One is the main flow with rapid transport and the other is associated with the bank and/or bed dead zones which exchange with the main flow according to some exchange coefficient related to local flow velocity and concentration within the flow.

A number of assumptions are made in the formulation of the ADZ model, the most fundamental being that the same dynamic structure used to describe a single dead zone can be used to describe the whole reach.

Considering a single dead zone of volume V_1 in a flow field of discharge Q, the dynamic mass balance is represented by an ordinary differential equation

$$\frac{d}{dt} V_1(t)x(t) = Q(t)u(t) - Q(t)x(t) - \kappa[V_1(t)x(t)] \tag{6.1}$$

Rate of change of mass	Mass in per unit time	Mass out per unit time	Mass lost per unit time

With the assumption of complete and instantaneous mixing throughout the dead zone, $x(t)$ is the concentration at time t both within the dead zone and in the output, and $u(t)$ is the concentration of solute entering the dead zone. Losses are assumed to be proportional to the mass present following a characteristic decay rate with parameter κ. This can be simplified to Equation 6.2 by assuming that the flow is steady and the volume constant

$$\frac{d}{dt}x(t) = -\left(\kappa + \frac{Q}{V_1}\right)x(t) + \frac{Q}{V_1}u(t) \tag{6.2}$$

This equation represents a model for the dispersion in a single dead zone. Within any given reach there will be a number of such dead zones of differing volumes, all contributing to the overall dispersion between any two points. By making the critical assumption that the total effect of all these dead zones can be described in the same way but replacing the individual dead zone volume V_1 by an effective ADZ volume V_e, representing the total volume within the reach associated with dead zone dispersion processes, an equation can be written of similar form to Equation 6.2. In moving to the reach level, account must also be taken of the effect of advection. This is simply represented by the introduction of a pure transport time delay, τ time units, defined as the time taken for the leading edge of the tracer cloud to be advected through the reach. The model now becomes

$$\frac{d}{dt}x(t) = -\epsilon x(t) + \lambda u(t - \tau) \tag{6.3}$$

For a conservative solute $\epsilon = \lambda = 1/T$, where $T = V_e/Q$ is the time constant associated with the passage of the solute through the ADZ element, termed the ADZ residence time.

In the study of any reach it will not be known at the outset whether the reach length and individual characteristics are of a form capable of representation by a single ADZ element of this kind, or whether more elements connected in some manner are needed. Such connections of any number of elements may be serial, where the subreaches could be of a similar nature with approximatley equal ADZ elements, or differing in nature with changing dead zone volumes and associated residence times. Alternatively, the connection might be parallel, where within one reach there are effectively two zones with different associated residence times. Indeed, if the modified forms of the ADE mentioned earier are a realistic representation of a channel with dead zones, it might be expected that parallel models would be commonly identified in the ADZ approach. However, this is rarely the case.

6.1.2.1 Model identification procedures

The technique used in the model calibration is of a flexible nature in that it does not require any *a priori* determination of the model structure, i.e. the number of ADZ elements needed to describe the reach. Instead, it is the observations that are used to suggest the best structure in a data based approach to modelling. The model identification is achieved using statistical calibration procedures based on recursive time series analysis techniques as developed by Young *et al.* (1980; 1984; 1985; 1986; 1989).

As the calibration procedure at present is computationally easier in discrete time terms and the data obtained from the field experiments come in sampled form, the general model is best considered as a numerical finite difference approximation of Equation 6.3. This can be written for a sampling interval of Δt time units as

$$x_k = -ax_{k-1} + b_0 u_{k-\delta} + b_1 u_{k-\delta-1} \qquad (6.4)$$

where the subscript k denotes the value of the variable at the kth sampling instant and δ, the advective time delay, is the nearest lower integral value of $\tau/\Delta t$. For $\tau/\Delta t$ close or equal to an integral value, the equation can be simplified by setting $b_1 = 0$. For this case the relationship between the continuous and discrete time model parameters, assuming the input is constant over the sampling interval, is

$$a = -\exp(\epsilon \Delta t); \quad b_0 = \lambda(1 + a) \qquad (6.5)$$

However, these relationships will be inaccurate due to discretization errors and so will only give an approximate solution to Equation 6.3.

A better way of representing the model to see how the identification procedures operate is to use the transfer function (TF) form. To derive this, use is made of the general backward shift operator z^{-p} where

$$z^{-p} x_k = x_{k-p}$$

Rearranging Equation 6.4 then gives

$$x_k = \left(\frac{b_0 + b_1 z^{-1}}{1 + a z^{-1}} \right) u_{k-\delta} = \left[\frac{(b_0 + b_1 z^{-1}) z^{-\delta}}{1 + a z^{-1}} \right] u_k \qquad (6.6)$$

In terms of a general model structure for any number of ADZ elements in series or parallel this becomes

$$x_k = \left(\frac{B(z^{-1})}{A(z^{-1})} \right) u_{k-\delta} = \left(\frac{B(z^{-1}) z^{-\delta}}{A(z^{-1})} \right) u_k \qquad (6.7)$$

where $B(z^{-1})$ and $A(z^{-1})$ are polynomials of the form

$$B(z^{-1}) = b_0 + b_1 z^{-1} + \ldots + b_m z^{-m}$$
$$A(z^{-1}) = 1 + a_1 z^{-1} + \ldots + a_n z^{-n}$$

The process of model order identification selects the integer values of n and m which will depend on the number and connection of the ADZ elements needed to describe the reach. This is achieved by a statistical time series analysis technique using a recursive–iterative refined instrumental variable method with a stochastic representation of the transfer function. For a full discussion of this, see the references by Young et al. (1980; 1984; 1989). The procedure involves recursively stepping through the data, updating the parameter estimates at each step for a variety of model structures, that is n and m values. Optimization is based on two criteria, one a goodness of fit term R_T^2, which approaches unity for good model fits, and another related to both fit and error on the parameter estimates, termed the Young information criterion (YIC) (Young 1989; Young and Lees, 1993). This latter term is very sensitive to over-paramaterization and its value will reflect when the identified model structure has more parameters than are really required to explain the observed input–output behaviour. The statistic is defined as

$$YIC = \ln\left(\frac{\sigma^2}{\sigma_x^2}\right) + \ln\text{(NEVN)} \qquad (6.8)$$

where σ^2 is the sample variance of the model residuals, σ_x^2 is the sample variance of the measured system output x_k about its mean value and NEVN is the normalized error variance norm (Young, Jakeman and McMurtie, 1980) defined as

$$\text{NEVN} = \left(\frac{1}{n+m+1}\right) \sum_{i=1}^{i=n+m+1} \left(\frac{\hat{\sigma}^2 p_{ii}}{\hat{a}_i^2}\right) \qquad (6.9)$$

Here $\hat{a}_i^2$ is the estimate of the ith parameter in the parameter vector, p_{ii} is the ith diagonal element of the matrix P_N, where P_N is related to the covariance matrix of the estimated parameter vector P_N^* by $P_N^* = \sigma^2 P_N$ (Young et al., 1980; 1984; 1989; Young and Lees, 1993), and N is the sample size (so that $\hat{\sigma}^2 p_{ii}$ is an estimate of the error variance associated with the ith parameter estimate after N samples).

The first term of Equation 6.8 provides a normalized measure of how well the model explains the data, becoming more negative as the variance of the model residuals becomes smaller in relation to the variance of the measured output. The second term gives a normalized measure of how well the parameter estimates are defined for the $(n + m + 1)$th order model, with those having smaller relative error variance being the statistically better defined estimates. As the model order increases, the first term tends to decrease whereas the second term may decrease at first, but will then increase markedly as the model becomes over-parametrized and the standard errors on the parameter estiamtes become large in relation to their estimated values. As the YIC is calculated as a logarithmic function, small increases in its value represent much larger actual changes in the variance of the parameters estimates. The best model structure able to describe the data is decided by selecting a model which has both a good fit to the data and low variance on the parameter estimates reflected in a negative YIC. The ADZ-Analysis package for the IBM PC is used for this time series analysis based on these procedures (Beven, Buckley and Young, 1991; Buckley et al, 1991).

6.2 DATA

6.2.1 Data description

The data analysed in this paper come from dye tracing experiments on two different rivers in England. The bulk of the data comes from extensive fieldwork carried out in 1987 and 1988 on two reaches of the River Severn in Shropshire, at Montford and Leighton (Beven and Carling, 1992). The Montford reach, west of Shrewsbury, is a relatively straight reach of 1 km length with a well defined pool−riffle sequence. The Leighton reach, downstream of Shrewsbury, is of 1.13 km length within a lowland meander sequence also containing pools and riffles. The other data set comes from the River Tonge near Bolton in north-west England. This reach is a straight section of 330 m length, constrained along its banks by concrete walls. Two weirs lie within the study section, one just below the injection site and the other about half-way along the reach, with the possibility of some abstraction just above the second weir. Two small industrial discharges are also present along the length and the bed material is of medium cobbles.

6.2.2 Data collection

The analysis requires the collection of at least two concentration versus time curves which contain the necessary information on dispersion within the reach to build the model. The dye tracing method used to collect the data involves gulp injection of a near-conservative tracer, rhodamine WT, at some upstream point, with the resulting dye cloud being measured at two sites downstream. The first site must be of sufficient distance from the injection point to give time for the dye to become fully mixed across the river. The dye fluorescence is measured by a Turner Designs fluorometer linked to an Epson-HX20 microcomputer, which logs the data and calculates the dye concentration. The system is set up to monitor the background and starts logging automatically on arrival of the tracer controlling the range changes on the fluorometer as it does so (Wallis, Blakeley and Young, 1987; Wallis, Young and Beven, 1989). Over the range of concentrations used the fluorometer can be assumed to be linear in its response, therefore requiring only an end-point field calibration. This method results in two concentration versus time curves at fixed positions in space which are analysed to build the ADZ model for that particular reach.

The data sets analysed in the following are particularly useful in that they contain a number of tracer tests across a range of discharges, allowing a predictive model, applicable over a wide range of conditions, to be constructed.

6.2.3 Data analysis

6.2.3.1 Initial modelling

The concentration curves collected are modelled using the ADZ-Analysis program, which calculates the ADZ model structure as described earlier (Beven, Buckley and Young, 1991; Buckley *et al.*, 1991). There is flexibility in this approach in that the structure is not fixed *a priori*, but is suggested by the data. Modelling in this way results in the identification of a best model structure and associated parameters for each tracer test, based on the statistical calibration procedures described briefly in Section 6.1 (see references by Young for full discussion). The criteria used in identifying the best model, the R_T^2 and the YIC are calculated for each model structure to enable the best one to be selected.

The results of the modelling, with R_T^2, YIC and parameter values for each data set are given in Table 6.1. In all instances the data could be adequately described with a better than 90% fit to the data by a first-order model, with only one or two tests yielding a second-order model as the best fit. In some instances these second-order models were also the best in YIC terms (see values marked * in Table 6.1). The data sets for which second-order models were identified as an equally good fit in terms of YIC or R_T^2 also have these results listed in the table. Figure 6.1 shows how well the ADZ model can reproduce the observed concentration curves.

6.2.3.2 Hydrological statistics and results

Analysis of the tracer data yields other important information necessary for defining the varying reach characteristics with discharge, including the advective time delay (τ), the mean travel time ($\bar{t}$) and the ADZ mean residence time (T). The advective time delay is the difference between the times of first arrival at the two sites, the mean travel time is the difference in

Table 6.1 Best estimated model structures and parameters

Date and file pair	Model structure	$Rt2$	YIC	a	b
River Severn					
2 June 1987	1,1,0	0.98290	− 12.5581	− 0.91425	0.08323
M1−M2	1,2,0	0.99253	− 10.9320	− 0.93463	0.25357
					− 0.18782
3 June 1987	1,1,0	0.97777	− 12.3181	− 0.92602	0.07952
M5−M6	1,2,2	0.98745	− 11.3553	− 0.94788	0.37711
					− 0.31817
13 August 1987	1,1,0	0.97304	− 12.5674	− 0.96073	0.02443
M9−M10					
4 February 1988	1,1,0	0.99543	− 14.9284	− 0.90017	0.08894
M13−M14	1,2,0	0.00868	− 13.5293	− 0.92053	0.22735
					− 0.15502
	2,2,0	0.99870	− 11.9758	− 1.40789	0.15088
				0.44717	− 0.12509
26 April 1988	1,1,0	0.99346	− 15.1722	− 0.94942	0.03745
M15−M16	1,2,1	0.99781	− 13.9219	− 0.95644	0.13746
					− 0.10430
	2,2,0	0.99803	− 8.78174	− 1.44150	0.06956
				0.46406	− 0.05243
27 April 1988	1,1,0	0.99273	− 15.0184	− 0.94872	0.04047
M17−M18	2,2,0*	0.99800	− 17.1105	− 1.87911	0.04843
				0.88182	− 0.04618
26 August 1988	1,1,0	0.99000	− 14.9938	− 0.97348	0.03046
M19−M20	2,2,2	0.99230	− 8.67775	− 1.83635	0.04343
				0.83969	− 0.03956
4 June 1987	1,1,0	0.99329	− 14.0559	− 0.85024	0.13459
L1−L2					
12 August 1987	1,1,0	0.95723	− 11.4748	− 0.95321	0.03225
L3−L4	2,2,0*	0.93618	− 13.5272	− 1.93031	0.03651
				0.93138	− 0.03576
27 October 1987	1,1,0	0.98659	− 13.5192	− 0.92314	0.07586
L7−L8	1,2,0	0.98854	− 8.21642	− 0.91823	0.04269
					0.03654
3 February 1988	1,1,1	0.98328	− 12.5798	− 0.88870	0.09447
L9−L10					
28 April 1988	1,1,1	0.99545	− 16.2581	− 0.95606	0.03891
L11−L12		0.99766	− 12.7027	− 1.74552	0.02605
				0.75357	− 0.01713
27 July 1988	2,1,0*	0.98601	− 12.0830	− 1.86997	0.00405
L13−L14				0.87314	
River Tonge					
28 April 1988	1,1,3	0.96991	− 10.4106	− 0.96725	0.03143
T1−T2	2,2,0	0.99798	− 6.89811	− 1.83015	0.00705
				0.83568	− 0.00196
20 May 1987	1,1,2	0.99234	− 10.0425	− 0.89883	0.12058
T7−T8					
24 June 1987	1,3,1	0.95957	− 4.97207	− 0.87846	0.04244
T15−T16					0.01927
					0.08990
16 March 1988	1,1,1	0.99780	− 17.0228	− 0.92160	0.07267
T17−T18					
6 September 1988	1,1,1	0.99331	− 14.9135	− 0.93773	0.07119
T21−T22					
6 October 1988	1,1,0	0.99783	− 17.2023	− 0.92643	0.09406
T23−T24					
6 October 1988	1,1,0	0.99736	− 16.9187	− 0.93504	0.07699
T25−T26					
6 October 1988	1,1,0	0.99762	− 17.1200	− 0.93792	0.06669
T27−T28					

* See text.

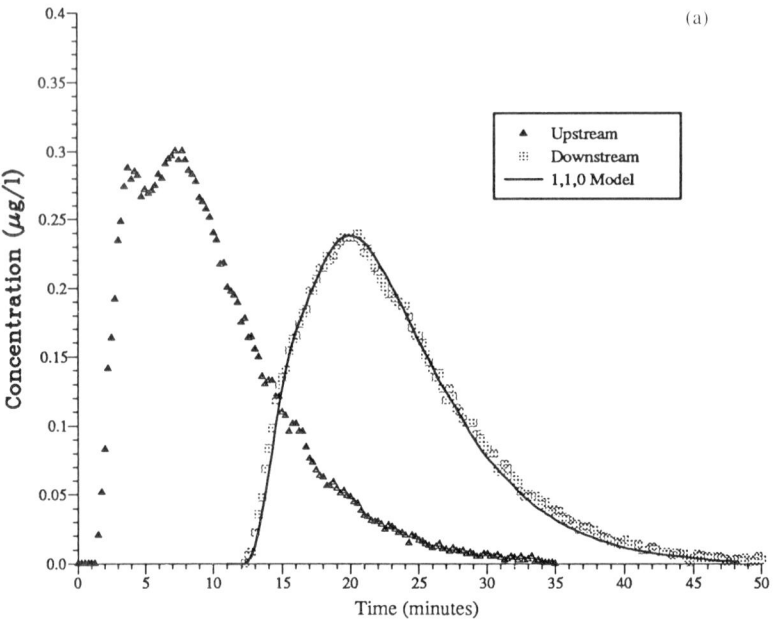

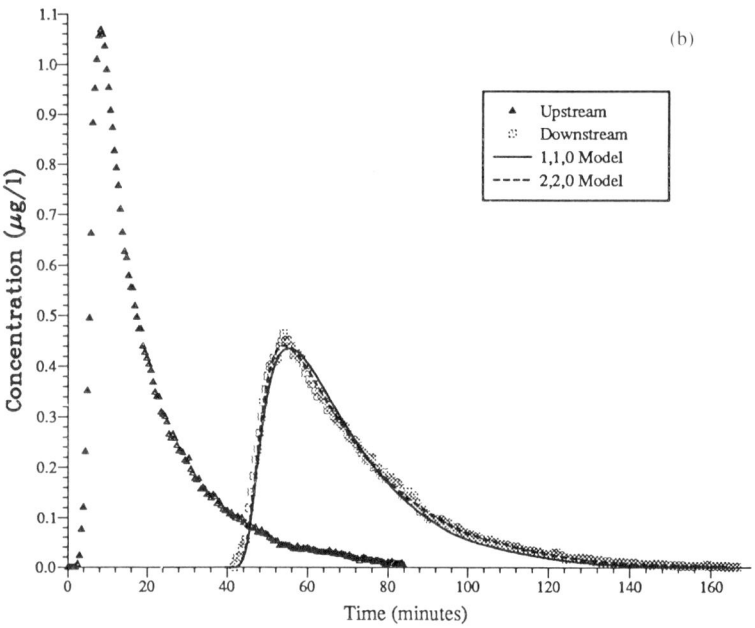

Figure 6.1 Examples of ADZ model fits during calibration. (a) Example of first-order model fit; (b) example of first- and second-order model fit. Data from the Montford reach of the River Severn

the centroid times of the two curves, and for a reach described by a first-order model and ADZ mean residence time is the difference between these two, i.e. $T = \bar{t} - \tau$. This is defined as the time the particle spends in the dead zone volume of the reach, which is that portion considered to be responsible for the dispersal effects. Dividing this T value by the mean travel time gives another important ADZ parameter, the dispersive fraction (DF). This can be viewed as the time a particle spends in the dead zone volume of the reach as a proportion of the total time it spends travelling through the reach or, alternatively, the fraction of the reach volume responsible for dead zone mixing processes.

A discharge value can be estimated, assuming mass conservation, by dividing the total mass of tracer injected by the area under either of the measured concentration curves. It is considered more reliable to use the area of the downstream curve as the tracer has had more time to become fully mixed across the flow at this site. However, over long reach sections the time taken to travel downstream may cause an overestimate of discharge due to the extra time available for adsorption or other losses. The estimates produced in this way have been shown to agree with gauging station estimates to within 10% (see Wallis, Young and Beven, 1989). An estimate of the loss between the upstream and downstream sites comes from

Table 6.2 Hydrological statistics calculated for all files

Files	Discharge Q (m³ s⁻¹)		Gain		τ (s)	$\bar{t}$ (s)	T (s)		DF
	Input	Output	A_0/A_i	TF			Data	Model	
M1−M2	5.6	5.6	1.008	0.971	4287	5602	1315	1339	0.235
M5−M6	7.4	6.4	1.159	1.075	3832	5738	1906	1561	0.332
M9−M10	8.0	12.7	0.631	0.622	4066	5338	1272	1498	0.238
M13−M14	107.1	117.5	0.912	0.891	665	808	143	143	0.177
M15−M16	19.5	25.3	0.771	0.740	2315	2948	633	578	0.215
M17−M18	20.4	24.7	0.825	0.789	2354	2996	642	570	0.214
M19−M20	22.0	19.8	1.115	1.149	2595	3550	955	1116	0.269
L1−L2	13.8	15.7	0.875	0.899	2472	2673	201	370	0.075
L3−L4	13.0	20.1	0.646	0.689	2868	3728	860	1252	0.231
L7−L8	105.0	101.9	1.031	0.987	1150	1590	440	375	0.277
L9−L10	223.8	258.4	0.866	0.849	852	1110	258	254	0.232
L11−L12	46.6	42.5	1.099	1.114	1811	2585	774	844	0.299
L13−L14	31.2	23.9	1.305	1.278	2076	3219	1143	1134	0.355
T1−T2	0.623	0.558	0.90	0.96	1048	1823	775	901	0.425
T7−T8	0.446	0.585	1.19	1.19	934	1827	893	703	0.490
T15−T16	0.585	0.780	1.33	1.25	655	1418	763	463	0.538
T17−T18	4.483	4.117	0.92	0.93	375	553	178	184	0.322
T21−T22	1.12	1.27	1.14	1.14	740	1191	451	467	0.379
T23−T24	3.069	3.462	1.13	1.14	454	620	166	196	0.268
T25−T26	2.646	3.093	1.17	1.19	490	684	194	223	0.284
T27−T28	2.474	2.665	1.08	1.07	494	727	233	234	0.320

Discharge. Calculated from the area under the curve divided by the total mass of dye injected. The input value is calculated using the upstream curve and the output value using the downstream curve.
Gain. A_0/A_i calculated from the ratio of the areas under the input and output concentration curves. TF steady-state gain comes from the sum of the b parameters over one plus the sum of the a parameters in the transfer function of the model.
T. Data calculated from $T = \bar{t} - \tau$; model calculated from $T = -\Delta t/\ln a_1$.
Dispersive fraction. Calculated by dividing the ADZ residence time T by the travel time $\bar{t}$.

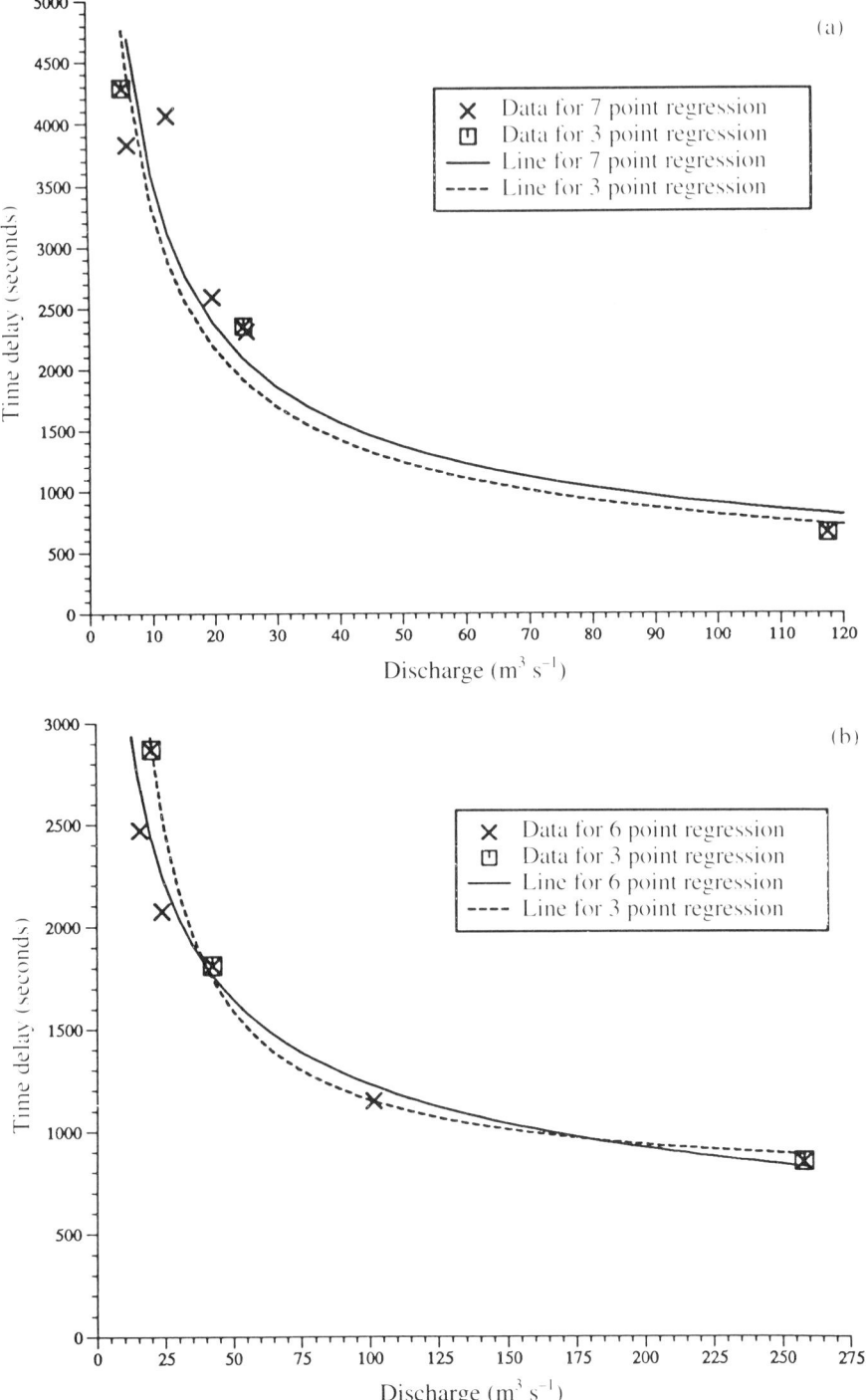

Figure 6.2 Regression lines of τ against discharge for (a) the River Severn at Montford and (b) the River Severn at Leighton

comparing the areas under the two concentration curves or from the steady-state gain of the model as calculated from the transfer function for a steady unit input. Table 6.2 details these values calculated for each of the tests. The two methods of calculation give good agreement, showing that the model identifies the gain in the data. In these experiments the gain values range from those near unity to 0.63 and 1.33 in the worst cases.

6.3 INCIDENT PREDICTION

6.3.1 Use of results for predictions

Having modelled the individual tracing tests for a reach, how is this information used in building the model for predictive purposes? The first stage is to define a general model structure for the particular reach, which is decided by the ADZ analysis of the data. In most reaches studied to date, where data from a number of tracing tests have been analysed, it has been found that a first-order model will often adequately define the reach, as is the case with the data sets for this study. Higher order models are only occasionally required (see * in Table 6.1), particularly if recession tails are high, and as yet no river has been studied which consistently requires higher than a first-order model throughout the measured discharge range.

The next stage is to estimate the model parameter values for a prediction. These are defined

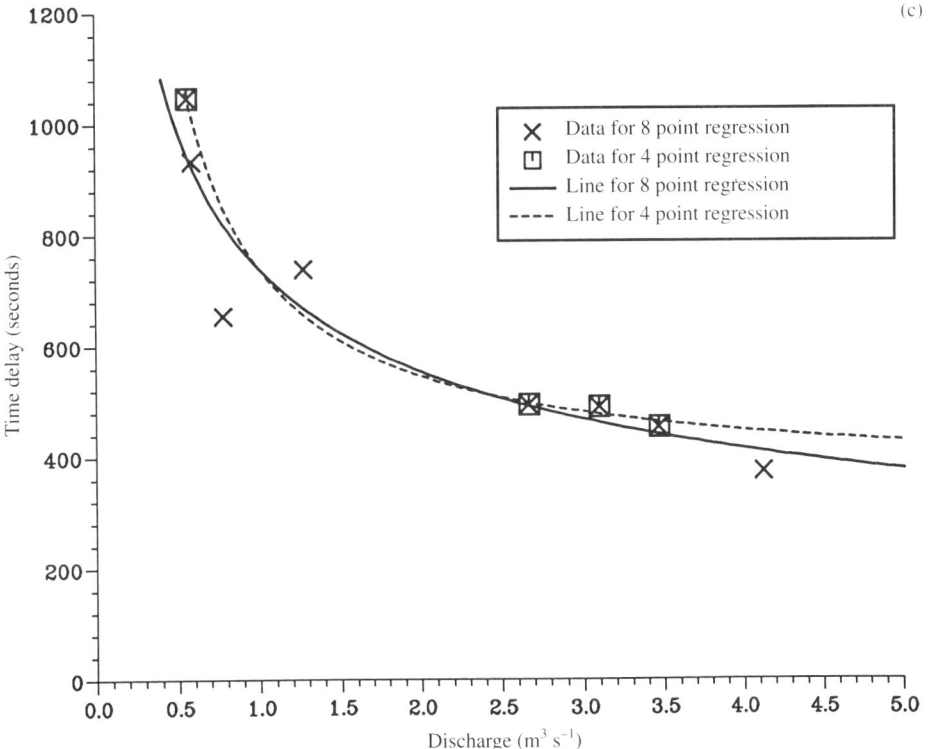

Figure 6.2 (*continued*) Regression lines of τ against discharge for (c) the River Tonge

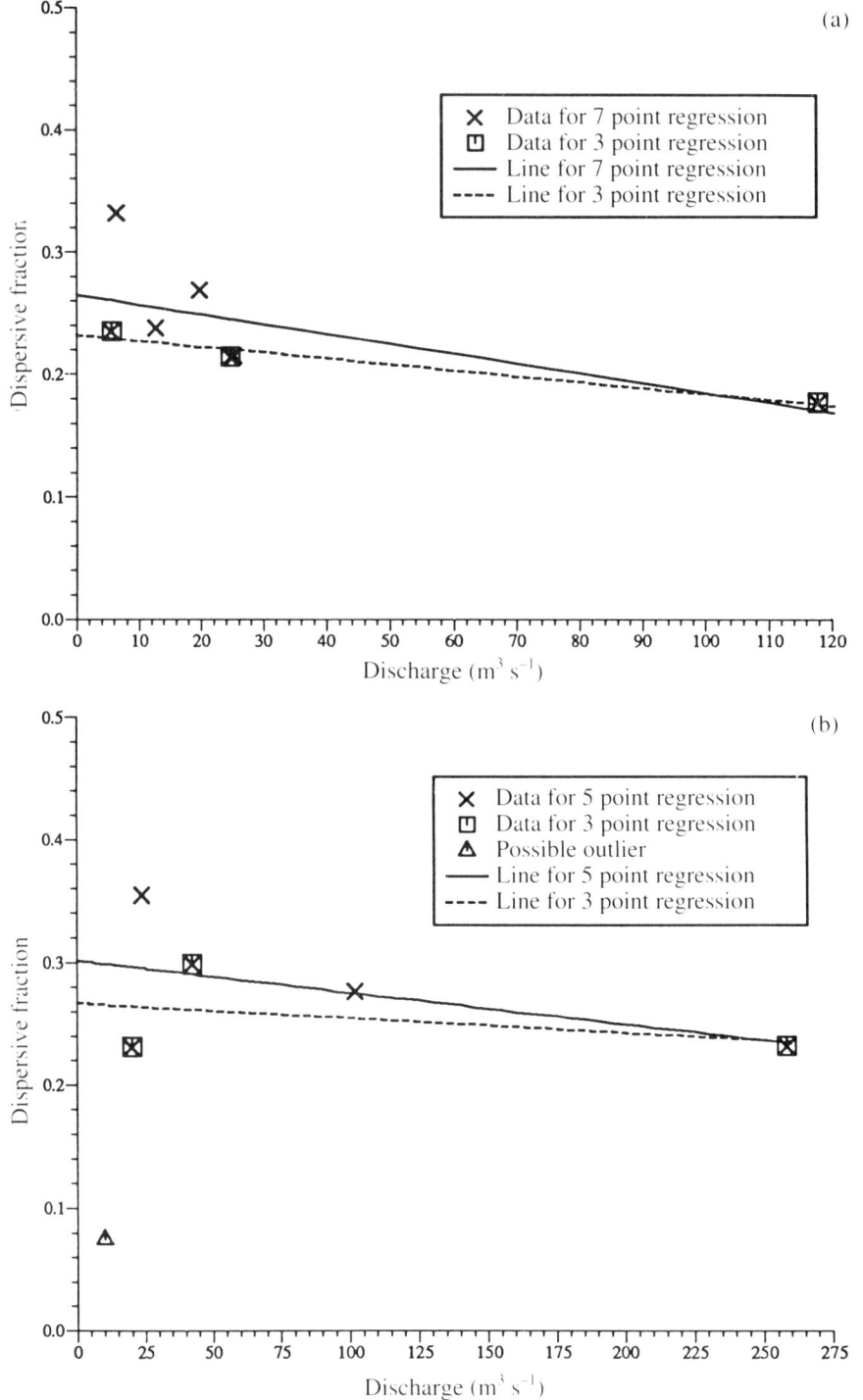

Figure 6.3 Regression lines of DF against discharge for (a) the River Severn at Montford and (b) the River Severn at Leighton

on a physical basis with reference to a residence time parameter, namely the ADZ mean residence time, T. The value for T is predicted from the dispersive fraction and the advective time delay, the two parameters used to define the changing reach characteristics with discharge. The mean travel time could be used instead of the advective time delay. Although both are liable to inaccuracies in measurement due to the effects of incomplete mixing, it should be easier to determine accurately the time delay, which only requires the definition of the first arrival, than the travel time, which requires good definition of the whole curve, particularly the tail section.

The nature of their changing relationship with discharge is estimated by linear regression analysis. Plotting the variables calculated against discharge shows that they all decrease with increasing discharge. Previous studies (Wallis, Young and Beven, 1989) have examined both power law and inverse linear relationships to describe changes in the parameter values with discharge. In this study the power law relationship

$$\tau = \alpha Q^\beta \qquad (6.10a)$$

or

$$\ln\tau = \ln\alpha + \beta\ln Q \qquad (6.10b)$$

was found to give the best fit to the data for all the study reaches (see Figures 6.2 and Table 6.3, which gives the values and standard errors of the regression parameters). Wallis, Young

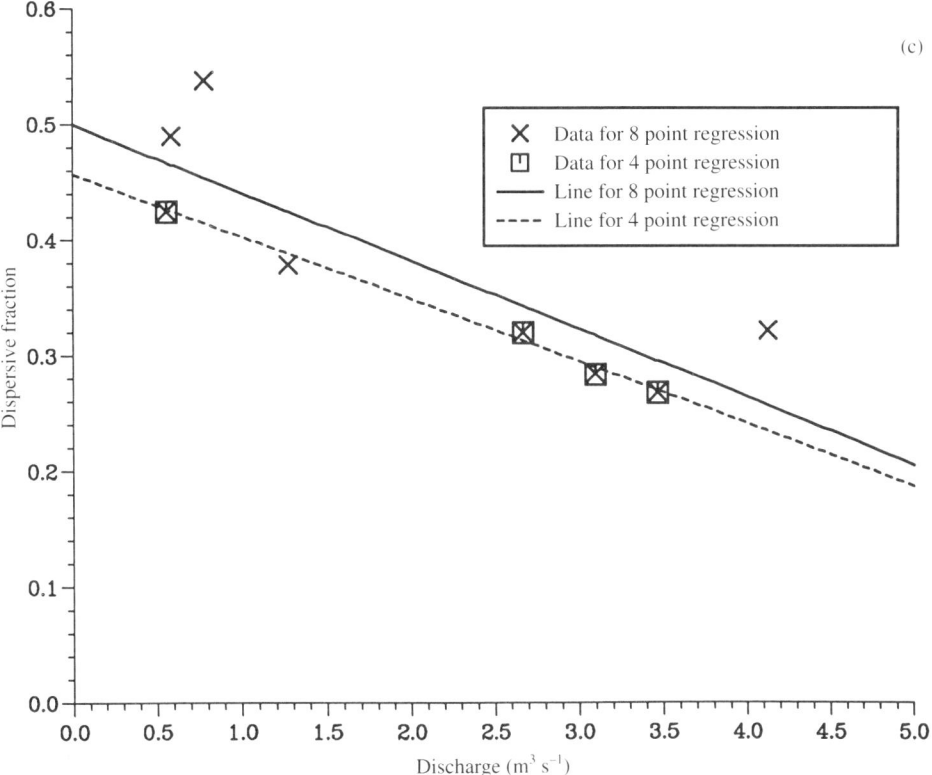

Figure 6.3 (*continued*) Regression lines of DF against discharge for (c) the River Tonge

Table 6.3 Regression results from full data sets

Site	$\ln\alpha$		β		$r^2_{\tau,Q}$		N_τ
	Value	SE	Value	SE	raw	adj	
Montford	9.564	0.210	−0.598	0.084	0.911	0.893	7
Leighton	9.052	0.109	−0.420	0.045	0.956	0.945	6
Tonge	6.606	0.118	−0.417	0.054	0.910	0.895	8

SE is the standard error; see Appendix for the formula used in calculation of the r^2 values.

and Beven (1989) have also suggested that the dispersive fraction may be constant with discharge. However, the data for these reaches indicates that the dispersive fraction may decrease with increasing discharge (see Beven and Carling, 1992), as shown in Figure 6.3. Such a result is not altogether surprising as it is reasonable to expect that the proportion of dead zones might increase at lower flows; however, it now means that a one-point estimate of its value for a reach is not necessarily valid and it is necessary to obtain an empirical expression for the variation with discharge.

As there are so few data points in Figure 6.3 it is difficult to quantify the nature of the variation in the dispersive fraction with discharge. A variety of regression equations were investigated, but the following simple linear regression, although not particularly good (see Table 6.3), provided the best compromise. Clearly, more data would be required to investigate the nature of the relationship in more detail.

$$DF = \gamma + \eta Q \tag{6.11}$$

Note that for the lowest discharge value in the Leighton data set there was an anomalously small dispersive fraction value. As other data sets held have not displayed such comparatively small values at their lowest discharges, this was treated as an outlying data point and omitted from the regression analysis. It would, however, be of interest to conduct repeated experiments at low flows to be more confident about the form of the relationship between the dispersive fraction and discharge in this region of the flow range.

Given these relationships between the dispersive fraction and discharge and time delay and discharge, values for τ, T, and $\bar{t}$ at any required discharge can be estimated and used to calculate the parameters of the model α_1 and b_0. However, all these estimates and the resulting parameter predictions will have uncertainty associated with them, which should be taken into account in any model predictions. An attempt can be made to quantify this uncertainty by considering the statistics used in the estimation procedure. The uncertainty arises from a number of factors: from the inherent variability associated with experimental observations in a natural system; from the residual errors associated with calibration of a particular model structure to each data set; and from the estimation error associated with the regression analysis of changes in the parameter values with discharge. In general, the residual errors associated with model calibration have been small, but the estimation error could be large as most of the data points lie at one end of the discharge range with only a few points at high flow. This means that the fit of the regression line is being strongly influenced by these one or two points. There is also a considerable degree of scatter in the data and a minimal number of data points. Thus in this instance the estimation error associated with the regression relationships dominates the source of error arising from fitting the model at each discharge.

γ		η		$r^2_{DF,Q}$		N_{DP}	$\rho_{df,\tau}$
Value	SE	Value	SE	raw	adj		
0.265	0.041	−0.0008	0.0004	0.423	0.308	7	0.650
0.302	0.051	−0.0003	0.0003	0.261	0.014	5	0.246
0.500	0.055	−0.0589	0.0145	0.732	0.687	8	0.705

6.3.2 Statistical uncertainty calculations

In attempting to quantify this uncertainty, from the prediction of one random variable from a linear regression given a value for the other random variable, the statistics of linear prediction and conditional probability are used. The equations and laws of this can be found in most statistics texts (for example, Benjamin and Cornell, 1970; Ang and Tang, 1975).

The prediction equations, including such statistical uncertainty calculations to derive the required values of time delay τ, dispersive fraction DF, and hence mean residence time T, and mean travel time $\bar{t}$, follow (see Appendix for notation). These equations are of a power law regression type, but equivalent equations can be derived for an inverse relationship.

From the regression lines, the conditional expected values for $\ln\tau$, given a value for $\ln Q$, and DF, given a value for Q, are given by Equations 6.12 and 6.13

$$E[\ln\tau|\ln Q] = \ln\alpha + \beta\ln Q \tag{6.12}$$

$$E[DF|Q] = \gamma + \eta Q \tag{6.13}$$

The variables $\ln\tau$ and DF are conditionally normally distributed about these mean values with the constant variance about the regression lines (Benjamin and Cornell, 1970) estimated from

$$\sigma^2_{\ln\tau|\ln Q} = \left(\frac{N_\tau}{N_\tau - 2}\right)(1 - r^2_{\tau,Q})\, s^2_{\ln\tau} \tag{6.14}$$

$$\sigma^2_{DF|Q} = \left(\frac{N_{DF}}{N_{DF} - 2}\right)(1 - r^2_{DF,Q})\, S^2_{DF} \tag{6.15}$$

The variance about an individual mean prediction of $\ln\tau$ given a value for $\ln Q = \ln Q_0$ is shown in Equation 6.16. This variance is dependent on the actual value of $\ln Q$ chosen for the prediction of $\ln\tau$, and will increase the further the desired value of $\ln Q$ is from the sample mean. That is, more confidence can be placed in values predicted near the mean of the data set than at the extremes.

$$\sigma^2_{\overline{\ln\tau}|\ln Q} = \frac{\sigma^2_{\ln\tau|\ln Q}}{N_\tau}\left(1 + \frac{\{\ln Q_0 - \overline{\ln Q}\}^2}{S^2_{\ln Q}}\right) \tag{6.16}$$

As the dispersive fraction is calculated from time variables T and $\bar{t}$, which are related to τ, its value will be conditional not only on the discharge, but also on the value of τ. For two correlated values, the observation of one should reduce the uncertainty in predicting the other, so use can be made of conditional probability equations in the prediction of DF. From the

Table 6.4 Regression results for reduced data sets

Site	$\ln\alpha$		β		$r^2_{\tau,Q}$		N_τ
	Value	SE	Value	SE	raw	adj	
Montford	9.527	0.252	−0.614	0.117	0.965	0.930	3
Leighton	9.309	0.082	−0.465	0.044	0.991	0.982	3
Tonge	6.684	0.030	−0.458	0.020	0.996	0.994	4

SE is the standard error; see Appendix for the formula used in calculation of the r^2 values.

predicted values of τ and the dispersive fraction, a value of the ADZ residence time is to be predicted for the model. This residence time T is directly related to the continuous time model parameter, $\epsilon = 1/T$ for a conservative tracer with $\kappa = 0$. However, solving the model in discretized form means that the approximation of the discrete time model parameter a_1 to the continuous time parameter [$a_1 = -\exp(-\epsilon\Delta t)$ where Δt is the sampling interval] will inherently include discretization errors.

To take account of this non-linear transform in predicting the parameters, a Monte Carlo technique is adopted to incorporate the stochastic error into the prediction process. The remaining equations including this and the conditional probability relations are given in the following (Benjamin and Cornell, 1970).

$$\ln\tau_i = E[\ln\tau|\ln Q] + R_{1,i} \times \sigma_{\ln\tau|\ln Q} \tag{6.17}$$

where $i = 1 \ldots N$ is the number of the Monte Carlo realization and R is a normally distributed random variate of zero mean and unit variance.

$$E[DF|\ln\tau] = E[DF|Q] = \rho_{DF,\tau}\left(\frac{\sigma_{DF|Q}}{\sigma_{\ln\tau|\ln Q}}\right)(\ln\tau_i - E[\ln\tau|\ln Q]) \tag{6.18}$$

$$\sigma_{DF|\ln\tau} = \sigma_{DF|Q} \times (1 - \rho^2_{DF,\tau})^{1/2} \tag{6.19}$$

$$DF_i = E[DF|\ln\tau] + R_{2,i} \times \sigma_{DF|\ln\tau} \tag{6.20}$$

$$\bar{t}_i = \frac{\tau_i}{1 - DF_i} \tag{6.21}$$

$$T_i = \bar{t}_i - \tau_i \tag{6.22}$$

From the calculation of time delay and ADZ mean residence time T, we can calculate the model parameters a_1 and b_0 at each Monte Carlo run as follows

$$a_1 = -\exp(-\Delta t/T_i) \tag{6.23}$$

$$b_0 = 1 + a_1 \tag{6.24}$$

where Δt is the time step at which the data are sampled.

The model outputs are then calculated for this realization from

$$x_k = \frac{b_0}{1 + a_1 z^{-1}} u_{k-\delta} \tag{6.25}$$

where $\delta = \tau_i/\Delta t$.

γ		η		$r^2_{DF,Q}$		N_{DP}	$\rho_{df,\tau}$
Value	SE	Value	SE	raw	adj		
0.232	0.008	−0.0005	0.0001	0.958	0.917	3	0.999
0.267	0.050	−0.0001	0.0003	0.170	−0.659	3	0.041
0.457	0.007	−0.0540	0.0030	0.994	0.991	4	0.971

Figure 6.4 shows an example set of Monte Carlo simulations calculated in this way.
As a better test of the predictive abilities of the model, a subset of the data was used in further regressions, allowing predictions to be made on files which had not been used in defining the starting values of the model. How many tracings are required to define adequately the relationships for a river is, as yet, unclear, which is another reason for testing the model on a subset to see how well it can still predict given only a very small number of values to define the regression relationships. The subset was taken as half the data set, with the points to be used randomly selected from groups chosen to span the discharge range. As very few tests were made at the higher discharges, the choice of points at this end was limited. The results of the regression analyses on the reduced data sets are given in Table 6.4.

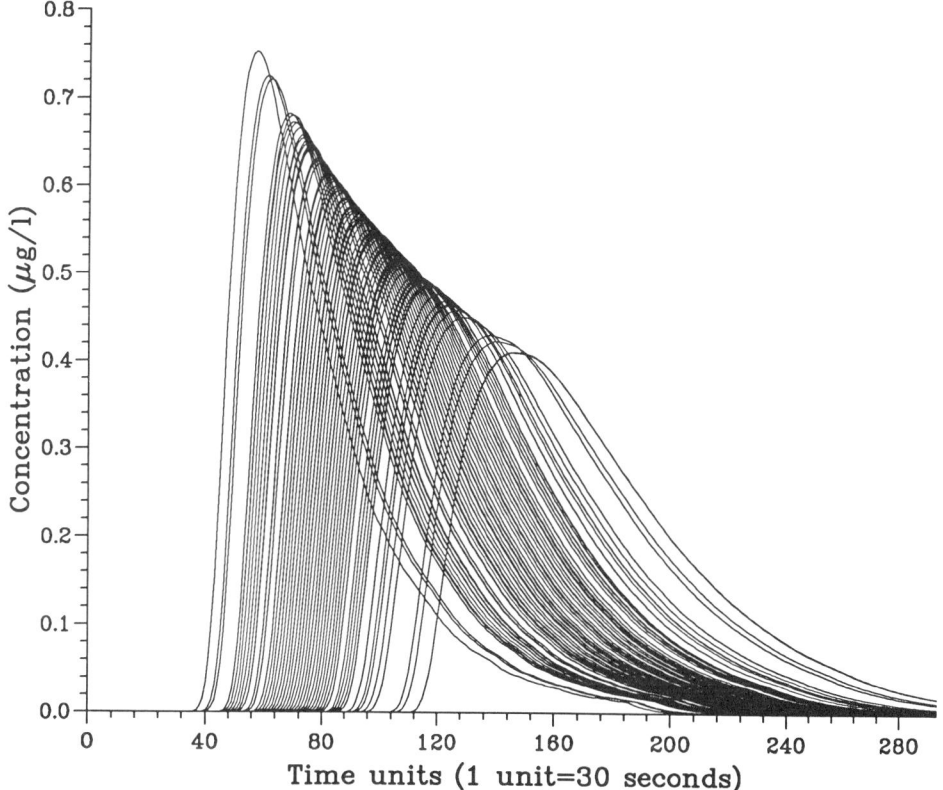

Figure 6.4 Example set of Monte Carlo simulations. Fifty runs (from 200) using the reduced data set regression results from files M15–M16

6.3.3 Representation of predictive uncertainty

The prediction process as described here yields an output concentration curve for each of the Monto Carlo simulations. In this instance 200 were sufficient to yield stable estimates of the confidence limits. From these some mean prediction must be derived, along with the associated uncertainty derived from the spread of these curves. This was initially achieved by taking the mean of the concentration values at each time step and representing the uncertainty by plotting two standard deviations to each side. This does not, however, give a good representation as it extends the output over all time steps including those where only a few runs gave values. A better method is to estimate a mean model from the average values of T and τ over the full number of Monte Carlo simulations, with the uncertainty plotted as the 5th and 95th percentiles of the ranked concentration values at each time step. Examples of the output for the simulations on each reach are discussed in the following section.

The prediction equations are based on using the normal distribution for confidence limits about the unknown population mean. These are calculated as $x \pm z\sigma_x$, where $\sigma_x = \sigma/N^{1/2}$ is the standard error of the mean (which is the standard deviation of the sampling distribution of the mean for a known population standard deviation σ), and z is the normal deviate for the chosen probability level. If the sample size is large, then the sampling distribution can still be assumed to be a standard normal distribution and a sample estimate of the population standard deviation S can be used instead of σ without significant error. However, when the sample size is small, with the population standard deviation to be estimated from this small sample, then the sampling distribution can no longer be assumed to follow the standard normal curve. Instead, the t distribution should be used with the t deviate $t = (x - \mu)/(S/N^{1/2})$, which is sensitive to the number of degrees of freedom. The t distribution, like the normal distribution, is symmetrical about the mean, but with a higher standard deviation, making it higher in the tails and lower at the peak. It has separate distributions for different degrees of freedom and rises at the peak, becoming smaller in the tails as N increases, and approaching the normal distribution for $N > 30$.

Tables exist to give the t-value for the required confidence level with the appropriate degrees of freedom (see Hamburg, 1979). The definition of a small sample set certainly includes the reduced data set regressions, which means that use of the t distribution would be statistically more rigorous. When plotting confidence limits about the regression lines the t-value for the three-point regression for one degree of freedom at the 5% confidence level is 12.706, considerably greater than the equivalent value of 1.96 for the normal statistic, which causes the limits to become extremely wide in some instances. Consequently, the normal distribution confidence limits calculated here, even using the full data sets, will generally be minimum estimates.

6.4 RESULTS

In this section we present results from a number of predictions for the different reaches at different discharges using regression relationships from both the full and reduced data sets. In all instances the predictive Equations 6.12−6.25 have been used. For the full data sets the results are compared with observations that were used in fitting the regression relationships. For the reduced data sets, comparisons are made only with observations not used in fitting the regressions. All predictions assume a conservative tracer and show the predictions from the expected parameter values and 90% confidence limits at each time step as determined from the Monte Carlo simulations.

Many of the predictions show that, given the measured upstream concentrations as input, both the full and reduced data sets can produce good predictions, in which the confidence limits determined by Monte Carlo simulation enclose the observed downstream data (see Figures 6.5–6.7).In some instances the calculated 90% confidence limits for the higher and lower discharges are very wide for the reduced data set (for example, Figure 6.5b), even though normally distributed estimates for the variation in τ and DF have been assumed. Figure 6.5b also shows the additional difficulties of predicting a concentration curve for which the gain is less than unity, i.e. non-conservative behaviour. No account has been taken of the uncertainty in gain in the present calculations. Given a fixed gain, short time delay simulations give rise to high peaks, whereas those for long delays will give low peaks. It is therefore possible that a simulation with the expected values of the parameters may lie close to the upper confidence limit at the peak (see Figure 6.5b).

Figure 6.6 shows an example where the prediction procedure has worked well, producing a reasonably accurate mean model with uncertainty bands that enclose the observed curve. The data set used in this prediction was one with a good fit to the τ and DF regressions and a gain value near unity.

The results for the River Tonge (Figure 6.7) show the limitations of estimating uncertainties from a small number of data points. In this instance, the regression correlation coefficients are high for all eight data points and even higher for the reduced set of data points. The result is very narrow and misleading confidence limits for the reduced data set.

6.5   DISCUSSION

The results presented here represent a first attempt to quantify the uncertainty associated with pollution incident prediction. The results have highlighted a number of difficulties and areas for improvement. In particular, it would be useful to be able to incorporate an estimate of the additional uncertainty due to variations in the gain caused by non-conservatism of the tracer. This may be difficult as there is no clear variation in gain with other variables. Apparent differences in gain may result from a number of causes, including adsorption on to vegetation surfaces, which might be seasonally dependent, and a lack of representative sampling in cases where the tracer is not fully mixed across the cross-section, something that the 'dead zone' concepts underlying the ADZ model suggest may happen fairly often.

The simulations for the River Tonge reach show the problem of using a small number of data points in assessing predictive uncertainty. In this instance the high correlations of the reduced data set result in a marked underestimate of the uncertainty. However, it is unusual to have even three sets of tracing experiments for a given reach; one, two or no measurements would be more common. The results of this paper suggest that even in such situations an attempt should be made to estimate the uncertainties associated with predictions of downstream concentrations, and it gives some guidelines as to the magnitude of the uncertainty in the parameter values to be expected.

It should also be noted that the results presented here are all for situations where there is a well defined input concentration curve. Consideration must also be given to the case where the input curve details are unknown. This may involve making predictions for the case where it must be assumed that there is a point source of pollutant and some initial mixing must be taken into account. It has been found that for such initial mixing, second- or third-order ADZ models are required to reproduce concentration curves in the reach immediately downstream of the input. In general, if such a model is estimated as nth order, then it can be approximated by n first-order ADZ models in series. If the time delay and dispersive fraction

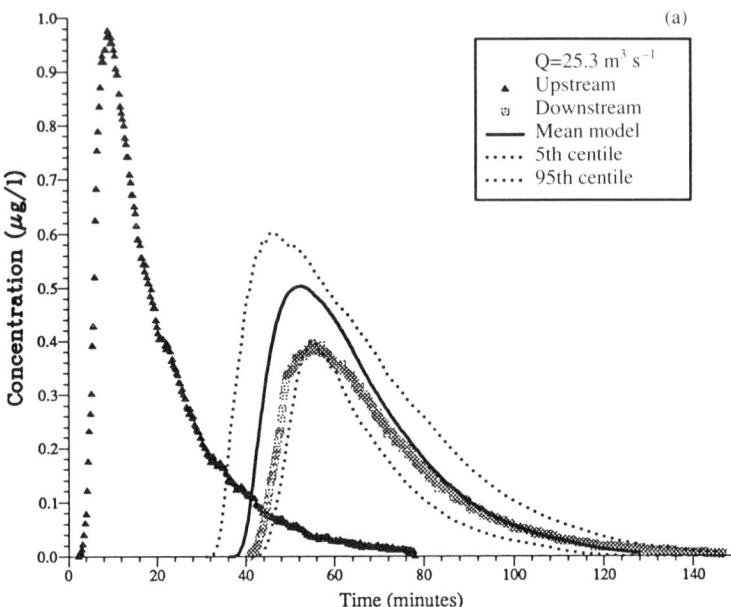

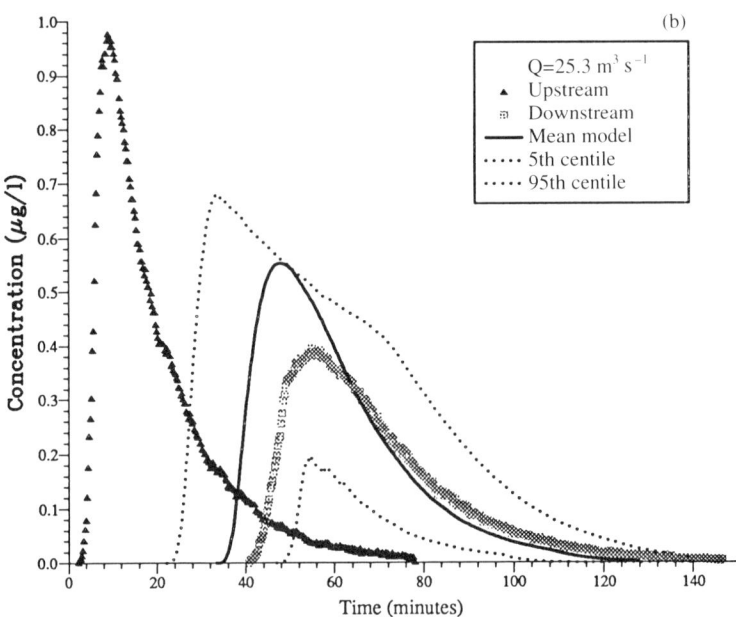

Figure 6.5 Prediction results at a low discharge for the Montford reach of the River Severn (a) using full data set regressions and (b) using reduced data set regressions

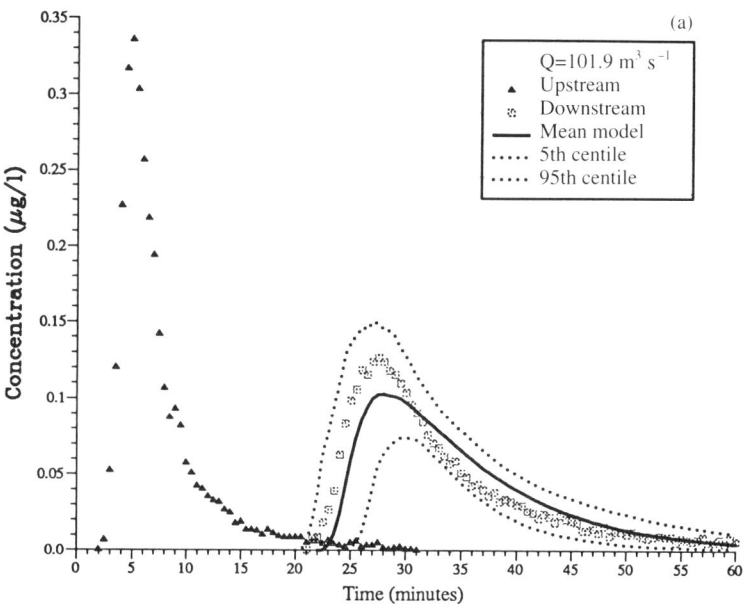

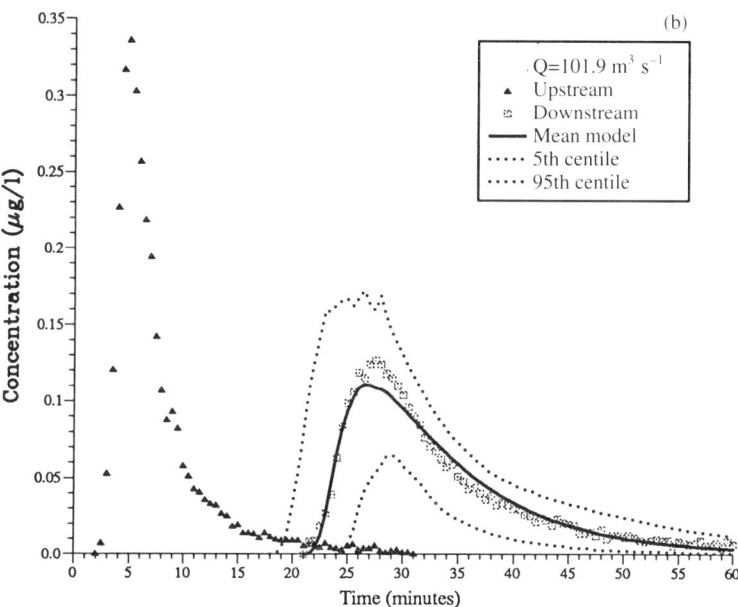

Figure 6.6 Prediction results at an intermediate discharge for Leighton reach of the River Severn (a) using full data set regressions and (b) using reduced data set regressions

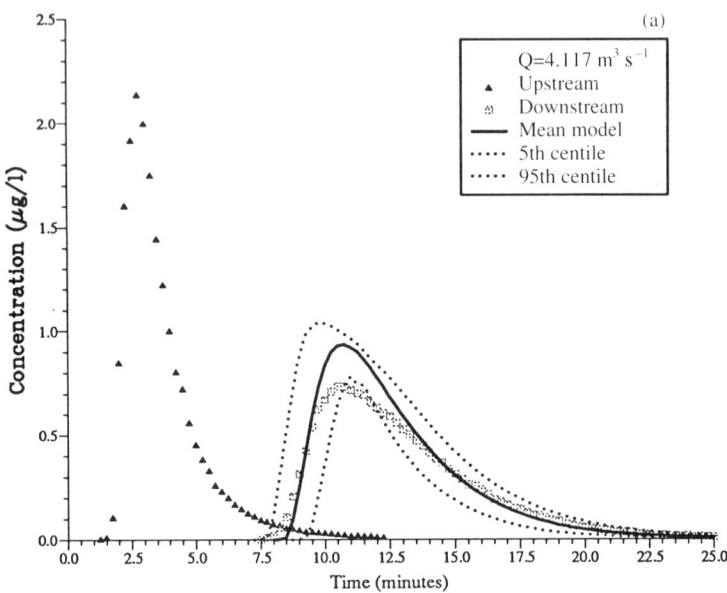

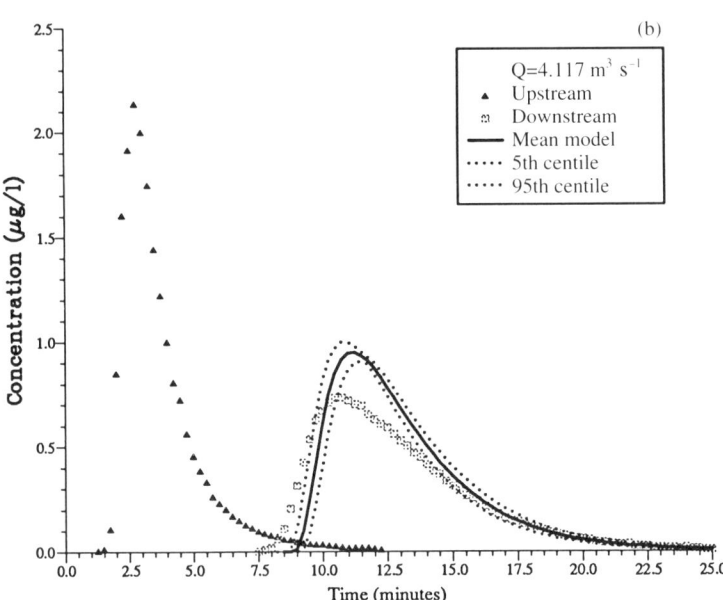

Figure 6.7 Predicted results at a high discharge for the River Tonge (a) using full data set regressions and (b) using reduced data set regressions

of a reach can be estimated, such a series of n first-order models can be constructed approximately, by dividing the total time constant for the reach by the model order n. It may also be necessary to propagate the uncertainty over a number of reaches. This may be performed in the same Monte Carlo framework by following the ith predicted concentration curve through the reaches, with a random choice of parameters for each reach. These techniques have been used in the ADZ-Network pollution incident prediction package available from CRES (Buckley and Beven, 1991).

ACKNOWLEDGEMENTS

We are grateful to Chris Hutchings, Nigel Vernon and Trevor Mission, who helped with the fieldwork. Steve Wallis and Nigel Vernon helped with the early work on the uncertainty estimation ideas presented here.

REFERENCES

Ang, A.H-S. and Tang, W.H. (1975) *Probability Concepts in Engineering Planning and Design.* Vol. 1. *Basic Pricniples.* Wiley, New York.

Beer, T. and Young, P.C. (1983) Longitudinal dispersion in natural streams. *J. Environ. Engin. Div. ASCE* **109**, 1049–1067.

Benjamin, J.R. and Cornell, C.A. (1970) *Probability, Statistics, and Decision for Civil Engineers.* McGraw-Hill, New York.

Beven, K. and Carling, P.A. (1992) Velocities roughness and dispersion in the lowland River severn. In: *Lowland Floodplain Rivers: Geomorphological Perspectives* (Eds P.A. Carling and G. Petts). Wiley, Chichester, 71–93.

Beven, K., Buckley, K.M. and Young, P.C. (1991) *ADZ-Analysis Manual I. CRES Report TR/90.* Lancaster University, Lancaster.

Buckley, K.M. and Beven, K. (1991) *ADZ-Network Manual. CRES Report TR/92.* Lancaster University, Lancaster.

Buckley, K.M., Beven, K., Young, P.C. and Benner, S.B. (1991) *ADZ-Analysis Manual II. CRES Report TR/91.* Lancaster University, lancaster.

Chatwin, P.C. and Allen, C.M. (1985) Mathematical models of dispersion in rivers and estuaries. Annu. Rev. Fluid Mech. **17**, 119–149.

Day, T.J. (1975) Longitudinal dispersion in streams. *Wat. Resour. Res.* **11**, 909–918.

Denton, R.A. (1990) Analytical asymptotic solutions for longitudinal dispersion with dead zones. *J. Hydr. Res.* **28**, 309–329.

Elder, J.W. (1959) The dispersion of marked fluid in turbulent shear flow. *J. Fluid Mech.* **5**, 544–560.

Fischer, H.B. (1966) A note on the one dimensional dispersion model. *Int. J. Air Wat. Pollut.* **10**, 443–452.

Fischer, H.B. (1967) The mechanics of dispersion in natural streams. *J. Hydr. Div. Proc. ASCE* **93**, 187–216.

Fischer, H.B. (1968) Methods for predicting dispersion coefficients in natural streams with applications to the lower reaches of the Green and Duwamish rivers Washington. *USGS Prof. Paper* **582-A**.

Fischer, H.B. (1973) Longitudinal dispersion and turbulent mixing in open channel flow. *Annu. Rev. Fluid Mech.* **5**, 59–78.

Fischer, H.B., List, E.J., Koh, R.C.Y., Imberger, J. and Brooks, N.H. (1979) *Mixing in Inland and Coastal Waters.* Academic Press, New York.

Hamburg, M. (1979) *Basic Statistics: A Modern Approach.* Harcourt Brace Jovanovich, New York.

Liu, H. (1977) Predicting dispersion coefficients of streams. *J. Environ. Engin. Div. ASCE* **103**, 56–69.

McQuivey, R.S. and Keefer, T.N. (1974) Simple method for predicting dispersion in streams. *J. Environ. Engin. Div. ASCE* **100**, 997–1011.

Nordin, C.F. and Sabol, G.V. (1979) Empirical data on longitudinal dispersion in rivers. USGS, Lakewood, 20–74.

Nordin, C.F. and Troutman, B.M. (1980) Longitudinal dispersion in rivers: the persistence of skewness in observed data. *Wat. Resour. Res.* **16**, 123−128.

Reichert, P. and Wanner, O (1991) Enhanced one dimensional modelling of transport in rivers. *J. Hydr. Engin. ASCE* **117**, 1165−1183.

Sabol, G.V. and Nordin, C.F. (1978) Dispersion in rivers as related to storage zones. *J. Hydr. Div. ASCE* **104**, 695−708.

Taylor, G.I. (1954) The dispersion of matter in turbulent flow through a pipe. *Proc. Roy. Soc. London Ser.* **A223**, 446−468.

Valentine, E.M. and Wood, I.R. (1977) Longitudinal dispersion with dead zones. *J. Hydr. Div. ASCE* **103**, 975−990.

Valentine, E.M. and Wood, I.R. (1979a) Experiments in longitudinal dispersion with dead zones. *J. Hydr. Div. ASCE* **105**, 999−1016.

Valentine, E.M. and Wood, I.R. (1979b) Dispersion in rough rectangular channels. *J. Hydr. Div. ASCE* **105**, 1537−1553.

Wallis, S.G., Blakeley, C. and Young, P.C. (1987) A microcomputer based fluorometric data logging and analysis system. *J. Inst. Wat. Engin. Scien.* **41**, 122−134.

Wallis, S.G., Young, P.C. and Beven, K.J. (1989) Experimental investigation of the aggregated dead zone model for longitudinal solute transport in stream channels. *Proc. Inst. Civ. Eng.; Part 2* **87**, 1−22.

Whitehead, P.G., Williams, R.J. and Hornberger, G.M. (1986) On the identification of pollutant or tracer sources using dispersion theory. *J. Hydrol.* **84**, 273−286.

Young, P.C. (1984) *Recursive Estimation and Time Series Analysis: An Introduction.* Springer Verlag, Berlin.

Young, P.C. (1985) Recursive identification, estimation and control. In: *Handbook of Statistics.* Vol. 5. *time Series in the Time Domain* (Eds E.J. Hannen *et al.*). North Holland, Amsterdam, 213−255.

Young, P.C. (1986) Time-series methods and recursive estimation in hydrological systems analysis. In: *River Flow Modelling and Forecasting* (Eds D.A. Kraijhenhoff and J.R. Moll). D. Reidel, Dordrecht, 129−180.

Young, P.C. (1989) Recursive estimation forecasting and adaptive control. In: *Control and Dynamic Systems* (Ed. C.T. Leondes). Academic Press, San Diego, 119−166.

Young, P.C. and Lees, M. (1993) The active mixing volume: a new concept in modelling environmental systems. In: *Statistics and the Environment* (Eds V. Barnett and R. Turknam). Wiley, Chichester, 3−43.

Young, P.C., Jakeman, A.J. and McMurtie, R. (1980) An instrumental variable method for model order identification. *Automatica* **16**, 281−294.

APPENDIX: NOTATION FOR UNCERTAINTY CALCULATIONS

r^2_{raw}	Correlation coefficient of the regression analysis
r^2_{adj}	Correlation coefficient of the regression analysis adjusted for the number of degrees of freedom. This is achieved using the equation $r^2_{\text{adj}} = 1 - (1 - r^2_{\text{raw}}) [(N-1)/(N-k)]$, where r^2_{adj} is the corrected value, r^2_{raw} is the uncorrected value, N is the number of observations and k is the number of constants in the regression equation (Hamburg, 1979)
$E[\ln\tau\,|\,\ln Q]$	Mean value of the logarithm of time delay, $\ln\tau$, as given by the regression equation for a chosen discharge
$E[DF\,|\,Q]$	Mean value of dispersive fraction, DF, as given by the regression equation for a chosen discharge
$\sigma^2_{\ln\tau\,|\,\ln Q}$	Constant variance about any $E[\ln\tau\,|\,\ln Q]$ estimate
N_τ	Number of data points used in the τ regression
$r^2_{\tau,Q}$	Correlation coefficient of the τ regression, adjusted for the number of degrees of freedom as above

$S^2_{\ln\tau}$	Sample variance of the logarithmic τ data values
$\sigma^2_{DF\|Q}$	Constant variance about any $E[DF\|Q]$ estimate
N_{DF}	Number of data points used in the DF regression
$r^2_{DF,Q}$	Correlation coefficient of the DF regression, adjusted for the number of degrees of freedom as above
S^2_{DF}	Sample variance of the DF data values
$\ln\tau_i$	Value of $\ln\tau$ at each Monte Carlo run
$R_{1,i}$ and $R_{2,i}$	Normally distributed random variables $\{N(0,1)\}$ for the ith realization
$\ln Q_0$	Logarithm of the discharge value at which the prediction is to be made
$\overline{\ln Q}$	Sample mean of the logarithmic Q data values
$S^2_{\ln Q}$	Sample variance of the logarithmic Q data values
$E[DF\|Q]$	Mean value of DF given a value for $\ln\tau$
$\rho_{DF,\tau}$	Correlation coefficient between DF and $\ln\tau$
$\sigma_{DF\|\ln\tau}$	Standard deviation of DF given a value of $\ln\tau$
DF_i	Value of dispersive fraction at each Monte Carlo run
$\overline{t}_i$	Value of mean travel time at each Monte Carlo run
T_i	Value of ADZ mean residence time at each Monte Carlo run

7 Preliminary Observations and Significance of Dead Zone Flow Structure for Solute and Fine Particle Dynamics

P.A. CARLING, H.G. ORR and M.S. GLAISTER

Institute of Freshwater Ecology, Ambleside, Cumbria, UK

7.1 INTRODUCTION

Research on dispersion processes through river reaches often has highlighted discrepancies between the observed and predicted residence times of solutes and particulates where the calculated values are obtained using Fickian-type dispersion models. Generally, the observed residence times are in excess of those predicted, and may be related to the effects of delayed advection induced by a variety of aggregated storage effects within a given reach. These effects are not accounted for by an application of those dispersion models which assume a uniform velocity field.

Delayed advection may be related to complex three-dimensional flow effects. At the simplest level, secondary (spiral) currents effectively increase the length of the transport path in relation to the primary current vector, but, in addition, the importance of spatial variation in drag exerted by a variation in boundary roughness and channel geometry has previously been largely neglected in natural streams (Cherkauer, 1973). For example, although it has long been known that near-bank flow dynamics in straight trapezoidal channels may impede particulate dispersion (e.g. Lane, 1937; p. 137), longitudinal changes in the geometry of natural channels induce flow contractions and expansions, as well as zones of flow separation. In addition, the channel roughness changes with variations in bed particle size, the presence of slumped bank material, submerged macrophytes and partially submerged riparian vegetation. The summation of these effects through any reach can be described statistically using the aggregated dead zone (ADZ) modelling algorithm (Wallis, Blackley and Young, 1987), but the quantification of deterministic constraints is poorly researched (Beven and Carling, 1992).

Detailed field studies on reaches of the River Severn in the UK (Figure 7.1) have indicated that significant fluid retention is associated with slow moving, recirculating or 'stagnant' water associated with near-bank zones, rather than with the near-bed or interstitial zone (for example, the bottom of deep pools). River water is shallow, rough−turbulent and well mixed in the vertical plane, whereas near-bank zones are characterized by greater flow resistance than flow over the mid-channel area and may exhibit flow separation cells, cordoned from the main flow by shear zones across which fluid is exchanged at variable rates. The physical dimensions and residence times of these individual 'dead zones' may vary with discharge,

Mixing and Transport in the Environment. Edited by K.J. Beven, P.C. Chatwin and J.H. Millbank
© 1994 John Wiley & Sons Ltd

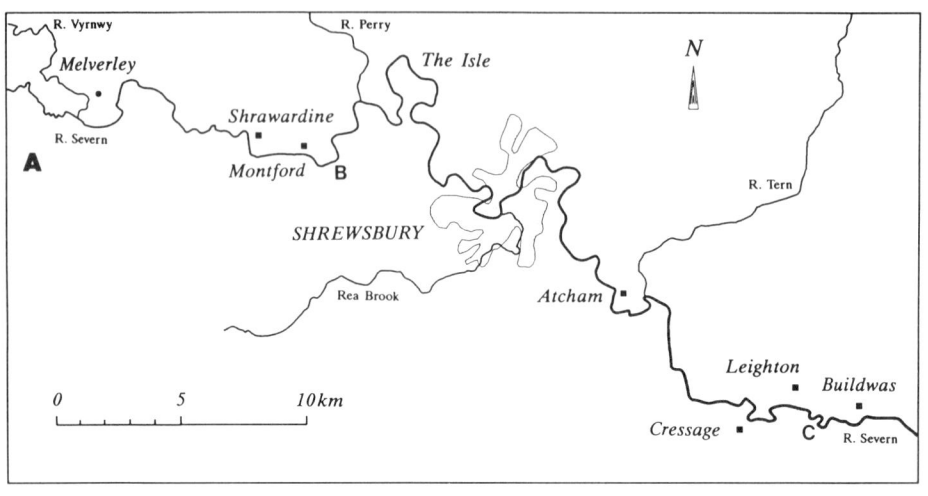

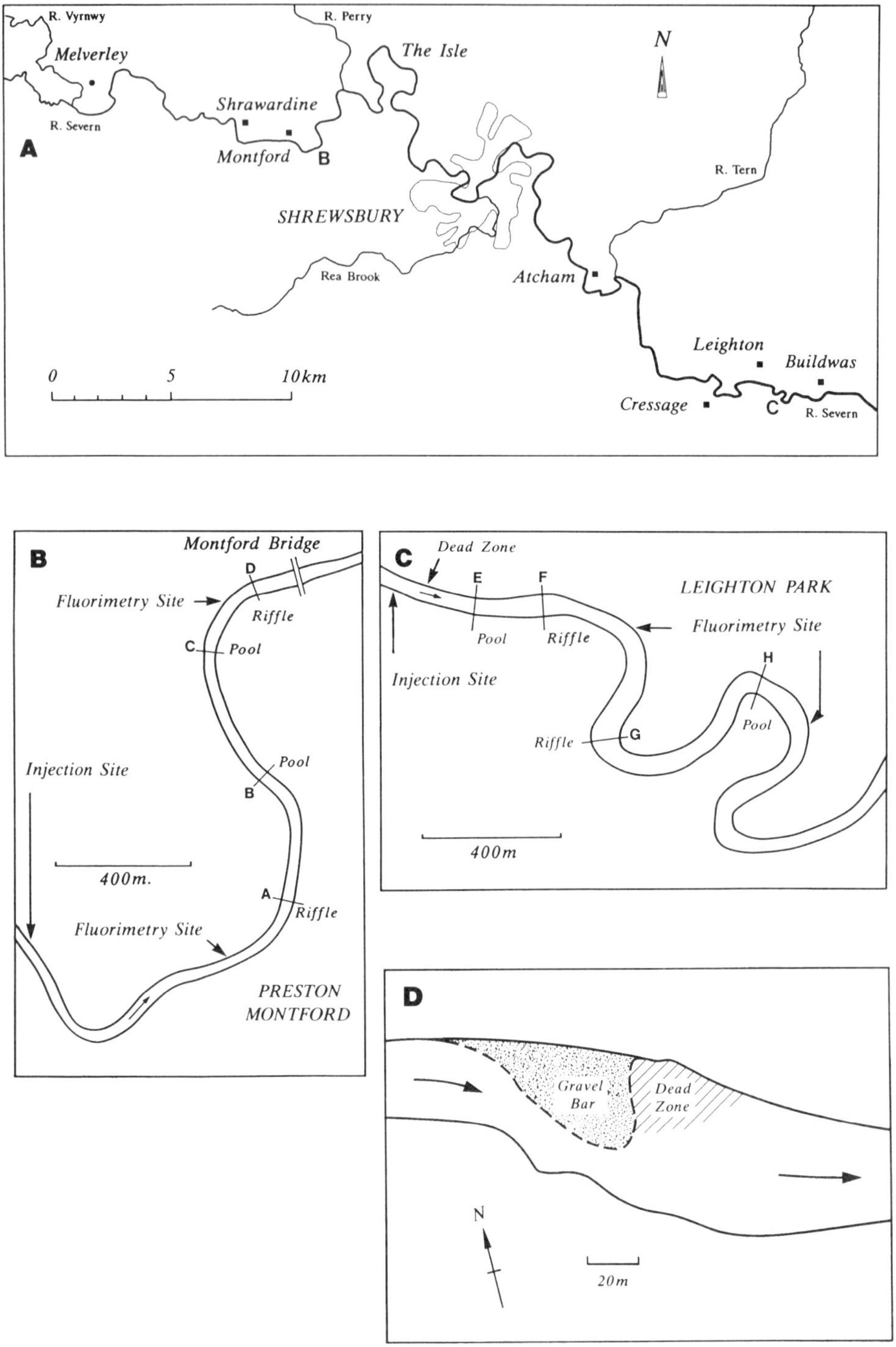

Figure 7.1 Location of the Montford and Leighton reaches of the River Severn, near Shrewsbury, UK

but overall it appears that their volumetric effect on the dispersion process is only weakly (negatively) dependent on discharge: the effective mixing volume decreases slightly with discharge (Beven and Carling, 1992). This is initially surprising as intuitively we expect a positive relationship, as mixing should be more thorough as discharge and velocity increase and channel-form roughness effects decrease (Parker and Peterson, 1980). Possible explanations include: (1) a reduction in the importance of grain roughness in the mixing process as water depths increase; (2) the replacement of low flow dead zone areas by new 'volume equivalent' high flow dead zones; (3) the growth of existing dead zones; or (4) the increased importance of strongly three dimensional flow effects on the overall mixing process, such that the mixing of the main flow volume is more thorough at high discharges.

Characterization of dead zone processes is important, not only with reference to bulk dispersion processes, but also with respect to processes active within dead zones themselves. Dead zones have been shown to act as inoculae with respect to maintaining plankton and algae populations in flowing river waters (Reynolds, 1988; Reynolds, Carling and Beven, 1991) and may also act as nursery areas, or refugia during floodflow, for fish (Harvey, 1987; Heggenes, 1988; Copp, 1989), aquatic plants and benthos (Fisher *et al.*, 1982; Gaschignard, Persat and Chessel, 1983; Biggs and Close, 1989). Further, if dead zones account for the delayed dispersion of particulates, it follows that some particles may accumulate on the bed of dead zones (Graham, 1990). This may lead to localized changes in sediment and water chemistry, with implications for benthic organisms, and the prospect of pulses of contaminants being released from storage when the discharge is sufficient to promote flushing (Berndtsson, 1990; Higgins, 1990). This latter scenario has been addressed using a preliminary model of deposition and resuspension within dead zones, which reproduces observed fine particle behaviour tolerably well when calibrated against limited hydraulic data (Tipping, Woof and Clarke, 1993).

In the absence of detailed data about the flow structure of dead zones, a research programme has been implemented to determine the relative volume of dead zones in relation to the effective mixing volume and to obtain basic physical characterization. This study is based on reaches of the River Severn near Shrewsbury (Figure 7.1), where considerable detailed data on the distribution of velocity at the scale of individual sections already exists (Carling, 1991; Reynolds, Carling and Beven, 1991; Beven and Carling, 1992; Carling *et al.*, 1992). These existing data are recognized as being suboptimal with respect to defining the detailed local hydraulic climate of dead zones, but are supplemented by preliminary turbulence measurements. An initial analysis of these data has provided insights useful for focusing future programme efforts both at the scale of individual dead zones and at the reach scale. In addition, the interpretation of remote sensing imagery and dispersion experiments facilitate the reach-scale interpretation of flow processes.

7.2 METHODS

7.2.1 Velocity

Velocity profiles throughout the water depth have been measured every 2 m across 11 cross-sections on the Montford and Leighton reaches of the River Severn near Shrewsbury, for a range of discharges ranging from very low flow during drought conditions (about $7 \, \text{m}^3 \, \text{s}^{-1}$) to bankfull ($250 \, \text{m}^3 \, \text{s}^{-1}$). The locations studied include meander crossings and pools, straight reaches with and without marginal vegetation and one major dead zone.

Data were collected using a rig consisting of six Ott current meters simultaneously integrating current speed over 60 s periods. By sequentially raising the rig, data were obtained throughout the water column to characterize the outer layer and near-bed velocity structure. Near-bed velocity data conforming closely to a log-normal profile (Wilkinson, 1984) have been processed to obtain information on the boundary layer depth-mean velocity (U), the near-bed shear stress ($\tau_0 = \rho u_*^2$), the friction factor [$f = 8(u_*/U)^2$], the roughness length (z_0) and Reynolds number (Ud/ν) where u_* is the shear velocity, ρ is the density, d is the flow depth and ν is the kinematic viscosity. The equation fitted to the data is that of von Kàrmàn (1934)

$$U_z = (u_*/\kappa) \, \ln(z - h/z_0) \tag{7.1}$$

where κ is von Kàrmàn's constant (~ 0.40) and h is the zero datum (about $0.5D_{50}$ for an open plane gravel bed; Jackson, 1981).

As the discharge varied insignificantly during a section traverse, pseudo-synoptic maps of isovel distribution have been produced for a total of 50 surveys. By assuming that isovels are distortable membranes (Bhowmik, 1982) it is possible to adduce qualitatively the nature and pattern of secondary currents. These maps have helped identify 'dead zone' areas and approximate their lateral and vertical dimensions at a section.

7.2.2 Turbulence

Turbulence data were collected on traverses normal to the bankline at a variety of sections and at a series of points across the major dead zone and adjacent main flow. Measurements were obtained using a Colnbrook electromagnetic current meter with a 55 mm sensor, which can measure two perpendicular components of velocity simultaneously. The signal was digitized at 10 Hz and was stored on an Epson portable computer. The current meter was mounted on a rigid frame which sat in a stable position on the river bed and data were obtained at a variety of relative heights (z/d) above the bed. Further details of the system and data processing are given by Heslop and Allen (1989; Heslop, Holland and Allen, Chapter 3).

7.2.3 Remotely sensed data

A series of high resolution airborne remote sensed images of reaches of the River Severn was obtained using the NERC Daedalus AADS-1230 thermal line scanner, which allowed water surface temperature differences of 0.2°C to be monitored at a spatial resolution of approximately 1.25 m. The system is calibrated by reference to temperature charts for black body sources. The data were processed using the I²S system at the Institute of Terrestrial Ecology at Bangor using false colour enhancement so that different colours represent a range of temperature bands, and were supplemented by synoptic *in situ* surface water temperature measurements taken across a transect of the river and vertical colour air photography.

7.2.4 Dispersion process

In the same reaches where current meter surveys were conducted at individual sections, additional solute dispersion surveys were conducted through given reaches using rhodamine WT and the computer controlled logging equipment described by Wallis, Blackley and Young (1987). Trials were conducted using a fine particle tracer to define the particulate dispersion characteristics of a reach exhibiting a well defined dead zone area and an attempt was made to measure solute residence time within the dead zone. The particle tracer used was an inert fine plastic fluorescent particle of similar size range and density to the natural suspended sediment ($3-300\,\mu$m). The variations in the particle tracing data were found to be too great for quantitative analysis, although tracer was detected within the dead zone, indicating that exchange had taken place across the shear zone.

The dye dispersion data were analysed using the ADZ model (Wallis, Guymer and Bilgi, 1990). Wallis, Young and Beven (1989) have shown that the model parameters can be interpreted in terms of the different residence times of individual reaches. The mean residence time, t, is the sum of the advective time delay, ς, and the tracer mean residence time, T, of an effective mixing volume V_e. Successful dispersion tests have been conducted on this reach of the Severn in the past (Beven and Carling, 1992), but for reasons which are not clear most of the current series of tests proved to be inadequately mixed for analysis. Limited interpretation of some runs, however, was possible and reference is also made to previously published dispersion data from the same river reaches (Beven and Carling, 1992).

In principle, a discharge, Q, can be calculated from dilution of the tracer, so the effective mixing volume is given by

$$V_e = QT \tag{7.2}$$

and an estimate of the total reach volume is given by

$$V = Qt \tag{7.3}$$

Consequently, the dispersive fraction parameter of the ADZ model (V_e/V) is given by

$$V_e/V = T/t = T/(T+\varsigma) \tag{7.4}$$

and the reach-averaged effective velocity, U_L is

$$U_L = V_e/tDW_e \tag{7.5}$$

where DW_e (the product of average depth and effective width) defines an effective reach-averaged cross-sectional area (A_e). Clearly from Equation 7.5, if the river flow is well mixed in the vertical plane, then a reduction in effective volume with discharge will be associated largely with changes in effective width, whereas velocity increases with discharge and should act to sustain the mixing volume. It is evident that where suitable data exist, reach-scale parameters such as effective velocity, residence time, effective area and dispersive fraction may be compared with section-scale data derived from current meter surveys.

7.3 RESULTS AND DISCUSSION

7.3.1 Section-scale flow structure

7.3.1.1 Velocity profile data

The current meter surveys showed that close to the vegetated banks, velocities may be very low compared with the main streamline, such that 'near-zero flow' regions extending from 2 to 4 m distant from the bank have velocities less than 3 cm s^{-1}. Reaches characterized by semi-submerged vegetation or physical re-entrants in the bank alignment often exhibited 'zero flow' zones as determined by current meter surveys and the minimal movement of floats within the dead zones (Figure 7.2). Similarly, within bendways, flow separation commonly

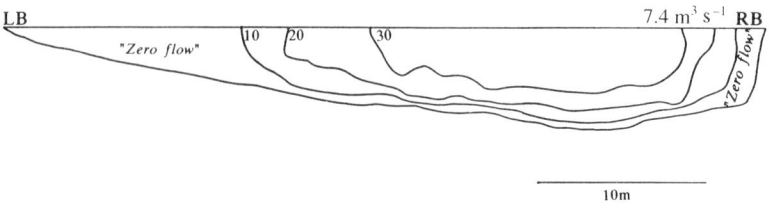

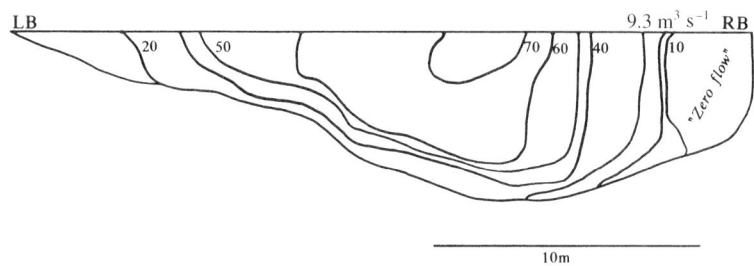

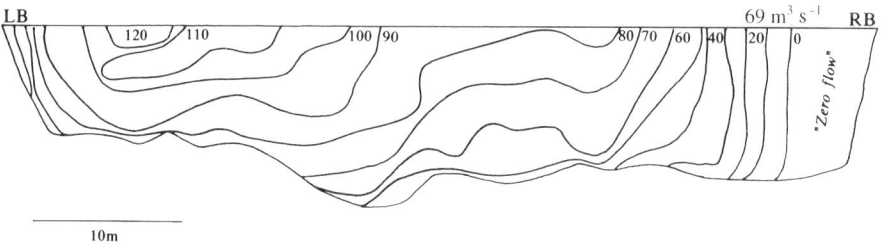

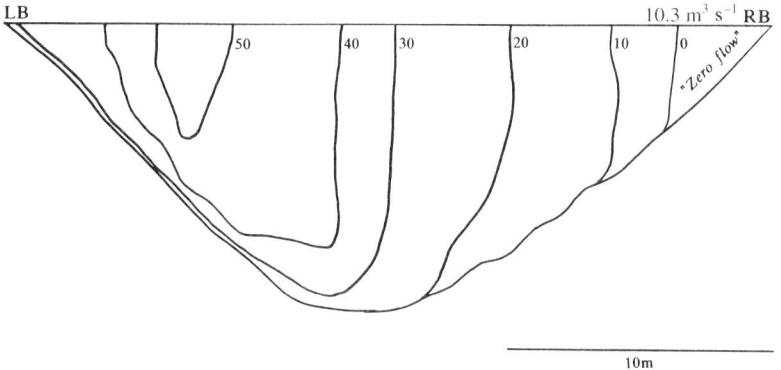

Figure 7.2 Examples of isovel distribution across sections characterized by dead zone development in association with the region of the bank batter. Contour values are in cm s^{-1}. Bulges in the isovels provide qualitative evidence for the presence of secondary currents. Vertical exaggeration ×5

occurs on the inside of the bend, but may also occur on the outside (for bends of tight curvature) when the main flowline cuts across the inner point bar (cf. Nanson and Page, 1983). The dead zones appeared to vary in size and presence depending on stage, with some being consistent in dimensions and location and others being transitory in nature. Delayed mixing is consequently affected by dead zones of variable residence time, although only the permanent features are likely to be of significance for the long-term storage and deposition of fine sediment.

Towards the centre of the channel the bed consists of a planar bed of cobbles. Here profiles conformed closely to Equation 7.1, consistent with steady uniform flow (Figure 7.3b) and roughness lengths of the order of 0.1 cm. In contrast, the structure of the velocity profile in the near-bank regions was different, because near the banks the bed consisted of cobbles, fines, slumped blocks of bank material and submerged aquatic vegetation; an altogether rougher composite surface. The primary effect of the rougher boundary was to decelerate the flow; reducing the average flow velocity so that dead zone near-bed shear stresses commonly did not exceed 0.4 N m^{-2} (Figure 7.3a) and the cobble bed was blanketed by fines. Elsewhere locally decelerating flow over a rough near-bank region was associated with increased shear velocities and large roughness lengths (Figure 7.3c and 7.3d).

The presence of submerged vegetation resulted in characteristic 'compound' or inflected profiles with slow flow within the vegetative mass and more rapid flow above (Figure 7.3e and 7.3f). In these latter cases it is often possible to isolate two roughness length values (Cionco, 1983) associated with the channel bed and vegetative surface, respectively, if Equation 7.1 or other functions are appropriate to one or other of the surfaces (Figure 7.3g). In the main flow over the cobble bed, however, the accurate determination of hydraulic parameters is not unduly sensitive to the choice of a zero datum value (Raupach and Thom, 1981), which has a limited optimization range (Jackson, 1981); this is not the case with many near-bank profiles. In the example given, the zero datum and roughness length values were determined by the iterative procedure of Inman (1963). Although this is not entirely satisfactory (Raupach and Thom, 1981), it represents a pragmatic solution to obtaining preliminary estimates of vegetative near-bank roughness from sparse data.

Decreased flow velocities and increased local bed shear close to the banks is reflected in the spatial variation in the value of the resistance coefficient (f) and Reynolds number. Synoptic data for pool and riffle traverses show the expected reduction of the value of f with high Reynolds number flows in mid-channel over gravel and higher resistance over macrophyte beds and within marginal submerged riparian vegetation where Reynolds number are lower (Figure 7.4). Beven and Carling (1992) showed that the relative roughness function for section-average data for riffles in the Severn (Hey, 1979) applied to the reaches presently investigated when near-bank data were excluded. Hey's section-averaged function is a version of the Colebrook−White function for the entire turbulent flow range

$$\frac{1}{f^{1/2}} = -2\log\left(\frac{k_s}{3.71d} + \frac{2.51}{Ref^{1/2}}\right) \tag{7.6}$$

Point data straddle this relationship with rougher beds lying above the curve and smoother beds lying below the curve, with near-bank data appearing as outliers (Figure 7.4).

7.3.1.2 Dead-zone turbulence

At this stage of the investigation the most useful turbulent quantities are the mean values

146

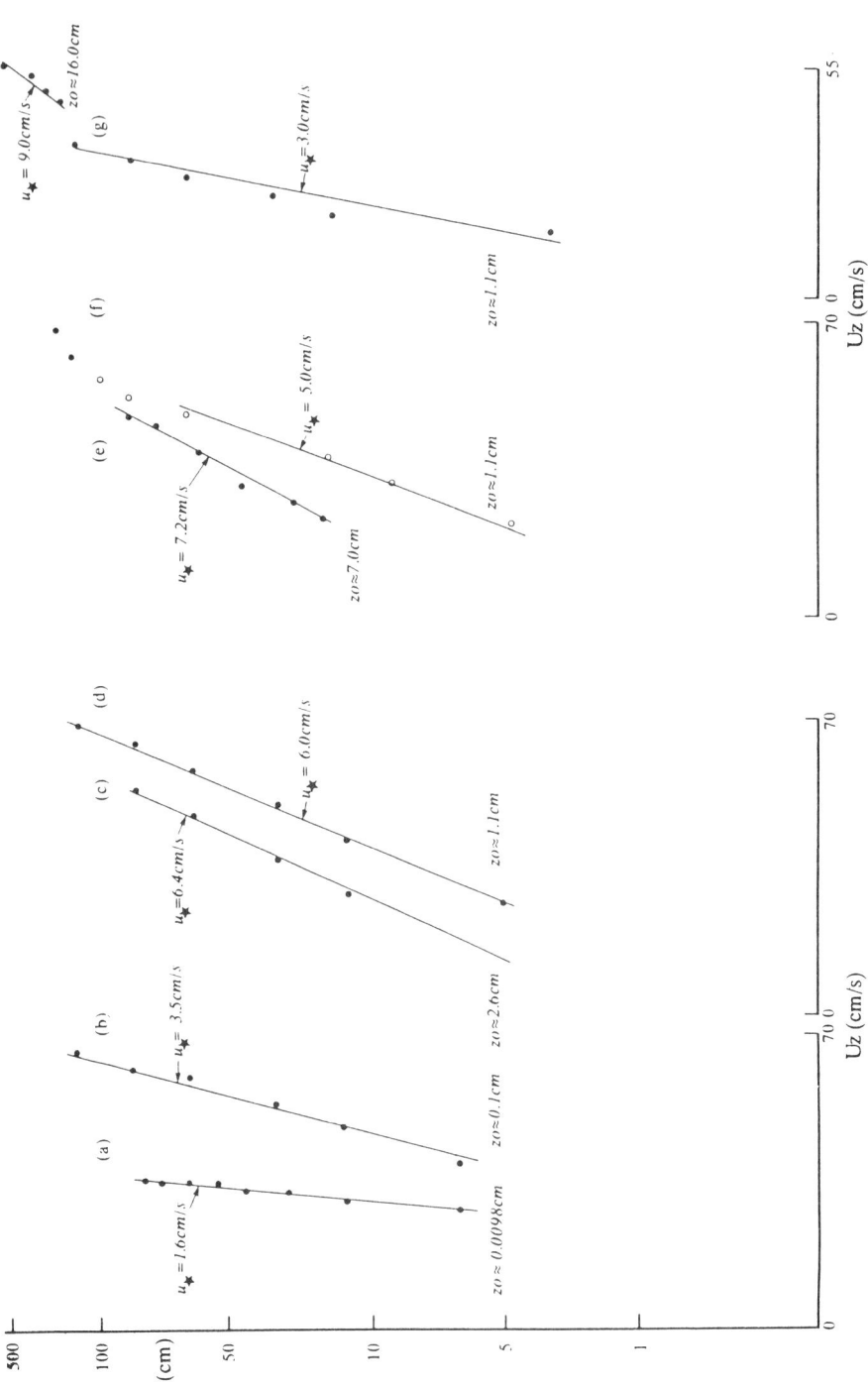

Figure 7.3 Near-bed velocity profiles in relation to curves representing Equation 7.1. (a) Flow over fine muddy sands within 'dead zone'; (b) flow over coarse gravel in mid-channel; (c) profile over slumped blocks 2 m from bank; (d) same location as (c) but over coarse gravel 4 m from bank; (e and f) compound profiles for flow through ($z+h<100$) tree branches and macrophytes in near-bank region, showing accelerative flow above branch layer ($z+h>100$); (g) well defined compound profile for flow through ($z+h<200$) and above ($z+h>200$) rank macrophyte layer

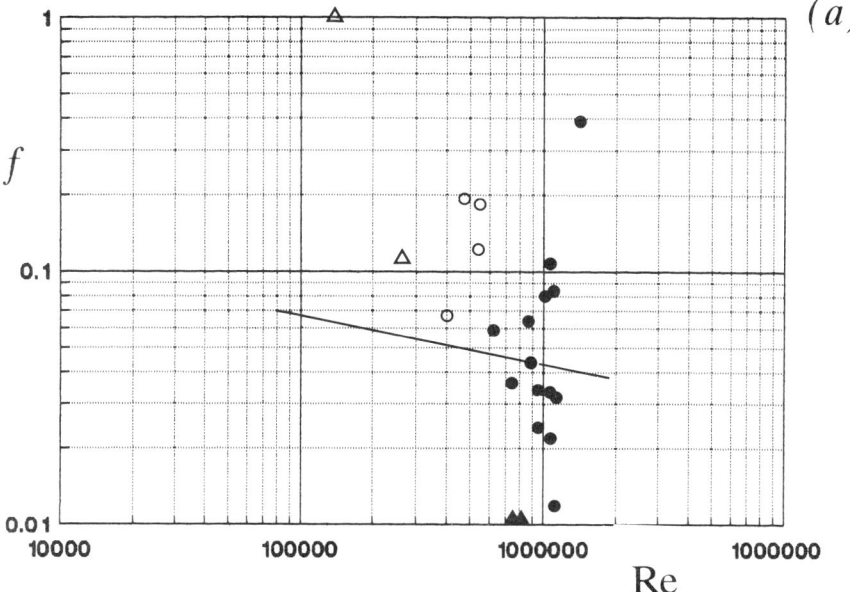

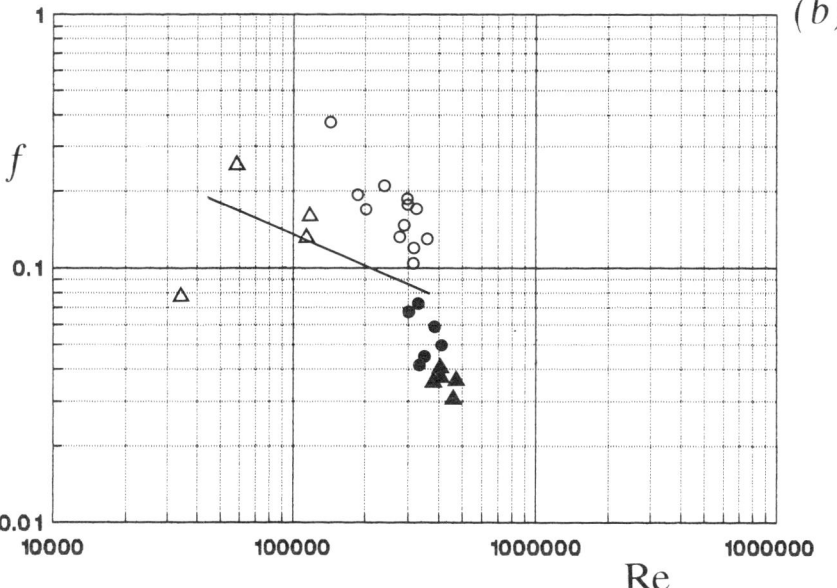

Figure 7.4 Synoptic point values of Darcy—Weisbach resistance coefficient for a traverse across (a) a pool on a straight reach at Leighton and (b) a riffle at the same location. The expected trend of a reduction in f (Equation 7.6) at higher Reynolds numbers and over smoother beds is demonstrated for reach-average data. Outliers are primarily associated with the near-bank regions. The points straddle the relative roughness function (Equation 7.6) for given point water depths and section-averaged bed grain size (pool $d_{84} = 0.025$ m; riffle $d_{84} = 0.05$ m). Symbols: o = coarse gravel and macrophytes; ● = fine gravel; ▲ = mud; ▵ = near bank

of U, V and W, together with the variances, $\overline{u^2}$, $\overline{v^2}$ and $\overline{w^2}$, of the instantaneous turbulent velocity excursions from the mean. The shear stress ultimately acts on the bed of the river and if sufficient in magnitude will entrain particles from the bed. Because near-bed dynamics are dominated by a consideration of τ_0 (the shear stress) it is common to scale turbulence data by the shear velocity $[u_* = (\tau_0/\rho)^{1/2}]$. As a generalization, Soulsby (1983) noted that the RMS of the turbulence intensities $[e.g.\ u' = (\overline{u^2})^{1/2}]$ are often of the same magnitude and scale with u_*

$$u' > v' > w' > u_* \tag{7.7}$$

This consistent relationship means that the tendency for suspended particles to settle or be resuspended from the bed may be expressed in terms of W, w' or u_* relative to the settling velocity of the unhindered particles in still water (w_s). It follows that the spatial and temporal variation in these turbulence parameters can, in part, be used to characterize dead zone behaviour.

In the same vein, the relative scales of eddies associated with different turbulent environments can be assessed if measurements are taken at similar heights above the bed in fluid subsystems with comparable length scales. The kinetic energy of the turbulence per unit volume due to large eddies is given by

$$E = \frac{\rho(\overline{u^2} + \overline{v^2} + \overline{w^2})}{2} \tag{7.8}$$

In Figure 7.5 turbulence intensity data in the y- and z-planes, collected along transects away from the river bank, have been scaled using u_* data obtained from synoptic velocity profiles. A number of straight or slightly curved sections on pools and riffles are represented (sections A and E) at which the tendency is for turbulence intensity to increase away from the bank. Both the impellor and electromagnetic flow meters often recorded sensibly zero flow within 2 m of the bank, although slight movement could be observed visually. At section A the gradient of increase in intensity with distance from the bank is steeper than at section E and may reflect the presence of partially submerged scrub trees inhibiting flow, which are not present at section E. The plot for section D has an opposite trend to the other data, being taken on the outside of a bend, and it reflects the higher turbulence generated by helical flow cells impinging on the bank.

The vertical velocity component is of particular interest as it may accelerate a particle towards the bed or hinder the rate of settlement. Figure 7.6 summarizes the available data for average values of W, taken with distance from the left bank. The quality of these data is not high, but some general observations are appropriate. At one site in particular the mean residual flow is towards the bed. No consistent pattern across the sections is evident, although the negative values in the slight bend at section A may reflect downwelling commonly associated with the outside bank of bends (Bathurst, Thorne and Hey, 1979; de Vriend and Geldof, 1983; Hicks, Yin and Steffler, 1990). More importantly, some indication is given of the average vertical velocities encountered, which, together with the variance, enables a determination of the range of fluid velocities encountered by settling particles in the main flowline close to the banks. Although the residence time of particles within the dead zones is of the order of $10^2 - 10^3$ s, vertical velocities are often adequate to prevent the deposition of particles of widely varying density (Carling et al., 1992).

To complete the turbulence analysis a comparison is required of parameters within a well developed dead zone, within the shear zone across which exchange of particles occurs and

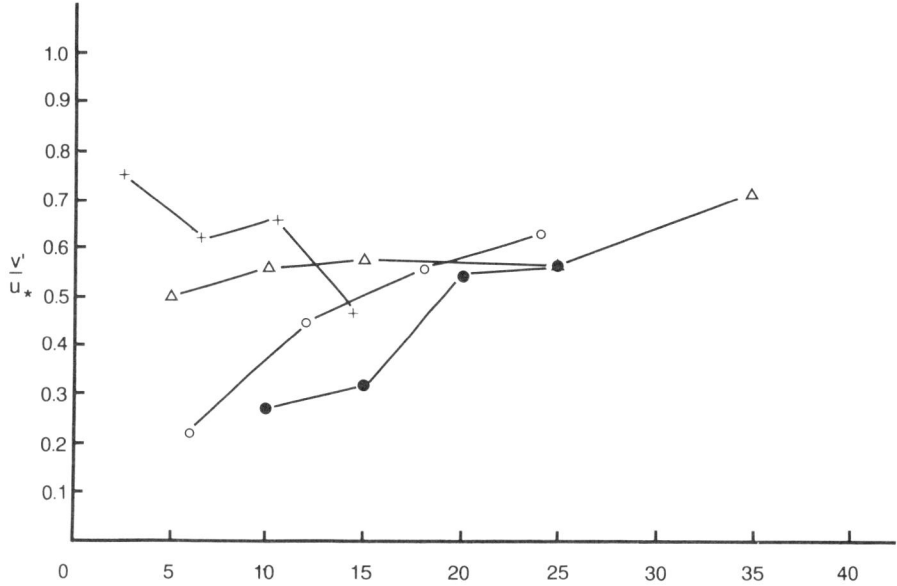

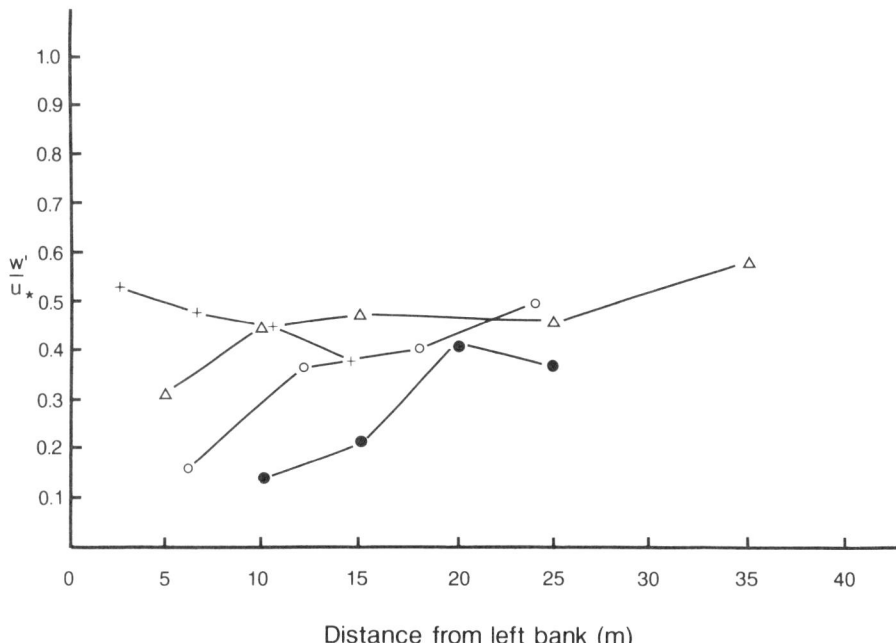

Figure 7.5 Increase in turbulence intensity with distance from river bank on three straight or slightly curved sections. Exception (+) is on the outside of a meander bend. All readings at 0.5 m above the bed except for ● at 0.65 m; ○, ● = section A; + = section D; △ = section E

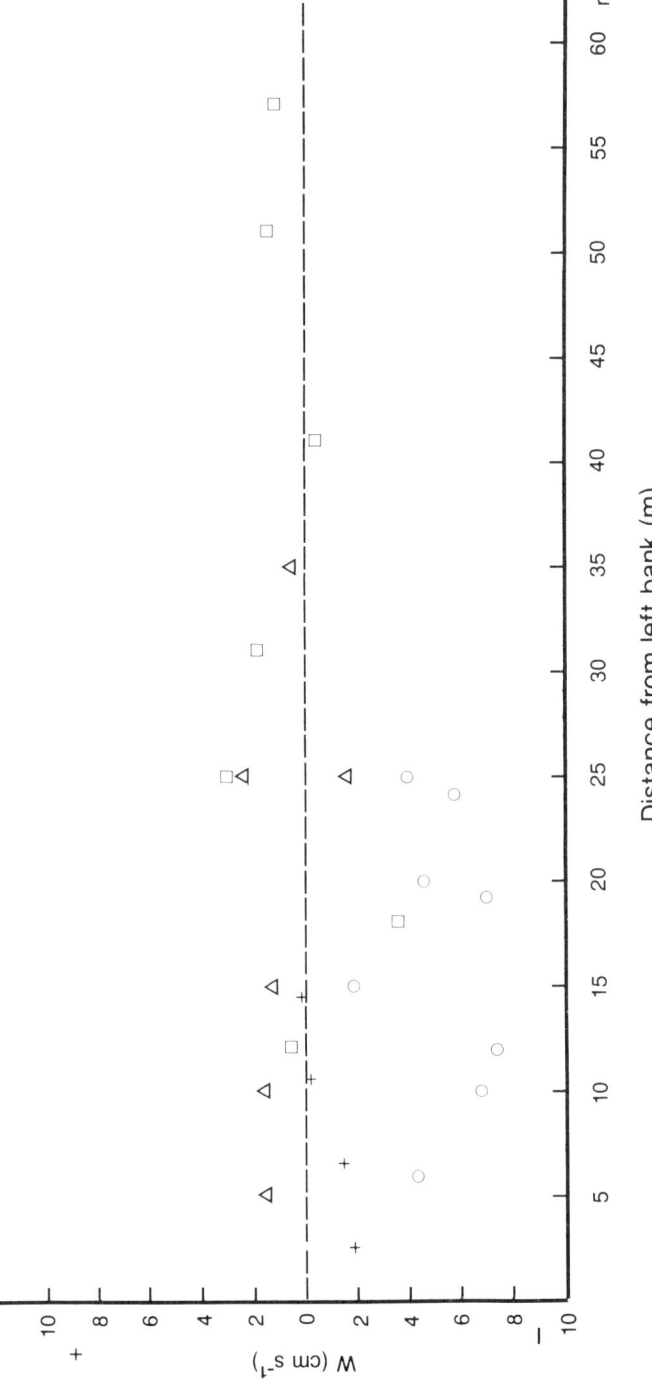

Figure 7.6 Variation in vertical turbulence component with distance from river bank. +, Upward residual; −, downward residual

Table 7.1 Turbulence data obtained during a discharge of 25 m³ s⁻¹ at three stations across section B−B′ (see Plate 6 for location)

Environ*	U	V	W	$(u^2)^{1/2}$	$(v^2)^{1/2}$	$(w^2)^{1/2}$	E	u_*	z/d
Control	0.007	0.02	−0.026	0.0018	0.003	0.0026	0.0100	0.0015	—
DeadZ	−0.035	0.141	0.105	0.0075	0.0066	0.0057	0.0650	0.0046	0.40
DeadZ	0.036	−0.07	−0.002	0.0099	0.0099	0.0072	0.1250	0.0046	0.556
ShearZ	0.497	0.51	0.028	0.095	0.079	0.09	11.6850	0.0412	0.20
ShearZ	0.812	0.124	0.07	0.097	0.078	0.06	4.8400	0.0413	0.20
ShearZ	0.511	0.112	0.123	0.081	0.052	0.043	5.5550	0.0412	0.464
ShearZ	1.036	−0.0007	−0.252	0.07	0.053	0.045	4.8650	0.0412	0.677
FreeZ	0.838	0.0468	0.0478	0.133	0.122	0.108	22.1200	0.0654	0.188
FreeZ	1.116	−0.325	−0.216	0.121	0.105	0.077	15.8000	0.0654	0.656

*Control is zero flow reading; FreeZ is main flow zone, etc.

within the adjacent mainflow (Table 7.1). As a traverse is made from the high velocity flow in mid-channel to the low velocity flow within the dead zone, a distance of a few tens of metres, the turbulence intensity decreases by three orders of magnitude, which is reflected in the scales of kinetic energy within each environment. Near-bed shear velocities are low within the dead zone and flow in the upstream direction is occasionally recorded.

The use of conventional flow meters to characterize the flow patterns within large-scale dead zones was found to be inappropriate across a wide spectrum of discharges. Velocity decreased across the shear zone to values below or equivalent to the threshold of detection (i.e. 0.03 m s⁻¹) at discharges of the order of 25 m³ s⁻¹ (Figure 7.7), although turbulence data reveal occasional values above this level. At the bankfull stage ($Q = 250$ m³ s⁻¹) flow within the 'dead zone' was extremely turbulent, exhibiting surface boils. Under these strongly three-dimensional flow conditions the accuracy of the current meter records is especially questionable. Nevertheless, these data (together with the turbulence data) were consistent and indicated that given a recirculating flow path within the dead zone the minimum transit time of any package of water through the dead zone area was typically between 400 and 1600 s during low flow. Dye tests during low flow are consistent with these data, two surveys giving estimated residence times of 860 and 1180 s, respectively. During high flow the bar upstream of the dead zone (Figures 7.1 and Plate 6) was submerged so that unsteady flow over the top of the bar increased the mixing rate. Close to bankfull, the minimum transit times estimated from the current meter survey ranged between 200 and 700 s. Two dye dispersion tests conducted during high discharge gave qualitative results, indicating the unsteady nature of the mixing process associated with bar-top inundation. The time delay for the dye to travel from an injection point in the centre of the dead zone to a point about 60 m downstream and outside the shear zone was 660 s, but the flux was unsteady and the centroid of the mass did not arrive until after 3300 s, giving an average velocity of about 2 cm s⁻¹ (Figure 7.8). In contrast, the time of travel of the centroid to a station in the mainflow 0.5 km downstream was only 4050 s (average velocity about 12 cm s⁻¹).

Shear velocities within the dead zone were calculated from the turbulence data and also estimated from the velocity profiles. Fine deposits were observed to blanket the bed of the dead zone when the shear velocity was 0.45 cm s⁻¹ and the discharge was 25 m² s⁻¹, whereas this fine blanket was absent within the shear zone ($u_* = 4$ cm s⁻¹) and within the mainflow ($u_* = 6$ cm s⁻¹). Inspection of the cross-channel distribution of values of u_* from other

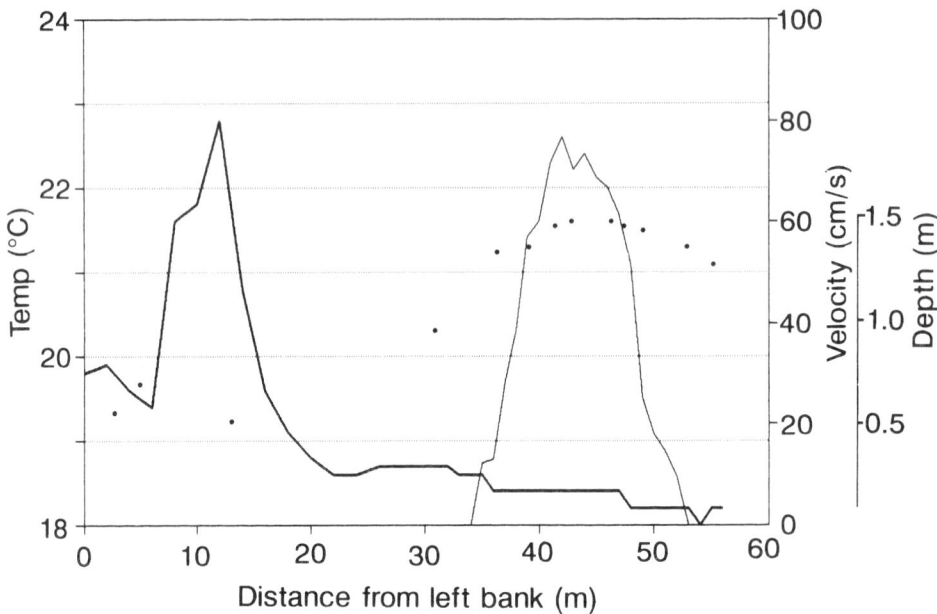

Figure 7.7 Surface water temperature (heavy line), depth-averaged velocity distribution; zero until 34 m (light line), and spot depth readings (●) across dead zone section B–B′ (see Plate 6b for location) on 14 August 1990. Macrophytes covered the bed as far as 37 m, but were dense between 9 and 17 m. Slight reverse flow was noted between the 0 and 9 m marks. Near-zero flow close to the right bank is dominated by submerged tree branches. Maximum depth about 1.5 m.

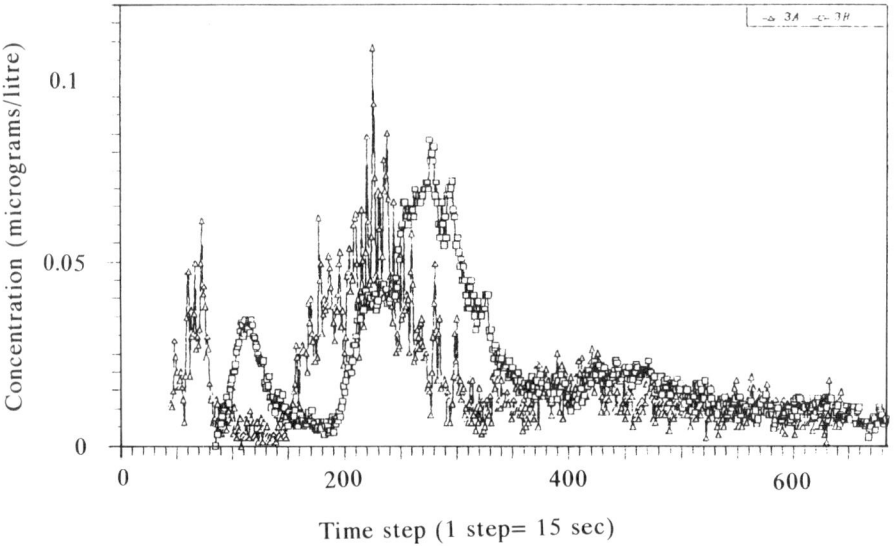

Figure 7.8 Dispersion of rhodamine WT tracer from centre of dead zone to fluorimetry station 3a 60 m downstream of shear zone (△) (point ⊕ Plate 6b) and station 3b (□) at a position 500 m downstream

sections indicated that the water was discoloured by resuspended sediment (3−300 μm) when u_* was typically 3−4 cm s^{-1}. It was concluded that the threshold for resuspension of the bulk of the fine deposits was about 2 cm s^{-1}. The fine particulates within this reach of the Severn typically range in size between 3 and 300 μm (averaging 7.5 μm) with an apparent density of 1.3−2.3 g cm^{-3} (averaging 1.8 g cm^{-3}), and settle in quiescent water in accord with Stokes' law (Carling *et al.*, 1992); consequently, the 2 cm s^{-1} threshold is consistent with resuspension of all particles less than about 200 μm; 300 μm particles are in suspension once the shear velocity exceeds 4 cm s^{-1}.

These results indicate that many marginal areas represent areas of net accumulation of fines until discharges exceed about 90 m^3 s^{-1}. The highest shear velocity recorded within the dead zone was 7 cm s^{-1} close to bankfull. These results are consistent with a published range of shear velocity data for this reach of the Severn (Carling, 1991), wherein it is shown that shear velocities in the mainflow, increasing from a minimum of about 2 cm s^{-1}, tend to become asymptopic to an cross-sectional average of 7−10 cm s^{-1}, with instantaneous peaks of 15−20 cm s^{-1}. Although deposition can occur only rarely within the main flow, dead zones remain potential repositories for fine sediments over a wide range of discharges. Modelling using these data for scaling have broadly reproduced the measured deposition and resuspension rates for given quiescent and flushing events within the dead zone driven by variations in discharge (Tipping, Woof and Clarke, 1993).

7.3.2 Reach-scale flow structure

7.3.2.1 Remote-sensed imagery

Surface water temperature profiles (Figure 7.7) showed that dead zones may be significantly warmer than the main flow during the summer period, reflecting the slow rate of exchange of fluid across enclosing shear zones. To extend this sectional interpretation to the reach scale, remotely sensed imagery provides a means whereby surface zones of flow exhibiting different thermal and hence different mixing character may be mapped.

The plan view in Plate 6a is a false colour thermal line scanner image of the large dead zone at Leighton during a summer low flow of 22 m^3 s^{-1}, whereas Plate 6b is a simplification drawn to highlight features of immediate interest. Slowly recirculating warm water occurs within the re-entrant delineated by a sharp thermal gradient towards cooler mid-channel water. Entrainment of cooler mainstream water into the dead zone is indicated by the indented isotherms within the re-entrant and signifies the presence of an anticlockwise gyre, whereas the ragged edge to the shear zone reflects the passage of vortex streets with interdigitating filaments of warm and cool waters.

Shown in section (Plate 6b) are three calibrated density slices and the surface water temperature measured at one section during the aerial survey. The ground truth is at variance with the calibration only in detail, although regional temperature gradients are correctly identified from the imagery. The infra-red detectors sense the temperature of only about the top 0.1 mm of the water, whereas the field thermistor was immersed to a depth of a few centimetres. Wind stress, ambient air temperature and water surface instability, coupled with the variable spatial and temporal rates of mixing within the water body, are responsible for the detail of the discrepancies.

By analogy with the velocity and mixing rate measurements taken within the dead zones and main flow during similar flow levels (25 m^3 s^{-1}), and from detailed visual observation of this reach, it is concluded that the $> 16.9°C$ contour (on this occasion) delineates waters (dead zones) with velocities much less than $3\,\mathrm{cm\,s^{-1}}$ and residence times of the order of hundreds of seconds. In contrast, the main flow is cooler, has velocities typically between 20 and $200\,\mathrm{cm\,s^{-1}}$, and residence times (for similar length scales) of the order of seconds.

Analysis of reaches with different macrophyte communities or channel morphologies should indicate those with a propensity for delayed advection associated with dead zone development. Figure 7.9a shows the cumulative temperature distributions for the main flow zone ($< 16.9°C$) of a number of channel reaches, whereas Figure 7.9b is a detail of adjacent dead zone environments. The lateral shift in the position of the 50th percentiles reflects differences in ambient flow temperature. For example, the surface of reaches characterized by deep pools are about $0.2°C$ cooler than reaches characterized by shallow riffle areas. This is a result of the slower current speeds within deep pools, which results in reduced vertical mixing (Simpson and Hunter, 1974; Weeks and Simpson, 1991) and enhanced evaporative cooling at the surface (Haigh and Pritchard, 1981); this contrasts with shallow fast-flowing riffle areas. Those areas $17°C$ or warmer account for between 6% and exceptionally 18% of the reach surface area (Figure 7.9). The meandering reach has the warmest waters owing to large expanses of shallowly inundated point bars and well developed flow separation zones on the inside of channel bends. Nevertheless, the straight reaches are characterized by a total dead zone area of $< 10\%$. This conforms with the velocity profile observations where a width up to 4 m was associated with delayed advection. The submerged bank batter in these straight deep reaches accounts for $3-4$ m of the river channel width between bank tops on each bank (e.g. see Figure 7.2 and Reynolds, Carling and Beven, 1991; their Figure 1). Assuming a flow depth of 1 m (Carling, 1991) for the low discharge at the time of the aerial survey, then simple geometrical calculation indicates that the flow volume over the bank areas is $< 10\%$ of the total sectional flow.

It may be concluded, therefore, that dead zone development at the reach scale is primarily associated with the region of the bank batter. The geometrical argument can also be used to indicate that as the water depth is increased, the relative area over the bank batter grows more rapidly than the area over the channel bed. For a reach which is well mixed with depth, the ratio of the effective width to wetted width is equivalent to the dispersive fraction. However, the dead zone width and hence the total volume over the bank batter at high discharge is not sufficient to account for the values of the dispersive fraction reported for these reaches of the Severn ($\sim 0.2-0.3$; Beven and Carling, 1992).

7.4 CONCLUSIONS

Near-bank dead zones play a significant part in fluvial mixing processes and the delayed advection of solutes and particulates, but cannot alone account for the low values of the dispersive fraction calculated for the Severn.

Preliminary hydraulic data can be used to demonstrate the distinctive character of dead zone dynamics. In particular, during low flows the kinetic energy can vary over three orders of magnitude along a traverse from the dead zone to the main flow. The dead zone areas are conservative; they maintain their relative volume and mixing rates as stage increases and remain important areas for the deposition of fines over a wide range of discharges. Closer attention to the detail of near-bank mixing processes should help to elucidate the nature of hydraulic interaction with the main flow. In particular, remote sensing provides an opportunity to derive reach-scale maps of water temperature variation which may be analogous to spatial variation in hydraulic structure and retentiveness.

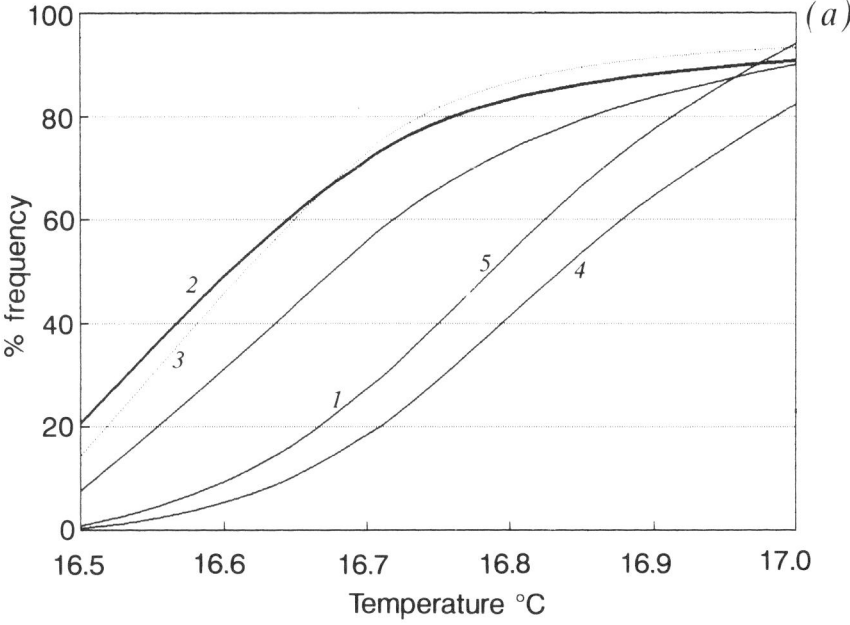

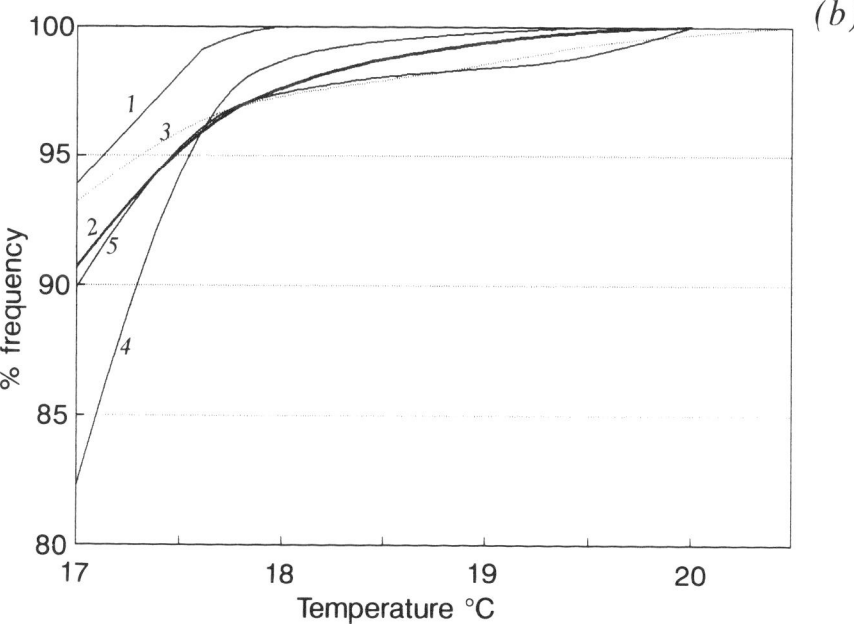

Figure 7.9 Cumulative surface water temperature distribution for selected reaches of the River Severn. (a) Main channel excluding dead zones < 17°C. (b) Dead zones > 17°C. 1 = Straight shallow reach, poor dead zone development; 2 = deep pool reach with marginal submerged macrophyte beds; 3 = straight reach with dead zones; 4 = shallow meandering section; and 5 = straight reach with large dead zones

Through this approach new perspectives on the physical interpretation of ADZ model output and conventional current metering surveys can be generated. Further investigation is planned to characterize the temperature structure and turbulent mixing character of dead zones as well as the exchange mechanisms and fluxes of suspended solids across shear zones. These measurements will be used to scale models of fine particulate dispersion, deposition and resuspension for application to river reaches characterized by delayed advection and dead zone development.

ACKNOWLEDGEMENTS

This investigation was funded by the DOE and NRA. Professor K.J. Beven, Hannah Green and Mr M.J. Holland are thanked for analyses and comment on turbulence and dispersion data. Drs C.M. Allen and S.E. Heslop provided unpublished turbulence data. Various staff and students of the Institute of Environmental and Biological Sciences and the IFE assisted in field work. Mr J. Marsh (formerly of the Plymouth Marine Laboratory) provided details of particulate tracers. Thermal line scanning data were obtained during the 1991 NERC Airborne Remote Sensing Campaign. The director and staff of the Preston Montford Field Studies Centre are thanked for continued hospitality over a number of years. Professor Beven and Dr P. Ashworth are thanked for commenting critically on earlier versions of the manuscript.

REFERENCES

Bathurst, J.C., Thorne, C.R. and Hey, R.D. (1979) Secondary flow and shear stress at river bends. *J. Hydr. Div. ASCE* **105**, 1277−1295.

Berndtsson, R. (1990) Transport and sedimentation of pollutants in a river reach: a chemical mass balance approach. *Wat. Resour. Res.* **22**, 1101−1108.

Beven, K.J. and Carling, P.A. (1992) Velocities, roughness and dispersion in a lowland River Severn. In: *Lowland Floodplain Rivers: Geomorphological Perspectives* (Eds P.A. Carling and G.E. Petts). Wiley, Chichester, 71−93.

Bhowmik, N.G. (1982) Shear stress distribution and secondary currents in straight open channels. In: *Gravel-bed Rivers* (Eds R.D. Hey, J.C. Bathurst and C.R. Thorne). *Wiley, Chichester, 31−55.*

Biggs, J.F. and Close, M.E. (1989) Periphyton biomass dynamics in gravel bed rivers: the relative effects of flows and nutrients. *Freshwater Biol.* **22**, 209−231.

Carling, P.A. (1991) An appraisal of the velocity reversal hypothesis for stable pool-riffle sequences in the River Severn, England. *Earth Surf. Process. Landforms* **16**, 19−31.

Carling, P.A., Tipping, E., Woof, C., Clarke, K., Orr, H. and Glaister, M. (1992) Fine Particles in Rivers, Related to Dead Zones. *Final Report NRA R&D Project 121*. National Rivers Authority, 64pp.

Cherkauer, D.S. (1973) Minimization of power expenditure in a riffle-pool alluvial channel. *Wat. Resourc. Res.* **9**, 1613−1628.

Cionco, R.M. (1983) On the coupling of canopy flow to ambient flow for a variety of vegetation types and densities. *Boundary-Layer Meteorol.* **26**, 325−335.

Copp, G.H. (1989) The habitat diversity and fish reproductive function of floodplain ecosystems. *Environ. Biol. Fishes* **26**, 1−27.

Fisher, S.G., Gray, L.J., Grimm, N.B. and Busch, D.E. (1982) Temporal succession in a desert stream ecosystem following flash flooding. *Ecol. Monogr.* **52**, 93−110.

Gaschignard, O., Persat, H. and Chessel, D. (1983) Répartition transversale des macroinvertébrés benthiques dans un bras du Rhone. *Hydrobiologia* **106**, 209−215.

Graham, A.A. (1990) Siltation of stone-surface periphyton in rivers by clay-sized particles from low concentrations in suspension. *Hydrobiologia* **199**, 107−115.

Haigh, G.A. and Pritchard, S.E. (1981) Quantitative analysis of aeriel infra-red data for heat loss surveys. *Harwell Rep. No. AERE-R 10390.*

Harvey, B.C. (1987) Susceptibility of young-of-the-year fishes to downstream displacement by flooding. *Trans. Am. Fish Soc.* **16**, 851−855.

Heggenes, J. (1988) Effects of short-term flow fluctuations on displacement of, and habitat use by, brown trout in a small stream. *Trans. Am. Fish Soc.* **117**, 336−344.

Heslop, S.E. and Allen, C.M. (1989) Turbulence and dispersion in larger UK rivers. In: *Proceedings of the 23rd Congress of the International Association of Hydraulic Research, Ottawa, Canada,* D75−D82.

Hey, R.D. (1979) Flow resistance in gravel-bed rivers. *J. Hydr. Div. ACSE* **105**, 365−379.

Hicks, F.E., Yin, Y.C. and Steffler, P.M. (1990) Flow near sloped bank in curved channel. *J. Hydr. Engin. ASCE* **116**, 55−70.

Higgins, R.J. (1990) Off-river storages as sources and sinks for environmental contaminants. *Regul. Riv.* **5**, 401−412.

Inman, D.L. (1963) Sediments, physical properties and mechanics of sedimentation. *Submarine Geology* (Ed. F.R. Shepard). Harper and Row, New York, 101−147.

Jackson, P.S. (1981) On the displacement height in the logarithmic velocity profile. *J. Fluid Mec.* **111**, 15−25.

Lane, E.W. (1937) Stable channels in erodible material. *Trans. ASCE* **63**, 123−142.

Nanson, G.C. and Page, K. (1983) Lateral accretion of fine-grained concave benches on meandering rivers. In: *Modern and Ancient Fluvial Systems* (Eds J.D. Collinson and J. Lewin). *Publ. Spc. Int. Assoc. Sedimentol. No. 6.*, Blackwell, Oxford, 13−143.

Parker, G. and Peterson, A.W. (1980) Bar resistance of gravel-bed streams. *J. Hydr. Div. ASCE* **106**, 1559−1573.

Raupach, M.R. and Thom, A.S. (1981) Turbulence in and above plant canopies. *Ann. Rev. Fluid Mech* **13**, 97−129.

Reynolds, C.S. (1988) Potomoplankton: paradigms, paradoxes and prognoses. In: *Algae and the Aquatic Environment* (Ed. F.E. Round). Biopress, Bristol, 285−311.

Reynolds, C.S., Carling, P.A. and Beven, K.J. (1991) Flow in river channels: new insights into hydraulic retention. *Arch. Hydrobiol.* **121**, 171−179.

Simpson, J.H. and Hunter, J.R. (1974) Fronts in the Irish Sea. *Nature* **250**, 404−406.

Soulsby, R.L. (1983) The bottom boundary layer of shelf seas. Chapter 5 In: *Physical Oceanography of Coastal and Shelf Seas* (Ed. B. Johns). Elsevier, Amsterdam, 189−266.

Tipping, E., Woof, C. and Clarke, K. (1993) Deposition and resuspension of fine particles in a riverine 'dead zone'. *Hydrol. Process.*, **7**, 263−277.

Vriend, H.J. de and Geldof, H.J. (1983) Main flow velocity in short river bends. *J. Hydr. Engin. ASCE* **109**, 991−1011.

von Kàrmàn, T. (1934) Turbulence and skin friction. *J. Aeronaut. Soc.* **1**, 1.

Wallis, S.G., Blackley, C. and Young, P.C. (1987) A micro-computer based fluorometric data logging and analysis system. *J. Inst. Wat. Engin. Scien.* **41**, 122−134.

Wallis, S.G., Guymer, I. and Bilgi, A. (1990) A practical engineering approach to modelling longitudinal dispersion. In: *Hydraulic and Environmental Modelling of Coastal, Estuarine and River Waters* (Eds R.A. Falconer, P. Goodwin and R.G.S. Matthew). Gower Technical, London, 291−300.

Wallis, S.G., Young, P.C. and Beven, K.J. (1989) Experimental investigation of the aggregated dead zone model for longitudinal solute transport in stream channels. *Proc. Inst. Civ. Engin.* **87**, 1−22.

Weeks, A. and Simpson, J.H. (1991) The measurement of suspended particulate concentrations from remotely-sensed data. *Int. J. Remote Sensing* **12**, 725−737.

Wilkinson, R.H. (1984) A method of evaluating statistical errors associated with logarithmic velocity profiles. *Geo-Mar. Lett.* **3**, 49−52.

8 Measurement of Transverse Mixing Using Digital Image Acquisition

J. HÖTTGES
Hydro-Ingenieure GmbH, Düsseldorf, Germany

U. ARNOLD
IKW GmbH, Berlin, Germany

and

G. ROUVÉ
Institute for Hydraulic Engineering and Water Resources, RWTH Aachen, Germany

8.1 INTRODUCTION

The mixing process in compound open channel flow is dominated by differential advection and turbulent mixing, in which the latter is the main influencing factor on transverse dispersion. Arnold, Höttges and Rouvé (1988) have shown that in the near-bank zone between the main channel and floodplain region large vortices are built, which exchange mass and momentum between these two regions.

Large numbers of flow situations were investigated to quantify the influence of channel geometry and roughness distribution on the mixing process of a non-buoyant solute. Digital image acquisition provides a powerful tool to measure the concentration of the tracer plume with a high measurement point density and a high rate of data aquisition, thereby allowing extended analysis of the measurement data, e.g. the evaluation of concentration gradients.

8.2 EXPERIMENTAL APPARATUS

The experiments were conducted in a 25.5 m long and 1 m wide straight channel with a compound cross-section (see Figure 8.1). The flow parameters varied and included channel slope and water depth, the ratio of the main channel and the floodplain width, and the main channel and floodplain roughness. The main channel roughness was simulated using a synthetic carpet, and wooden sticks of varying density produced the effect of flood plain vegetation.

A non-buoyant tracer (Uranine) was injected continuously and depth-averaged at 11 different lateral positions and seven different distances to the measurement cross-section (see Figure 8.2). The injection apparatus consists of a small inner tube, adjusted to the water surface, and an outer wing with small holes at its downstream edge, which distributes the tracer over the water depth and is moved down to the channel bed at each injection position. The injection flow-rate is adapted to the water velocity, which is measured with a hydrometric vane and is controlled by a gear pump varying in a range from 5 to 50 l/h.

Mixing and Transport in the Environment. Edited by K.J. Beven, P.C. Chatwin and J.H. Millbank
©1994 John Wiley & Sons Ltd

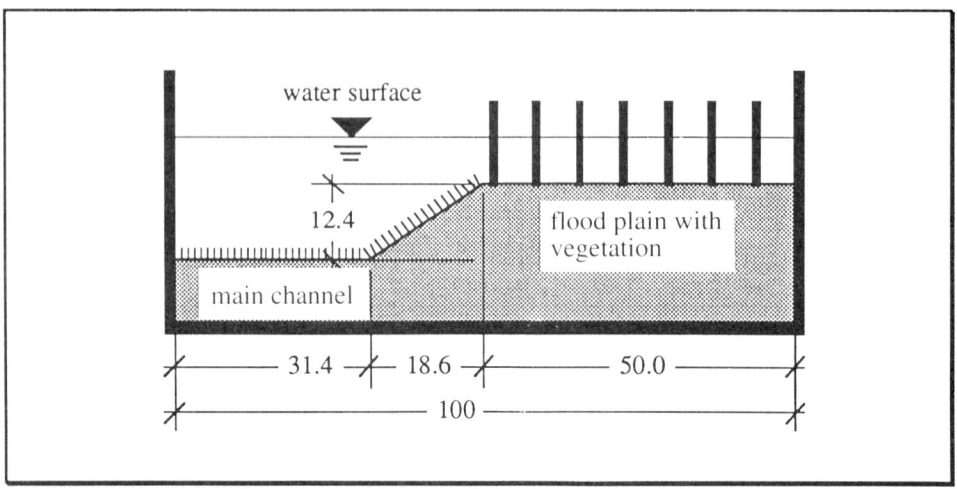

Figure 8.1 Cross-section of straight channel (all measurements in cm)

8.3 MEASUREMENT TECHNIQUES

The concentration in a cross-section of the channel is measured using a CCD camera and UV light to activate the tracer Uranine ($C_{20}H_{10}Na_2O_5$; $M = 376.28$ g/mol). A pair of black-light tubes is installed at a distance from each other and from the water surface that provides an optimum balance between reflection near the surface and light absorption and diffusion in the deeper layers. The camera is mounted in the top of a black tent to keep out daylight. The video signal passes through an adjustable amplifier and is digitized using an 8-bit frame grabber run on a PC. A program (C language) written by the authors controls the frame grabber using a subroutine library which is loaded with the image-processing hardware.

The light intensity digital values range between 0 and 255 and are displayed in real time on an additional monitor. A light intensity cross-sectional profile is taken by averaging over 10 lines with 128 points and 400–1000 pictures recorded at 10 Hz. Each measurement is preceded by a reference measurement in order to eliminate the influence of the increasing background concentration caused by the low volume of the water cycle (only 30 m^3).

The concentration at each measurement point is computed by interpolation based on the calibration measurements. The light intensity depends linearly on the concentration, provided that a minimum concentration at the beginning and a maximum concentration in the plume core are observed. To gain an interpolation function for every measurement point, a calibration measurement is taken at the beginning and end of a whole measurement series, with concentration values computed from the used tracer and the volume of the water cycle.

This method has been checked with a fluorimeter. Thus external influences are reduced, e.g. the slowly decreasing intensity of the UV light tubes. The measurement method requires a lot of practical experience due to numerous disturbing influences (e.g. external light, the reflection of the water surface, the depth-averaging of light intensity and tracer recirculation).

For geometrical transformation of the data to their true position, small light stripes are fixed at each of the side walls of the measurement region. Figure 8.2 (bottom right) gives an example of a tracer plume from an injection in the main channel near the bank. It contains seven concentration profiles, taken from 1.0 to 4.0 m.

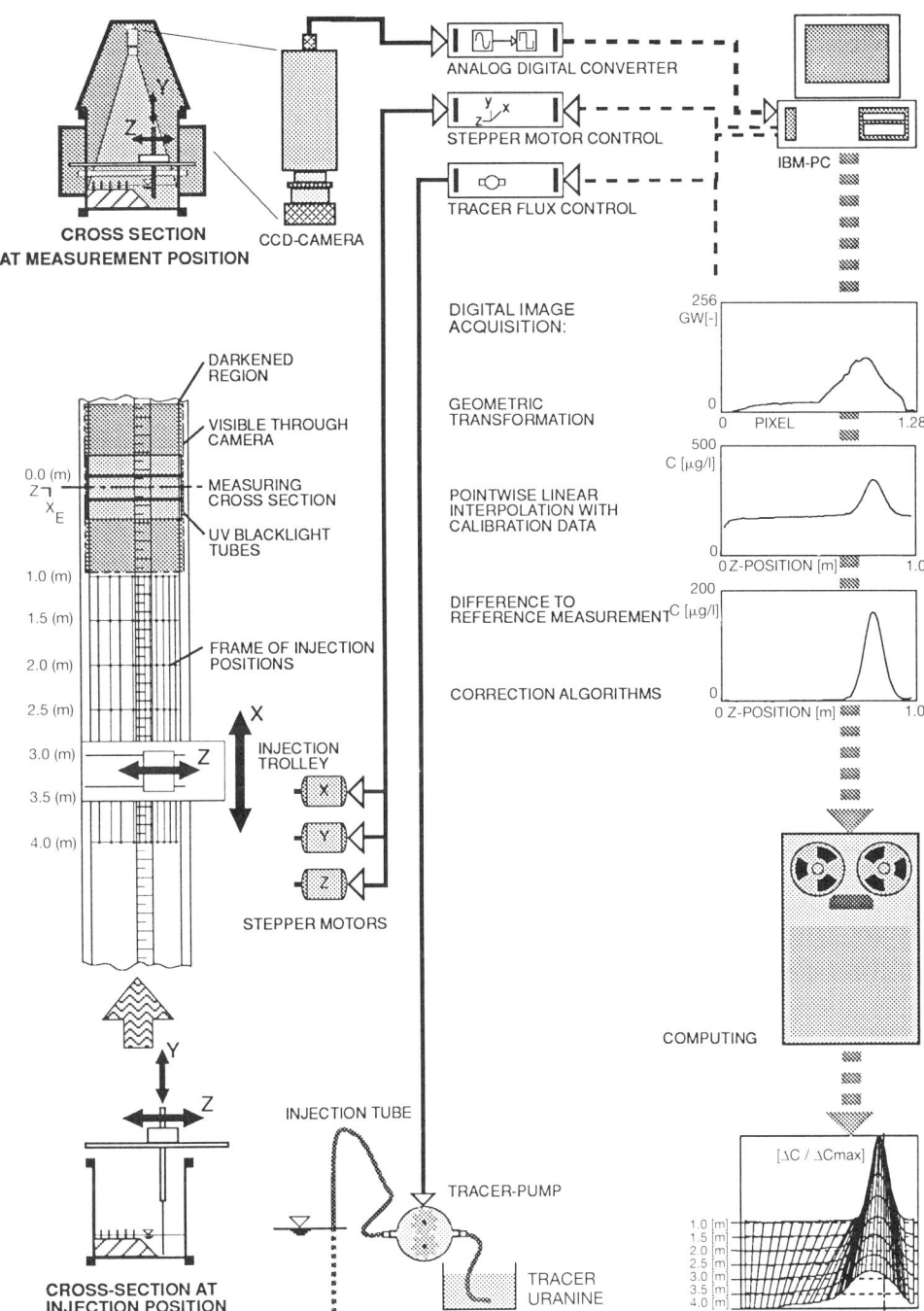

Figure 8.2 Experimental setup and concentration analysis procedure

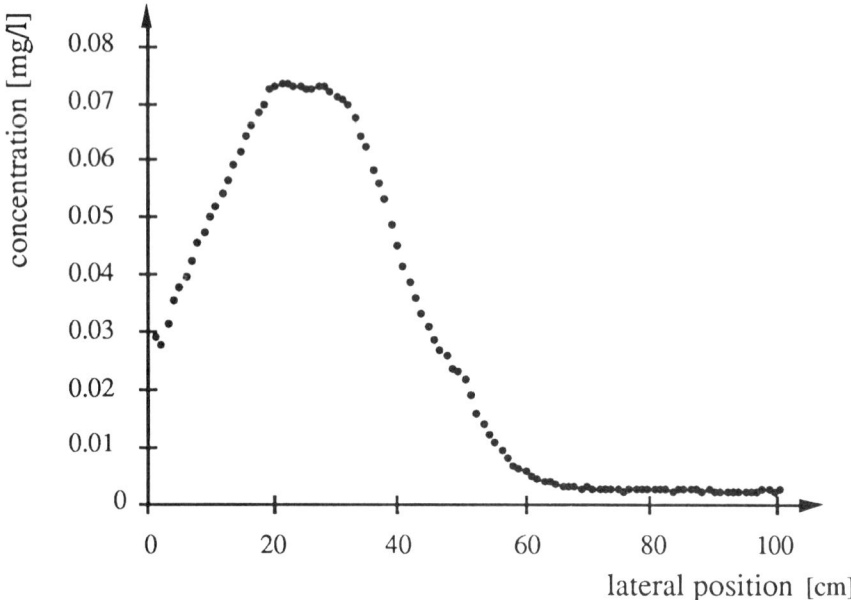

Figure 8.3 Rough concentration data

Tracer injection is automatically controlled by the PC and a stepper motor driven trolley. It takes place at seven different locations to give a picture of the whole tracer plume and at 11 different transverse positions to investigate the mixing conditions at different transverse sections of the channel. The corresponding velocity measurements yield the flow field necessary to calculate the tracer flux.

For very turbulent flow, water surface waves result in decreased quality of the data. In this instance it is useful to damp the waves with a 10 cm transparency, which is held at its front edge 5 cm above the water surface and touches the water surface from above. The entire measurement procedure is controlled by an IBM PC, including the control of the stepper motors and the tracer flux.

8.4   ANALYSIS

Data analysis consists of data verification and improvement, the derivation of parameters influencing the mixing process and data presentation. The rough concentration data is mostly of good quality, as can be seen in a typical example in Figure 8.3. Nearly all the post-processing steps, which have been developed for older digital image acquisition hardware, could be removed. Owing to the high sensitivity of the calculation of concentration gradients to data quality, data correction methods (especially smoothing and boundary data checking) are still necessary. In addition, a tracer flux integral is calculated, which can be used to check the tracer pump and the differences in light intensity received from the tracer cloud in different water depths.

For the calculation of the mixing coefficient two methods have been used. The generalized change of moment method (Holley and Abraham, 1973) has been modified (Arnold, 1987;

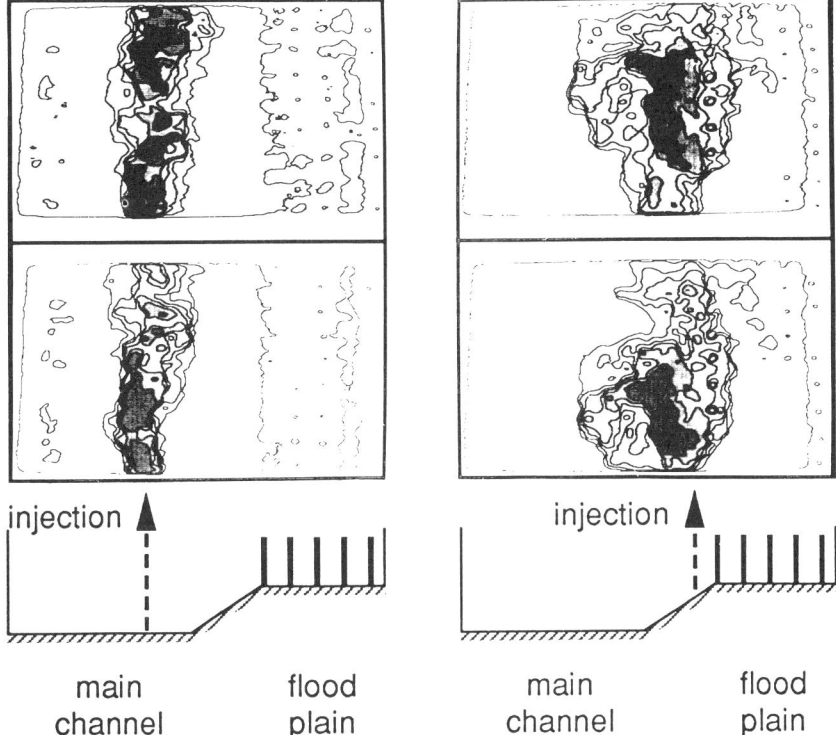

Figure 8.4 Tracer plume from a tracer injection in the near-bank zone and in the main channel; time difference between both pictures $\Delta t = 0.3\,\mathrm{s}$

Höttges, Arnold and Rouvé, 1990) to deal with a shifting maximum position. As a result of the inhomogeneity of the mixing process throughout the cross-section, a predefined distribution function for the mixing coefficient is used, which is based on a subdivision of the compound cross-section into different subregions, the assumption of an apparent wall between the floodplain and main channel (Pasche, Rouvé and Evers, 1985), and different turbulence assumptions (Arnold, 1987).

The other method is based on the original advection diffusion equation

$$hu\,\frac{\partial c}{\partial x} = \frac{\partial}{\partial z}\left(he_z\frac{\partial c}{\partial z}\right) \tag{8.1}$$

which is transformed to give the transverse mixing coefficient as

$$e_z(z) = \frac{\displaystyle\int_{z_0}^{z} hu\frac{\partial c}{\partial x}\mathrm{d}\eta}{h(z)\dfrac{\partial c}{\partial x}(z) - h(z_0)\dfrac{\partial c}{\partial z}(z_0)} \tag{8.2}$$

with h = water depth, u = longitudinal depth-averaged velocity, c = concentration, x = longitudinal coordinate, z, z_0 = transverse coordinate and e_z = transverse mixing coefficient. The formula may be understood as a comparison of the derivation of the tracer flux and the concentration gradient. The problems with this formula are caused by the gradients. They have to be smoothed in a manner sufficient to remove noise, but small enough to keep the original gradient values. Two methods are possible:

(1) A two-dimensional polynomial (x-direction, second degree; z-direction, third degree) regression can be applied to replace each value by a smoothed value. The method is very time consuming because a new polynomial has to be calculated for every measurement point. The method provides transverse and longitudinal gradients, which can easily be calculated from the polynomials.
(2) A one-dimensional Fourier transform for every measurement cross-section yields a frequency spectrum, which can be smoothed by simply deleting the higher frequencies. The smoothed concentration profile is obtained by inverse Fourier transform. In addition, the transverse gradients can also be easily calculated from the transform.

Additional attempts were made to calibrate the mixing coefficient by optimization based on a finite element program, which solves Equation 8.1 numerically. The average difference between measurements and the computed results was reduced to about $6-10\%$ of the maximum concentration value.

8.5 RESULTS

Many flow variations have been investigated, providing a large amount of concentration and velocity measurement data. The data are available in ASCII format on PC disks.

A preliminary analysis has shown that the evaluation of the mixing coefficient has strong scattering. Thus it is not possible to calculate an exact value. Additionally, the Fickian assumption of a gradient-driven tracer flux does not hold in the near-bank zone for a vegetated floodplain, where large vortices are the main factor responsible for tracer transfer between the main channel and floodplain region. Figure 8.4 shows large vortices in this region compared with a uniform tracer plume in the main channel.

It can be seen that the maximum of the concentration profiles moves, in many instances, to the floodplain, i.e. to the lateral position of lower mixing. This is not caused by a real one-way transfer of tracer, but by an effect that can be easily explained with a pile of sand: if someone digs the sand away from one side, the top moves slowly to the other side, although no grain moves in that direction.

ACKNOWLEDGEMENTS

The research activities described in this paper are sponsored by the German Research Foundation under grant no. Ro 365/30-5.

REFERENCES

Arnold, U. (1987) Zur bilddaten- und modellgestützten Bestimmung der Schadstoffausbreitung in naturnahen Fließgewässern. *Rep. Inst. Wasserbau RWTH Aachen,* Vol. 67 [in German].
Arnold, U., Höttges, J. and Rouvé, G. (1988) Combined digital image and finite element analysis of mixing in compound open channel flow. In: *Proceedings of the Third International Symposium on Refined Flow Modelling and Turbulence Measurements,* 26−28 July 1988, IAHR, Tokyo, Japan (eds. Y. Iwasa, N. Tamai and A. Wade), 569−576.

Höttges, J., Arnold, U. and Rouvé, G. (1990) Profiles of mixing coefficient in compound open channel. In: *Fourth International Symposium on Refined Flow Modelling and Turbulence Measurements*, 20−23 September 1990, IAHR, Wuhan, China, Hemisphere Publishing Corporation, New York (eds. Z. Liang, C.J. Chen and S. Cai).

Holley, E.R. and Abraham, G. (1973) Laboratory studies on transverse mixing in rivers. *J. Hydr. Res.* **11**, 219−253.

Pasche, E. Rouvé, G. and Evers, P. (1985) Flow in compound channels with extreme flood-plain roughness. In: *Proceedings of the 21st IAHR Congress, Melbourne, Australia*, 19−23 August 1985, Vol. 3, 383−389.

Part II

TRANSPORT AND DISPERSION IN TIDAL SYSTEMS

9 Tidal Mixing in the Gulf of California

J.H. SIMPSON and A. J. SOUZA

University of Wales Bangor, School of Ocean Sciences, Gwynedd, UK

and

M.F. LAVIN

CICESE, Departamento de Oceanografia Fisica, Ensenada, Baja California, Mexico

9.1 INTRODUCTION

The Gulf of California is a 1500 km long narrow inlet on the west coast of Mexico. It exhibits a strong M_2 tidal response (Hendershott and Speranza, 1971) with vigorous currents, consequent frictional effects and a large dissipation of energy. The impact of this input of tidal energy (~ 5 GW) in stirring the water column and thus controlling its structure have been demonstrated for the relatively shallow regions of the northern gulf (Argote, Amador and Morales, 1985; Durazo, 1989) and the analogy with similar systems such as the Irish Sea have been clearly established.

What makes the Gulf of California particularly, and perhaps uniquely, interesting is the fact that its topographic configuration facilitates the operation of tidal stirring on the deep stratification of the ocean. The deep waters of the Pacific have free access into the southern gulf down to depths in excess of 2000 m, with Pacific intermediate and deep waters occupying most of the volume below 500 m south of the Guaymas Basin (Bray, 1988b). To the north of this basin the bottom depth decreases rapidly, with pronounced sills in the region of the midriff islands (Figure 9.1) separating the deep southern basin from shallow regions to the north. This shoaling, combined with the proximity to the M_2 amphidrome, is responsible for vigorous tidal flows, which are further enhanced by the presence of the islands which squeeze the tidal flow by severely reducing the channel cross-section. Direct observations of currents in the vicinity of the islands are few, but results from shipborne acoustic Doppler current profile and current meter moorings (Badan-Dangon, Hendershott and Lavin, 1991) near the sills indicate flows in excess of $1.5\,\mathrm{m\,s^{-1}}$ at spring tides.

The impact of such strong tidal flows on the structure of the water column and the exchange flow over the sills has been little investigated, although efforts have been made to observe the operation of hydraulic control and the possible generation of hydraulic jumps and lee wave phenomena in the stratified flow over the sills (Badan-Dangon, 1989). There is good evidence from synthetic aperture radar imagery that propagating packets of internal waves do have their origin in the flow over the sills (Fu and Holt, 1984), and it has been proposed (e.g. Paden, 1990) that the mixing associated with these internal wave phenomena may operate at a much higher level of efficiency than stirring by bottom stresses, especially in the high amplitude region near the source.

Mixing and Transport in the Environment. Edited by K.J. Beven, P.C. Chatwin and J.H. Millbank
© 1994 John Wiley & Sons Ltd

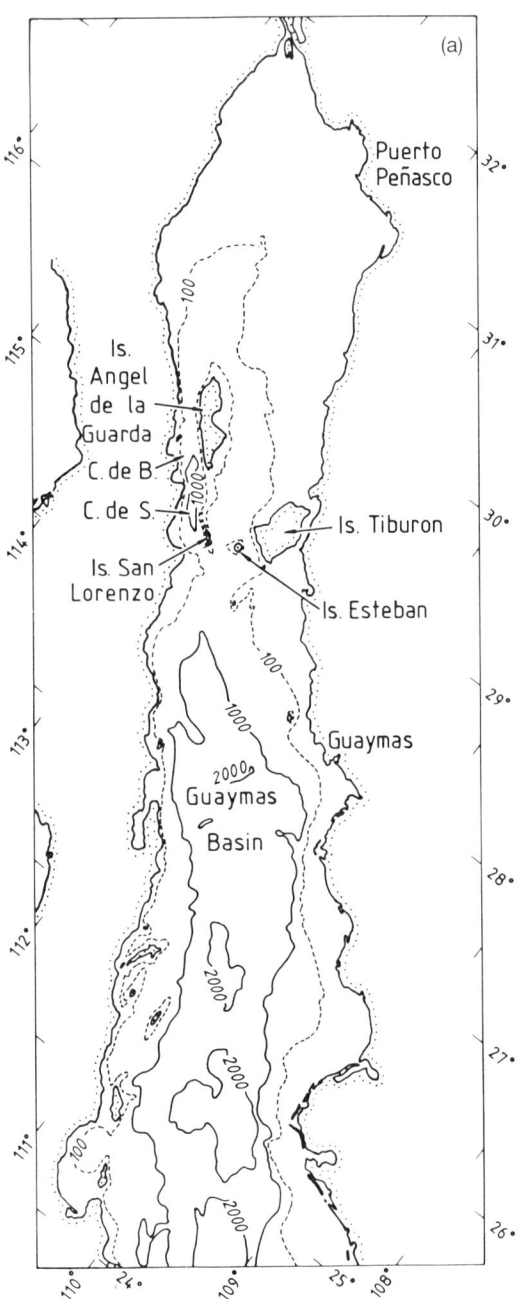

Figure 9.1 Study area in the Gulf of California. (a) Bathymetry in the region of the northern and cental Gulf of California

In an innovative study using EOF (Empirical Orthogonal Functions) techniques, Paden (1990) has analysed AVHRR (Advanced Very High Resolution Radiometer) thermal infra-red data to demonstrate the influence of tidal stirring on SST (Sea Surface Temperature) distributions. This approach indicates the presence of a marked fortnightly modulation of surface temperature gradients, which is most pronounced in the vicinity of the midriff islands, where there is a significant lowering of surface temperatures with an associated frontal region.

In this present chapter, we report efforts to diagnose the influence of tidal mixing by observing this spring−neap tidal modulation in the structure of the whole water column. The nature and intensity of the tidal mixing processes are not only of considerable physical interest

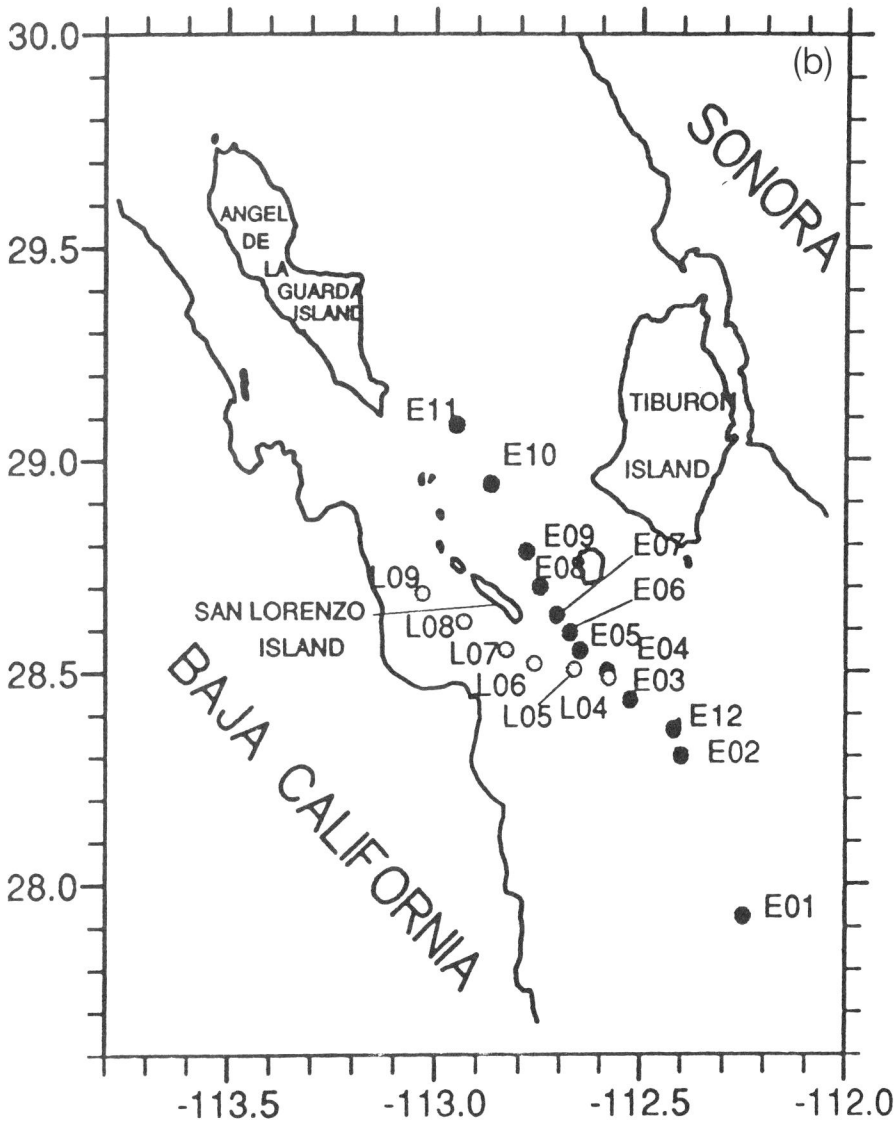

Figure 9.1 (*continued*) (b) Station positions showing line E (E1−E11) and line L (E1−L9)

in relation to the forcing of the circulation in the gulf, but they may also be the key to understanding the intensity and distribution of primary production, the unusually high levels of which are arguably supported by a supply of nutrients from deep waters.

9.2 SURVEYS

The observations were made from the USNS *De Steiguer* during the period 17−26 July 1990 as part of a collaborative programme to examine a range of physical and biological processes in the Gulf of California. The observational strategy was designed to determine the impact of tidal mixing directly, by contrasting surveys programmed to exploit the large spring−neap ratio in tidal range (Figure 9.2). As dissipation increases with u^3 (u = amplitude of the tidal velocity), the change in range means that tidal stirring is effectively switched off for a period around the neap tides. It is clear from the results of Paden and previous studies of the neap and spring cycle (Simpson and Bowers, 1981) that maximum and minimum stratification will occur, not at the neap and spring tides, but with a time lag of two to three days.

Two near-identical surveys of the key sections in the area of the midriff islands were therefore conducted over periods of 2.5 days at times as close to a three day lag from neap and spring tides as operational constraints allowed (observational periods shown in Figure 9.2). Both sections started in the deep Guaymas Basin; one section continued across the San Esteban sill (the line E in Figure 1b) and the second over the San Lorenzo sill into the Salsipuedes Channel (line L).

Measurements of the temperature and salinity profiles were made twice at each of these 18 stations using a NBIS mark III CTD (Conductivity Temperature and Depth profile). A rosette bottle sampling system attached to the CTD facilitated frequent calibration of the sensors and provided water samples for chemical analysis and estimates of plant pigment concentrations. Calibrations were consistent throughout the survey so we have confidence that all the presented data are accurate to $\pm 0.01\,°C$ and ± 0.01 psu (practical salinity units) in temperature and salinity, respectively. Satisfactory pairs of profiles were obtained for all stations except for the first profile at E1, which was lost due to initial problems with the CTD logging system. A flow-through system was also available for the continuous measurement of temperature, salinity and fluorescence of chlorophyll.

9.3 RESULTS

The temperature and salinity data for line L from the two surveys is presented in Figure 9.3 in the form of quasi-synoptic plots. Results for the sections L and E are closely similar except that line E does not have the deep water basin of the Salsipuedes Channel. The general picture is of a highly stable water column in deep water to the south-east, with indications of reduced stratification and lower surface temperatures in the sill region, which continues north-west of the sill into the deep Salsipuedes Channel. This reduction in stratification is noticeably more marked in the post-spring tide survey, with an increased spreading of the isotherms so that, for example, the total range of temperature over the centre of the sill decrease from 18 to 13°C after the post-spring tide survey.

To highlight the changes, we have subtracted the two temperature fields to yield the difference plot of Figure 9.4. This shows a substantial reduction in the temperature of a large volume of near-surface waters, particularly over and to the south-east of the sill. There is

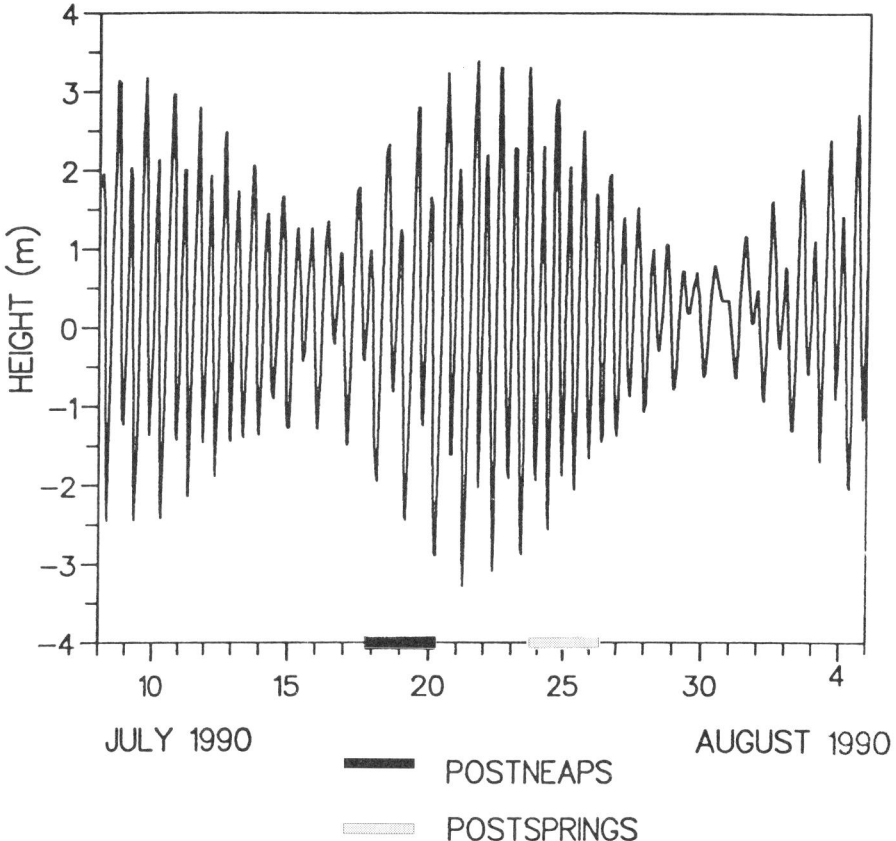

Figure 9.2 Tides at Peurto Peñasco (in the northern Gulf of California) during the observation period showing the times of the post-neap and post-spring tide surveys

a corresponding warming (by $\sim 1\,°C$) of the lower part of the water column over the sill and down to 500 m. On both sections E and L there is a large reduction in temperature variance (σ_T^2)

$$\sigma_T^2 = \overline{(T - \overline{T})^2}; \qquad \overline{T} = \frac{1}{h} \int_{-h}^{0} T\, dz \qquad (9.1)$$

(where T = temperature, h = water depth, and z = vertical coordinate, positive upwards from the sea surface) at stations over and to the south-east of the sill, with the RMS temperature decreasing by as much as $1.4\,°C$ after the spring tide from a post-neap tide value of $\sim 6.6\,°C$ (Figure 9.5a). To the north-west of the sill there is little change in the variance of temperature.

Perhaps the most striking contrast between the surveys is in the surface temperatures, which were greatly modified on both sections. It is evident from Figure 9.4 that, on line L, the SST was reduced by up to $4\,°C$. Similarly, on line E (Figure 9.6a), surface records show the post-spring tide temperature in the vicinity of E4 plunging to $25\,°C$. At the same time it appears that the high surface gradient region south-east of the sill (a front) has intensified and moved further away from the sill. Parallel measurements of chlorophyll fluorescence

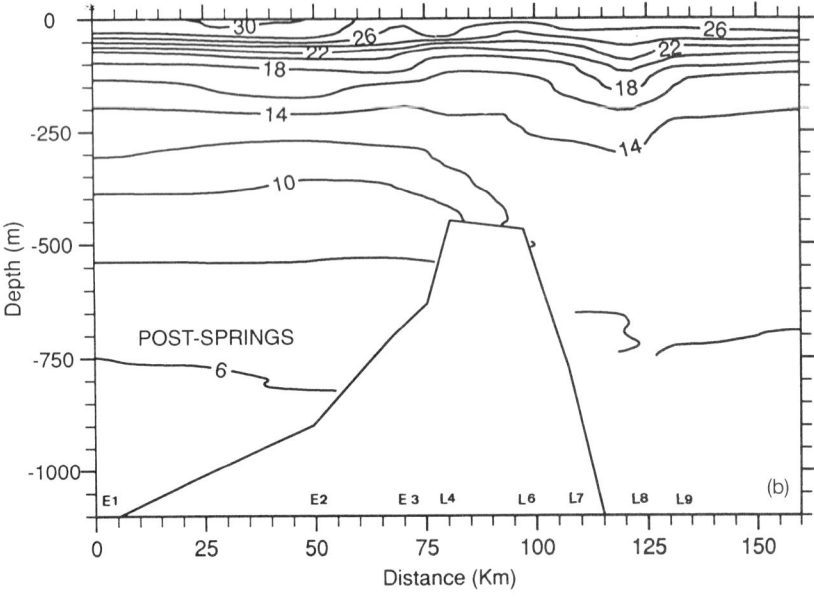

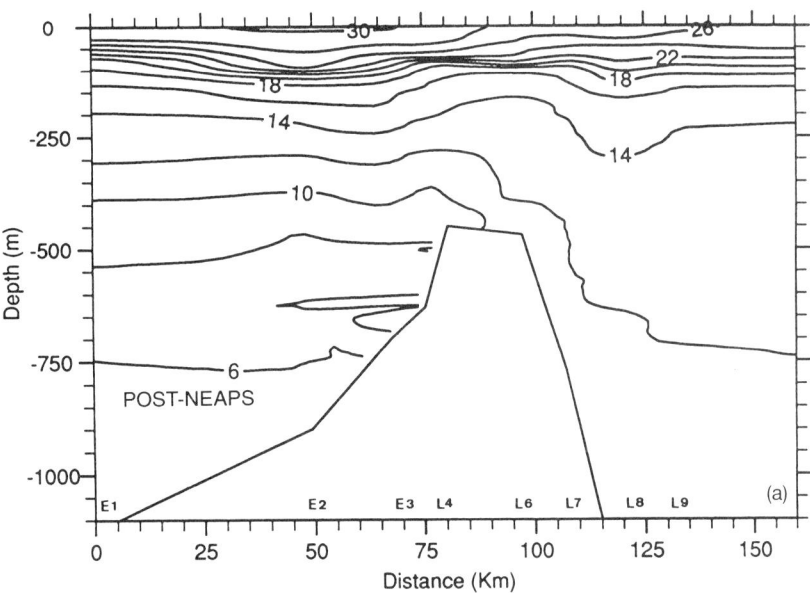

Figure 9.3 Post-neap and post-spring tide sections E1–L9. (a) Temperature variation post-neaps; (b) temperature variation post-springs

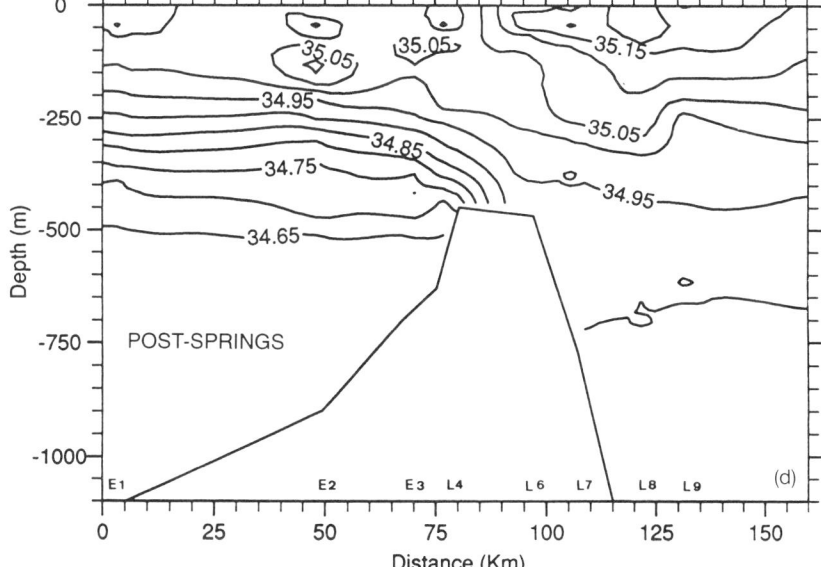

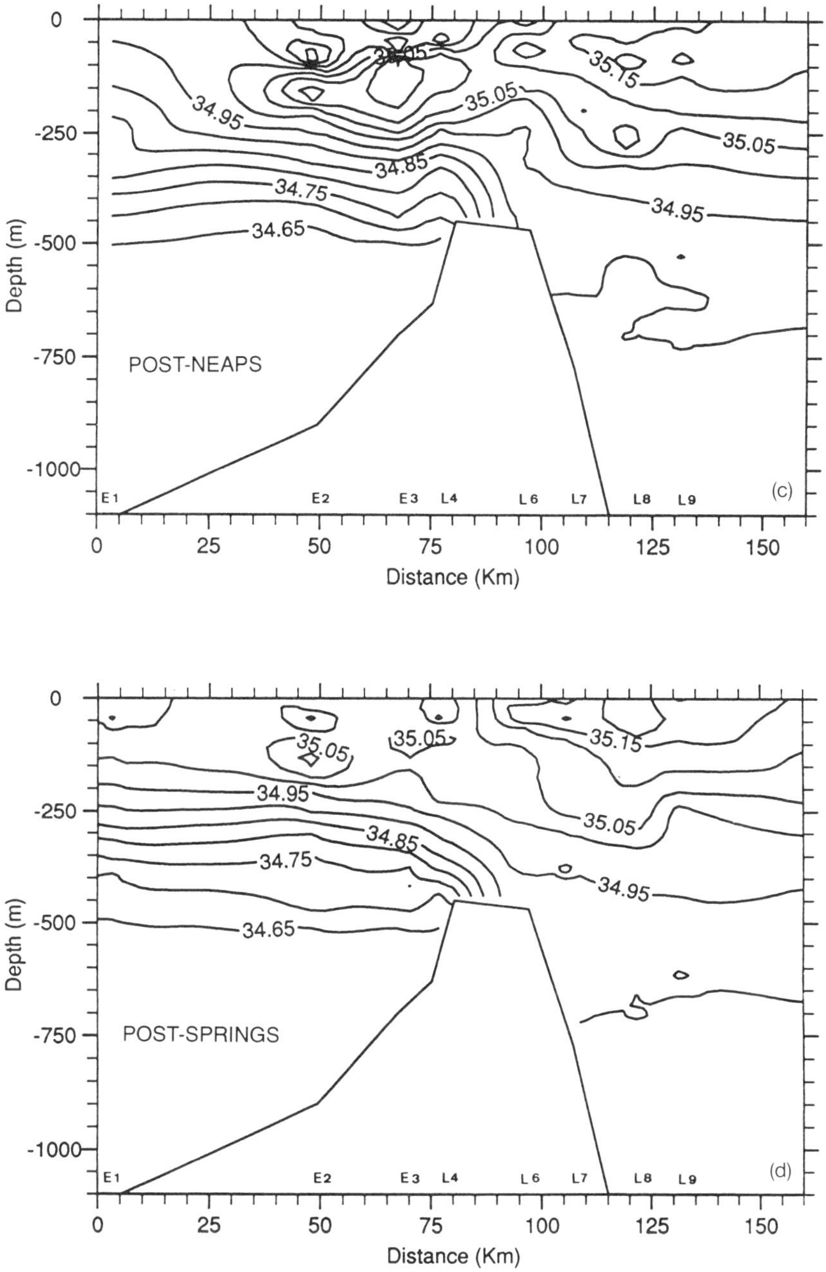

Figure 9.3 (*continued*) (c) salinity variation post-neaps; (d) temperature variation post-springs. Distance along both sections is measured north-westward from a datum at E1

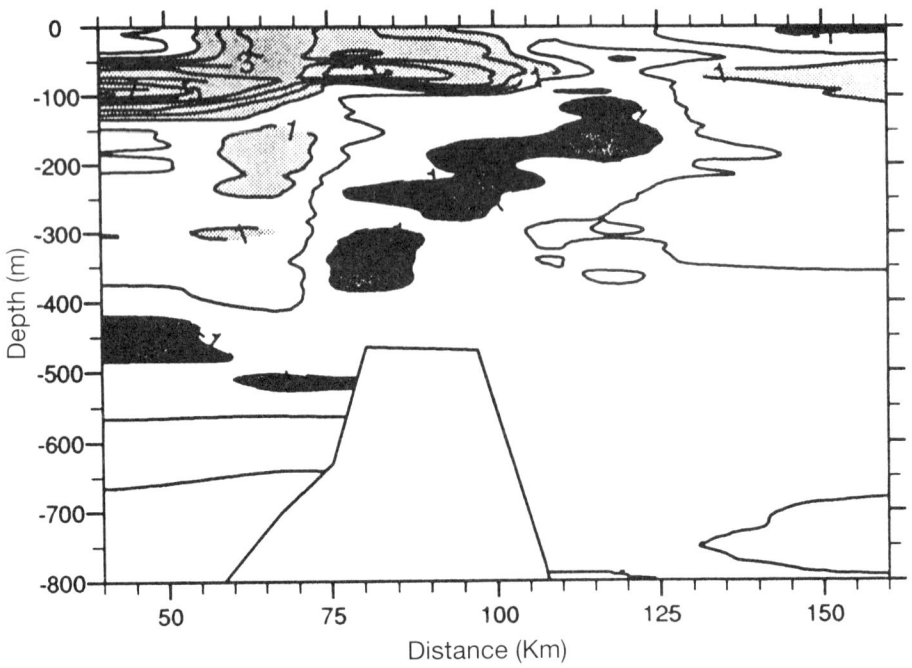

Figure 9.4 Plots of the temperature difference between post-spring and post-neap tide surveys. Black areas indicate warming of more than 1°C; light shading corresponds to cooling of more than 1°C. Distance is measured north-westward from E1

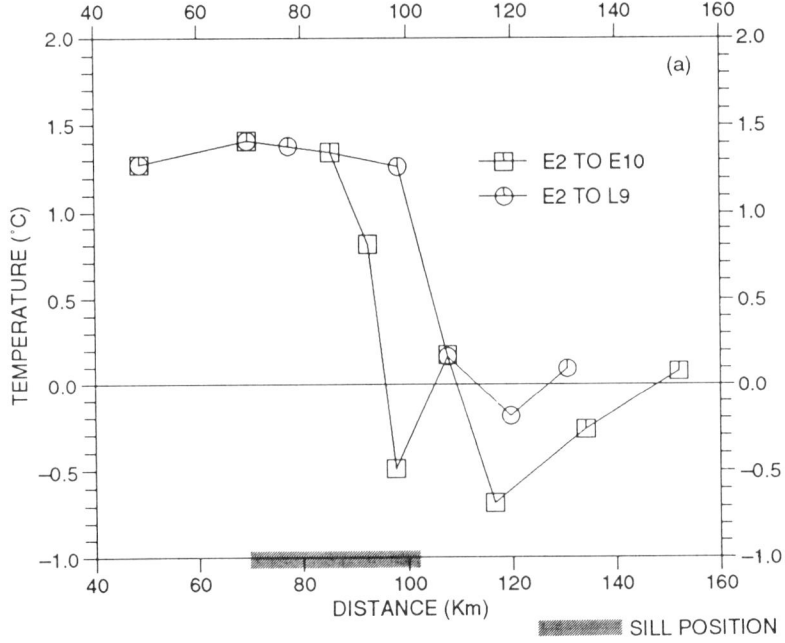

Figure 9.5 Summary of changes on the two sections E and L. (a) Station difference of temperature standard deviation

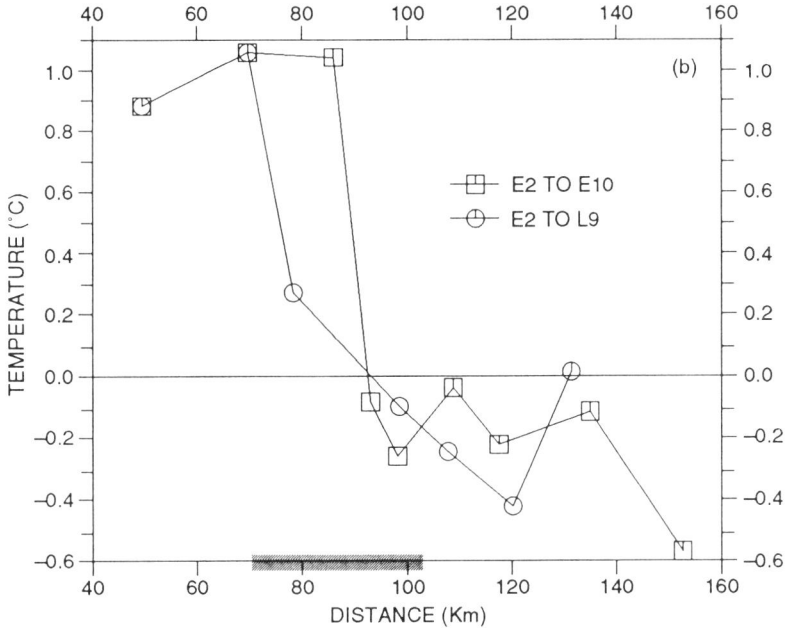

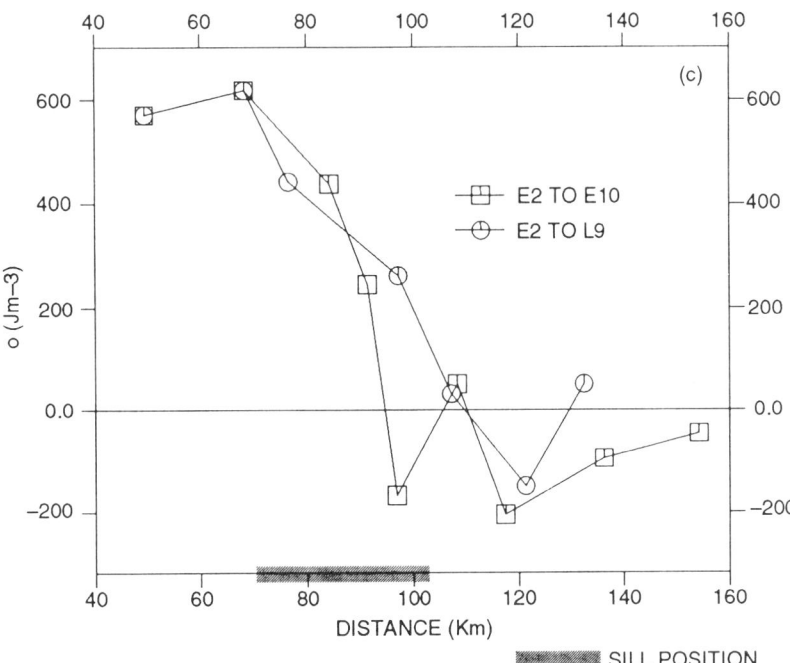

Figure 9.5 (*continued*) (b) Station difference of depth-mean temperatures. (c) Potential energy anomaly difference ϕ. In all cases, the difference is the neaps value minus the springs value

(Figure 9.6b) show a pronounced increase in the post-spring tide survey. The enhancement of fluorescence is by as much as a factor of three over an area coinciding with the region of strongest temperature reduction.

Pronounced changes are also apparent in the salinity fields. The strong salinity maximum extending south-westwards from the northern gulf at about 150 m in the post-neap tide sections (Figure 9.3c) is much lower in intensity after the spring tides (Figure 9.3d), especially in the vicinity of the sill. As with temperature, there is a marked reduction in the variance of salinity after spring tides at stations over and to the south-east of the sill. For example, at station E3 the standard deviation of salinity decreased from 0.205 to 0.160 psu between the two surveys, representing a 20% reduction in variability.

9.4   INTERPRETATION

The contrast in the surveys described here is strongly suggestive of the operation of vertical mixing brought about by intensified tidal stirring during the spring tide period. The reduced variability of water properties over the sills and to the south-east, with warming of the deep layers and cooling near the surface, point to the impact of large vertical fluxes driven by stirring. The magnitude of these fluxes is greatly in excess of the surface exchange, as is clear from the strong surface temperature reduction. This reinforcement of the SST minimum in the vicinity of the islands at a time of steady surface heating seems to be particularly clear and unambiguous evidence for the operation of enhanced vertical mixing. The area of reduced SST is consistent with that observed in infra-red data (Paden, 1990), and correlates closely with the observed strong increase in chlorophyll fluorescence. The introduction of nutrients to the surface layers by vertical mixing may be expected to stimulate production in the vicinity of the temperature minimum. Surface chlorophyll levels may also increase as a result of the upward mixing of plankton previously confined to lower layers in the photic zone.

Interpreting all the changes as being due to the effects of vertical mixing, we could use the changes in density [calculated by the UNESCO (1981) equation] distribution to calculate the change in potential energy anomaly (ϕ)

$$\phi = \frac{1}{h} \int_{-h}^{0} (\bar{\rho} - \rho) \, gz \, dz; \quad \bar{\rho} = \frac{1}{h} \int_{-h}^{0} \rho \, dz \qquad (9.2)$$

(where ρ = water density, and g = gravitational acceleration) due to the episode of spring tide mixing. A large decrease in ϕ of up to $500 \, \mathrm{J \, m^{-3}}$, apparent on both sections (Figure 9.5c) over and to the south-east of the sills, is a quantitative indication of the amount of work performed by stirring processes over the interval between the surveys. This large decrease in stability contrasts with indications of a slight increase in stratification on the north-western side of the sills in both basins.

If we ascribe these changes in the density field to mixing, then the apparent efficiency at stations in the region of greatest change of ϕ can be estimated from

$$\epsilon = \frac{h\Delta\phi}{k\rho\overline{u^3}\Delta t} \qquad (9.3)$$

where $k = 0.0025$ is the bottom drag coefficient, u is the tidal stream speed and $\Delta\phi$ is the change in ϕ over a time interval Δt (Simpson, 1981). For $u \simeq 1 \, \mathrm{m/s}$, we have $\epsilon \sim 0.1$, an estimate which is at least an order of magnitude greater than the values usually associated

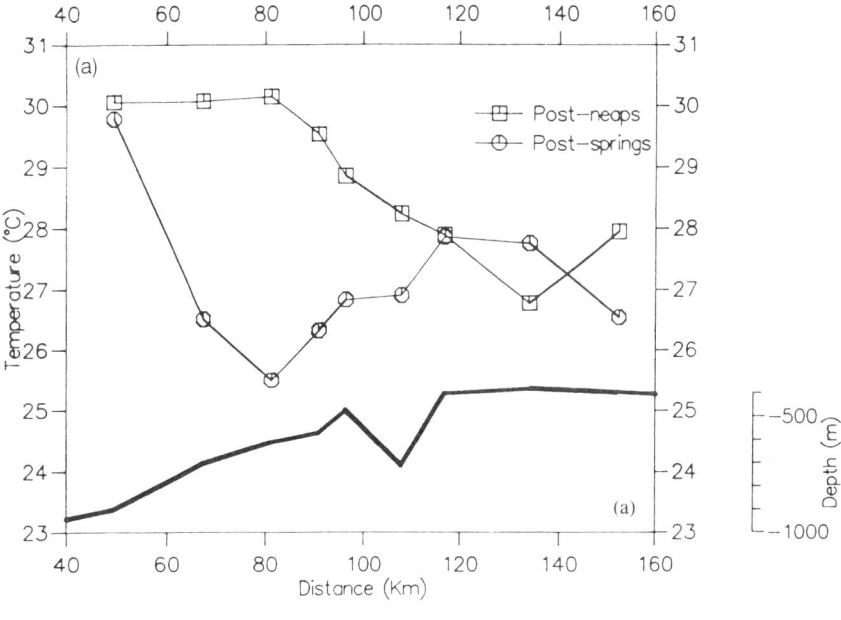

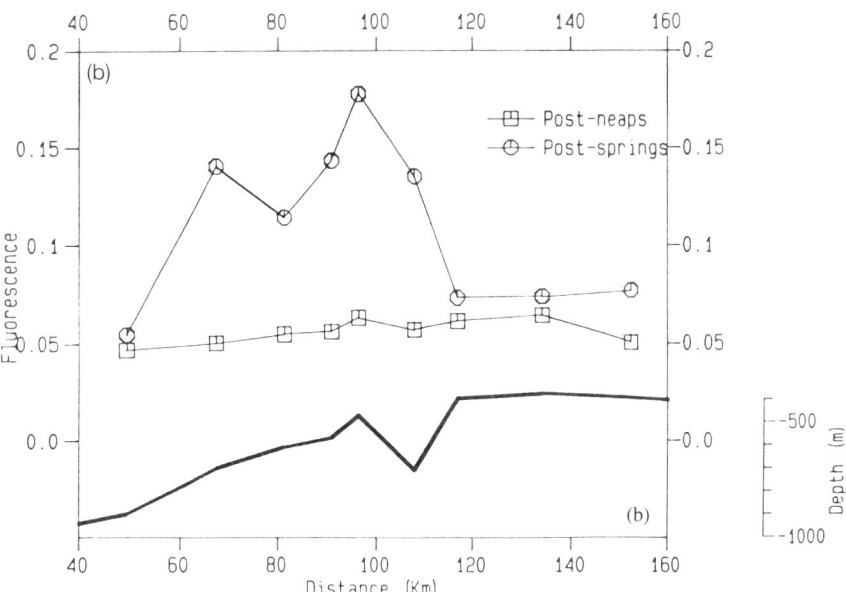

Figure 9.6 Surface measurements on line E. (a) Temperature; (b) chlorophyll fluorescence

with bottom stress stirring. The inference must be that another mixing mechanism is operating and the most likely candidate seems to be the action of hydraulic jumps and lee waves, as suggested by Badan-Dangon (1989).

This simple interpretation assumes that all the observed changes are due to vertical mixing. If this is true, then all the depth mean properties would be preserved. This is clearly not the case as there are significant changes in the column mean temperature (Figure 9.5b) at the stations in the south-east of the survey region where T increases by up to $1°C$, with indications of a slight temperature increase at the north-west end of the sections. The asymmetry of the net cooling of the near-surface waters suggests the operation of advection away from the sill to the south-east. Without such a flow it is difficult to account for strong cooling in deep water 50 km from the sills. The configuration of the subsurface salinity maximum (Figure 9.3c and 9.3d) also seems to support the idea of a near-surface flow away from the sill transporting the high salinity core, which originates in the northern gulf, to the south-east.

9.5 DISCUSSION AND CONCLUSIONS

Before attempting to draw conclusions from our results it is important to recognize the limitations of our strategy of contrasting post-neap and post-spring tide situations. The obvious attraction of this approach is that in the springs—neaps cycle, nature offers us something approaching a controlled experiment in which the modulation of tidal stirring should highlight its impact in terms of the mixing effects produced. To make an unambiguous interpretation of differences, however, we need to be able to assume that other processes remain sensibly steady during the interval between surveys. In the case of air—sea exchanges of heat, water and momentum, this is probably a reasonable assumption as meteorological conditions were consistent throughout the survey period and are characteristic of the summer regime in this part of the Gulf of California. The interpretation could, however, be complicated by advective effects in the form of large-scale internal adjustments of the structure within the gulf. The change in depth-mean temperature strongly suggests that some movement of this kind has occurred. Whether this is simply reactive to the intensified mixing, or is due to external forcing, is not clear from the present observations.

A further limitation lies in the fact that the surveys cannot be made in a truly synoptic manner, so that corresponding station surveys are generally not observed at the same phase of the tide. This means that there may be differences between profiles arising from horizontal displacements or from the effects of the internal tide. The former is probably not important in the present context, where horizontal differences on the scale of the tidal excursion are mostly small. A comparison of profile pairs, taken at well matched and opposing tidal phases, does not suggest any systematic influence of the internal tide. Its influence, and that of other internal waves, although they are probably small, cannot be ruled out without further observations.

Notwithstanding these caveats, the strategy of contrasting post-neap and post-spring tide surveys has yielded direct evidence of the operation of tidal mixing and indications of its role in this region. The conceptual picture that emerges (Figure 9.7) is of vigorous stirring in the vicinity of the sill promoting the mixing of the saline northern gulf water with underlying waters, including a component of the cold, low salinity Pacific intermediate water which is present just below the sill depth (Bray, 1988b). A slow northward flow of this intermediate water is needed to supply demand by sill mixing and a component of it will be carried over the sill in the flow to compensate for the outflow of water from the northern Gulf of California.

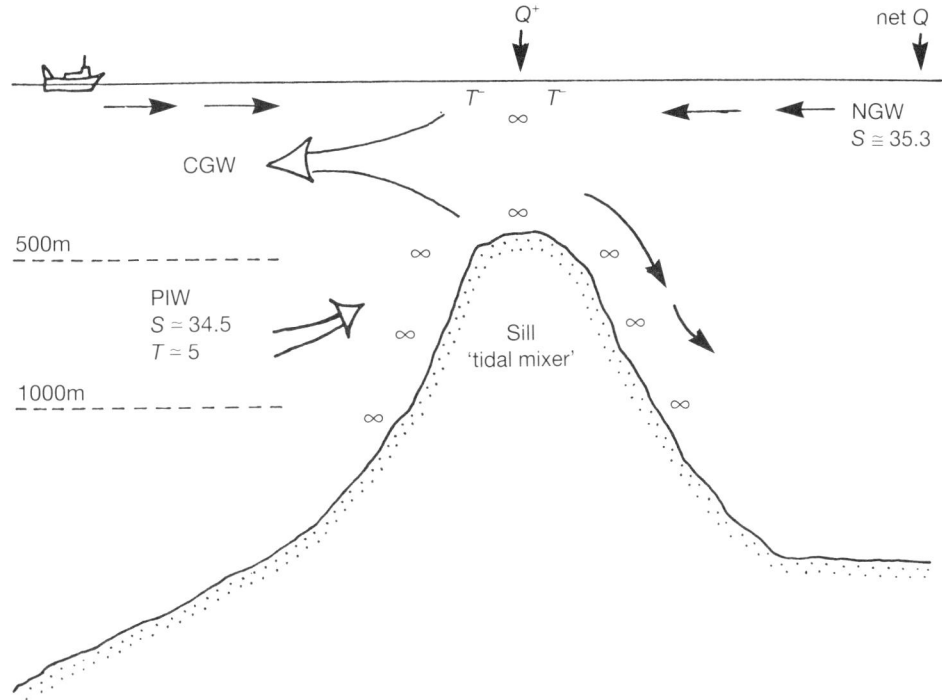

Figure 9.7 Schematic diagram of mixing and circulation in the Gulf of California. Strong stirring over the sills acts to mix Pacific intermediate water (PIW) with northern gulf water (NGW) to produce central gulf water (CGW)

Some surface water is also recruited from the south-east into the mixing process as in the three-layer circulation observed in laboratory mixing experiments (Hachey, 1934).

It has been shown (e.g. Lavin and Organista, 1988) that, although there is strong net evaporation in the northern gulf of about 1 m/year, the effect of this water loss on density is less than that of the net heat gain, so that the gulf system tends to circulate as a 'thermal estuary', with lighter water flowing out near the surface. Tidal mixing around the islands modifies this pattern and is responsible for the production of the central gulf water, which flows south-eastwards 100 m or more below the surface.

This combination of a thermal 'estuarine' circulation with circulation induced by deep mixing at the sills is consistent with the previous results of Bray (1988a) based on water mass analysis and inferences derived from the heat budget of the northern gulf (Lavin and Organista, 1988). Our results, moreover, strongly suggest that we are dealing with a system which experiences marked fortnightly modulation due to spring—neap tide variations in the activity of the sill mixing engine. We expect this to lead to corresponding pulsations in the circulation and structure over a wide area and there seems to be a good case for a campaign of observations by moored instruments on the south-east side of the sills to test this notion.

By lowering the surface temperature in the island region and also in the shallower regions further north, tidal stirring is acting to increase the input of heat to the gulf and at the same time reduce evaporation. This may contribute significantly to the predominant influence of heat on buoyancy which distinguishes the gulf from other evaporative basins such as the Mediterranean and the Red Sea, which have an outflow of denser water.

ACKNOWLEDGEMENTS

This study would not have been possible without the generous support of Ken Richter of the US Naval Oceanographic Service at San Diego and the help of Charlie and Clarice Yentsch, who helped to bring the interested parties together. Graham Allen provided valuable assistance with the observations and data analysis. This research was partially funded by CONACYT, Mexico.

REFERENCES

Argote, M.L., Amador, A. and Morales, C. (1985) Variacion estacional de la estratificacion en la region norte del Golfo de California. In: *Memorias de la Reunion Anual.* (Eds J. Urrutica and J.F. Valdez-Garcia) Union Geofisica Mexicana, 334−338.

Badan-Dangon, A. (1989) The flow over San Lorenzo sill. *WHOI Summer Geophysical Fluid Dynamics*, 164, 1−7, Woodshole Oceanographic Institute.

Badan-Dangon, A., Hendershott, M.C. and Lavin, M.F. (1991) Underway Doppler current profiles in the Gulg of California. *Trans. Am. Geophys. Union* 72, 209−218.

Bray, N.A. (1988a) Thermohaline circulation in the Gulf of California. *J. Geophys. Res.* 93, 4993−5020.

Bray, N.A. (1988b) Water mass formation in the Gulf of California. *J. Geophys. Res.* 93, 9223−9240.

Durazo, R. (1989) Frentes termicos de verano en el alto Golfo de California. *MSc Thesis.* CICESE, 66pp.

Fu, L.-L. and Holt, B. (1984) Internal waves in the Gulf of California: observations from a spaceborne radar. *J. Geophys. Res.* 89, 2053−2060.

Hachey, H.B. (1934) Movements resulting from the mixing of stratified waters. *J. Biol. Board Can.* 1, 133−143.

Hendershott, M.C. and Speranza, A. (1971) Co-oscillating tides in long, narrow bays: the Taylor problem re-visited. *Deep-Sea Res.* 18, 959−980.

Lavin, M.F. and Organista, S. (1988) Surface heat flux in the northern Gulf of California. *J. Geophys. Res.* 93, 14,033−14,038.

Paden, C. (1990) Tidal and atmospheric forcing of the upper ocean in the Gulf of California. *PhD Thesis.* University of California, 82pp.

Simpson, J.H. (1981) The shelf-sea front: implications of their existance and behaviour. *Phil. Trans. Toy. Soc. London Sser. A302*, 531−546.

Simpson, J.H. and Bowers, D.G. (1981) Models of stratification and frontal movement. *Deep-Sea Res.* 28, 727−738.

UNESCO (1981) Tenth Report of Joint Panel on Oceanographic Tables and Standards. *UNESCO Technical Papers in Marine Science No. 36.* UNESCO, Paris.

10 Suspended Sediment Dynamics: Measurement and Modelling in the Dover Strait

S. JONES and C.F. JAGO
University of Wales Bangor, School of Ocean Sciences, UK

and

D. PRANDLE and D. FLATT
Proudman Oceanographic Laboratory, Bidston Observatory, Birkenhead, UK

10.1 INTRODUCTION

The resuspension, transport and deposition of suspended particulate matter (SPM) play a crucial part in a range of marine processes, including benthic fluxes, biological productivity, biogeochemical cycling and pollutant dispersal. There are, however, surprisingly few data sets enabling detailed investigation of SPM dynamics because it has hitherto been difficult to monitor particle concentration, composition and behaviour over appropriate time and length scales.

This chapter describes an innovatory approach to the problem which exploits both technological advances that extend our observational capabilities, and modelling developments that advance our predictive capabilities. Thus optical and acoustic techniques have been used which, by measuring current velocity and SPM concentration on several temporal and spatial scales, provide extensive and detailed data sets. As such data are necessarily limited in space and time, the challenge is then to produce a model which first simulates the observations and then can be extrapolated in depth and time to provide comprehensive predictions of SPM dynamics.

10.2 AIMS

10.2.1 Acoustic Doppler current profiler measurement of SPM concentration

In this study, both optical beam transmittance and acoustic backscatter strength have been used to measure SPM concentration. Thus a self-recording transmissometer provides a long time series of data, but this is essentially a 'point' measurement at a fixed height above the bed. The advance provided by the acoustic Doppler current profiler (ADCP) is that, in theory, it can provide a near-bed to near-surface profile of current velocity and SPM concentration (and hence SPM flux); however, the feasibility of ADCP measurement of SPM concentration has not been much explored. One aim of this work has been to investigate the potential of the ADCP in this respect.

Mixing and Transport in the Environment. Edited by K.J. Beven, P.C. Chatwin and J.H. Millbank
© 1994 John Wiley & Sons Ltd

10.2.2 Development of SPM flux model

Suspended particulate matter has been monitored using instruments deployed in two configurations: short-term depth profiling over tidal cycles at anchor stations and long-term 'point' moorings over periods of up to one year. The resulting data set, combined with the *in situ* measurement of particle settling rates, has provided a detailed assessment of SPM dynamics at the experimental site (although only a portion is described here). The second aim of this work was to develop a particle tracking model to simulate observed SPM concentration time series, and to apply it predict SPM fluxes.

10.3 FIELD OBSERVATIONS

As part of an intensive study of fluxes through the Dover Strait, a combination of current meter, transmissometer and bottom-mounted ADCP was deployed at a fixed site for one calendar month. The site was also investigated during a RRS *Challenger* cruise (Prandle, 1990), when profiles of temperature, salinity and optical beam transmittance were recorded at half-hourly intervals over a period of 25 hours, using an NBIS Mk 3 CTD System equipped with auxiliary sensors. Water samples were analysed for SPM concentration and composition, and further samplers were deployed for the 'quasi-*in situ*' determination of particle settling velocity. This combination of measurements has enabled a detailed assessment of SPM dynamics at the site.

10.3.1 Study area

The study site is situated 4 km off Cap Gris Nez on the eastern (French) shore of the Dover Strait (Figure 10.1). A grab-sample survey in the area indicated that the bottom sediment consists of sand—coarse gravel in the vicinity of the site, and sand, gravel and gravelly mud to the north-east and south. Tides in the region have been intensively studied (Prandle, Loch and Player, 1993), and are characterized by a strong, predominantly unidirectional, tidal flow through the Dover Strait. The site is subject to periodic fresh water influences due to tidal advection of stratified coastal water (Brylinski and Lagadeuc, 1990).

10.3.2 Moored instrumentation

The ADCP was a 1 MHz instrument built at the Proudman Oceanographic Laboratory (POL) following development work at both the Proudman Oceanographic Laboratory (POL) and the Institute of Oceanographic Sciences Deacon Laboratory (IOSDL) (Griffiths and Flatt, 1987). It was programmed to record 24 bins, each of 1.4 m depth, with the first bin situated at 3.1 m above the sea bed. In addition to velocity, it records the received signal strength and setting of an automatic stepping gain control, which optimizes the received signal strength for current measurement (acting independently on groups of four adjacent bins). It was attached to a bottom frame, from which it was acoustically released on recovery.

In addition to the bottom-mounted ADCP, a U-shaped mooring was deployed with an Anderaa RCM7 current meter (incorporating pressure, conductivity and teperature sensors) and an optical beam transmissometer. The transmissometer, positioned 5 m above the bed (thus corresponding to the second ADCP bin) was designed and built at University College North Wales (UCNW). The optics are similar to the commercially produced Sea Tech

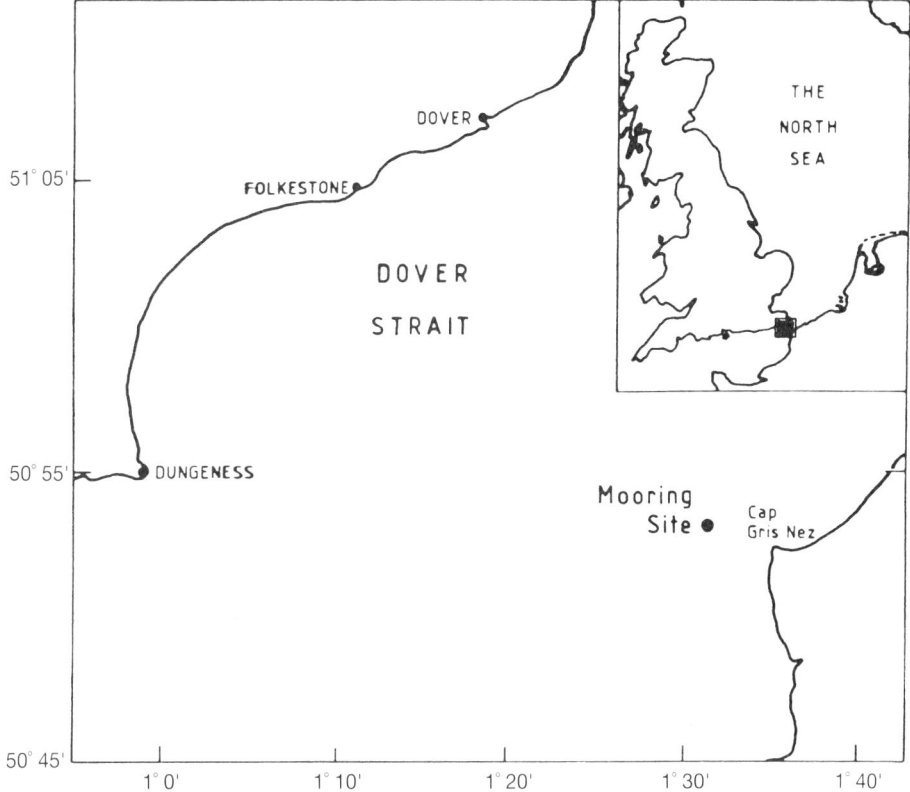

Figure 10.1 Location map of study site

instrument mounted on the RS *Challenger* CTD frame (Bartz, Zaneveld and Pak, 1983), comprising a 660 nm LED source and 25 cm optical path. Instantaneous (compensated for ambient light) values are logged onto a solid-state memory card at one minute intervals for up to three months.

The moored instrument was intercalibrated with the CTD-mounted instrument on site before and after deployment; no significant drift was measured. A further check was performed during the 25 hour anchor station, when half-hourly CTD measurements at 5 m above the bed corresponded almost exactly to the moored instrument readings.

10.3.3 Gravimetric calibration of transmissometers

Throughout the cruise period, GO-FLO bottle samples were collected during CTD casts and known volumes filtered through preweighed GF/F glass microfibre filters for the gravimetric determination of total SPM concentration. The organic carbon content was then determined by loss on ignition (mean value 35%).

Beam attenuation measured by the CTD transmissometer was calibrated against SPM concentration (the measured variation in organic content did not significantly influence the calibration). A highly significant regression was obtained ($R^2 = 0.564$; $p < 0.001$), despite considerable scatter (Figure 10.2). The scatter is due to a combination of unavoidable

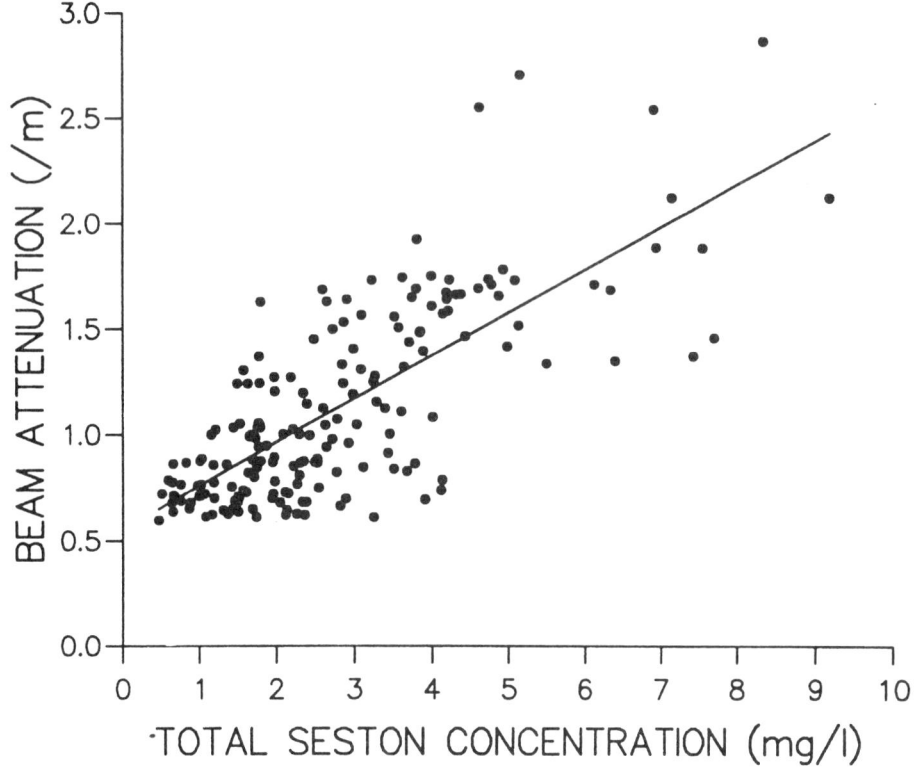

Figure 10.2 Calibration of transmissometer

inaccuracies in the gravimetric technique (especially at these relatively low concentrations), and localized or temporal variations in particle population characteristics (i.e. size/shape/refractive index; Campbell and Spinrad, 1987; Moody, Butman and Bothner, 1987). The calibration includes samples taken across the Dover Strait; restriction to those taken in the immediate vicinity of the mooring did not significantly alter the regression coefficients.

10.3.4 Settling velocity

At regular intervals during the CTD station, samples were collected for the determination of settling velocity spectra. The instrument used was developed at UCNW from an original design by Professor I.N. McCave (University of Cambridge), and collects effectively undisturbed samples of water in a horizontal tube (Jago, Jones and Boon, in preparation). This is brought back on board and mounted vertically, allowing the suspended particles to settle. Subsamples (550 ml) are withdrawn from the bottom at fixed time intervals and passed through preweighed 0.4 μm Nuclepore membrane filters, enabling the settling velocity distribution to be calculated (Owen, 1976). A bottom-release mechanism allows sampling at 1 m above the sea bed: a pressure release is also incorporated which allows sampling at approximately 7 m below the sea surface. A total of three sets of near-bed and near-surface settling velocity spectra were determined at times when the current flow was at a maximum.

10.4 RESULTS

10.4.1 Currents

Detailed analysis of the current structure and residual flow has been presented elsewhere (Knight *et al.*, 1992; Prandle and Player, in press). At 5 m above the bed, current speeds reached up to $1.4\,\mathrm{m\,s^{-1}}$ during spring tides, and $0.9\,\mathrm{m\,s^{-1}}$ during neaps (Figure 10.3a). Strong diurnal asymmetry was observed post-springs and during neaps. The predominant direction lies along the axis of the Dover Strait (Figure 10.4a); perpendicular to this a persistent (non-oscillatory) offshore flowed is observed. The unfiltered progressive vector plot (Figure 10.4b) shows a marked lunar variation, with offshore residuals during spring tides and through-Strait residuals during neap tides.

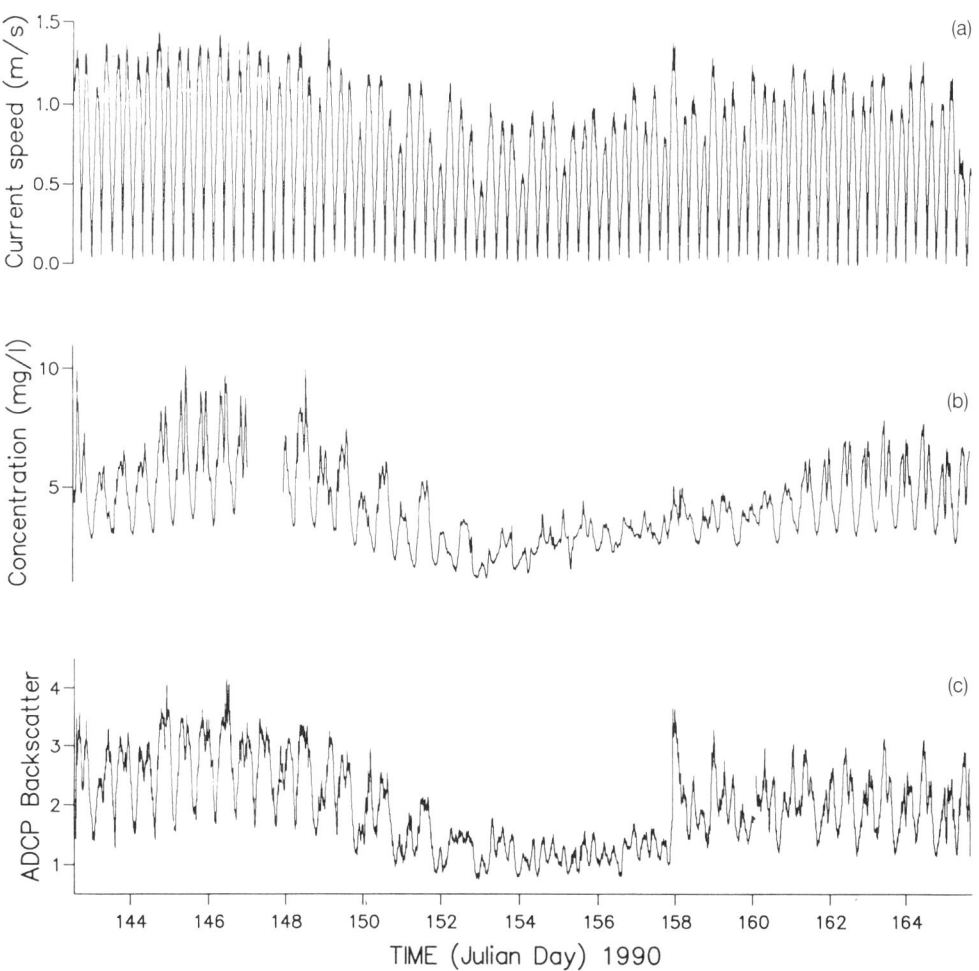

Figure 10.3 Time series of mooring observations 5 m above the sea bed. (a) Current speed; (b) SPM concentration from transmissometer; (c) ADCP backscatter strength (arbitrary units)

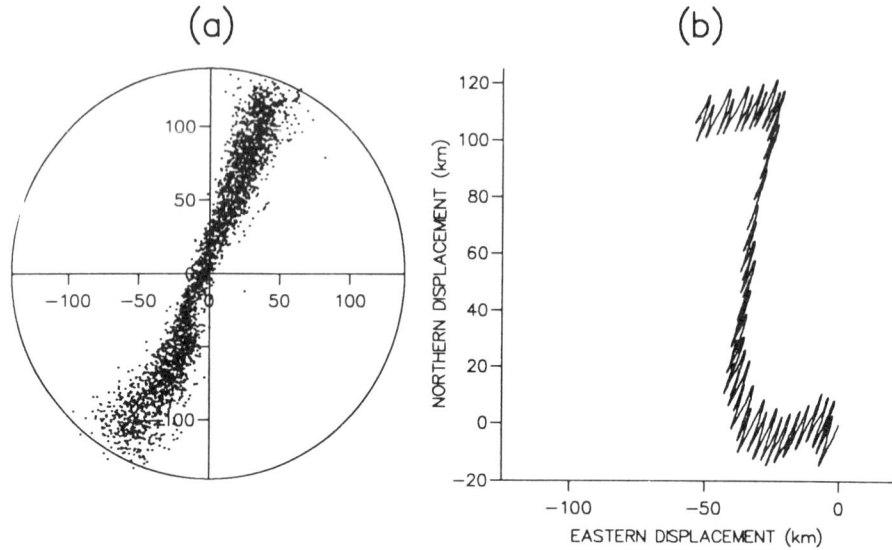

Figure 10.4 Analysis of currents 5 m above the bed. (a) Scatter plot; (b) progressive vector plot

10.4.2 SPM concentration

The time series recorded by the moored transmissometer, converted into SPM concentration, is shown in Figure 10.3b. Data were lost for a short period during the deployment due to temporary blockage of the light path.

The time series indicates several components of variability. A significant spring/neap variation in 'background' concentration is observed due to increased levels of regional resuspension during spring tides. Superimposed on this are two higher frequency contributions: a quarter-diurnal variation associated with local tidal resuspension, varying in strength over the spring/neap cycle, and an additional strong semidiurnal signal present only during spring tides. This additional signal causes alternate slack water minima to be offset in magnitude by up to 2.5 mg l^{-1}, resulting in a 'twin-peaked' signal which has been found to be characteristic of shelf sea regions exhibiting tidal resuspension (Jago et al., 1993). It can be explained most successfully as due to the effect of tidal advection past the mooring of a horizontal SPM concentration gradient (Weeks and Simpson, 1991), resulting in differences between SPM levels at high and low slack water. This gradient, which will be discussed in more detail in the next section, must be due either to spatial heterogeneity in resuspension levels (because of local variability in current speeds or bottom sediments), or to the regional influence of fine sediment sources (such as river mouths).

The potential of the 1 MHz ADCP for measuring SPM concentration profiles was also investigated. In addition to Doppler phase shifts (and hence currents), the strength of signal backscattered from within each bin was recorded. This is a function of bin height (as energy loss occurs via beam spreading and absorption during propagation) and, importantly, of the concentration and scattering cross-section of SPM within each bin. The scattering cross-section depends on particle size and composition, as well as signal frequency. Purpose-built acoustic backscatter probes operate at frequencies of 3 MHz and above, and profile SPM concentrations in the bottom 1−3 m of the water column (Libicki, Bedford and Lynch, 1989; Thorne et al., 1991). A 1 MHz transducer mounted on a bed-frame and looking up into the water column has been used to measure SPM concentration profiles up to 10 m above the bed (Lynch *et al.*, 1991).

The ADCP seems to be ideally suited for adaptation to monitor SPM concentrations, although there has been surprisingly little work in this area (Gordon and Rogers, in press; Thomson, Gorden and Dymond, 1989; Bos, 1991). If combined current/SPM profiles can be measured by a single instrument, the potential for monitoring particulate fluxes is enormous.

The ADCP data were analysed as follows. For each bin, the recorded signal strength was corrected to account first for the appropriate applied stepping gain, then for scattering/absorption loss in intermediate bins during propagation out and back. The value obtained is then an independent measure of scatter from that particular bin. Further analysis of the signal to obtain an estimate of particle concentration (for a given scattering cross-section) (e.g. Lynch, 1985) has not yet been acheived. However, it was possible to obtain empirical calibrations of adjusted signal strength against SPM concentrations measured by the transmissometer, both at 5 m depth over the full deployment and for the 25 hour CTD station throughout much of the water column.

Recorded signal strength from the ADCP bin corresponding to 5 m above the sea bed (bin 2) is presented in Figure 10.3c. It corresponds well with the transmissometer record, indicating that the ADCP is effectively responding to variations in SPM concentration. Both instruments indicate similar spring−neap cycles in background concentration. The higher frequency variations are also remarkably similar, with the same resuspension/horizontal advection signals present.

There are also, however, some interesting differences: these could be accounted for by differences in the response of the two instruments to particles of various sizes and compositions. For example, for a given composition and mass concentration, finer particles produce higher beam attenuation, whereas coarser particles scatter 1 MHz sound more effectively.

To illustrate the similarities and differences between the two records, sections of both time series are presented with current data in Figure 10.5. The graph axes have been arbitrarily adjusted to give the best correspondence between transmissometer/ADCP time series. Note that the combination of advection/resuspension components leads to a shift in the concentration maximum with respect to current speed, so that pairs of concentration peaks are shifted towards each other. It is clear that, although both instruments show a similar 'twin-peaked' response, there are parts of the record where the ADCP resuspension signal is broader (and more closely in phase with the current speed). This can be explained by considering the effect of two populations of sediment: a fine population which forms the background horizontal concentration gradient and is additionally locally resuspended, producing the record observed by the transmissometer, and a coarse population which is tidally resuspended, but not subject to a horizontal concentration gradient (hence not shifted in phase), observed by the ADCP in addition to the fine population.

If this explanation is correct, and further supporting evidence will be presented in Section 10.4.4, there are exciting implications for suspended sediment monitoring. In theory, these observed differences could be used to infer variations in SPM composition, and further experiments to investigate this possibility are in progress. For the purposes of this paper, however, the relatively minor observed deviations due to variations in the composition have been ignored and regression performed to relate primary variability in ADCP backscatter strength to the 'generalized' SPM concentration calculated for the transmissometers. It is not, of course, possible to state that one instrument or the other is 'correct', as each provides a biased measurement in suspensions of varying composition.

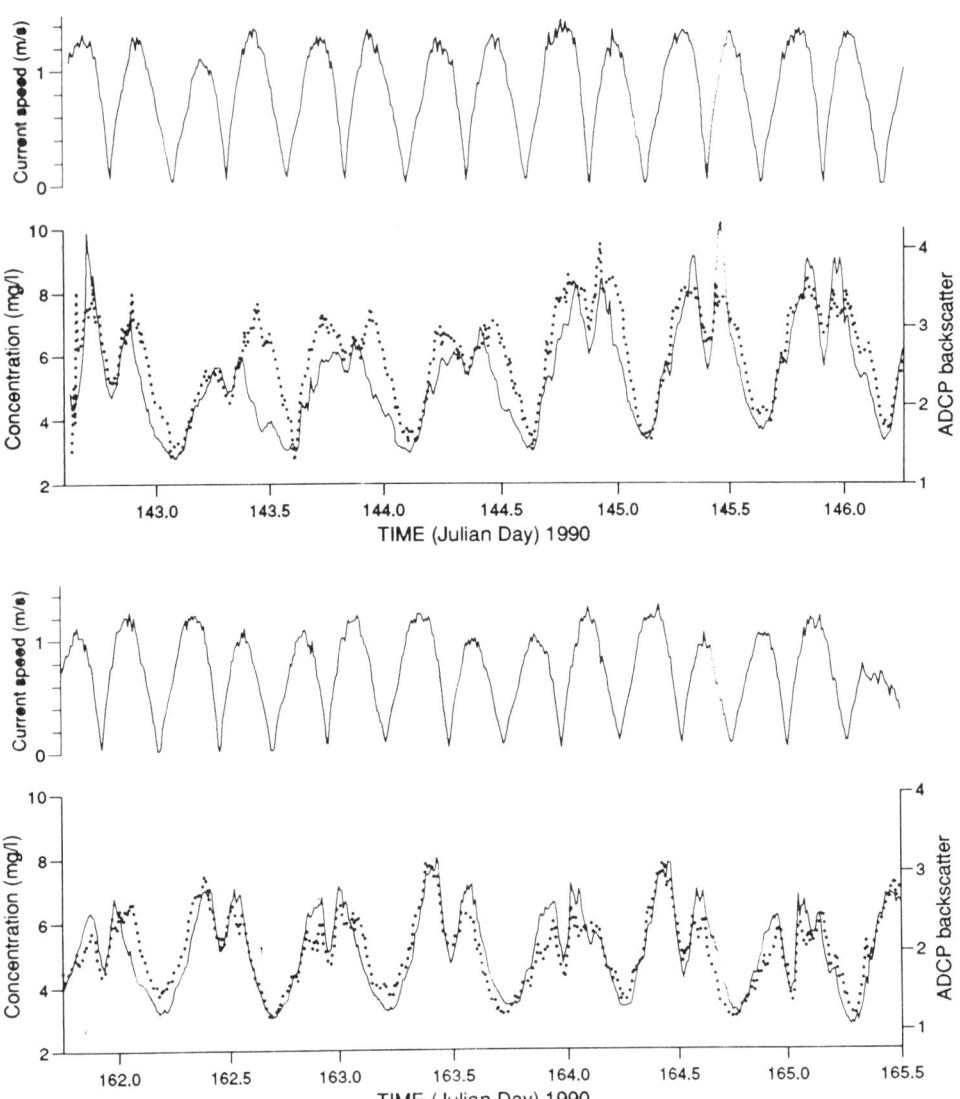

Figure 10.5 Comparison of concentration 5 m above the bed measured using moored transmissometer with ADCP backscatter strength (arbitrary units)

So far only data from bin 2 have been discussed. Clearly, the most important application of the ADCP is in measuring SPM concentration profiles using signal strength from all bins. Figure 10.6 illustrates selected comparisons between the ADCP and corresponding 'bin-averaged' data from the CTD transmissometer. Highly significant ($R^2 > 0.6$; $p < 0.001$) correlations were obtained for most bins up to 22 m above the bed (bin 14), beyond which the signal to noise ratio deteriorated.

Contoured time series of SPM profiles measured by CTD transmissometer and ADCP are presented in Plate 7. The two show encouraging similarity, allowing for differences in temporal

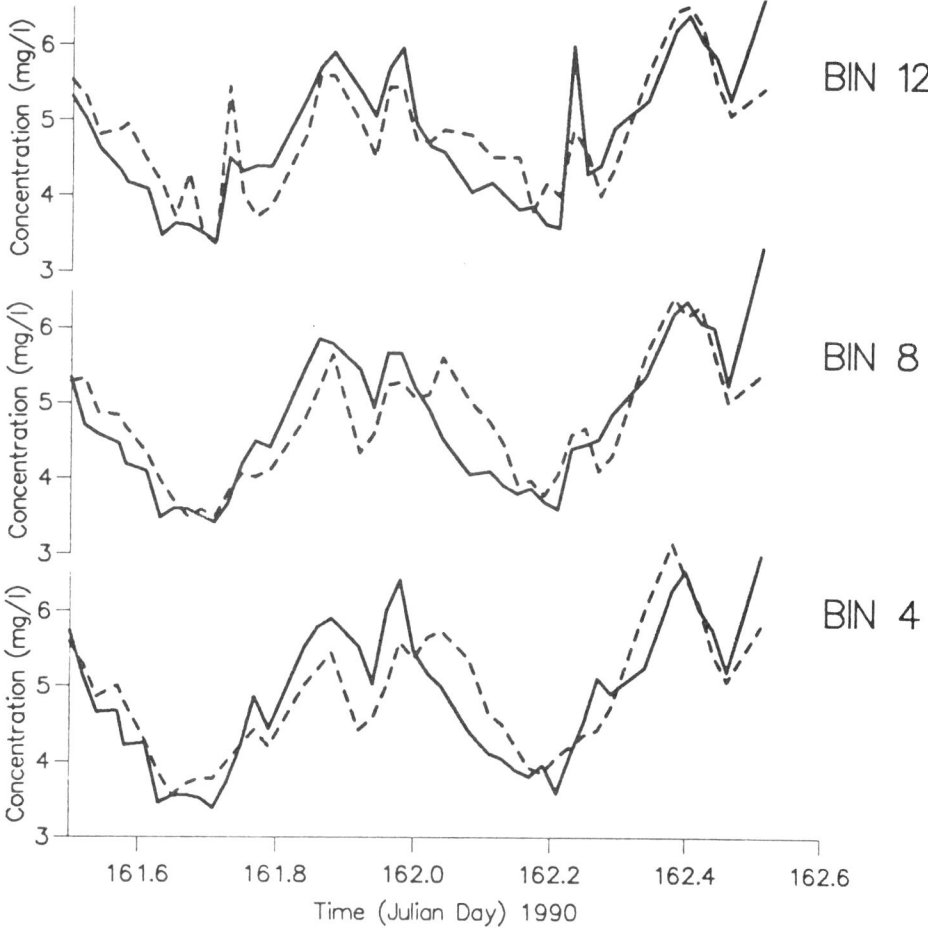

Figure 10.6 Comparison of half-hourly CTD-mounted transmissometer readings (averaged over depth intervals corresponding to ADCP bins) with ADCP backscatter data from selected bins. ADCP data have been converted to concentration using least-squares regression against transmissometer

and vertical resolution (and hence apparent duration and extent of some features). Both show strong resuspension during maximum current flow, superimposed on an advecting background horizontal concentration gradient. The ADCP shows enhanced resuspension around the two periods of current maxima on day 162.05, 162.35 compared with the transmissometer, perhaps in response to coarser-grained SPM. The two instruments also show enhanced SPM concentrations in the upper part of the water column just after the slack water ebb. Salinity and temperature profiles indicate that this corresponds to the advection of stratified coastal water, causing a warm, turbid fresh water layer to encroach on the mooring site.

The CTD calibration for bin 2 corresponded closely to that obtained from the moored transmissometer over the entire deployment period. This implied that other bin calibrations could be safely extended. However, when this was attempted it was found that for bins above bin 4, the automatic gain adjustment varied anomalously during deployment, thereby rendering the CTD calibration invalid. This software error has since been rectified. Fortunately, however, it was found that no such anomalies occurred between day 160 and mooring recovery, which encompassed the calibration period. Therefore it was possible to convert ADCP data to SPM concentration for the last five days of the experiment. The same pattern of resuspension/horizontal advection was observed during this time, with the turbid surface layer increasing in influence, presumably as tidal stirring reduced after spring tides.

10.4.3 Horizontal concentration gradients

As horizontal concentration gradients cannot be predicted by a one-dimensional model, validation of the model described in Section 10.5 with observed data required several simplifying assumptions. In fact, although the intention is eventually to include tidal advection of prescribed horizontal gradients, these have not yet been incorporated. Therefore the effects of regional heterogeneity had to be isolated from those due to local resuspension.

It is proposed that the observed time series in SPM concentration can be described by a series of discrete components. For rectilinear flow, the total concentration at the mooring site (at a fixed height) is then assumed to vary according to:

$$C(x,t) = C(x,0) + C_b(t) + C_r(t)$$

$$= C_0 - \partial C_b/\partial x \Big|_0^t U_x dt + f[U_x(t)]$$

where $C_r(t)$ describes resuspension/vertical diffusion of bed sediment as a function of the current speed U_x and $C_b(t)$ describes the change at x due to tidal advection of a (depth uniform) 'background' fine population past the mooring site. $\partial C_b/\partial x$ is assumed to be constant over the time period under consideration, which implies that resuspension/diffusion is spatially uniform in the vicinity of the mooring site, and hence that there are horizontal gradients only in relatively slowly settling (fine) SPM concentration. This assumption cannot be justified in all instances: as will be shown, it appears to be correct in this particular example.

The advective component can therefore be effectively decoupled from the local resuspension component without making any assumptions about the form of $f[U_x(t)]$. However, numerical separation of this semidiurnal component from the observed time series is still not straightforward. The SPM time series do not lend themselves to normal tidal analysis procedures because the resuspension component is always positive: filtering will therefore always include a tide-averaged resuspension component in addition to any lower frequency contributions.

Analysis was therefore performed manually by comparing tidal displacement s_T, calculated along the predominant current direction, with observed SPM concentration (C) for separate 25 hour sections of the time series. Oscillatory flow was only observed along this direction; the continuous offshore flow normal to this has been ignored in the model.

The advected component was computed as $-(\partial C/\partial x)s_T$, with values of the (constant) concentration gradient adjusted for each 25 hour section until a best fit to the observed semidiurnal variation was achieved. Excellent fits were obtained during spring tides. Figure 10.7 illustrates concentration gradients inferred by this method for the ADCP time series: a positive value indicates concentrations increasing towards the North Sea. The same pattern, but with slightly higher gradients, was obtained for the transmissometer record. It is clear that positive gradients developed during spring tides, which gradually reduced post-springs

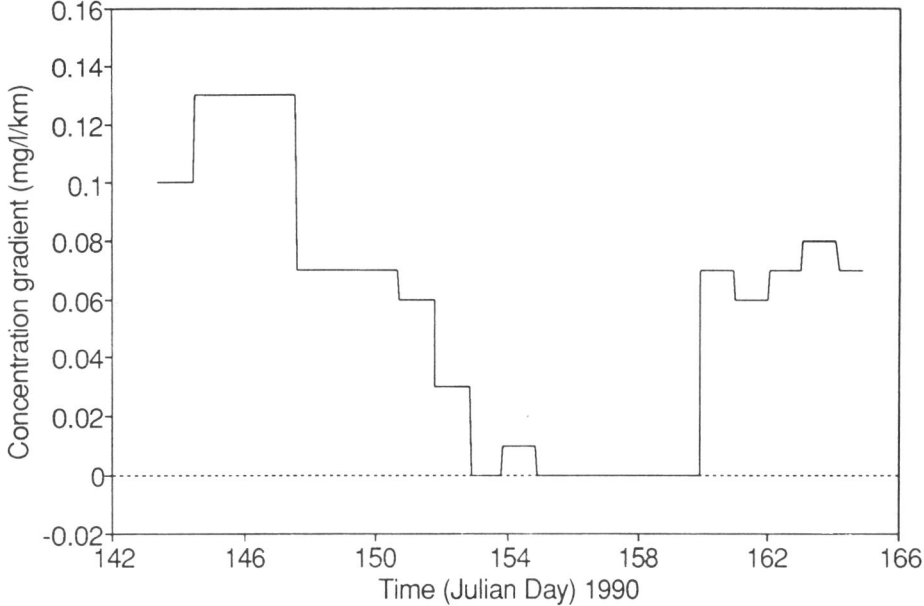

Figure 10.7 Horizontal (through-Strait) concentration gradient inferred for 25 hour sections from semi-diurnal advection signals observed in the ADCP concentration time series and tidal displacement computed 5 m above the bed

to zero during the neap tides. From the uniformity of such gradients on successive spring tides, the assumption of within-tide uniformity is justified.

Incidentally, tidal advection of weak, vertically uniform gradients in temperature and salinity was also observed during spring tides (by both the CTD and by the moored RCM current meter), implying an increase in temperature and reduction in salinity towards the North Sea. Thus higher turbidity is associated with warmer, fresher water. This association agrees with that found for the stratified surface layer, indicating that local fresh water sources are also sources of fine SPM. Interestingly, the apparent source of the background gradient (to the north-east) was unrelated to the observed stratified water, which encroached on the mooring site from the south-east immiedately after the least turbid, most saline, coolest background values were obtained.

10.4.4 SPM fluxes

The net horizontal flux of SPM through the Dover Strait is of considerable interest for modellers of North Sea water quality (Eisma, 1987). Combined current/concentration measurements allowed the calculation of 'point' fluxes over the full mooring deployment using the 5 m time series, and of a net flux profile between 3 and 22 m for the last five days of the deployment using ADCP profile data. Only fluxes along the major current axis (i.e. through the Dover Strait) have been considered here.

Fluxes were first computed for the single-point time series, which was split into three (7×25 hour) sections, corresponding to two spring tides and one neap. The ADCP (bin 2) SPM record was used because it was the most complete. Positive values indicate net flux into the

North Sea. Net fluxes of 0.00014 and 0.0019 kg m^{-2} s^{-1} were obtained over the two spring tide periods, while a value of 0.00039 kg m^{-2} s^{-1} was obtained during neaps. Interestingly, neap tide fluxes were twice the spring tide values, despite considerably lower concentrations, because the residuals were much higher during neaps. If these values are assumed to be typical for spring/neap tides throughout the year, an annual 'point' flux of 8.8 t m^{-2} is indicated, neglecting the potentially important influence of storms. Less justifiably, if this value is assumed to apply throughout the water column and across the Dover Strait, a total net influx of 9.4 × 10^{-6} tonnes per year into the North Sea is obtained, matching the value estimated by Eisma (1981).

Such extrapolation does, of course, represent a gross over-simplification and ignores the vertical structure and horizontal heterogeneity. Figure 10.8 indicates the vertical distribution of SPM flux calculated from ADCP profile data for the 5 × 25 hour spring tide period (days 160–165). Note that net fluxes out of the North Sea are indicated in the upper layers, but that far greater amounts are transported towards the North Sea nearer the bed. Observational data to investigate variability across the Dover Strait have not yet been analysed.

This clearly shows the potential of the ADCP to monitor SPM flux profiles. However, both the important near-bed region and the upper part of the water column can only be included by extrapolation of an appropriate model.

10.4.5 Settling velocity

The settling velocity tubes enabled independent SPM concentration measurements in addition to settling velocity spectra. Figure 10.9 compares such concentrations measured at 1 and 25 m above the bed (sampled near to the maximum current flow), with those recorded at 5 m above the bed by the transmissometer and ADCP, respectively. Much higher concentrations were obtained near the bed than at 5 m above the bed, which corresponded more closely to the samples at 25 m. The near-bed region is clearly extremely important, and cannot be properly monitored using the present ADCP/transmissometer measurement system.

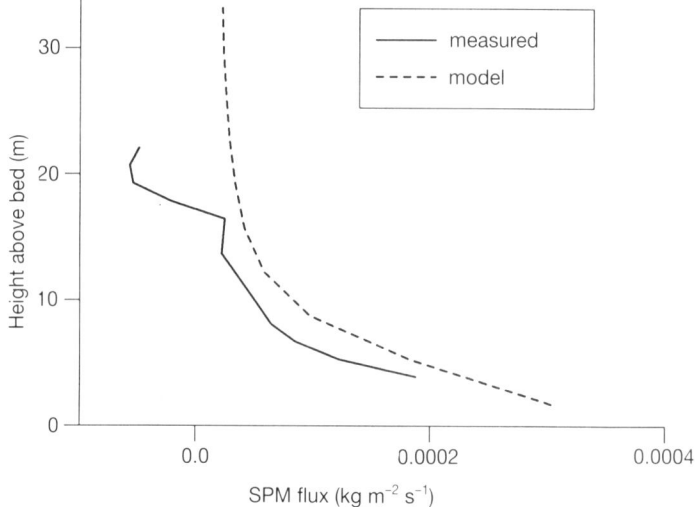

Figure 10.8 Net SPM flux profile computed from ADCP current/concentration profiles measured over last five days of deployment. Model prediction for same period superimposed

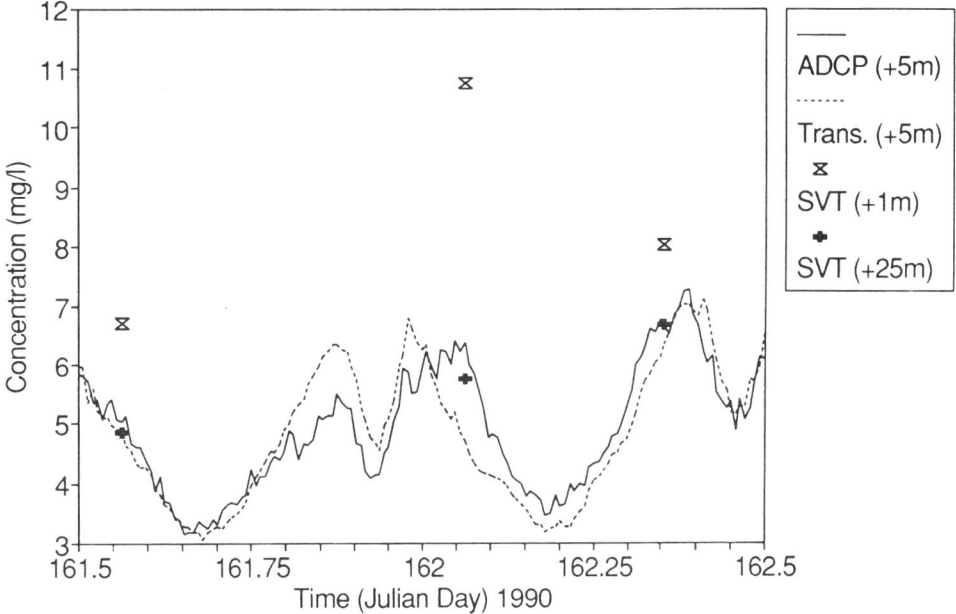

Figure 10.9 Comparison of SPM concentrations measured using settling velocity tubes with moored instrument data. Heights in legend are relative to the sea bed

The settling velocity spectra indicate some interesting temporal and vertical variability (Figure 10.10). All samples showed a broad mode of about 0.1 mm s^{-1}. This corresponds (for inorganic particles) to a grain diameter of about 10 μm, or fine silt. The near-bed samples also showed a more rapidly settling mode of about $3-10$ mm s^{-1}; this corresponds to a grain diameter of $60-150$ μm, or fine sand. The proportion of this sand fraction varied considerably, with very high concentrations observed during the highest peak flow (day 162.0). In contrast, in the near-surface samples very little sand was observed; the additional, very slowly settling, mode may have been biogenic (e.g. almost neutrally buoyant phytoplankton).

These observations support the hypothesis about differences in ADCP/transmissometer records. The highest proportion of sand was measured *during* the period of maximum concentration according to the ADCP, but *after* that measured by the transmissometer. Therefore the ADCP is apparently more sensitive to sand-sized SPM.

10.5 RANDOM WALK PARTICLE TRACKING SIMULATION APPLIED TO SEDIMENT RESUSPENSION

The random walk particle tracking approach to dispersion problems can be readily applied to particle resuspension (Allen, 1982).

The dispersion equation (limited for convenience to a single horizontal direction and only vertical dispersion) is as follows

$$\frac{\partial C}{\partial T} + U \frac{\partial C}{\partial X} + W \frac{\partial C}{\partial Z} = \frac{\partial}{\partial Z} \left(E \frac{\partial C}{\partial Z} \right)$$

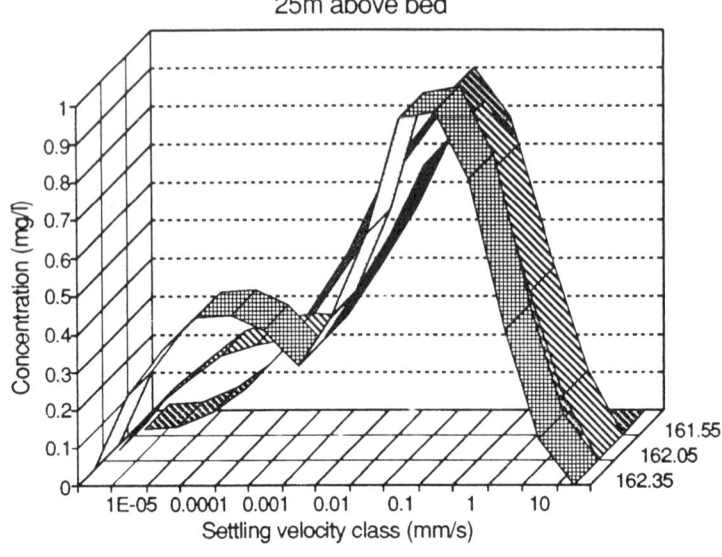

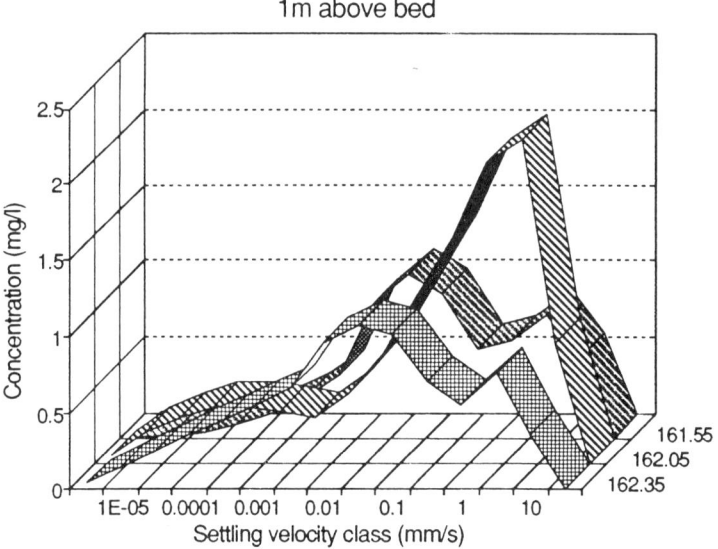

Figure 10.10 Settling velocity spectra for samples indicated in Figure 10.10. (a) 25 m above sea bed; (b) 1 m above sea bed. Note log scale. Labels indicate time of sampling (day no.)

where C is the concentration of a given particle type, T is time, U is the horizontal velocity along the X axis, W is the vertical velocity along the Z (upwards) axis and E is a coefficient of vertical eddy diffusivity.

It is further convenient to generalize the vertical components of the above equation by adopting

$$w = \frac{W}{D}, \quad e = \frac{E}{D^2}, \quad z = \frac{Z}{D}$$

where D is the water depth, hence $z = 0$ corresponds to the bed and $z = 1$ the surface.

10.5.1 Random walk simulation

In simulating sediment resuspension in tidal flows, the vertical velocity W can be assumed to be the settling velocity W_s for the particle in quiescent water (Lick, 1982). A distribution in settling velocity class is assumed (as measured in the settling velocity tubes described in Section 10.3.4), raning from 100 to 10^{-4} mm s^{-1} (coarse sand to fine clay for inorganic particles).

Fischer *et al.* (1979) indicate how turbulent mixing in a continuum can be represented by successive randomly directed step motions at time intervals Δt of length L where

$$L = (2E\Delta t)^{1/2}$$

Numerical simulation of sediment resuspension simply involves the release at the bed of a number of particles, NP, at each time step to reproduce erosion. Here this number is proportional to some power P of the specified tidal velocity U, i.e.

$$NP = \text{INT} \left(\alpha | U(t) |^P \right)$$

The form of this function is consistent with other published work, with P commonly ranging from 1.0 to 10.0 (Lavelle, Mojfeld and Baker, 1983; Shi, Larsen and Downing, 1985; Dyer and Soulsby, 1988; Davies, 1992). In this simple formulation α and P are assumed to be independent of particle settling velocity.

In subsequent time steps, all suspended particles are transported vertically by $\pm L_s - W_s\Delta t$; (where W_s is the settling velocity), the vertical height $Z(t)$ is stored together with the horizontal position $X(t)$ [updated each time step as $X(t+\Delta t) = X(t) + U(t)\Delta t$].

At the surface any computed movement beyond $Z(t) = 1$ is 'reflected' back into the water column. Conversely, at the bed any computed movement below $z = 0$ causes the particle to be removed from further motion, i.e. it is assumed to be deposited.

The model can accommodate vertical and time variations in both U and E and likewise W_s can be adjusted to vary with particle concentrations (e.g. to include the effects of aggregation). However, for the examples cited here E is assumed constant, W_s is invariant with concentration (a reasonable assumption given the low concentrations encountered) and U varies only with time. $\partial C / \partial X$ is also assumed to be zero.

10.5.2 Simulations

Preliminary model development to date has concentrated on the simplest application for non-depth-varying values of U and E. This is clearly not suitable for the prediction of the observed variation in density structure due to tidal advection of stratified coastal water. Therefore no attempt has been made at this stage to model the SPM concentrations observed in the upper part of the water column.

Instead, the model has been applied to simulate observed SPM concentrations in the mixed region near the bed (and, in fact, throughout the water column for much of the time). Attention has focused on tidal resuspension and subsequent vertical mixing of the SPM. The ADCP record at 5 m above the bed was chosen to validate the model because it showed reduced sensitivity to advective variation in background 'fines' (which is yet to be incorporated in the point model) and enhanced sensitivity to locally resuspended coarse material. Further modelling work is being conducted to allow 'tuning' of the fine:coarse components to simulate both transmissometer/ADCP time series; inferences about instrument sensitivity to time-varying particle composition should then be possible.

Particle concentrations were calculated over the 22 day study period for the following parameters: $D = 33$ m, $E = 0.1$ m^2 s^{-1}, $W_s = 10$ and 0.1 mm s^{-1}, $P = 3$ and total number of particles $= 2000$. The two settling velocity classes selected correspond (to first order) to the two modes observed during peak flow near the bed (Section 10.4.5). The horizontal current was specified from ADCP measurements: a level of 5 m above the bed was selected.

Using a least-squares fitting procedure the calculated time series for the two settling velocity classes (corresponding to coarse and fine particle populations) were calibrated to best fit the observations, as shown in Figure 10.11. The measured time series has been modified to remove the advected background component described in Section 10.4.3 using the gradients shown in Figure 10.7. It was found that a coarse:fine particle proportion equivalent to a mean concentration ratio of 3:1 produced an optimum simulation. The correspondence between observed and calculated suspended sediment time series provides encouragement for the further integration of models with observations.

Figure 10.12 shows the net flux of SPM predicted by the model over a 28 tide, spring−neap period during the mooring deployment (days 151−164). Model output has been calibrated using the same constants used in Figure 10.11; fluxes are calculated for each settling velocity class separately, then summed. The variability in depth profiles for the two particle settling velocities selected shows (i) the difficulty in estimating the fluxes of separate components of SPM (and hence particle-dependent contaminants) from direct measurement and (ii) the

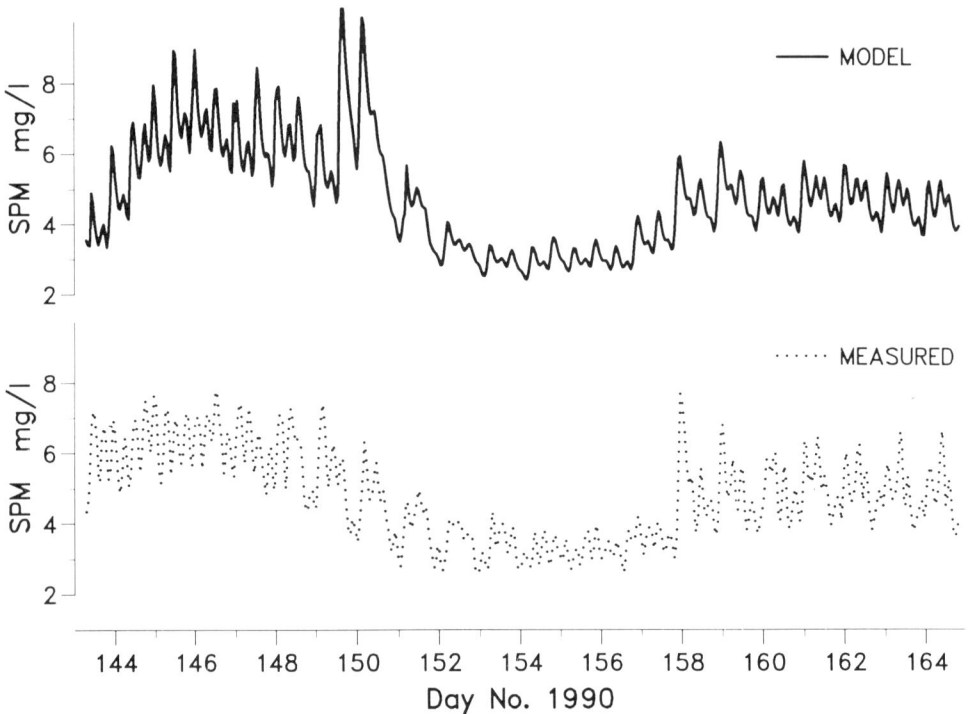

Figure 10.11 Comparison of model and measurement of SPM concentration (using ADCP) 5 m above the bed. Horizontal advection components have been removed from the ADCP data (using gradients shown in Figure 10.8)

diversity of results that can be achieved from such models. Interestingly, depth integration yields a net spring/neap flux of $3.4\,t\,m^{-1}$ width for the coarse population, and a similar $3\,t\,m^{-1}$ width for the fine population, despite markedly different vertical distributions.

The combined model flux for days 160–165 has been superimposed on the flux profile calculated from ADCP data (Section 10.4.4) in Figure 10.8. It shows reasonable agreement nearer the bed, but deviates in the upper half of the water column. This disagreement is not unexpected, as the model (i) assumes a constant vertical velocity and (ii) ignores the influence of stratified coastal water and residual transport along horizontal concentration gradients. However, it is clear that even in its simplest form described here the model can simulate SPM flux profiles to first order, and hence could be used for the vertical/temporal extrapolation of observations. It also serves to highlight the likely importance of the near-bed region for net flux estimates. The model predicts an annual depth-integrated flux of $5.9 \times 10^6\,t$ under quiescent conditions on simple extrapolation across the Dover Strait; further modelling development is underway to allow more realistic estimates.

10.6 CONCLUSIONS

An extensive data set of SPM concentration and current velocity has been obtained over a period of one month at a site in the Dover Strait. Simultaneous deployment of an optical beam transmissometer and bottom-mounted ADCP has allowed detailed intercomparison of their performance in the measurement of SPM concentration and provided data for input to, and validation of, a particle tracking model of SPM dynamics. Preliminary investigation has shown that

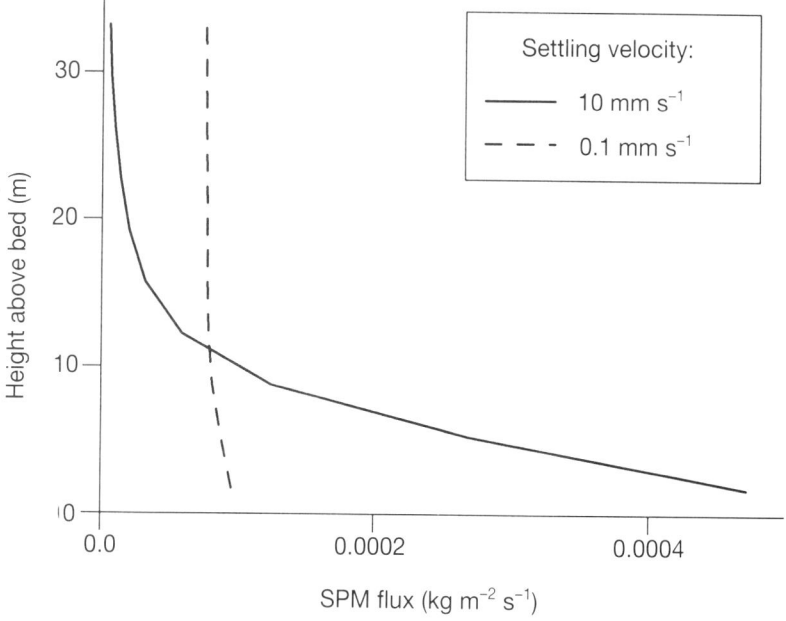

Figure 10.12 SPM concentration flux profiles calculated from the model over the 21 day deployment period for separate settling velocity classes

(1) Time series of acoustic backscattering strength recorded by a 1 MHz ADCP bear close correspondence with those of SPM concentration measured by a transmissometer. Both instruments indicate spring−neap variations in tidal resuspension and advection of horizontal gradients of SPM. Once calibrated, the ADCP can measure profiles of SPM concentration and, when combined with velocity profiles measured by the same instrument, SPM flux.

(2) In detail, the acoustic and optical signals may differ; this can be attributed to their differing response to temporal variations in the composition of SPM. The ADCP appears to be more sensitive to locally resuspended coarse (sand-sized) particles, the transmissometer to horizontally advected finer particles. Thus combining acoustic and optical techniques offers important potential for investigating the temporal variability of SPM composition (provided that appropriate *in situ* calibration is performed).

(3) These direct observations are necessarily limited in space and time, and require extrapolation using a suitable model for monitoring or predictive purposes. A particle tracking model, using two particle populations with settling velocities corresponding to *in situ* observations, has been successfully applied to simulate SPM concentrations at 5 m above the bed over a lunar cycle.

(4) Flux profiles calculated by the model are in reasonable agreement with those measured by the ADCP over a five day period. The model exhibits a marked sensitivity to particle settling velocity, indicating contrasting flux profiles of coarse and fine SPM. Extrapolation of both observations and modelling allows a rough estimate of an annual SPM flux into the North Sea (due to tidal pumping) of 1×10^7 t. However, this ignores spatial heterogeneity across the Dover Strait and the influence of storms and biological processes, which are almost certainly significant.

It has been shown that a combination of innovative measurement techniques and numerical models has tremendous potential for monitoring SPM fluxes.

REFERENCES

Allen, C.M. (1982) Numerical simulation of contaminant dispersion in estuary flows. *Proc. Roy. Soc. London A* **381**, 179−194.

Bartz, R., Zaneveld, J.R.V. and Pak, H. (1978) A transmissometer for profiling and moored observations in water. *SPIE Ocean Optics V* **160**, 102−108.

Bos, W.G. (1991) A comparison of two Doppler Current Profilers. *IEEE J. Ocean. Eng.* **16**, 374−381.

Brylinski, J.M. and Lagadeuc, Y. (1990) The inshore/offshore waters interface off the French Coast in Dover Strait: a frontal area. *Compt. Rend. Acad. Sci. Paris Ser. 2* **311**, 535−540.

Campbell, D.E. and Spinrad, R.W. (1987) The relationship between light attenuation and particle characteristics in a turbid estuary. *Est. Coast. Shelf Sci.* **25**, 53−65.

Davies, A.G. (1992) Modelling the vertical distribution of suspended sediment concentration in combined current flow. In: *Dynamics and exchanges in estuaries and the coastal zone* (Ed. D. Prandle). AGU Coastal and Estuarine studies **40**, 441−466.

Dyer, K.R. and Soulsby, R.L. (1988) Sand transport on the continental shelf. *Ann. Rev. Fluid Mech.* **20**, 295−324.

Eisma, D. (1981) Supply and deposition of suspended matter in the North Sea. *Spec. Publ. Int. Assoc. Sedimentol.* **5**, 415−428.

Fischer, H.B., List, E.J., Kok, R.C.Y., Imberger, J. and Brooks, N.H. (1979) *Mixing in Inland Coastal Waters*. Academic Press, New York, 483pp.

Gordon, R.L. and Rogers, J. Observations of water and suspended material fluxes in San Diego Bay. *J. Hydr. Engin.*, in press.

Griffiths, G. and Flatt, D. (1987) A self-contained acoustic Doppler current profiler. ⏗peration. In: *Fifth International Conference on Electronics for Ocean Technology, Edinᵤ UK, 24−26 March 1987. IERE Publ. No. 72.* IERE, London, 41−47.

Jago, C.F., Bale, A.J., Green, M.O., Howarth, M.J., Jones, S.E., McCave, I.N., Millward, G.E., Morris, A.W., Rowden, A.A. and Williams, J.J. (1993) Resuspension processes and seston dynamics, Southern North Sea. *Phil. Trans. Roy. Soc. London Series A* **343** (1669), 475−491.

Jago, C.F., Jones, S.E. and Boon, D. Quasi-in-situ settling velocities of suspended particulate matter in the North Sea, in preparation.

Knight, P.J., Howarth, M.H., Flatt, D. and Loch, S.G. (1992) Current profile and sea-bed pressure and temperature records May 1990−July 1991, Dover Strait. *Proudman Oceanogr. Lab. Rep. No. 22*, 234pp.

Lavelle, J.W., Mojfeld, H.O. and Baker, E.T. (1983) An in situ erosion rate for a fine-grained marine sediment. *J. Geophys. Res.* **89**, 6453−6552.

Libicki, C., Bedford, K.W. and Lynch, J.F. (1989) The interpretation and evaluation of a 3 MHz acoustic backscatter device for measuring benthic boundary layer sediment dynamics. *J. Acoust. Soc. Am.* **85**, 1501−1511.

Lick, W. (1982) Entrainment, deposition and transport of fine-grained sediments in lakes. *Hydrobiologia* **82**, 31−40.

Lynch, J.F. (1985) Theoretical analysis of ABSS data for HEBBLE. *Mar. Geol.* **66**, 277−289.

Lynch, J.F., Gross, T.F., Brummley, B.H. and Filyo, R.A. (1991) Sediment concentration profiling in HEBBLE using a 1-MHz acoustic backscatter system. *Mar. Geol.* **99**, 361−385.

Moody, J.A., Butman, B. and Bothner, M.H. (1987) Near-bottom suspended matter concentration on the continental shelf during storms: estimates based on in-situ observations of light transmission and a particle size dependent transmissometer calibration. *Cont. Shelf Res.* **7**, 609:628.

Owen, M.W. (1976) Determination of the settling velocities of cohesive muds. *Hydr. Res. Stat. Wallingford, Rep. No. IT 161.*

Prandle, D. (1990) RRS "Challenger" Cruise 66B/90, 3−17 June 1990. Measuring the flux of contaminants through the Dover Straits. *Proudman Oceanogr. Lab. Cruise Report* **9**, 17pp.

Prandle, D. and Player, R. Residual currents through the Dover Strait measured by H.F. Radar. *J. Phys. Oceanogr.*, in press.

Prandle, D., Loch, S.G. and Player, R. (1993) Tidal flow through the Straits of Dover. *J. Phys. Oceanogr.* **23**(1), 23−37.

Shi, N.C., Larsen, L.H. and Downing, J.P. (1985) Predicting suspended sediment concentration on continental shelves. *Mar. Geol.* **62**, 255−275.

Thomson, R.E., Gordon, R.L. and Dymond, J. (1989) Acoustic Doppler current profiler observations of a mid-ocean hydrothermal plume. *J. Geophys. Res.* **94**, 4709−4720.

Thorne, P.D., Vincent, C.E., Hardcastle, P.J., Rehman, S. and Pearson, N. (1991) Measuring suspended sediment concentrations using acoustic backscatter devices. *Mar. Geol.* **98**, 7−16.

Weeks, A. and Simpson, J.H. (1991) The measurement of suspended particulate concentrations from remotely sensed data. *Int. J. Remote Sensing* **12**, 725−737.

11 Tidal Flushing of Semi-enclosed Bays

D. BOOTH

Département d'Océanographie, Université du Québec à Rimouski, Quebec, Canada

11.1 INTRODUCTION

No-one can deny the importance of coastal regions, whether for human activity or for the ecosystem as a whole. Bays, and in particular semi-enclosed bays, offer particular habitats and attractive conditions for enterprises such as aquaculture. The conditions within semi-enclosed bays depend, at least in part, on the exchange with the sea outside, i.e. flushing. The work described here focuses on small bays at the entrances of which tidal motion dominates the Eulerian flow. The river flow is small compared with the tidal flow, at least in summer.

In practice, evaluating the flux of some dissolved or suspended substance through a narrow entrance is not easy. Ideally, one should integrate through time and over the entrance cross-section the product uc, where u is the longitudinal current velocity and c the substance concentration. Spatial gradients make such an integration difficult, particularly when c has to be measured by discrete sampling.

A Lagrangian method of evaluating the exchange might consist in an attempt to follow a mass of water for several days and at the same time measuring the concentration within that mass. Drifting buoys, the only practical Lagrangian method for routine measurements on scales of between 10 m and 10 km in shallow water, lack the precision required. Furthermore, buoys do not respond to vertical motion and are subject to wind and wave drag. Slippage through the water is generally minimized by maximising the surface area of the drogue compared with the surface area of the supporting buoy. Any small slippage, however, may take the buoy across streaklines. Long-term drifts must therefore be considered with caution.

Although the present studies do rely on drifting buoys, no attempt was made to track them for periods longer than a tidal period. Indeed, tracking a group of 25 buoys for just 13 hours at these sites proved to be a challenge. Nevertheless, the results throw light on the mechanisms of exchange and it is these results and their implications that are presented here.

If we consider just one tidal cycle, then the net exchange is simply that part of the incoming flood water that does not immediately exit with the ebb tide. A more thorough definition would require recourse to the notion of 'age distribution' (Bolin and Rodhe, 1973), but for clarity we consider just one tide.

In a long narrow bay, most of the flood water is likely to return out on the ebb. Wide bays, however, are prone to asymmetries between the flood and ebb flow patterns. This is particularly so for bays with tidal jets (e.g. Bruun, Mehta and Johnsson, 1978). Topography, including simple slopes (Wolanski and Imberger, 1987), can affect the difference between the jet and the ebb flows. Moreover, the generation of eddies on either side of a jet (Kashiwai, 1984) complicates the circulation and can induce a net exchange. Wind stress and baroclinic pressure gradients also influence the exchange.

Mixing and Transport in the Environment. Edited by K.J. Beven, P.C. Chatwin and J.H. Millbank
©1994 John Wiley & Sons Ltd

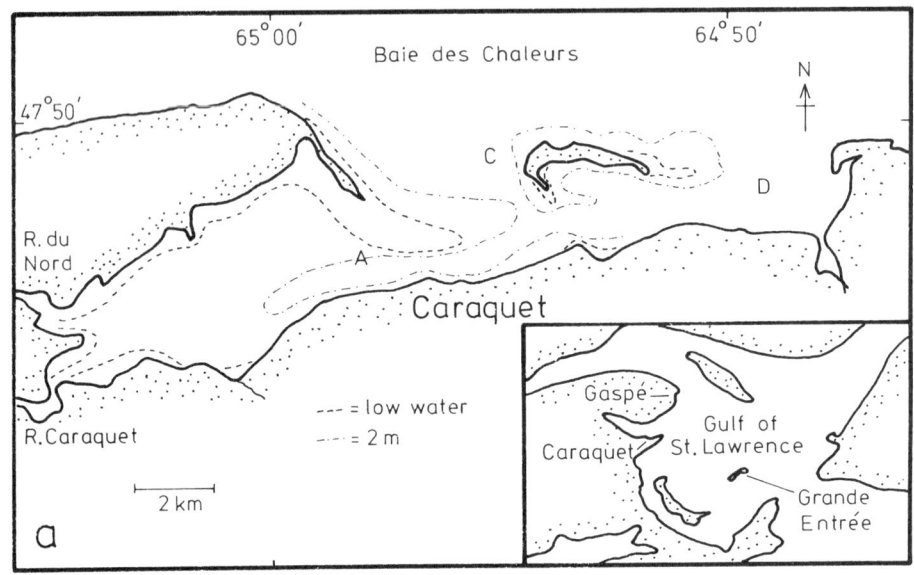

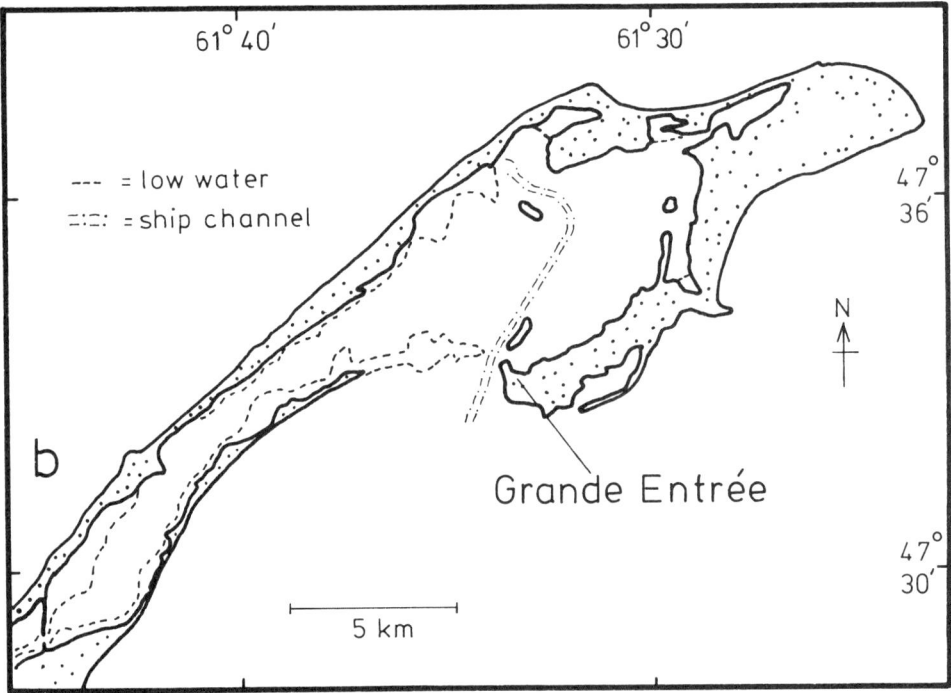

Figure 11.1 Study sites. (a) Caraquet Bay (A, C and D indicate mooring positions); (b) Lagune de la Grande Entrée

Between 1988 and 1990, three different embayments on the coast of the Gulf of St Lawrence were studied. In each instance, several deployments of drifting buoys were made while other recording instruments were in operation. Each study was undertaken in an attempt to evaluate, if only qualitatively, the flux of shellfish larvae or planktonic algae through the entrance.

11.2 SITES

11.2.1 Caraquet Bay

Caraquet Bay is a shallow bay on the south shore of Baie des Chaleurs (Figure 11.1a). Its horizontal dimensions are 8 × 3 km. It is protected by a dune, which at low tide stretches almost completely across the bay's width. Apart from the entrance channel, depths at low water are about 2 m. The River Caraquet, the river entering the south-west corner of the bay, is the principal river. Two smaller rivers contribute to the total runoff of about 13 m^3 s^{-1} in spring and 3 m^3 s^{-1} in summer. Tides are mixed with a large tidal range of 2.1 m. The bay is a centre of oyster cultivation. Collectors are set out for spat. Ten months later the seed oysters are scattered on the bottom. The bay is ice-bound in winter. The bay was visited in 1988 during July, the pelagic period of the oyster larvae.

11.2.2 Lagune de la Grande Entrée, Iles de la Madeleine

The Lagoon is the northernmost and longest lagoon of the archipelago of the Iles de la Madeleine (Figure 11.1b). It is about 25 km long and 1–5 km wide. It is connected to Lagune du Havre aux Maisons to the south by a narrow channel, but the principal contact with the

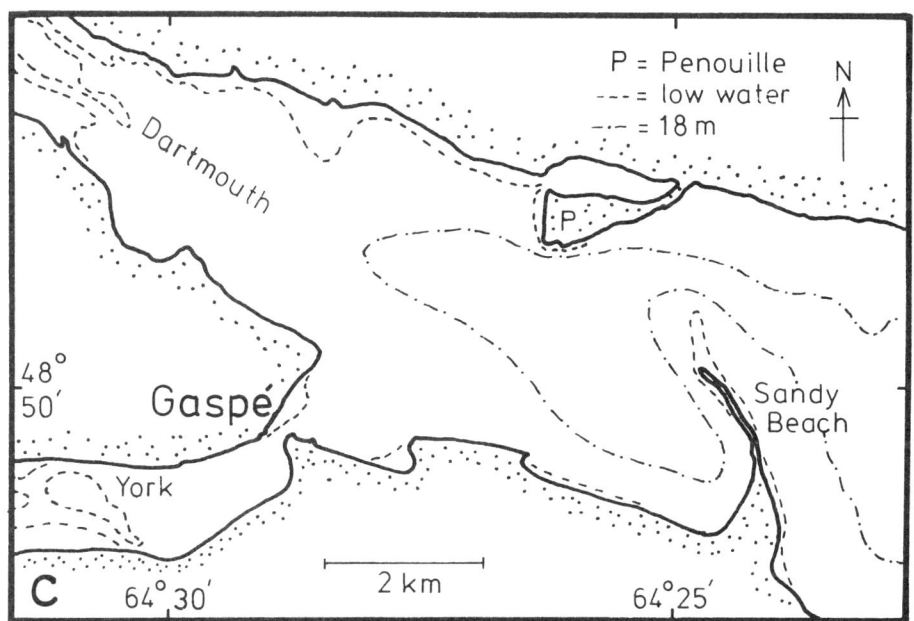

Figure 11.1 (*continued*) (c) Havre de Gaspé

sea outside is by its main entrance. A ship channel, which is dredged regularly, splits the lagoon into two, the north-eastern part being the deeper with depths of 5–6 m. Mussels are cultivated in this deeper area. There are no rivers, but icemelt provides fresh water in spring and there is some evidence in STD profiles that groundwater seeps from the bottom of the lagoon. The tides are mixed diurnal–semidiurnal with a large tidal range of 0.9 m. As part of a joint study with the Institut National de la Recherche Scientifique, Rimouski, to study the potential of the lagoon for mussel culture, the lagoon was visited twice, in May and September 1989.

11.2.3 Havre de Gaspé

The head of Baie de Gaspé is protected by the bar of Sandy Beach to form Havre de Gaspé, which has dimensions of 8 × 3 km (Figure 11.1c). The entrance channel slopes upwards

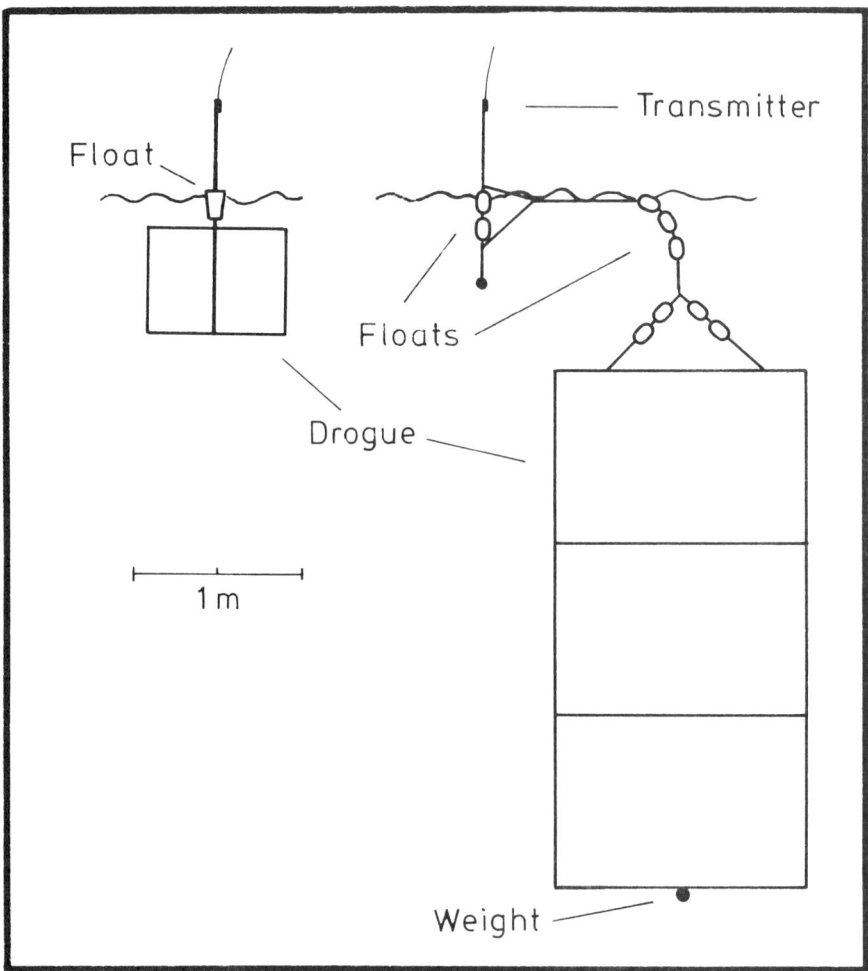

Figure 11.2 Two drifting buoys, both with window-blind drogues. The smaller buoy was used in 1988 and 1989, the larger one in 1990

into the harbour, inside which depths in the outer half are uniform at about 20 m. Two rivers enter the harbour, the Darmouth and the York, the latter entering a shallow tidal basin before emerging into the harbour. The York drainage basin is the larger by 26%. Total freshet runoff reaches 250 $m^3 s^{-1}$, but summer values are about 25 $m^3 s^{-1}$ (Carrière, 1974). The tides are again mixed, with a large tidal range of 1.9 m. The harbour was visited during June 1990. The question of interest was the possible flux of toxic algae into the harbour.

11.3 MATERIALS AND METHODS

Drifting buoys used at Caraquet and in Lagune de la Grande Entrée were almost identical (Figure 11.2). Slippage due to wind drag alone on these buoys was about 4×10^{-3} of the wind velocity. In 1988 the buoys were deployed in groups to aid visual contact, but the following year micro-transmitters were attached to some buoys to allow location by radio direction finding. For the work at Gaspé, new buoys with larger drogues and small tethered floats were constructed (Figure 11.2). All the buoys were now equipped with transmitters. Slippage due to wind drag on these buoys was about 8×10^{-4} of the wind velocity. Drogues on these buoys were centred at depths of either 2.5 or 9 m.

Buoys were followed by boat and positioned with a Motorola Miniranger microwave system with two or three slave units on shore referenced to geodesic points. The system was checked over known distances. Position accuracies were at best of the order of 2 m. Twenty or 25 buoys were usually deployed and their positions fixed about once an hour. In all, 26 deployments were made and of these four provided series of over 12.5 hours.

Temperature and salinity profiles were made with an Applied Microsystems STD12 equipped with a shallow water pressure transducer. Eulerian currents and time series of temperature and salinity were obtained with current meters of the types Aanderaa RCM4S/7 and InterOcean S4. Seven current meters were deployed in Caraquet and at Gaspé. In the lagoon, a total of 19 current meters were deployed for a joint study. Also deployed at each site were Aanderaa pressure recorders WLR5/7 and an Aanderaa meteorological station. The instruments were typically deployed for periods of about four to six weeks.

11.4 RESULTS

11.4.1 Caraquet Bay

During July 1988 the bay was generally well mixed apart from two regions: the channel, where a halocline was occasionally found, and within the estuarine plume of the River Caraquet. The estuarine layer in the inner part of the bay was sometimes too thin to be detected with the STD, but was nevertheless visible by a glimmering sheen in the water. The vertical shear across 10 or 20 cm was made more obvious by drifting fragments of sea grass. The behaviour of the plume, which was delineated by slicks and the accumulation of weed, is of interest. On the ebbing tide the plume expanded out from the river mouth to cover the southern part of the bay. In the channel, the plume became narrow with its northern edge near the inner side of the channel. On the flood, the plume was pushed back. On one occasion, a temporary V-shaped front (Largier, 1992) was observed at the inner end of the channel. By high water the plume was back in the river mouth.

A back and forth drift of the buoys drogued at 0.5 m resembled the oscillatory behaviour of the plume. In general, buoys released during the ebb in the inner part of the bay drifted

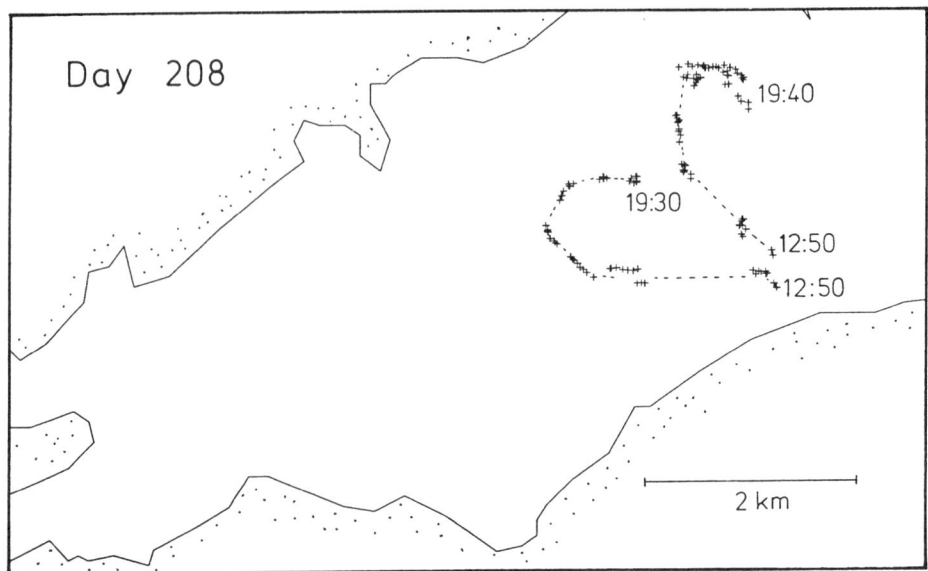

Figure 11.3 Caraquet Bay, 1988, day 208. Two groups of buoys deployed on the north side of the channel. Crosses represent position fixes per buoy, except at deployment where they indicate limits of group. Wind: $1-5$ m s^{-1} from the south-west and south. Observed tidal elevations: 1140, -0.5 m; 1710, 0.0 m; 2210, -0.5 m (times in UTC)

seawards towards or into the channel and returned on the flood with a net drift of about $2-4$ cm s^{-1} towards the east or south-east. Apart from the area just north of the channel, the flood−ebb flow patterns were almost symmetrical (Sephton and Booth, 1992).

Two sets of buoy results are of interest for the present subject, those of days 208 and 211. On the former day, two groups of buoys were deployed near the edge of the channel about an hour after low water (Figure 11.3). Both groups moved in an anticyclonic manner. Separation of the two groups occurred in the main near the edge of the channel just after deployment. At this time, the wind was from the south with a speed of about 1 m s^{-1}. Three days later, one group of 20 buoys was released near the northern river mouth two hours after high water (Figure 11.4). The group entered the channel and elongated while drifting along a temporary slick near the edge of the channel. By low water the group was midway down the channel where it turned, as if to return on the flood. Abruptly, however, the group deviated northwards and crossed the edge of the channel. The wind was about 1 m s^{-1} from the north. The slick disappeared. The group then drifted directly towards the north shore, becoming dispersed on the way.

During the five day period preceding day 211, temperatures outside the bay decreased (Figure 11.5), with a corresponding increase in salinity. In the inner part of the bay, the change was more gradual and less pronounced. Tidal variations of both temperature and salinity in the channel were enhanced during this period.

11.4.2 Lagune de la Grande Entrée

Currents in the lagoon were weak, particularly in the north-eastern part where currents rarely exceeded 0.1 m s^{-1}. In the narrower south-western part, tidal currents reached 0.4 m s^{-1},

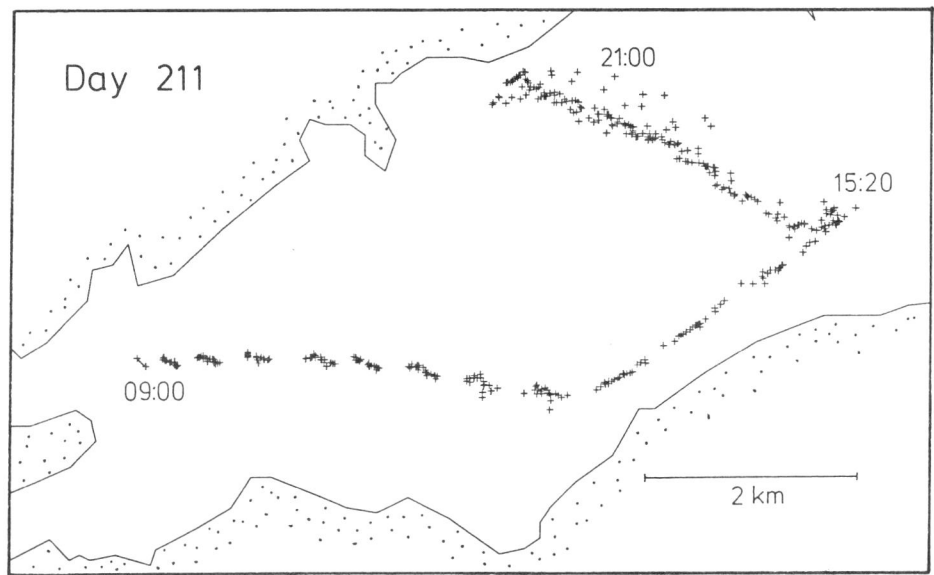

Figure 11.4 Caraquet Bay, 1988, day 211. One group of 20 buoys deployed near the mouth of North River. Symbols are as for Figure 11.3. Wind: $0-3$ m s^{-1} from the north-west until low water, then veering to south. Observed tidal elevations: 0650, 1.1 m; 1450, -0.8 m; 1930, 0.2 m (times in UTC)

but it is around the entrance that the tidal currents were the strongest, with speeds of 0.8 m s^{-1} (Navarro, 1991).

The lagoon was generally well mixed. Causes of stratification include icemelt, solar heating and exchange through the entrance, but the habitual wind quickly breaks up any stratification.

Of the eight buoy experiments conducted in the lagoon, four involved buoys deployed at the entrance on the flood tide. Efforts to follow the buoys for a complete tidal cycle were unsuccessful, but the results did show the guiding effect of both the dredged and secondary channels (Figures 11.6 and 11.7). The tidal excursion in the dredged channel for moderate tides penetrated to the north side of the lagoon. Some buoys did begin to return along the same path; others drifted away further into the lagoon. One buoy, deployed in the entrance on day 260, was found in the channel two days later. A feature common to all deployments in the entrance was the separation of the flow between the ship channel and the natural channel to the north-west. On day 260, two buoys deployed 25 m and 30 seconds apart chose different channels, thus drifting into different regions. Wind drag on both days undoubtedly contributed to the buoy displacements, but deployments inside the north-eastern part of the lagoon indicated that wind induced much of the circulation there.

11.4.3 Havre de Gaspé

Inside the harbour, Eulerian currents at mid-depth peaked typically to 0.2 or 0.3 m s^{-1}. In the channel, currents were stronger with peak speeds of 0.7 m s^{-1} at a depth of 18 m, the flooding current being swifter than the ebb. For all meters below a depth of 3 m, the time series of salinity and temperature were inversely correlated on a time-scale of days with a lag between mid-bay and the channel of one or two days. Both in the channel and inside the harbour the vertical density structure varied from one with a single pycnocline at about

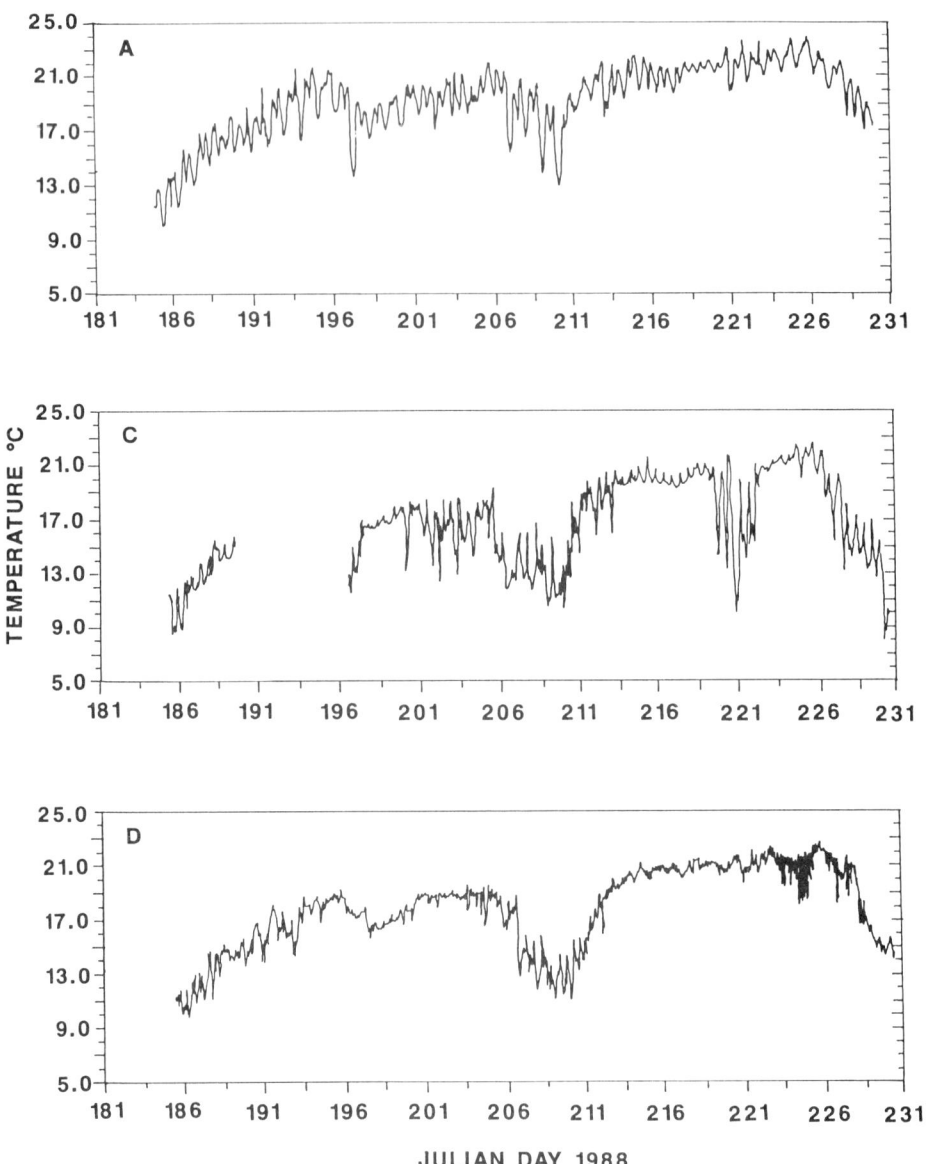

Figure 11.5 Temperatures recorded at 3 m depth at positions A, C and D (see Figure 11.1)

2 m depth to one resembling a step formation with two or three pycnoclines. The plume emanating from the River York, as indicated by surface patterns, was irregular and tidal.

A subsample of drogue tracks are presented for days 181 and 183. Complete results are presented in Pettigrew, Booth and Pigeon (1991). On the former day, 25 buoys were released between two hours and one hour before high water (Figure 11.8). Tides were neaps with a range of 0.5 m. The buoys with shallow drogues (centred at 2.5 m) clearly followed a cyclonic gyre in the eastern part of the harbour. Ten other buoys (not shown) released in mid-bay

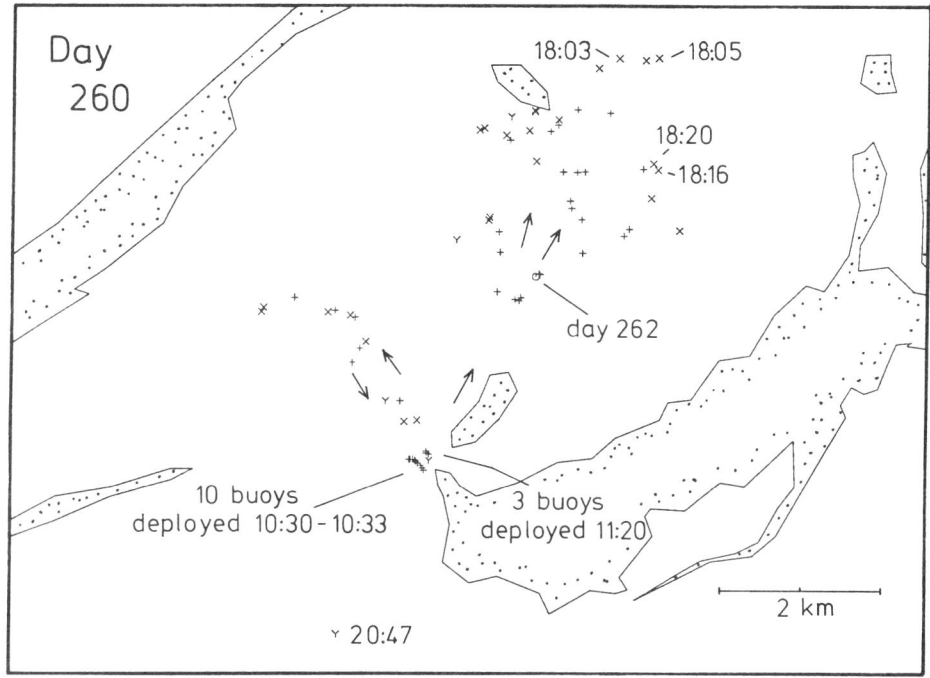

Figure 11.6 Grande Entrée, 1989, day 260. Two groups of buoys with a total of 13 buoys deployed in the entrance. Position fixes: + = before 1435; × = 1552–2045; ⋎ = 2024 to 2206; o = 1630, day 262. Wind: 4–6 m s^{-1} from the south-west. Predicted tides: 0745, 0.3 m; 1410, 1.0 m; 2005, 0.6 m (times in UTC)

also followed this motion. Six buoys released near the south shore (three of which are shown in Figure 11.8a) ended up after the tidal cycle in the channel, thus reflecting the behaviour of the York plume. Two of five buoys with deep drogues (centred at 9 m) also picked up this gyre, but the other three moved westwards to circulate in an anticyclonic gyre (Figure 11.8b). Buoys 4 and 5 of Figure 11.8b were deployed 480 m and nine minutes apart. STD casts at buoy 4 revealed surface and bottom layers, both with a thickness of about 2 m. Some small steps were recorded at mid-depth. Winds were light with speeds less than 5 m s^{-1} and direction from the north-west or north.

Two days later the buoys were deployed across the channel and across the bay between four and three hours before high water (Figure 11.9). The morning tidal range was 0.3, but the evening high tide was 0.4 m higher than the preceding tide. The five deep drogued buoys (Figure 11.9b) and two of the shallow drogued buoys moved in a cyclonic gyre in the eastern part of the harbour. Of the buoys deployed in the channel (not shown), most drifted outwards and then back in. Two of these, those nearest Penouille, moved directly inwards. Likewise, six buoys (not shown) deployed in the outer and mid parts of the bay moved north-westwards. Those deployed just south of Penouille, however, diverged, two moving towards the channel, the others inwards (Figure 11.9a). Buoys 5 and 6 of Figure 11.9a, deployed within three minutes of each other and 260 m apart were separated by 5 km after 12 hours. Two hours after deployment a slick was observed near buoy 7 of Figure 11.9a. The water column was stratified with a pycnocline at depths of between 2 and 3 m. Winds were light to moderate, attaining speeds of 7 m s^{-1} from the south-east during the early afternoon.

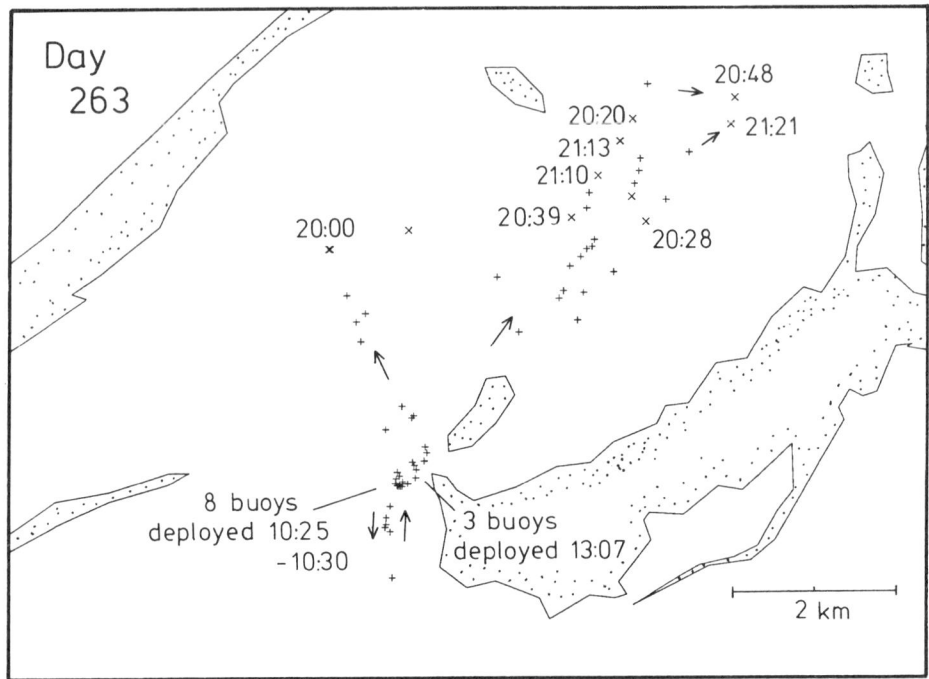

Figure 11.7 Grande Entrèe, 1989, day 263. Two groups with a total of 11 buoys deployed in the entrance. Position fixes: + = before 1705; x = after 1705. Wind: 5−0 m s⁻¹ from the south-west. Predicted tides: 0950, 0.2 m; 1705, 0.9 m; 1845, 0.7 m (times in UTC)

11.5 DISCUSSION

The flushing time of an embayment estimates the time required to renew or replace a certain percentage of the water therein. It is an imprecise term, often used without qualification, but for which rigour requires use of the 'age distribution' of the water (Bolin and Rodhe, 1973; Zimmerman, 1988). In the simplest scenario, however, flushing is determined simply by the fraction of the tidal prism that remains inside, the rest being washed out with the ebb. This obviously does not account for the return of ebb water. It indicates the amount of water replaced each tide, irrespective of the previous history of the water. With this definition, the asymmetry in the tidal circulation, including the displacements due to small-scale motions, describes the mechanisms of flushing.

11.5.1 Tidal jets

Because of a construction at the entrance, the flood tide enters the embayment as a jet. The flood water may penetrate far into the embayment, as in the case of Caraquet Bay and Lagune de la Grande Entrée. The ebb tide, on the other hand, withdraws water in a more radial manner (Bruun, Mehta and Johnsson, 1978). The side channels often found diagonally inside the entrance (e.g. Drapeau, 1988) may be a result of erosion by this ebb flow. The flow is, however, controlled by topography. Even a simple slope has the effect of guiding the retreating flow, thus reducing the asymmetry between the flood and ebb flows (Wolanski and Imberger,

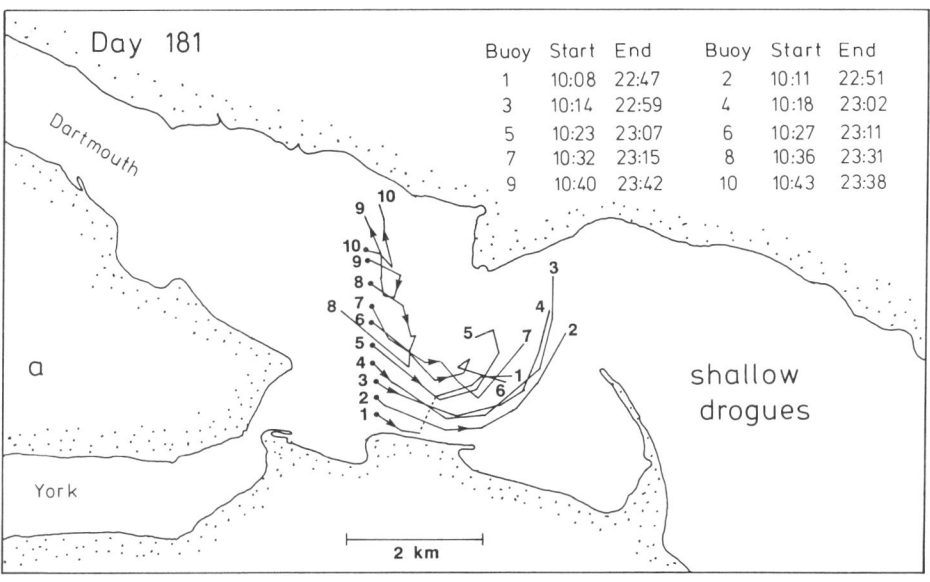

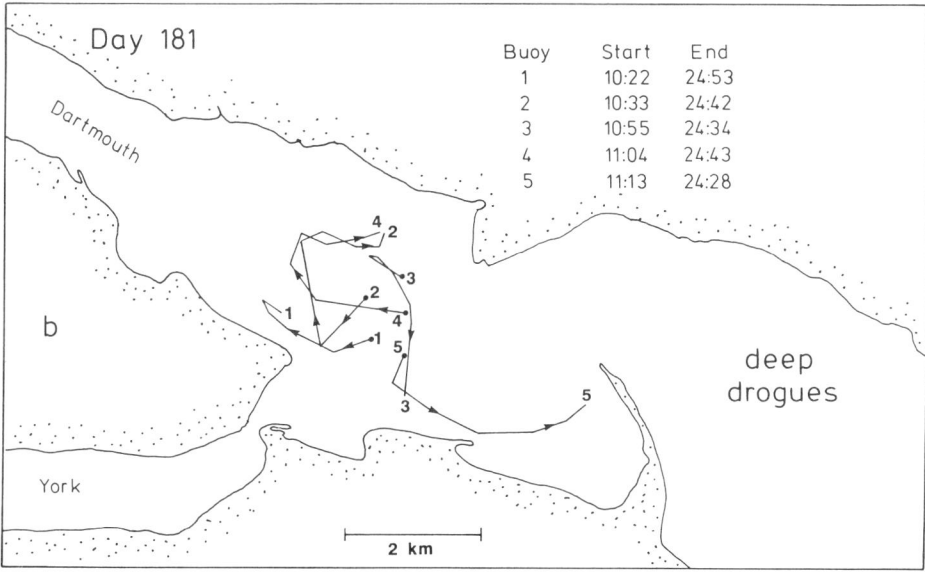

Figure 11.8 Havre de Gaspé, 1990, day 181. Fifteen of the 25 buoys deployed. (a) Shallow drogues; (b) deep drogues. Wind: $1-6 \text{ m s}^{-1}$ from the north-west and north. Predicted tides: 1115, 1.1 m; 1730, 0.6 m; 2420, 1.2 m (times in UTC)

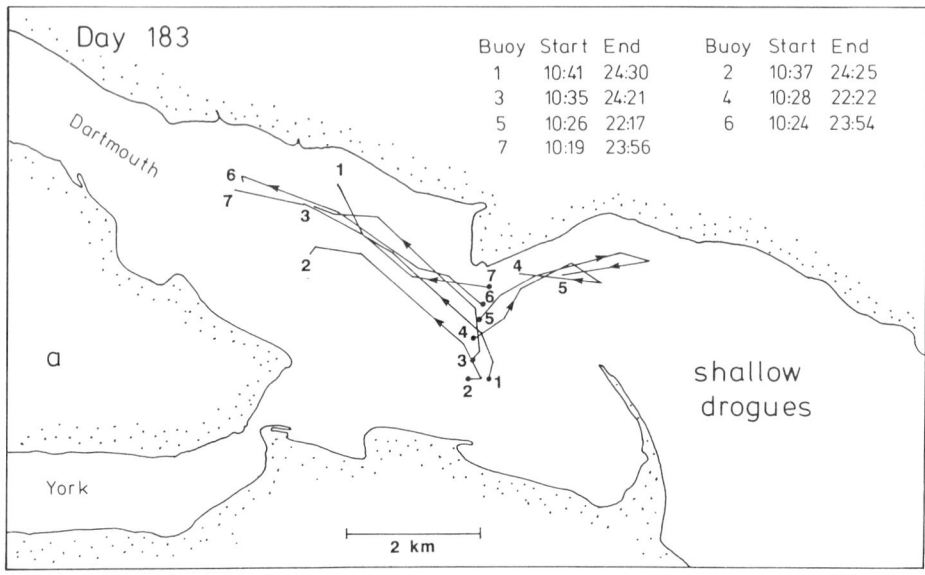

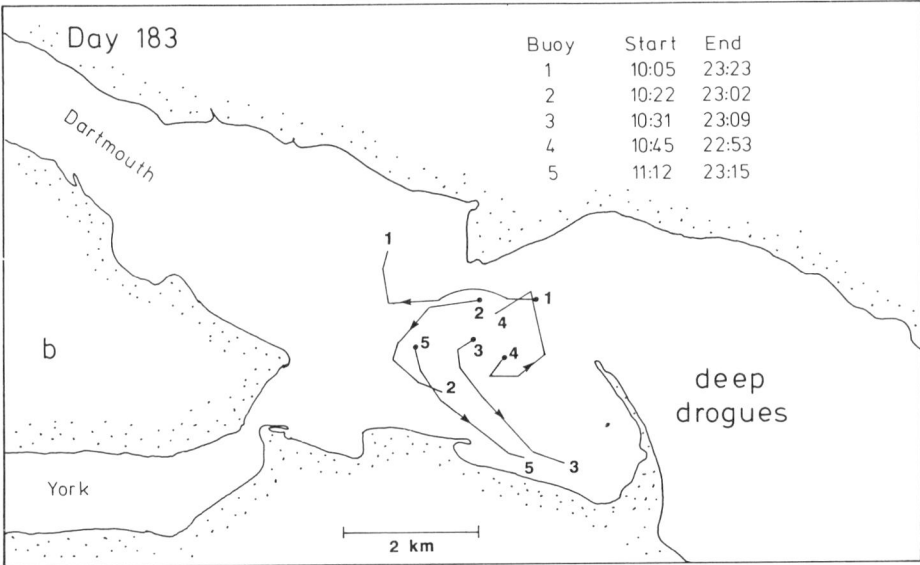

Figure 11.9 Havre de Gaspé, 1990, day 183. Twelve of the 25 buoys deployed. (a) Shallow drogues; (b) deep drogues. Wind: $1-7\,\mathrm{m\,s^{-1}}$ from the north-west and later the south-east. Predicted tides: 1305, 0.9 m; 1840, 0.6 m; 0210, 1.3 m (times in UTC)

1987). In the case of a long channel, the guiding is extreme. Both Caraquet Bay and Lagune de la Grande Entrée showed topographic guiding. The channels thus limit their tidal flushing.

11.5.2 Tidal eddies

Eddies, generated by discontinuities on either side of the jet (Kashiwai, 1984), play a part in returning water to the entrance region. Eddies were observed at both Gaspé and Caraquet. Evidently, not enough data were collected to deduce the residual circulation, but it is clear that eddy motion contributed to the tidal circulation. According to Kashiwai (1984), the behaviour of the eddies, and thus the exchange through the entrance, depends on the tidal prism (the volume of water between low and high tidal levels) and the dimensions of the channel. His results suggest that the tidal exchange due to the vortex displacement should be greater in Lagune de la Grande Entrée than at the other two sites, in the main because the length of the entrance opening is short. The results apply to flat-bottomed basins and do not account for topographic effects. In the case of Caraquet, for example, Kashiwai's analysis indicates that a permanent eddy should fill the bay, but an eddy motion was only observed in the north-eastern part of the bay. The existence of eddies, however, allows another approach to the question of flushing. Using Aref's blinking vortex flow (Khahker, Rising and Ottino, 1986) to represent alternating tidal eddies inside and outside the bay, it can be shown for Caraquet that dispersion by these eddies is expected to be weak, except during spring tides. This kinematic model is, however, again rather unrealistic.

11.5.3 Wind stress

Wind stress is undoubtedly effective in removing water from the captive region around the entrance. In the north-eastern part of Lagune de la Grande Entrée, for example, tidal currents were weak compared with other more irregular variations. During moderate winds ($10 \, \text{m s}^{-1}$), displacements there during half a tidal cycle can be 3 km, which is greater than the ebb captive region based on a tidal prism of $64.1 \times 10^6 \, \text{m}^3$ (Drapeau, 1988) and a typical depth of 5 m. Although bottom water may be trapped in the channel and although counter flows against the wind may return flood water to the captive region, it is likely that much of the flood water can be swept away by strong winds, thus providing efficient flushing. Wind stress is expected to be as important in Caraquet Bay, but because of the greater depth, less so in Havre de Gaspé.

11.5.4 Density gradients

When the density of the flooding water differs from that of the water inside, another process occurs, that of interleaving or intrusion. In Caraquet Bay, and to a lesser extent in Havre de Gaspé, frontal convergences at the surface indicated such occurrences. Inside both these embayments, salinities and temperatures reflected the changes occurring outside. Bottom layers were also found in Lagune de la Grand Entrée and in Havre de Gaspé. During July 1987 a low salinity layer was found in the lagoon (Booth, 1991). Although later measurements of unstable water columns suggested seepage from the lagoon sediments, the salinity of surface waters outside the lagoon are at this season reduced by the freshet from the St Lawrence River (Petrie, 1990). The outer surface waters remain denser than waters inside the lagoon due to their lower temperatures.

In all three locations, therefore, there occur conditions susceptible to gravity intrusions, the propagation speed of which is determined by the difference in density, the thickness of the layer, and the ambient density profile, and is in general proportional to the internal Froude number (Simpson, 1982). This raises the possibility of estimating the removal of flood water from the captive region by density currents based simply on the hydraulics of the intrusion. For Lagune de la Grande Entrée, a simple such estimate gave a volume left behind by a tide of 0.1 times the tidal prism (Booth, 1991). Mixing at the head of the intrusion (Simpson and Britter, 1979) has the effect of increasing the flow inside the intrusion and dispersing some of the intrusion water into the ambient water. Entrainment must also be included in the small-scale motion (Luketina and Imberger, 1989). Thus increased scatter in the true Lagrangian displacements should be expected in such regions, but it is the irreversibility of these flows that make them effective in flushing.

11.5.5 General remarks

When attempting to estimate flushing some recourse is usually made to mixing between the flood water and the ambient water inside, whether or not this is done sectionally along the inlet (e.g. Zimmerman, 1976). Because of the complexity of the mechanisms involved, mixing is represented by some empirical expression, an exchange coefficient, for example. Exchange may, however, occur in pulses rather than as a continuous leakage. Although it is difficult to distinguish vertical and horizontal fluxes, time series of temperature and salinity from both Caraquet and Gaspé did suggest pulsed exchanges.

Exchange through a strait is starkly illustrated by the numerical two-dimensional model of Awaji *et al.* (1980), where the interleaving striated convolutions between the original two water bodies became increasingly complex. Although some diffusion is included in the model, turbulence is not necessary to achieve dispersion. The waters are stirred rather than mixed. Dispersion without turbulent mixing is clearly demonstrated, for example, by the simple models of fluid flow presented by Khakhar, Rising and Ottino (1986) and Pasmanter (1988; 1991). Although the series of buoy data is too short by far to indicate the degree of dispersion, except at a local scale, there were indications that dispersion was not just by turbulence alone.

In Caraquet Bay, one group of buoys had a net tidal displacement comparable with the tidal excursion (Figure 11.4), and this because of its displacement across the edge of the channel. Water that remained in the channel would have returned along the channel and into the western part of the bay. Thus not only the dispersion of the group, but also its residual displacement depended on a process at some small scale. The sudden change in direction of the buoys was startling at the time of observation, but it was clear visually that water was flowing over the shallow edge of the channel. We might expect that the lateral pressure gradients were associated with the side-slope of the channel, which according to the original field chart of the published nautical chart extends over about one-tenth of the channel width, i.e. a few tens of metres. For the case of uniform density, the lateral pressure gradient would imply upwelling and thus divergence at the surface. It is not clear if density played a part, but this was a period when the density outside the bay was changing and tidal variations in the channel were enhanced. During this same period oyster larvae were flushed in and out of the bay (Sephton and Booth, 1992). As brackish water entered the channel on the ebbing tide and high salinity water pushed it back on the flood, it is most probable that the density gradient across the edge of the channel alternated with the tide, thus changing the direction of any underflowing intrusions. Whatever the process, it appears that a separation in the flow was introduced by the topography.

Separation of trajectories was observed at all three sites. At Caraquet on day 208, two groups of buoys separated near the edge of the channel. In Lagune de la Grande Entrée, the inward flow through the entrance separated between the two channels. At Gaspé there were separations of flows both at the surface and at mid-depth, the latter between two gyres. The surface separation occurred near Penouille and may have been associated with a plume.

Separation does not necessarily imply divergence, but divergence, whether one- or two-dimensional, does imply separation and requires a source from a flow on another axis, often the vertical. Divergences occur near coastal upwelling, fronts (Simpson and James, 1986), in Langmuir circulations, upstream of topography, and in front of advancing intrusions. Net Lagrangian displacements are known to be sensitive to the tidal phase of release and thus the position of release, particularly near the times of slack water (Gomez-Reyes, 1989). In regions where the length scale of the horizontal shear is much smaller than the tidal excursion, this sensitivity is enhanced. Shear is induced by both density gradients and topography, both of which can have small horizontal scales in estuarine bays. Thus these bays are likely sites for flow separation and sensitivity to position, conditions that can lead to chaotic displacements.

The fact that these flows are either temporal or have a component in the third spatial dimension supports the prospect of chaotic displacements (Ottino, 1989), if only at a local or intermittent level. A criterion for chaos that is lacking from the experimental results is an exponentially increasing separation. Although dispersion studies in the sea do indicate an exponential curve (Okubo, 1974), this is normally attributed to the turbulent mixing. It is tempting to take the buoy results as indicating the presence of certain regions where dispersion is not just a result of turbulence, but rather a sensitivity to position, i.e. regions of what Zimmerman (1986) calls deterministic dispersion. Small-scale turbulent motion might be expected to enhance the dispersive effects of sensitive regions by carrying particles across the shear, a process known as shear diffusion in a Eulerian framework (Sanderson and Okubo, 1987). Dispersion must be particularly enhanced when the initial separation leads tracers into different regimes of flow as, for example, two different gyres, two different layers, or as in the case of Lagune de la Grande Entrée, two regions with different responses to local forcing. By causing a dispersion of flood water out of the ebb captive region, sensitivity to position within a limited region enhances flushing. The results lead us to the suggestion that regions of sensitivity to position — i.e. regions of strong shear or divergence — must be considered in any evaluation of flushing.

11.6 CONCLUSIONS

Despite the 26 buoy experiments, the volume of data is still diminutive. The few clues in the data, together with attempts to imagine the flow field, suggest the following scenario based on the tidal flushing being determined by the asymmetry between the flood and ebb flows. The kinetic energy of the entrance jet is a major factor in this asymmetry, but topography tends to decrease the asummetry. Wind stress can enhance flushing by removing water from the captive region around the entrance. Density currents also enhance flushing by propagating into the ambient water. These are the bulk motions. Added to these, and on top of the background turbulence generated by stress, appear to be certain regions where the displacement is sensitive to position. Sensitivity is likely not only in regions of high shear, such as at the pycnocline, but also at divergences, whether they be one or two dimensional. Divergences are likely to occur near vertical motions (including those of Langmuir circulation), near topographic obstacles, near the boundaries of intrusions whether jet-like or plume-like, and between eddies. As the sensitivity to position provides a means for the small scale to interact directly with large scales, the resulting dispersion is expected to be efficient and thus important in the determination of flushing. Quantifying these effects is the challenge that awaits.

ACKNOWLEDGEMENTS

Many colleagues contributed to this work, which was undertaken while the author was employed by the Canadian Department of Fisheries and Oceans. In particular, I thank Roger Pigeon, Sylvain Cantin, Bernard Pettigrew, Rémi Desmarais and Danny Boulanger for their dedication to the experimental work.

REFERENCES

Bolin, B. and Rodhe, H. (1973) A note on the concepts of age distribution and transit time in natural reservoirs. *Tellus* **25**, 58–62.

Booth, D. (1991) Flusing of Lagune de la Grande Entrée. In: *Proceedings of Workshop on Summer Mortality of Mussels on the Magdelaine Islands, April 1990*, CAPZ, Quebec.

Bruun, P., Mehta, A. and Johnsson, I. (1978) *Stability of Tidal Inlets — Theory and Engineering*. Elsevier, Amsterdam.

Carrière, J.B. (1974) A physical oceanographic study of the Havre and Baie de Gaspé. *MSc Thesis*. Dalhousie University, Halifax.

Drapeau, G. (1988) Stability of tidal inlet navigation channels and adjacent dredge spoil islands. In: *Hydrodynamics and Sediment Dynamics of Tidal Inlets* (Eds D.G. Aubrey and L. Weishar) *Lecture Notes in Coastal and estuarine Studies 29*. springer, New York, 226–244.

Gomez-Reyes, E. (1989) Tidally driven Lagrangian residual velocity in shallow bays. *PhD Thesis*. University of New York, Stony Brook.

Kashiwai, M. (1984) Tidal residual circulation produced by a tidal vortex Part 1 — life history of a tidal vortex. *J. Oceanogr. Soc. Jpn* **40**, 279–294.

Khakhar, D., Rising, H. and Ottino, J. (1986) Analysis of choatic mixing in two model systems. *J. Fluid Mech.* **172**, 419–451.

Largier, J.L. (1992) Tidal intrusion fronts. *Estuaries* **15**, 26–39.

Luketina, D. and Imberger, J. (1989) Turbulence and entrainment in a buoyant surface plume. *J. Geophys. Res.* **94**, 12619–12636.

Navarro, N. (1991) Océanographie physique de la lagune de la Grande Entrée, Iles de la Madeleine, Golfe de St. Laurent. *MSc Thesis*. University du Québec, Rimouski.

Okubo, A. (1974) Some speculation on oceanic diffusion diagrams. *Rapp. P.-v. Réun. Cons. Int. Explor. Mer* **167**, 77–85.

Ottino, J. (1989) The mixing of fluids. *Sci. Am.* Jan, 56–67.

Pasmanter, R. (1988) Anomalous diffusion and anomalous stretching in vortical flows. *Fluid Dyn. Res.* **3**, 320–326.

Pasmanter, R. (1991) Deterministic diffusion, effective shear and patchiness in shallow tidal flows. In: *Non-linear Variability in Geophysics* (Eds D. Schwetzer and s. Lovejoy). Kluwer, Dordrecht.

Petrie, B. (1990) Monthly means of temperature, salinity and sigma-t for the Gulf of St. Lawrence. *Can. Tech. Rep. Hydrogr. Ocean Sci.* **126**, 137pp.

Pettigrew, B., Booth, D. and Pigeon, R. (1991) Oceanographic observations in Havre de Gaspé during the summer 1990. *Can. Data Rep. Fish. Aquat. Sci.*, **100**, 94pp.

Sanderson, B. and Okubo, A. (1987) Comments on the "shear effect" and diffusion in the Lagrangian framework. *J. Oceanogr. Soc. Jpn* **43**, 183–196.

Sephton, T. and Booth, D. (1992) Physical ocenographic and biological data from the study of the flushing of oyster (*Crassostrea virginica*) larvae from Caraquet Bay, New Brunswick. *Can. Manuscr. Rep. Fish. Aquat. Sci.*, **2162**, 61pp.

Simpson, J.E. (1982) Gravity currents in the laboratory, atmosphere and ocean. *Annu. Rev. Fluid Mech.* **14**, 213–234.

Simpson, J.E. and Britter, R. (1979) The dynamics of the head of a gravity current advancing over a horizontal surface. *J. Fluid Mech.* **94**, 477−495.

Simpson, J.H. and James, I. (1986) Coastal and estuarine fronts. In: *Baroclinic Processes on Continental Shelves* (Ed. C. Mooers), *Coastal and Estuarine Sciences 3*. American Geophysical Union, Washington, 63−93.

Wolanski, E. and Imberger, J. (1987) Friction-controlled selective withdrawal near inlets. *Est. Coast. Shelf Sic.* **24**, 327−333.

Zimmerman, J. (1976) Mixing and flushing of tidal embayments in the western Dutch Wadden Sea, part II: analyses of mixing processes. *Neth. J. Sea Res.* **10**, 397−439.

Zimmerman, J. (1986) The tidal whirlpool: a review of horizontal dispersion by tidal and residual currents. *Neth. J. Sea Res.* **20**, 133−154.

Zimmerman, J. (1988) Estuarine residence times. *Hydrodynamics of Estuaries* (Ed. B. Kjerve), Vol. I. *Estuarine Physics*. CRC Press, Boca Raton, 75−84.

12 Measurements of Tidal Currents and Estimated Energy Fluxes in the Bristol Channel

R.J. UNCLES and M.B. JORDAN
Plymouth Marine Laboratory, Plymouth, UK

12.1 INTRODUCTION

The Severn Estuary and Bristol Channel constitute a large, partially enclosed body of water in the south-west of the British Isles (Figure 12.1). The section UV (Figure 12.1) can be considered as the seaward boundary of the Bristol Channel and the section YZ as the seaward boundary of the Severn Estuary.

The area has received considerable attention from coastal oceanographers because of the theoretical interest arising from its large tides and the practical interest in tidal power generation (Shaw, 1987). Three-dimensional research models of the circulation in the Bristol Channel have been presented by Owen (1980), Stephens (1986) and Wolf (1987). Two-dimensional models have been used for investigation of the tidal and residual circulations (Uncles, 1981; 1982; 1983a; 1983b). Circulation models have also been used for studies of water quality (Radford, Uncles and Morris, 1981), ecological (Uncles and Joint, 1983) and meteorological problems (Proctor and Flather, 1989).

Whatever the application of circulation models, validation is required based on field observations of currents. It was for this reason that we made observations of currents in the Bristol Channel and Severn Estuary from 1975 to 1977. The long term (one month) data sets from this period of measurements continue to be used by theoreticians (Stephens, 1986; Wolf, 1987) although the data have never been published nor interpreted other than as points on observed−modelled validation graphs (Uncles, 1981) or as numerical values in observed−modelled comparison tables (Stephens, 1986).

The intention of this chapter is to present and interpret the data from this campaign of observations. During the campaign emphasis was placed on observing currents at as many stations, and over as wide an area, as was feasible with the resources available. In consequence, the observations at most stations extended over only a few tidal cycles, although at five stations the observations extended over one month. These data delineate the vertical structure of the semidiurnal currents over the region. In conjunction with a simple vertical structure model and a depth-averaged, two-dimensional numerical tidal model of the region, the data enable estimates to be made of the depth-averaged M_2 tidal currents. These depth-averaged currents are used to draw co-phase and co-amplitude lines for the M_2 surface elevations and to derive estimates of the associated fluxes of tidal energy across the sections UV, WX and YZ (Figure 12.1).

Mixing and Transport in the Environment. Edited by K.J. Beven, P.C. Chatwin and J.H. Millbank
© 1994 John Wiley & Sons Ltd

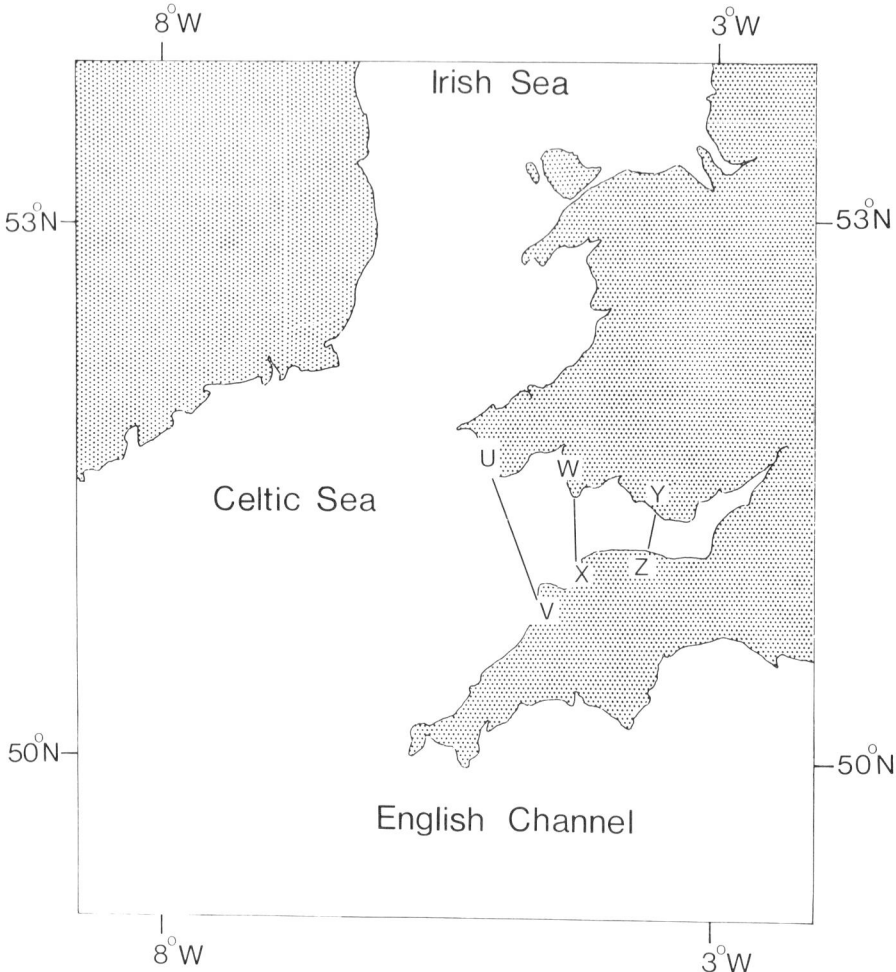

Figure 12.1 Location map of south-west British Isles and approaches. The Bristol Channel can be defined as the region between sections UV and YZ and the Severn Estuary as the region landward of section YZ

Most of the early observations of currents in the Bristol Channel and Severn Estuary were made for navigational purposes using float and line measurements of surface tidal streams, and are summarized by Heaps (1968; p. 503). Data from a single current meter mooring in the western Bristol Channel, deployed during September 1975, were used by Robinson (1979) to estimate the residual transport of M_2 tidal energy into the region. Heathershaw *et al.* (1978) reported observations of residual currents at 11 stations in a small area of the eastern Bristol Channel (Swansea Bay). Current meter data from Swansea Bay have also been reported by Collins, Ferentinos and Banner (1979).

More recently, observations of surface currents in a small area of the Bristol Channel have been made using OSCR (Ocean Surface Current Radar) (Hammond *et al.*, 1987) and near-bed residual currents have been derived from seabed drifter studies (Collins and Ferentinos, 1984). We will restrict our attention here to the observations made by us during 1975–7, rather than attempt to review all the water current data available. Our attention is further restricted to an analysis of the data in terms of semidiurnal currents and elevations. This enables us to present a reasonably brief synthesis, which nevertheless covers the overwhelmingly dominant tidal signal in the region. Results are not presented for overtides, although these are implicitly taken into account in the data reduction technique (Appendix). An analysis of M_4 tidal properties, largely based on this data set, is given by Uncles (1991).

12.2 METHODS

12.2.1 Methods of observations

Measurements were made using both recording current meters (RCMs) and direct reading current meters (DRCMs). In general, RCMs were deployed in the deeper water of the Bristol Channel and DRCMs from anchored ships in the Severn Estuary. At each RCM station two or three meters were suspended from a wire line which was kept taut by subsurface buoyancy and which formed part of a standard 'U'-shaped mooring (Howarth, 1975).

Long-term stations 1C−3C (Figure 12.2) each deployed two meters, and stations 1B and 3B three meters, although of these latter six meters only three supplied data for the full 32 day periods. Three meters were deployed at the short-term RCM stations when the average depth exceeded 40 m and two meters on the shallower moorings (Figure 12.2). Currents were profiled throughout the water column at the short-term DRCM stations (Figure 12.2).

The RCM measurements were made at 10 minute intervals and the quoted instrument inaccuracies for speed and direction were ± 5 cm s^{-1} and $\pm 5°$. For the DRCM data measurements were made at half-hourly intervals and the quoted instrument inaccuracies were ± 5 cm s^{-1} and $\pm 10°$.

12.2.2 Data reduction

All velocity data were decomposed into northerly and easterly components and analysed separately. For the long-term data, tidal constituents were derived using standard, least-squares, harmonic analysis (Doodson, 1954). For the short-term stations, the semidiurnal component of the current was derived using a method similar to that described by Rossiter and Lennon (1965), except that the data were analysed in blocks of 12 hours rather than 24 hours (see Appendix).

The quoted, non-systematic instrument inaccuracies generally have a minor influence on the accuracy of estimated semidiurnal currents. This is because of the random nature of the instrument speed and direction errors and the large number of instrument readings within a semidiurnal tidal period.

12.2.3 Tidal ellipse

In the Argand plane the semidiurnal currents (with u the easterly and v the northerly component of current) can be represented in the form

$$w = u + iv = w_+ + w_-$$

$$= |w_+|\exp[i(2\pi t/T + \theta_+)] + |w_-|\exp[-i(2\pi t/T - \theta_-)]$$

where $|w_+|$ and $|w_-|$ are the amplitudes of the positive (anticlockwise) and negative

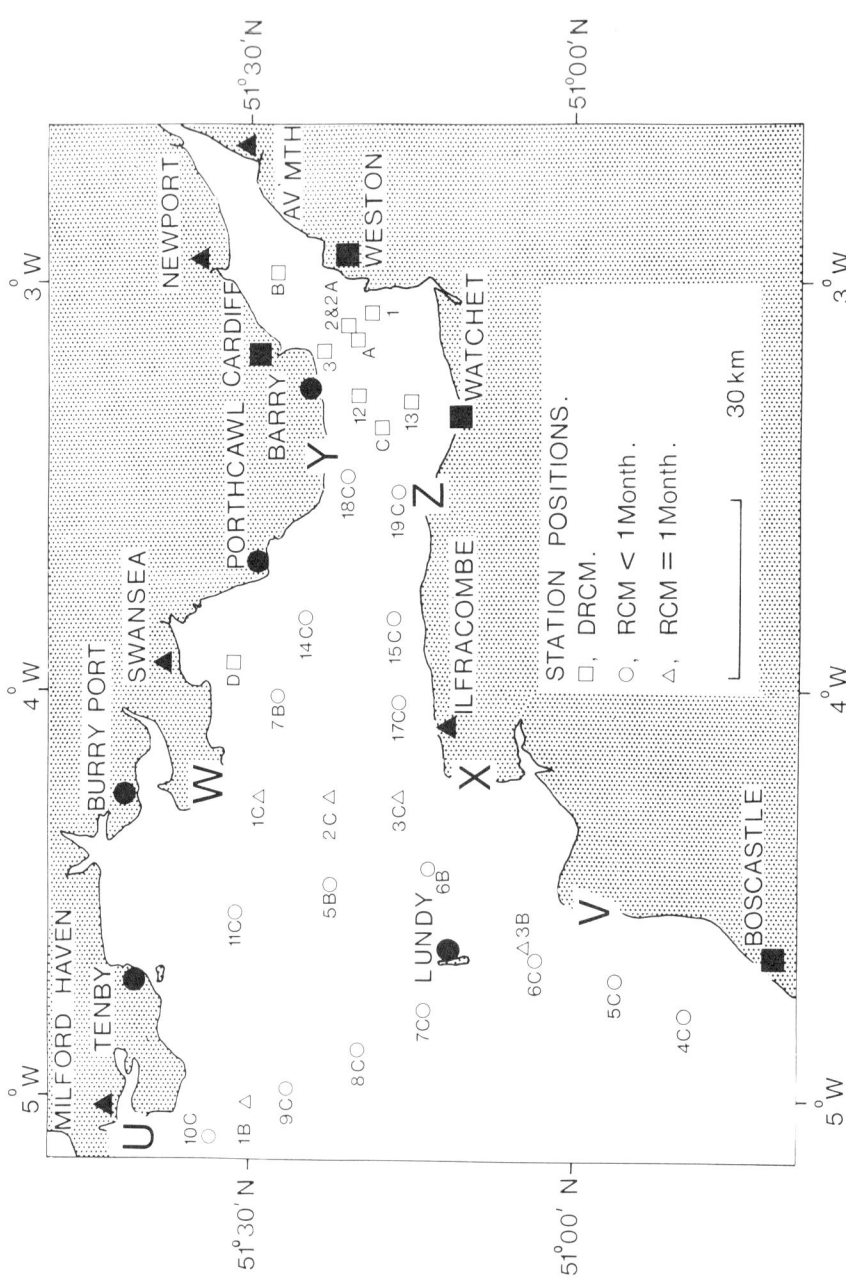

Figure 12.2 Positions and names of stations at which currents were observed. (□) DRCM station; (○) short-term RCM stations (less than one month of data); and (△) long-term RCM stations (one month of data). Also shown are coastal stations for which harmonic analyses of tide gauge data are available: (■, ●, ▲) denote harmonic analyses for data extending over two weeks, one month and one year, respectively

(clockwise) circular motions, and θ_+ and θ_- their phases; t is time and T the period of tidal cycle investigated. The ellipse major axis, a, and minor axis, b, and orientation relative to true east, δ_a, may be written

$$a = |w_+| + |w_-|$$
$$b = |w_+| - |w_-|$$

and

$$\delta_a = (\theta_+ + \theta_-)/2,$$

in which positive values of b imply that the current vector rotates in a positive sense, and conversely for negative values of b. The ellipticity, ϵ, may be defined as

$$\epsilon = b/a$$

The ellipse phase, g_a, which is usually defined with respect to the equilibrium tidal constituent phase at the Greenwich meridian, is computed from

$$g_a = (\theta_- - \theta_+)/2$$

An analysis of the tidal ellipse in terms of real Cartesian coordinates is given by Pugh (1987; pp. 425–429).

The small influence of non-systematic instrumental inaccuracies on the tidal ellipse estimations can be illustrated by considering the effect of the quoted DRCM directional inaccuracy of $\pm 10°$. As an example, consider $a = 120 \, \text{cm s}^{-1}$, $\epsilon = 0.09$, $g_a = 80°$ and $\delta_a = -10°$, and an unrealistically poor sampling rate of one reading per hour, then tidal ellipse errors resulting from the directional inaccuracy of $\pm 10°$ are only $0.2 \, \text{cm s}^{-1}$, 0.01, $0.2°$ and $1.3°$, respectively.

12.2.4 Vertical structure model

Some features of the observed vertical distribution of tidal flow can be interpreted using a simple linear model of the form

$$\partial w/\partial t = i\omega(w_+ - w_-)$$
$$= -ifw + F + (\alpha/h) \cdot \partial^2 w/\partial \eta^2 \tag{12.1}$$

with

$$F = F_+ + F_- = -g\partial\zeta/\partial x - ig\partial\zeta/\partial y \tag{12.2}$$

and subject to

$$\partial w/\partial \eta = 0 \qquad \text{at } \eta = 1 \tag{12.3}$$

and

$$\partial w/\partial \eta = (k/\alpha)w \quad \text{at } \eta = 0 \tag{12.4}$$

In equation 12.1, ω is the frequency of the semidiurnal tide ($2\pi/T \simeq 1.4 \times 10^{-4} \text{s}^{-1}$), f is the Coriolis parameter ($1.14 \times 10^{-4} \text{s}^{-1}$), η is the relative height above the sea bed and the vertical eddy viscosity, N_z, is assumed to be of the form $N_z = \alpha h$ with α constant and h the mean depth (Heaps, 1973; p. 104).

Equation 12.2 represents the surface elevation forcing in the Argand plane, g is the acceleration due to gravity ($981 \, \text{cm s}^{-2}$), ζ is the local semidiurnal surface elevation and (x, y)

denote easterly and northerly coordinate axes. Equations 12.3 and 12.4 are the conditions of zero stress at the surface and linear frictional stress at the bottom (with coefficient k), respectively.

The solutions to Equations 12.1 to 12.4 are

$$w_+ = \{F_+ / i(\omega + f)\}[1 + A\cosh\{\beta(1 - \eta)\}] \tag{12.5}$$

and

$$w_- = \{F_- / i(\omega - f)\}[-1 + B\cosh\{\gamma(1 - \eta)\}] \tag{12.6}$$

with

$$A = -(k/\alpha)[\beta\sinh(\beta) + (k/\alpha)\cosh(\beta)]^{-1}$$

$$B = (k/\alpha)[\gamma\sinh(\gamma) + (k/\alpha)\cosh(\gamma)]^{-1}$$

$$\beta = (1 + i)[h(\omega + f)/2\alpha]^{1/2}$$

and

$$\gamma = (i - 1)[h(\omega - f)/2\alpha]^{1/2}$$

Integrating Equations 12.5 and 12.6 over η yields a solution for the depth-averaged tidal ellipse. Simple vertical structure models of this form are explored in detail by Prandle (1982).

12.2.5 Depth-averaged semidiurnal currents

To deduce depth-averaged tidal ellipses from the RCM observations it is necessary to relate depth-averaged properties to the vertical distributions of current and, therefore, to obtain estimates of α and k. Estimates of F (Equation 12.2) have been derived from a vertically-integrated numerical model of the semidiurnal tide in the region (Uncles, 1981) and can be used to calculate w from Equations 12.5 and 12.6.

Treating station 5B as a test case (Figure 12.2), then a good approximation to the observed vertical distributions of a and b at station 5B is obtained using $\alpha = 0.2$ cm s^{-1} and $k = 0.3$ cm s^{-1} in the vertical structure model.

Modelled values of $|w_+|$ and $|w_-|$ as functions of η for station 5B are shown as continuous and broken lines, respectively, in Figure 12.3a, together with $|\bar{w}_+|$ and $|\bar{w}_-|$, the amplitudes of the depth-averaged circular currents, which are shown by vertical, broken lines. Modelled currents equal modelled depth-averaged currents, $|w_\pm| = |\bar{w}_\pm|$, for values of η very close to 0.4 (Figure 12.3a). Also shown as points in Figure 12.3a are logarithmic curves of least-squares fit to the modelled $|w_+|$ and $|w_-|$

$$|w_\pm| = \lambda_\pm [\ln(1 - \eta/\mu_\pm) - \{\mu_\pm/(1 - \mu_\pm)\}\ln\{1 - \eta/(1 - \mu_\pm)\}] \tag{12.7}$$

which satisfy $|w_\pm| = 0$ at $\eta = 0$ and $\partial|w_\pm|/\partial\eta = 0$ at $\eta = 1$. Equation 12.7 represents leading order solutions to the equation of horizontal momentum with a parabolic profile of vertical eddy viscosity (McGregor, 1972; p. 106).

Values of $|\bar{w}_\pm|$ computed from Equation 12.7 are in excellent agreement with values of $|\bar{w}_\pm|$ deduced from Equations 12.5 and 12.6, as are values computed from Van-Veen profiles of least squares fit to the modelled currents, $|w_+|$ and $|w_-|$

$$|w_\pm| = \lambda'_\pm \eta^{0.19} \tag{12.8}$$

(see, for example, Bowden and Hamilton, 1975; p. 285). Therefore, assuming that the vertical

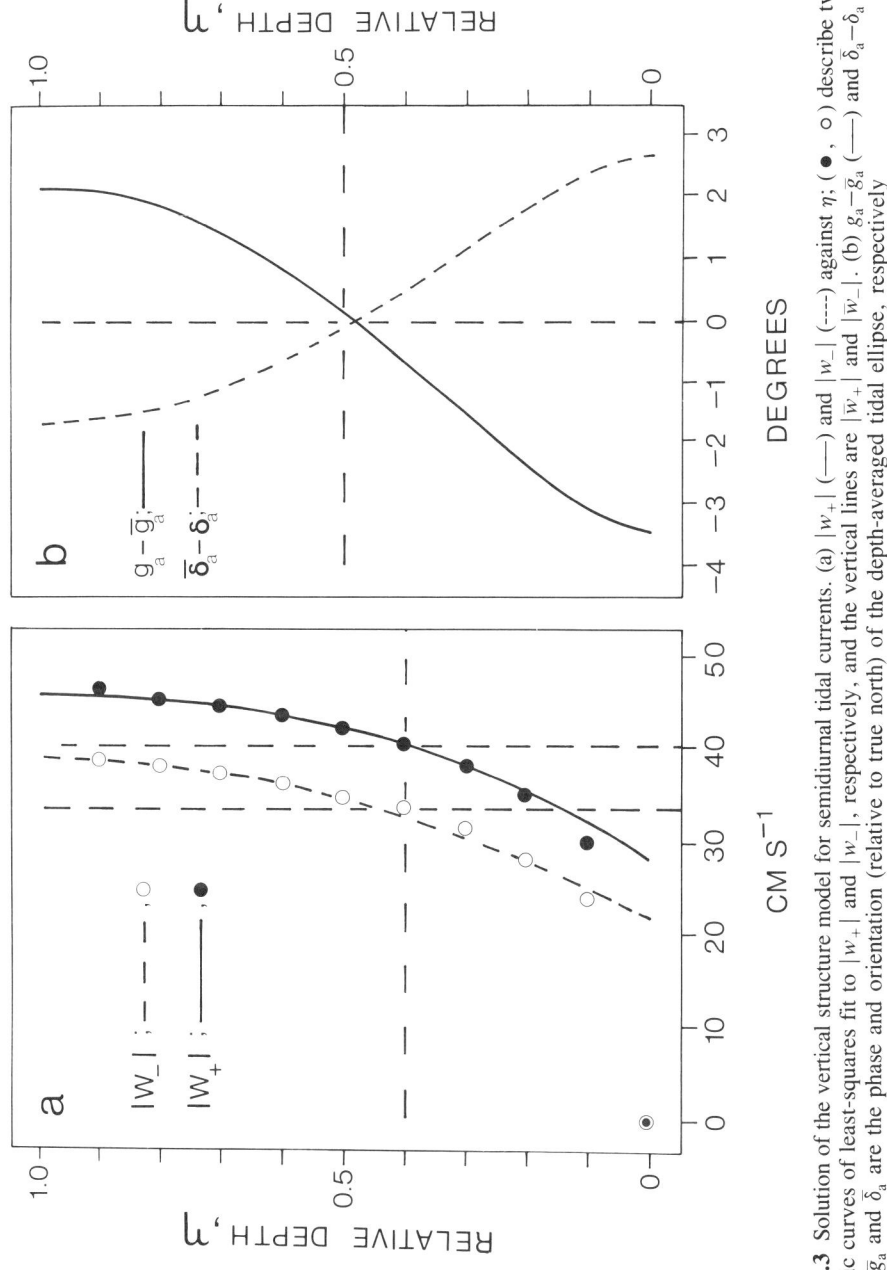

Figure 12.3 Solution of the vertical structure model for semidiurnal tidal currents. (a) $|w_+|$ (——) and $|w_-|$ (---) against η; (●, o) describe two-parameter logarithmic curves of least-squares fit to $|w_+|$ and $|w_-|$, respectively, and the vertical lines are $|w_+|$ and $|w_-|$. (b) $g_a - \overline{g}_a$ (——) and $\overline{\delta}_a - \delta_a$ (---) against η, where $\overline{g}_a$ and $\overline{\delta}_a$ are the phase and orientation (relative to true north) of the depth-averaged tidal ellipse, respectively

structure model provides a reasonable representation of the actual current profile, then Equation 12.7 can be used to estimate depth-averaged currents, $|\bar{w}_\pm|$, when observed data at two or three points through the column are available, and Equation 12.8 can be used when data at only one point are available.

Figure 12.3b shows modelled values of $(g_a - \bar{g}_a)$ and $(\delta_a - \bar{\delta}_a)$ as functions of η for station 5B, where $\bar{g}_a$ and $\bar{\delta}_a$ represent the phase and orientation (relative to true north) of the depth-averaged ellipse. Modelled phases and orientations equal depth-averaged values, i.e. $g_a = \bar{g}_a$ and $\delta_a = \bar{\delta}_a$, for values of η very close to 0.5 (Figure 12.3b). Again, assuming that the vertical structure model provides a reasonable representation of the actual current profile, then interpolation or extrapolation of g_a and δ_a to mid-depth can be used to define $\bar{g}_a$ and $\bar{\delta}_a$ when observed data at two or three points through the column are available. When data at only one point are available, then values of g_a and δ_a must be assumed to be representative of $\bar{g}_a$ and $\bar{\delta}_a$. Fortunately, for these cases (stations 3B and 10C; Figure 12.2), data are available near mid-depth.

These relationships between the modelled vertical distribution of current and its depth-averaged properties have been computed for $\alpha = 0.2 \text{ cm s}^{-1}$ and $k = 0.3 \text{ cm s}^{-1}$, although they also apply with reasonable accuracy for cases tested in the range $\alpha = 0.04-1.00 \text{ cm s}^{-1}$ and $k = 0.1-1.0 \text{ cm s}^{-1}$.

The effects of quoted instrument inaccuracies on the estimated semidiurnal tidal ellipses are small. The methods used here to deduce depth-averaged current properties by interpolation (or slight extrapolation) to mid-depth for phase and direction and by fitting velocity profiles for semi-major and semi-minor axes do not amplify these non-systematic errors.

12.2.6 Co-amplitude and co-phase lines for M₂ tides

The M_2 surface elevation at any point in the Bristol Channel is

$$\zeta = \zeta_0 \cos(\omega t - g_\zeta) = \zeta_c \cos(\omega t) + \zeta_s \sin(\omega t)$$

The co-amplitude lines (ζ_0 = constant, half the magnitude of the co-range lines) subtend an angle Ψ' with the x (east) direction, and have a perpendicular gradient $\partial\zeta_0/\partial n'$, where (Robinson, 1979; p. 177)

$$\tan\Psi' = -\left[\zeta_c\frac{\partial\zeta_c}{\partial x} + \zeta_s\frac{\partial\zeta_s}{\partial x}\right]\left[\zeta_c\frac{\partial\zeta_c}{\partial y} + \zeta_s\frac{\partial\zeta_s}{\partial y}\right]^{-1} \tag{12.9}$$

and

$$\frac{\partial\zeta_0}{\partial n'} = \zeta_0^{-1}\left[\left(\zeta_c\frac{\partial\zeta_c}{\partial x} + \zeta_s\frac{\partial\zeta_s}{\partial x}\right)^2 + \left(\zeta_c\frac{\partial\zeta_c}{\partial y} + \zeta_s\frac{\partial\zeta_s}{\partial y}\right)^2\right]^{1/2} \tag{12.10}$$

The co-phase lines (g_ζ = constant, commonly referred to as co-tidal lines) subtend an angle Ψ with the x axis and have a perpendicular gradient $\partial g_\zeta/\partial n$, where

$$\tan\Psi = -\left[\zeta_c\frac{\partial\zeta_s}{\partial x} - \zeta_s\frac{\partial\zeta_c}{\partial x}\right]\left[\zeta_c\frac{\partial\zeta_s}{\partial y} - \zeta_s\frac{\partial\zeta_c}{\partial y}\right]^{-1} \tag{12.11}$$

and

$$\frac{\partial g_\zeta}{\partial n} = \zeta_0^{-2}\left[\left(\zeta_c\frac{\partial\zeta_s}{\partial x} - \zeta_s\frac{\partial\zeta_c}{\partial x}\right)^2 + \left(\zeta_c\frac{\partial\zeta_s}{\partial y} - \zeta_s\frac{\partial\zeta_c}{\partial y}\right)^2\right]^{1/2} \tag{12.12}$$

The depth-averaged momentum equations can be derived by integrating Equation 12.1 over η, subject to Equation 12.3. If a linearized version of the quadratic frictional term is used instead of Equation 12.4 (Proudman, 1953; p. 309) and, as an approximation, is expressed in terms of the depth-averaged currents, the equations can be written

$$F = \bar{w}_+[\nu\Phi_+(\bar{\epsilon}) + i(\omega+f)] + \bar{w}_-[\nu\Phi_-(\bar{\epsilon}) - i(\omega-f)] \tag{12.13}$$

where

$$\nu = \frac{8\bar{a}D}{3\pi h} \tag{12.14}$$

in which Φ_+ and Φ_- take into account current ellipticity (such that $\Phi_+(0) = 1$) and where D is the usual quadratic bottom drag coefficient, with $D = 2.5 \times 10^{-3}$ (Proudman, 1953; Pingree and Maddock, 1977; p. 343).

Values of $\partial\zeta/\partial x$ and $\partial\zeta/\partial y$ can be derived from Equations 12.2 and 12.13 for each current meter station at which depth-averaged M_2 currents have been estimated. These surface slopes enable Ψ', $\partial\zeta_0/\partial n'$, Ψ and $\partial g_\zeta/\partial n$ to be computed iteratively from Equations 12.9 to 12.12, once estimates of ζ_c and ζ_s are available. Initial values of ζ_c and ζ_s were derived from the depth-averaged hydrodynamic model of the region, and the contours of ζ_0 and g_ζ constrained to satisfy the high quality data for ζ_0 and g_ζ at Swansea and Ilfracombe (Figure 12.2). Results of a second iteration (ζ_c and ζ_s being derived from the first iteration) differed insignificantly from the first, and the contours were accepted as being a valid representation of the M_2 co-amplitude and co-phase lines to within the accuracy of the current meter data. The technique used here is described in detail by Doodson and Corkan (1932) and Robinson (1979).

12.2.7 Energy balance for M_2 tides

In the Argand plane the rate of residual transport of tidal energy per unit width due to the M_2 tide is

$$\bar{E}_h = \rho g h \langle \bar{w}\zeta \rangle \tag{12.15}$$

in which $\langle \cdot \rangle$ denotes an average over the M_2 tidal period, and where tidal variations in the water density, ρ, and the total depth of water are neglected. Equation 12.15 also ignores transport due to the direct astronomical forcing and Earth tides, both of which are very small for the Bristol Channel (Robinson, 1979; p. 193). If $\bar{\delta}_E$ denotes the direction of propagation of $\bar{E}_h$ and δ_A the direction of a normal to the plane, A, across which the transport is computed, then the perpendicular residual flux across any point of A is, from Equation 12.15

$$\bar{E}_f = \bar{E}_h\cos(\bar{\delta}_E - \delta_A)/h$$
$$= \tfrac{1}{2}\rho g \zeta_0 \bar{a}[\cos^2(g_\zeta - \bar{g}_a) + \bar{\epsilon}^2\sin^2(g_\zeta - \bar{g}_a)]^{1/2}\cos(\bar{\delta}_E - \delta_A) \tag{12.16}$$

where

$$\tan\bar{\delta}_E = [\sin\bar{\delta}_a\cos(g_\zeta - \bar{g}_a) + \bar{\epsilon}\cos\bar{\delta}_a\sin(g_\zeta - \bar{g}_a)]$$
$$\times [\cos\bar{\delta}_a\cos(g_\zeta - \bar{g}_a) - \bar{\epsilon}\sin\bar{\delta}_a\sin(g_\zeta - \bar{g}_a)]^{-1} \tag{12.17}$$

and the rate of residual transport of energy across A is

$$E = \int_A \bar{E}_f \, dA \tag{12.18}$$

12.3 RESULTS

12.3.1 Long-term RCM stations

The major tidal constituents can be computed for nine meters, the ellipse phase for each constituent, g_a, being defined with respect to the equilibrium tidal phase at the Greenwich meridian. Values of a, b, g_a and δ_a (in this instance relative to true north) for the M_2 tidal ellipses are given in Table 12.1. Here, η is the mean fractional height of the meter in the water column, which was derived from computer simulations of each mooring using the definition

$$\eta = \langle z(t)/H(t) \rangle$$

where z is the height of the meter above the bottom, H is the instantaneous depth of water and $\langle \cdot \rangle$ is the time-average of the period of observations.

Significant semidiurnal oscillations are also generated by the S_2, N_2 and K_2 tidal constituents, although K_2 is related to S_2 in the analyses and has the same ellipticity and orientation. Expressing their major axes as percentages of the M_2 major axis for each meter, and computing means and standard deviations for the nine meters, gives 36 ± 3, 19 ± 4 and $10 \pm 1\%$, respectively. Similarly, the orientations of the S_2 and N_2 major axes differ in absolute magnitude from the orientation of the M_2 major axis by only $3° \pm 3°$ and $3° \pm 2°$, respectively, showing that the direction of maximum flood is essentially the same for spring and neap tides.

For those meters where the M_2, S_2 and N_2 minor axes, $|b|$, exceed about 1 cm s^{-1} the ellipse vectors rotate in the same sense (ϵ has the same sign throughout). Results for smaller values of $|b|$ are confused and probably dominated by experimental and numerical errors.

Averaging data for all current meters, the ratio of the S_2 ellipticity to the M_2 ellipticity ($\epsilon_{S_2}/\epsilon_{M_2}$) is about 70%. The N_2 ellipticity shows no definite relationship to the M_2 values over the data set. Therefore the ellipticity of the semidiurnal current varies between spring and neap tides. An investigation of the semidiurnal tidal ellipses at spring and neap tides for station 1B (Figure 12.2, where ellipticity is largest for the long-term RCMs) shows that the spring–neap variation in ϵ is greatest for the near bottom meter (the neap tide ellipses having larger values of ϵ), but is between about $\pm 20\%$ of the M_2 values. The ellipse orientations are essentially the same at spring and neap tides.

Table 12.1 M_2 tidal ellipses for long-term RCM stations, a, b, g_a and δ_a are the major axis, minor axis, phase (relative to the Greenwich meridian) and orientation (relative to true north). η is the relative height of the RCM above the sea bed

Station	η	a(cm s^{-1})	b(cm s^{-1})	g_a	δ_a
1C	0.60	93	0	093°	088°
	0.12	70	+10	089°	087°
2C	0.65	87	−1	091°	080°
	0.10	55	+10	087°	082°
3C	0.60	109	+2	095°	069°
	0.12	84	+6	091°	068°
1B	0.61	85	+14	077°	100°
	0.13	62	+21	051°	103°
3B	0.36	77	+10	105°*	060°

* Phase suspect due to timing error.

12.3.2 Short-term and DRCM stations

The Fourier analysis method of data reduction (Appendix) yields properties of the semidiurnal tidal ellipses which are applicable to the periods of observations, and in particular to the state of the spring—neap cycle.

Average values of a, b, g_a, δ_a (relative to true north) and T (the semidiurnal tidal period) for the short-term RCM stations are given in Table 12.2 for each position, η.

12.3.3 Vertical distribution of currents

Tables 12.1 and 12.2 show that the ellipse major axes decrease with increasing depth

Table 12.2 Semidiurnal tidal ellipses for short-term RCM stations. a, b, g_a and δ_a are applicable to the periods of observations; g_a is defined from Equation 12.A4 and δ_a is relative to true north. T is the tidal periodicity, averaged over each period of observation

Station	η	$a(\text{cm s}^{-1})$	$b(\text{cm s}^{-1})$	g_a	δ_a	T (hours)
4C	0.84	31	0	117°	045°	12.33
	0.50	27	+5	109°	046°	12.33
5C	0.46	43	+6	112°	030°	12.33
	0.19	39	+12	109°	031°	12.33
6C	0.70	60	+2	113°	052°	12.68
	0.42	52	+9	100°	061°	12.68
	0.13	37	+11	092°	064°	12.68
7C	0.71	57	+7	103°	069°	12.68
	0.42	50	+10	095°	069°	12.68
	0.13	40	+12	088°	068°	12.68
8C	0.69	62	+10	083°	074°	12.51
	0.36	60	+21	079°	089°	12.51
	0.11	48	+29	063°	092°	12.51
9C	0.69	75	+4	078°	088°	12.51
	0.36	63	+18	070°	099°	12.51
	0.11	54	+31	058°	100°	12.51
10C	0.37	96	+16	043°	116°	12.35
11C	0.52	81	−2	085°	083°	12.35
	0.21	67	+7	083°	081°	12.35
14C	0.38	130	−4	093°	099°	12.26
	0.21	126	−4	093°	098°	12.26
15C	0.39	167	−1	092°	084°	12.26
	0.16	148	+1	092°	087°	12.26
17C	0.38	131	−1	087°	089°	12.41
	0.22	119	0	087°	083°	12.41
18C	0.38	150	+2	101°	104°	12.23
	0.16	129	+5	101°	100°	12.23
19C	0.36	137	0	104°	102°	12.18
	0.20	124	0	104°	099°	12.18
5B	0.64	104	+6	092°	078°	12.33
	0.40	94	+8	090°	079°	12.33
	0.15	75	+12	088°	079°	12.33
6B	0.64	98	+6	093°	061°	12.37
	0.11	71	+15	086°	061°	12.37
7B	0.50	79	−2	092°	090°	12.57
	0.22	67	−3	092°	086°	12.57

(decreasing η) at each station. The ellipse phases also decrease with increasing depth. Therefore the maximum flood occurs earlier towards the bottom, although the phase differences between meters are insignificant for mean depths of water less than about 35 m, where the currents are also fast. Both of these features arise from increasing frictional forces towards the bottom (Proudman, 1953; Defant, 1961).

Tables 12.1 and 12.2 also show an increase in the minor axes with increasing depth at those stations where the ellipticity, ϵ, exceeds a few per cent. This effect, coupled with decreasing a, produces large increases in ϵ with depth.

At most stations the variation in ellipse orientation through the column is slight (less than 4°). If those stations are omitted for which differences in ellipse orientation between the top and bottom meters are only 1° then, in general, the rotation of the ellipse orientation with depth is negative (clockwise) where mean depths exceed 40 m, and positive where mean depths are less than about 35 m. Thus shallow water and strong currents in the eastern Bristol Channel tend to be associated with a positive rotation of the ellipse axis with increasing depth, and conversely for deep water and weaker currents in the western Bristol Channel.

12.3.4 Depth-averaged M_2 currents

Depth-averaged M_2 and semidiurnal tidal ellipses for the RCM stations were computed from data in Tables 12.1 and 12.2 using the theoretical relationships between vertical distributions of current and their depth-averaged properties (Equations 12.5 to 12.8 and Figure 12.3). To compare the depth-averaged currents on a geographical basis it is necessary to reduce the results for the short-term RCM and DRCM stations to a common tidal state. This is taken to be the M_2 tide owing to its domination of the semidiurnal tidal regime (see Appendix).

The estimated, depth-averaged M_2 tidal ellipses are plotted in Figure 12.4. Figure 12.4a shows the estimated values of $\bar{a}$ throughout the Bristol Channel, together with the amplitudes of the M_2 surface elevations at coastal stations, ζ_0. Both $\bar{a}$ and ζ_0 generally increase in magnitude from the seaward boundary to the estuary.

Values of $\bar{\epsilon}$ are shown in Figure 12.4b. The ellipticity is largest at the seaward boundary, UV, and has its maximum observed value of 31% in the central part of the Bristol Channel, where lateral flows due to surface slopes and Coriolis effects are able to develop, away from the influence of coastal boundaries.

In general, the current vectors rotate in a positive sense ($\bar{\epsilon} > 0$) throughout the Bristol Channel. A significant exception occurs at stations D, 7B and 14C south of Swansea (Figure 12.2), for which the negative rotation appears to be a consequence of local bay and coastline topography.

Concerning the accuracy of $\bar{\epsilon}$, the long-term station 3B (north of position V) and its neighbouring short-term station 6C both yield $\bar{\epsilon} = +13\%$. Similarly, the long-term station 1B (south-east of position U) and its neighbouring short-term station 9C both yield $\bar{\epsilon} = +21\%$. This indicates that values of $\bar{\epsilon}$ deduced from the short-term stations are reasonable estimates of the M_2 current ellipticities.

The estimated phases, $\bar{g}_a$, and orientations, $\bar{\delta}_a$, of the M_2 current ellipses are shown in Figure 12.4c, together with the phases of M_2 surface elevations at coastal stations, g_ζ. The orientations are shown as arrows which are predominantly parallel to the Bristol Channel's medial line. The phases $\bar{g}_a$ can be compared with g_ζ for those stations in the vicinity of coastal tide gauge stations. Up-estuary of Ilfracombe (Figure 12.4c) $\bar{g}_a$ and g_ζ generally increase along the Bristol Channel, showing the partially progressive nature of the tidal wave, although the phases of the currents also depend on local topography (Proudman, 1953; p. 315).

233

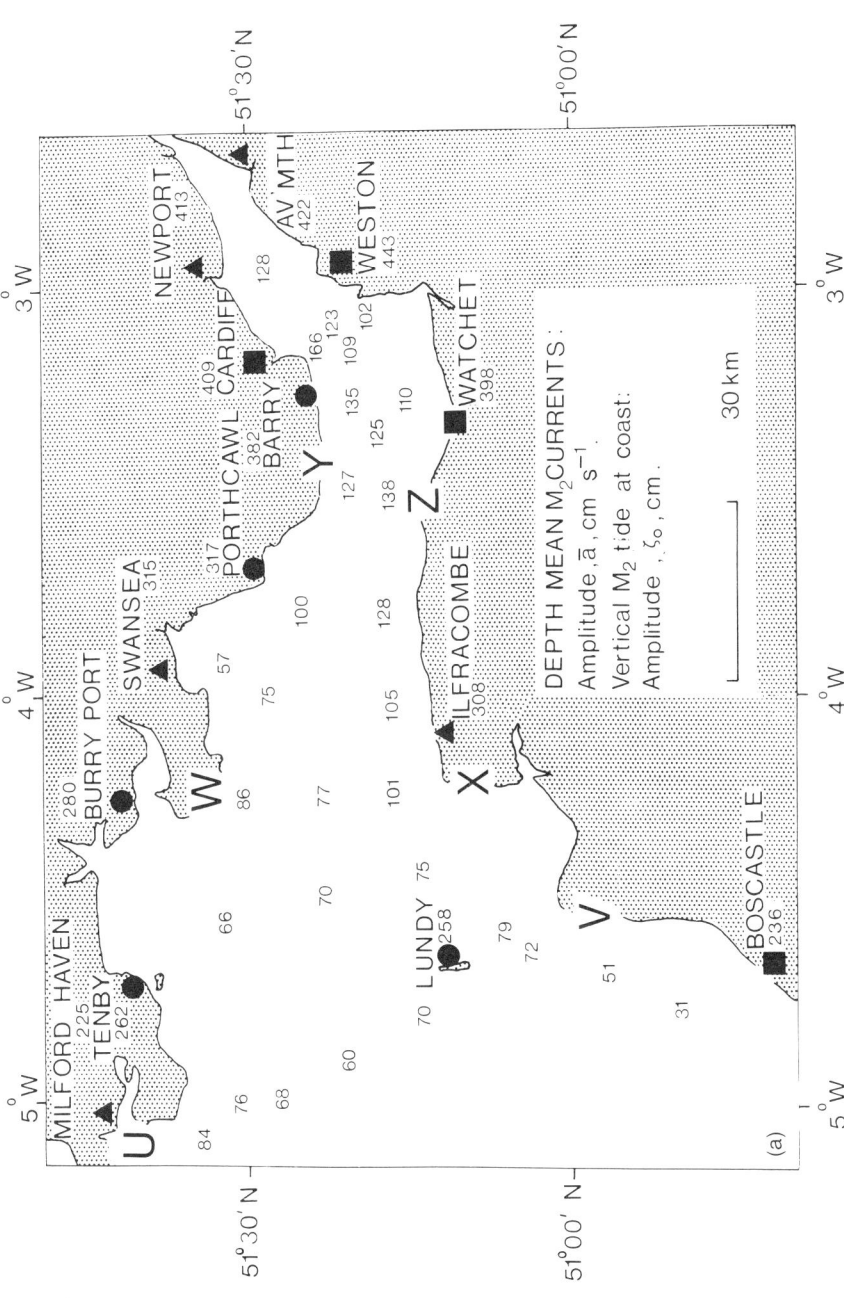

Figure 12.4 Estimated depth-averaged current ellipses for the M_2 tide. Data for stations 2 and 2A (Figure 12.2) are averaged. (a) Semi-major axis of the M_2 current ellipse, $\bar{a}$ (cm s^{-1}), and amplitudes of the M_2 surface elevations at coastal stations ζ_0 (cm) *(continues overleaf)*

234

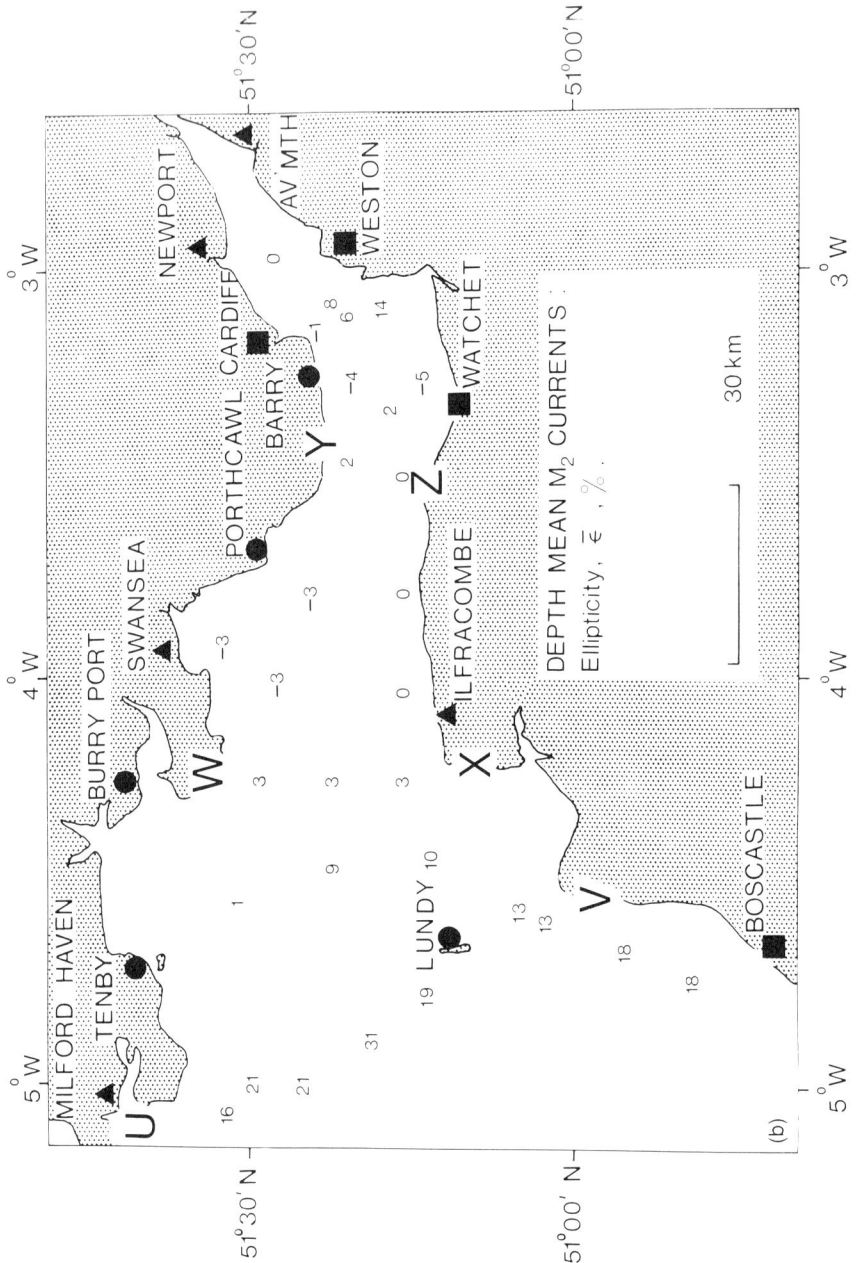

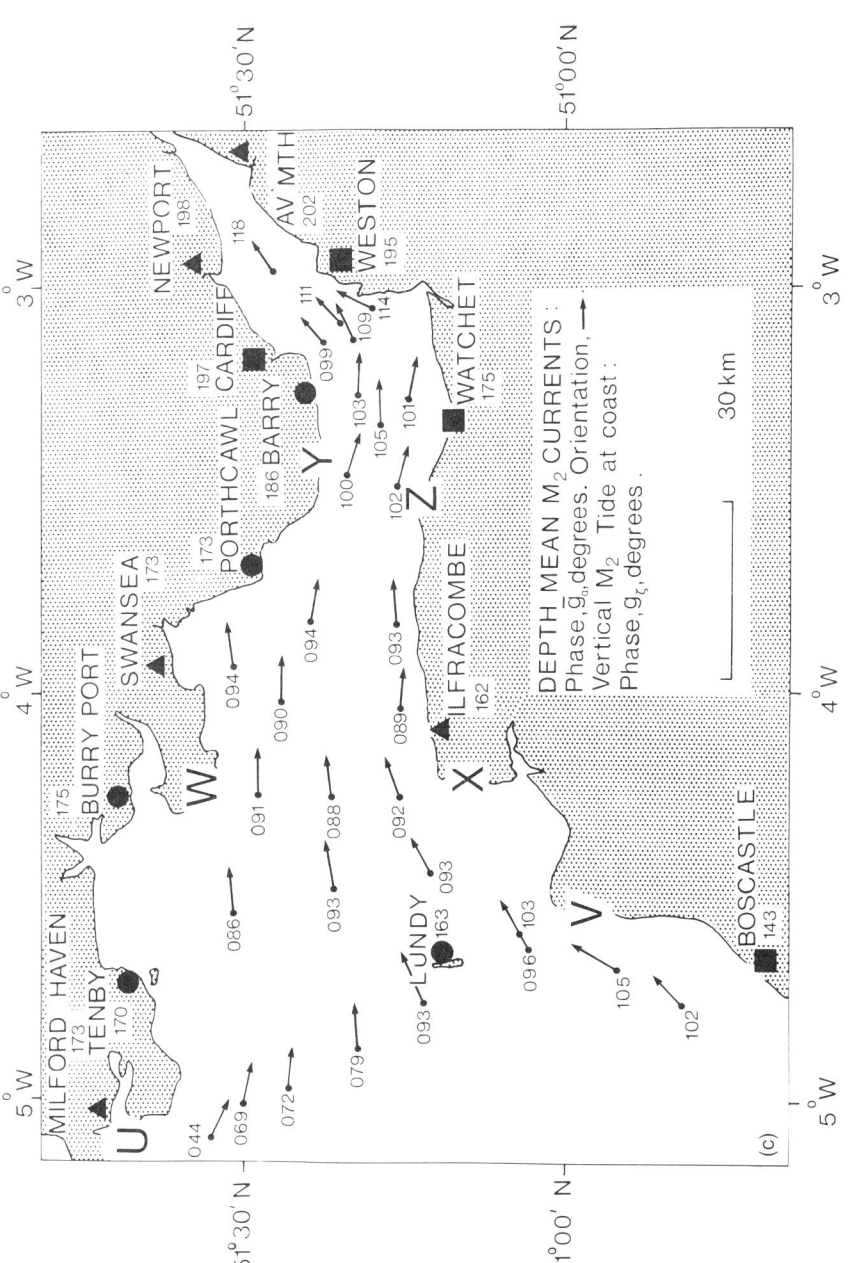

Figure 12.4 (*continued*) (b) Ellipticity of the M_2 currents, $\bar{\epsilon}(\%)$. (c) Phase $\bar{g}_a$ (degrees, relative to Greenwich) and orientation of the M_2 current ellipse; also shown are the phases of the M_2 surface elevations at coastal stations

The phase differences between elevation and current, $(g_\zeta - \bar{g}_a)$, are roughly $70° - 80°$, whereas a one-dimensional standing tidal wave would satisfy $(g_\zeta - \bar{g}_a) = 90°$.

Seaward of section WX the distribution of phases, $\bar{g}_a$, is more complex. Between Boscastle and Ilfracombe (Figure 12.4c) g_ζ increases from $143°$ to $162°$, whereas along the northern coastline g_ζ is roughly $170°$, so that high water occurs earlier along the southern coastline. This is in contrast with the behaviour of $\bar{g}_a$ along the seaward boundary UV, for which maximum flood occurs earlier in the north of the Bristol Channel. Also, the elevation—current phase difference, $(g_\zeta - \bar{g}_a)$, varies from rougly $40°$ near Boscastle to $70°$ near Ilfracombe, whereas in the north of the Bristol Channel $(g_\zeta - \bar{g}_a)$ is typically $80° - 100°$.

It follows that the tidal wave is much more progressive in the south of the Bristol Channel and can be identified as the source of residual mass transport and energy which flows landwards across the seaward boundary due to Stokes drift (see, for example, Tee, 1976, p. 614; Uncles and Jordan, 1980).

12.3.5 Co-amplitude and co-phase lines for M_2 tides

Results for co-amplitude lines, $\zeta_0 = $ constant (co-range lines are more commonly plotted) and co-phase lines, $g_\zeta = $ constant (commonly referred to as co-tidal lines) are given in Figures 12.5a and 12.5b, respectively. The co-amplitude lines are approximately one-dimensional and increase in magnitude progressing landwards along the Bristol Channel's medial line. The co-amplitude lines diverge in the bays to satisfy boundary conditions at the coast (Proudman, 1953; p. 267) and, up-channel of Ilfracombe, show slightly enhanced elevations on the southern coastline due to the Coriolis effect.

The co-phase lines (Figure 12.5b) also have a simple form, being approximately parallel to the northern coastline seaward of Ilfracombe, while gradually adjusting to a more cross-channel direction in the Severn Estuary.

The computed co-amplitude and co-phase lines for the Bristol Channel are in qualitative agreement with, although they show much greater detail than, those presented by Doodson and Corkan (1932; p. 51) and Robinson (1979; p. 180). The general patterns of the lines are very similar to those computed by Miles (1979; p. 79) and Uncles (1981) using two-dimensional numerical models of the region.

12.3.6 Energy balance for M_2 tides

The fluxes, $\bar{E}_f$, were computed from Equation 12.16 using the values of ζ_0 and g_ζ given in Figure 12.5 and are plotted in Figure 12.6 for stations along the sections UV, WX and YZ. Also shown in Figure 12.6 are graphical representations of the sections, in which depths have been averaged over distances of roughly 3 km for ease of computation.

Error bars on the fluxes represent 95% confidence intervals, which have been constructed by making the reasonable (but rather arbitrary) assumptions that the corresponding intervals for $\bar{a}$, $\bar{\epsilon}$, $\bar{g}_a$, $\bar{\delta}_a$, ζ_0 and g_ζ and ± 5 cm s^{-1}, $\pm 10\%$ (fractional), $\pm 3°$, $\pm 5°$, ± 5 cm and $\pm 1°$, respectively. If these errors are doubled, then the confidence intervals for $\bar{E}_f$ are also doubled.

Figure 12.6 shows that the largest landward fluxes of energy occur on the south side of the Bristol Channel, which is qualitatively similar to that expected from a simplified description of the tidal regime in terms of the interference of landward travelling, and damped (reflected) seaward travelling, progressive Kelvin waves (Hunter, 1972; p. 32). At the seaward boundary,

Figure 12.6a, $\overline{E}_f$ is negative (seaward) in the north of the Bristol Channel, implying that the reflected tidal wave is dominant there.

Calculating the rate of residual transport of M_2 tidal energy across UV, WX and YZ using Equation 12.18 yields $(8.6 \pm 0.8) \times 10^6$ kW, $(6.1 \pm 0.7) \times 10^6$ kW and (3.0 ± 0.4) 10^6 kW, respectively. The errors are 95% confidence intervals and are based on the computed uncertainties in $\overline{E}_f$ shown in Figure 12.6 together with the assumption that errors are negligible in the use of linear interpolation and extrapolation for defining $\overline{E}_f$ over the sections. These errors are doubled if those for $\overline{E}_f$ are doubled.

Unfortunately, the use of linear interpolation and extrapolation is unlikely to be an accurate approximation owing to the large spatial variability exhibited by the currents (Figure 12.4a) and the computed values of E are probably only accurate to within 20−30%. An attempt has been made to improve the accuracy of the value for section UV by omitting the contribution to E over the length of Lundy Island (Figure 12.5b), as the current must be essentially parallel to UV in this small region.

A rough estimate of the average perpendicular component of the Stokes drift over each section, u_s, can be made using (Uncles and Jordan, 1980)

$$u_s = E/\rho gAh$$

where h is here an average depth over A. Values of u_s are 0.5, 1.0 and 3.0 cm s^{-1} for sections UV, WX and YZ, respectively.

The estimated value of the M_2 energy transport, $E = 8.6 \times 10^6$ kW, for the seaward boundary is reasonably consistent with a value of 7.2×10^6 kW for the Bristol Channel derived by Robinson (1979; p. 193), which was based on data from just one mooring near the centre of the Bristol Channel's seaward boundary. Both estimates are much smaller than Miller's value of 28×10^6 kW (Miller, 1966; p. 2487), which therefore appears to be in error.

Using a simple model, Bennett (1975) computed the rate of residual transport of M_2 tidal energy across the sections WX and YZ to be 11.3×10^6 and 8.4×10^6 kW, respectively, both of which are larger than the present values of 6.1×10^6 and 3.0×10^6 kW. However, the value of 3.0×10^6 kW for the dissipation of M_2 tidal energy in the Severn Estuary is consistent with a value of 2.9×10^6 kW for the section of estuary up-estuary of Porthcawl derived by Uncles and Jordan (1980) using a numerical one-dimensional hydrodynamic model of greater physical complexity than the analytical model used by Bennett (1975).

12.4 CONCLUSIONS

The wide coverage of stations used to observe tidal currents in the Bristol Channel and Severn Estuary has enabled us to delineate several important aspects of the current regime. In general our results confirm those of earlier modelling studies, which were based on much less data. The data presented here will be of value in future modelling studies of the region, especially for three-dimensional calculations. Some salient features of the results are as follows:

(1) Solutions of a simple, linear vertical structure model with constant eddy viscosity are in qualitative agreement with the observed vertical distributions of semidiurnal tidal current. The major axis and phase of the current ellipse decrease with increasing depth and the ellipticity increases. Differences in these ellipse properties through the column are smaller in the eastern Bristol Channel, where the currents (and the vertical mixing) are larger, in accordance with the simple model.

238

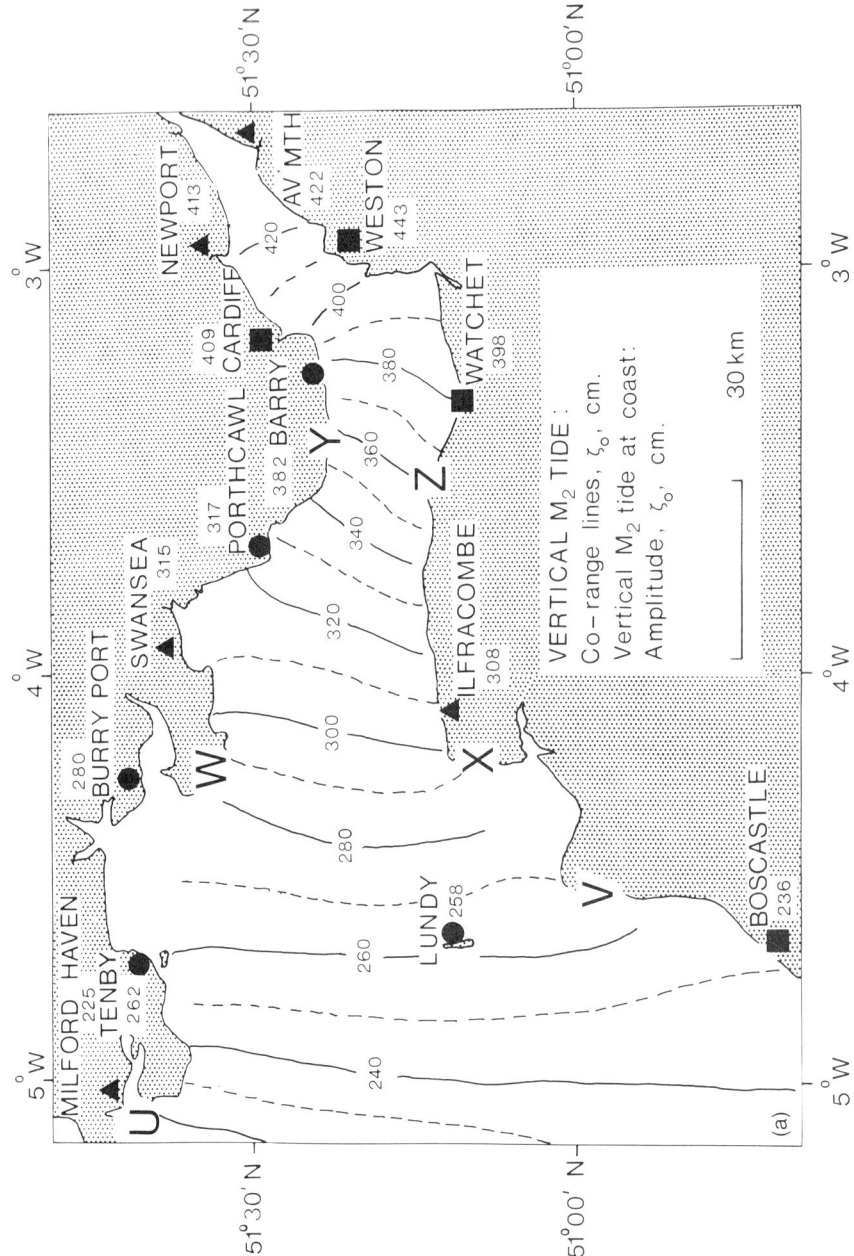

(a)

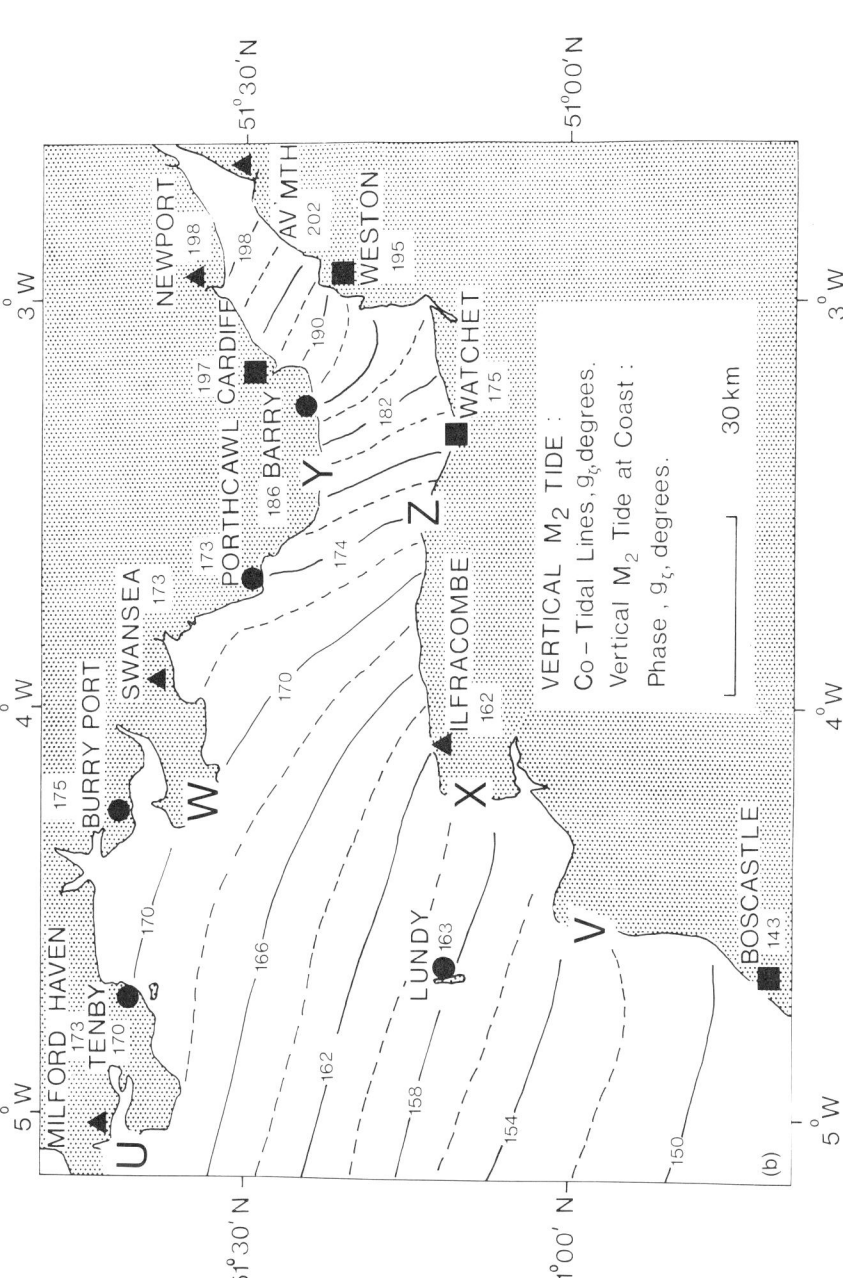

Figure 12.5 Co-amplitude and co-phase lines for the M_2 tide. (a) Amplitude of the M_2 surface elevations ζ_0 (cm), together with observed values for coastal stations. (b) Phase of the M_2 surface elevations, g_ζ (degrees, relative to Greenwich meridian), together with observed values for coastal stations

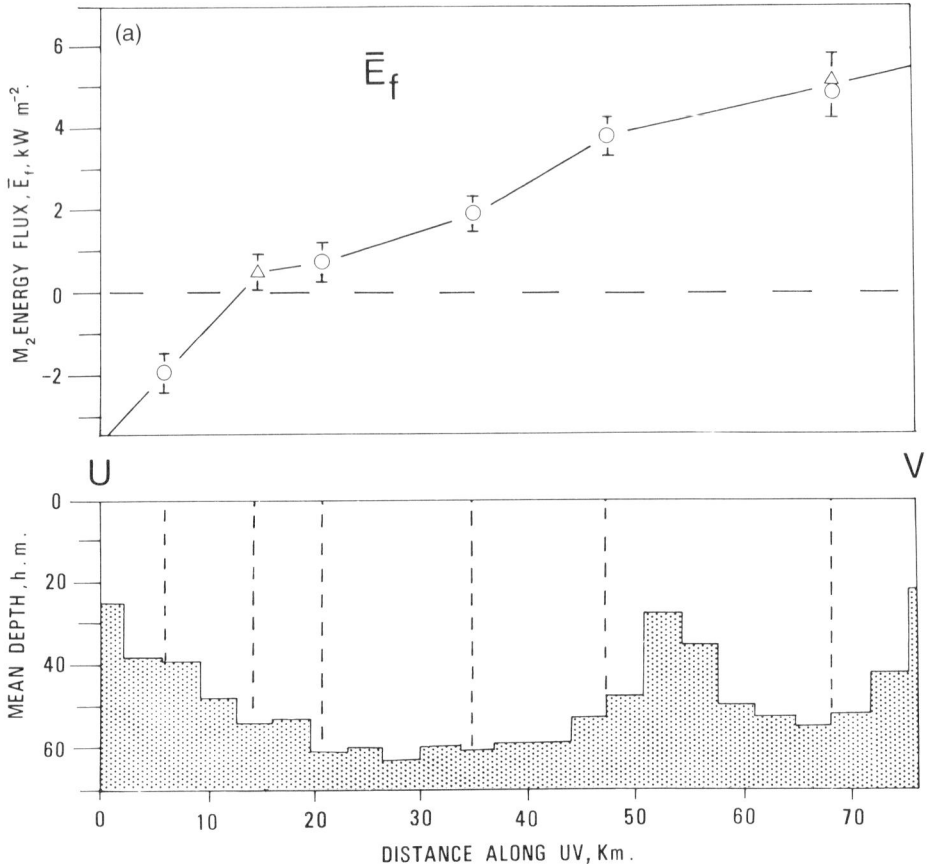

Figure 12.6 Residual fluxes of M_2 tidal energy $\bar{E}_f$, across sections UV, WX and YZ, together with depths, h, over the sections. (a) $\bar{E}_f$ and h over section UV

(2) The horizontal distribution of M_2 tidal current shows that its ellipse major axis increases up-channel, its value at the seaward boundary of the Severn Estuary being roughly twice that at the western boundary of the Bristol Channel. The data show enhanced speeds in the vicinity of headlands and reduced speeds in bays.

(3) The current ellipticity is largest in the central part of the Bristol Channel at the seaward boundary, reaching an observed maximum of 31%. The ellipticity generally decreases rapidly progressing up-channel, owing to the decreasing width of the Bristol Channel, although it can become significant ($>10\%$) through local topographical influences. The ellipticity is generally positive (anticlockwise rotation) over the region, except in the north-eastern Bristol Channel, where bay and coastline topography appear to dominate the flow.

(4) Up-estuary of Ilfracombe the M_2 ellipse phases generally increase progressing along the Bristol Channel, showing the partially progressive nature of the tidal wave, the phase differences between elevation and current being roughly $70°-80°$. Down-estuary of Ilfracombe the situation is more complex; maximum flood occurs first in the northern entrance to the Bristol Channel and becomes successively later progressing landwards

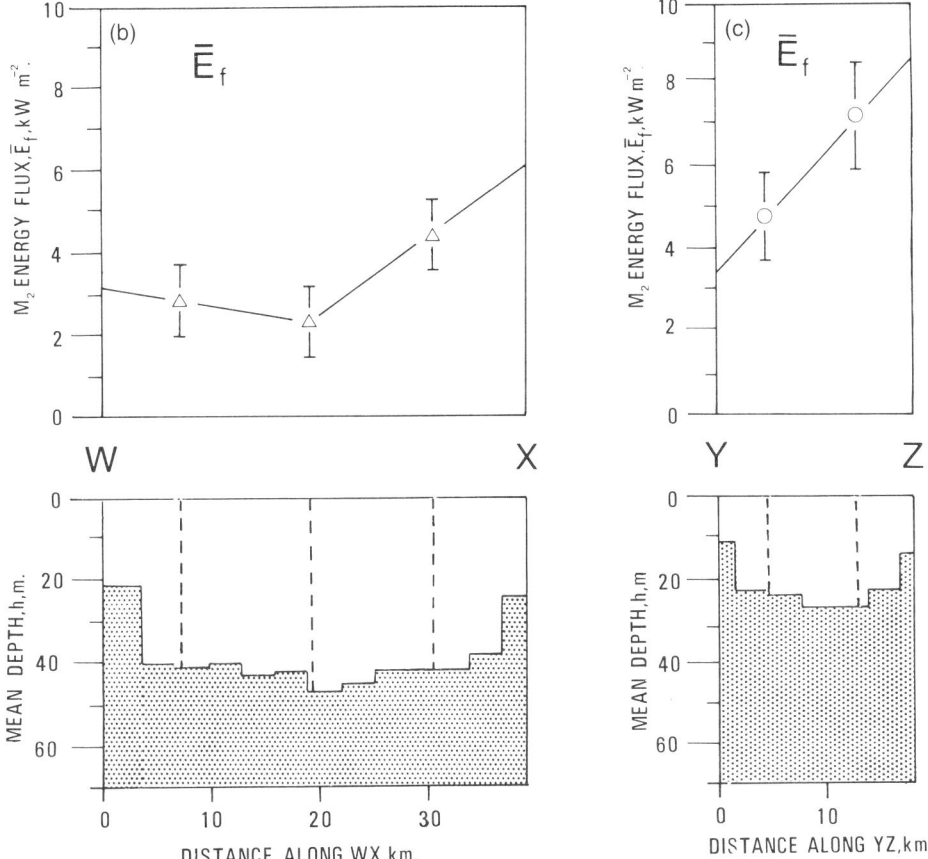

Figure 12.6 (*continued*) (b) $\bar{E}_f$ and h over section WX; and (c) $\bar{E}_f$ and h over section YZ

along the northern coastline, the reverse being true along the southern coastline down-estuary of Ilfracombe. In contrast, high water for the M_2 tide occurs first at the southern entrance to the Bristol Channel, so that the tidal wave in the southern part of the Bristol Channel appears to be undergoing an adjustment from near-progressive at the seaward boundary, to near-standing in the Bristol Channel.

(5) The co-amplitude lines for M_2 surface elevations in the region are essentially one-dimensional and increase in magnitude progressing along the Bristol Channel's medial line. The lines diverge in bays and show slightly enhanced elevations along the southern coastline due to the Coriolis effect. The co-phase lines also have a simple form, being essentially parallel to the northern coastline down-estuary of Ilfracombe and adjusting to a more cross-channel direction in the Severn Estuary, their orientation in the Severn depending mainly on the relative magnitudes of the frictional and Coriolis forces.

(6) The estimated rate of residual transport of M_2 tidal energy across the seaward boundary of the Bristol Channel amounts to 8.6×10^6 kW, the landward fluxes of energy being larger on the south side of the Bristol Channel than in the centre, and being seawards near the northern coastline. Associated with this residual flux of energy is a Stokes drift,

which has a cross-sectionally averaged value of roughly 0.5 cm s^{-1}. The computed rate of residual transport of M_2 tidal energy across the seaward boundary of the Severn Estuary is $3.0 \times 10^6 \text{ kW}$, and the associated Stokes drift is roughly 3 cm s^{-1}, the maximum flux occurring in the south of the Bristol Channel. The cross-sectional dependence of the energy flux is consistent with a tidal regime dominated by the interference of landward travelling, and damped (reflected) seaward travelling, progressive Kelvin waves.

REFERENCES

Bennett, A.F. (1975) Tides in the Bristol Channel. *Geophys. J. R. Astron. Soc.* **40**, 37–43.

Bowden, K.F. and Hamilton, P. (1975) Some experiments with a numerical model of circulation and mixing in a tidal estuary. *Est. Coast. Mar. Sci.* **3**, 281–301.

Collins, M.B. and Ferentinos, G. (1984) Residual circulation in the Bristol Channel, as suggested by Woodhead sea-bed drifter recovery patterns. *Oceanol. Acta* **7**, 33–42.

Collins, M., Ferentinos, G. and Banner, F.T. (1979) The hydrodynamics and sedimentology of a high (tidal and wave) energy embayment (Swansea Bay, northern Bristol Channel). *Est. Coast. Mar. Sci.* **8**, 49–74.

Defant, A. (1961) *Physical Oceanography*. Vol. 2. Pergamon Press, Oxford, 598pp.

Doodson, A.T. (1954) The analysis of tidal observations for 29 days. *Int. Hydrol. Rev.* May, 3–31.

Doodson, A.T. and Corkan, R.H. (1932) The principal constituent of the tides in the English and Irish Channels. *Phil. Trans. R. Soc. London A* **231**, 29–53.

Hammond, T.M., Pattiaratchi, C.B., Eccles, D., Osborne, M.J., Nash, L.A. and Collins, M.B. (1987) Ocean Surface Current Radar (OSCR) vector measurements on the inner continental shelf. *Cont. Shelf Res.* **7**, 411–431.

Heaps, N.S. (1968) Estimated effects of a barrage on tides in the Bristol Channel. *Proc. Inst. Civ. Engin.* **40**, 495–509.

Heaps, N.S. (1973) Three-dimensional numerical model of the Irish Sea. *Geophys. J. R. Astron. Soc.* **35**, 99–120.

Heathershaw, A.D., Carr, A.P., Blackley, M.W.L. and Hammond, F.D.C. (1978) Swansea Bay (SKER) report. *IOS Rep. No. 74*, 47pp.

Howarth, M.J. (1975) Current surges in the St Georges Channel. *Est. Coast. Mar. Sci.* **3**, 57–70.

Hunter, J.R. (1972) An investigation into the circulation of the Irish Sea. *University College of North Wales Marine Science Laboratories, Oceanogr. Rep. 72-1*, 166pp.

McGregor, R.C. (1972) The influence of eddy viscosity on the vertical distribution of velocity in the tidal estuary. *Geophs. J. R. Astron. Soc.* **29**, 103–108.

Miles, G.V. (1979) Estuarine modelling — Bristol Channel. In: *Tidal Power and Estuary Management* (Eds R.T. Severn, D. Dineley and L.E. Hawker). Scientechnica, Bristol, 296pp.

Miller, G.R. (1966) The flux of tidal energy out of the deep oceans. *J. Geophys. Res.* **71**, 2485–2489.

Owen, A. (1980) A three-dimensional model of the Bristol Channel. *J. Phys. Oceanogr.*, **10**, 1290–1302.

Pingree, R.D. and Maddock, L. (1977) Tidal residuals in the English Channel. *J. Mar. Biol. Assoc. UK* **57**, 339–354.

Prandle, D. (1982) The vertical structure of tidal currents. *Geophys. Astrophys. Fluid Dyn.* **22**, 29–49.

Proctor, R. and Flather, R.A. (1989) Storm surge prediction in the Bristol Channel — the floods of 13 December 1981. *Cont. Shelf Res.* **9**, 889–918.

Proudman, J. (1953) *Dynamical Oceanography*. Methuen, London, 409pp.

Pugh, D.T. (1987) *Tides, Surges and Mean Sea-level (A Handbook for Scientists and Engineers)*. Wiley, Chichester, 472pp.

Radford, P.J., Uncles, R.J. and Morris, A.W. (1981) Simulating the impact of technological change on dissolved cadmium distribution in the Severn Estuary. *Wat. Res.* **15**, 1045–1052.

Robinson, I.S. (1979) The tidal dynamics of the Irish and Celtic Seas. *Geophys. J. R. Astron. Soc.* **56**, 159–197.

Rossiter, J.R. and Lennon, G.W. (1965) Computation of tidal conditions in the Thames Estuary by the initial value method. *Proc. Inst. Civ. Engin.* **31**, 25–56.

Shaw, T.L. (1987) Environmental aspects of tidal power barrages in the Severn Estuary. In: *Tidal Power: Proceedings of the Symposium Organised by the Institution of Civil Engineers, London, 30–31 October 1986*, Thomas Telford, London, pp. 235–254.

Stephens, C.V. (1986) A three-dimensional model for tides and salinity in the Bristol Channel. *Cont. Shelf Res.* **6**, 531–560.

Tee, K.T. (1976) Tide-induced residual current, a 2-d non-linear numerical tidal model. *J. Mar. Res.* **34**, 603–628.

Uncles, R.J. (1981) A numerical simulation of the vertical and horizonal M$_2$ tide in the Bristol Channel and comparisons with observed data. *Limnol. Oceanogr.* **26**, 571–577.

Uncles, R.J. (1982) Computed and observed residual currents in the Bristol Channel. *Oceanol. Acta* **5**, 11–20.

Uncles, R.J. (1983a) Hydrodynamics of the Bristol Channel. *Mar. Pollut. Bull.* **15**, 47–53.

Uncles, R.J. (1983b) Modeling tidal stress, circulation and mixing in the Bristol Channel as a prerequisite for ecosystem studies. *Can. J. Fish. Aquat. Sci.* **40** (suppl. 1), 8–19.

Uncles, R.J. (1991) M$_4$ tides in a macrotidal, vertically mixed estuary: the Bristol Channel and Severn. In: *tidal Hydrodynamics* (Ed. B.B. Parker). Wiley, Chichester, 341–355.

Uncles, R.J. and Joint, I.R. (1983) Vertical mixing and its effects on phytoplankton growth in a turbid estuary. In: *Proceedings of the Symposium on the Dynamics of Turbid Coastal Environments. Can. J. Fish. Aquat. Sci.* **40** (Suppl. 1), 221–228.

Uncles, R.J. and Jordan, M.B. (1980) A one-dimensional representation of residual currents in the Severn Estuary and associated observations. *Est. Coast. Mar. Sci.* **10**, 39–60.

Wolf, J. (1987) A 3-D model of the Severn Estuary. In: *Three Dimensional Models of Marine and Estuarine Dynamics* (Eds. J.C.J. Nihoul and B.M. Jamart). *Elsevier Oceanogr. Ser.* **45**, 609–624.

APPENDIX

Data reduction at short-term stations

All velocity data were decomposed into northerly and easterly components and analysed separately. Data were extracted from a smoothed record at 30 minute intervals and the record divided into blocks of 25 such values. Each block was accompanied by simultaneous data for the surface elevations at Avonmouth, which were used to specify the period of the tidal cycle investigated, T, where $T > 12$ hours, and the quantity, r, which is the ratio of the averaged observed surface elevation amplitude at Avonmouth during each period of short-term measurements to the M$_2$ elevation amplitude of 422 cm (Figure 12.5a).

Each block of velocity and tide gauge data was assumed to satisfy periodicity over T and quadratic interpolation was used to define data at intervals of $T/50$ between 0 and T. Fourier analysis of the interpolated data then yielded the semidiurnal currents at the station and the semidiurnal surface elevations at Avonmouth (subscript A) for each block of data in the form

$$v = v_0 \cos\{2\pi(t - t_v)/T\} \tag{12.A1}$$

$$u = u_0 \cos\{2\pi(t - t_u)/T\} \tag{12.A2}$$

and

$$\zeta_A = \zeta_{A,0} \cos\{2\pi(t - t_{A,\zeta})/T\} \tag{12.A3}$$

where v and u represent northerly and easterly components of the current and ζ_A the surface elevation at Avonmouth. Using the Argand plane representation

$$w = u + iv = |w_+|\exp\{i(2\pi t/T + \theta_+)\} + |w_-|\exp\{-i(2\pi t/T - \theta_-)\}$$

then the time of maximum flood, t_a, within a block (tidal cycle) of data is

$$t_a = T(\theta_- - \theta_+)/4\pi$$

which can be related to the time of high water at Avonmouth for the semidiurnal tide $(t_{A,\zeta})$ using

$$\Delta t = t_a - t_{A,\zeta}$$

The Fourier analysis method of data reduction yields a, b, δ_a, t_a and Δt for each tidal cycle of observations. Averaging these data for each station yields properties of the semidiurnal current ellipses which are applicable to the periods of observations, and in particular to the state of the spring–neap cycle. The tidal state is defined from Equation 12.A3, and is quantified using the parameter, r, where

$$r = \zeta_{A,0}/\zeta_{A,0,M2}$$

with $\zeta_{A,0,M2} = 422$ cm. Average values of r for the short-term stations are used. Δt is converted to a phase and used to define the phase of the observed current ellipse

$$g_a = 202° + 28.984\Delta t \tag{12.A4}$$

where $28.984°$ hour^{-1} is the frequency of the M_2 tide and $202°$ is the phase of the M_2 surface elevations at Avonmouth. Defined in this way, g_a can easily be compared with M_2 phases for the long-term RCM stations and with phases for the surface elevations, although it represents only a first estimate of the phase of the M_2 tidal currents. Similarly, a first estimate of the M_2 tidal currents can be obtained by computing (a/r) and (b/r).

Estimating depth-averaged M_2 currents

Considering the short-term stations, the orientation of the tidal ellipse, $\bar{\delta}_a$, is nearly the same at spring and neap tides and can therefore be equated to the orientation of the M_2 ellipse.

The ellipticity, $\bar{\epsilon}$, varies over a spring–neap cycle and may not equal the M_2 value for a specific set of observations. However, data for the long-term station 1B showed that $\bar{\epsilon}$ is unlikely to differ from the M_2 value by more than $\pm 20\%$ and this is probably an overestimate for those stations where ellipticity is significant (greater than about 0.1). The observed values of $\bar{\epsilon}$ are therefore taken to be equal to the M_2 values.

Estimating the major axis and phase of the depth-average M_2 currents from estimated values of the depth-averaged semidiurnal currents is not straightforward because they are not related in a simple way to the semidiurnal surface elevations at Avonmouth. The presence of quadratic frictional forces in the Bristol Channel means that depth-averaged variables $\Delta \bar{t}$ and $\bar{a}/\zeta_{A,0}$ are not constant at a station but vary with the state of the spring–neap cycle.

At neap tides the quadratic frictional forces are small and phase differences along the Bristol Channel are reduced (being zero for a frictionless channel; Proudman, 1953; p. 230), the reverse being true at spring tides. Similarly, the frictional damping of the tidal wave along the Bristol Channel is larger at spring tides than at neap tides. These features manifest themselves in the generation of overtides, which although not considered explicitly here, can be taken into account by defining quantities $r^*(x,y,r)$ and $\Delta \bar{g}_a(x,y,r)$ such that

$$\bar{a}(M_2) = \bar{a}(\text{observed})/r^*$$

and

$$\bar{g}_a(M_2) = \bar{g}_a(\text{observed}) + \Delta \bar{g}_a$$

r^* and $\Delta \bar{g}_a$ are computed by running a depth-averaged hydrodynamic model of the region with seaward input elevations ($n \times$ the M_2 value) corresponding to $1.4 \times M_2$, $1.0 \times M_2$

and $0.6 \times M_2$, i.e. $n = 1.4$, 1.0 and 0.6, and using the modelled currents and elevations to tabulate

$$r* = \bar{a}(n \times M_2)/\bar{a}(1.0 \times M_2)$$

and

$$\Delta \bar{g}_a = 28.984[\Delta \bar{t}(1.0 \times M_2) - \Delta \bar{t}(n \times M_2)]$$

against $r = \zeta_{A,0}/\zeta_{A,0,M2}$ for each station. The tabulated values satisfy $r* = 1$ and $\Delta \bar{g}_a = 0$ for $r = 1$, and define $r*$ and $\Delta \bar{g}_a$ for all r values through quadratic interpolation. It is stressed that the model is used only to provide small corrections to the major axes ($\bar{a}/r*$, $r* \approx r$) and phases. These corrections are insensitive to whether or not the model predicts the precise absolute currents or elevations. The corrections are very small for stations in the Severn Estuary (up-estuary of stations 18C and 19C; Figure 12.2) owing to their close proximity to Avonmouth, and increase seawards. However, the maximum magnitude of $\Delta \bar{g}_a$ is only $-5°$ (≈ 10 minutes; stations 4C, 5C and 6C), and $r*$ differs from r by 8% at most (stations 6C and 7C).

13 Seasonal Modulations of the Principal Semidiurnal Lunar Tide

D. PUGH

Institute of Oceanographic Sciences Deacon Laboratory, Godalming, Surrey, UK

and

J.M. VASSIE

Proudman Oceanographic Laboratory, Birkenhead, UK

13.1 INTRODUCTION

Tidal variations of sea levels and currents occur principally at periods of one day and half a day — the diurnal and semidiurnal tidal species. Because of ocean resonances, the semidiurnal variations are usually dominant, as is the case around the British Isles. However, there are modulations in the species amplitudes over periods of two weeks (the spring−neap cycle of the semidiurnal tides), one month, six months, one year and 18.6 years. These modulations are due to the addition of lunar and solar tides, and to modulations in the moon's orbit around the earth and the earth's orbit around the sun. They are well accommodated in standard tidal analysis and prediction procedures.

Response analyses of tides have shown clearly that there are other small forcing factors which modulate the tides, notably the radiational solar diurnal and solar semidiurnal tides. The major lunar semidiurnal constituent, M_2, is subject to a small annual perturbation in the astronomical forcing due to the solar influence on the lunar orbit. However, the observed changes to M_2 through the year are greater than this astronomical forcing would imply. The purpose of this chapter is to investigate the nature and origin of these enhanced variations. There are seasonal modulations in other lesser harmonic tidal constituents, but here we concentrate only on the major semidiurnal tide.

Both the astronomical and non-astronomical variations are only of the order of 1% of the M_2 amplitude. Evidently the interest in the seasonal modulations is mainly scientific, but there are also immediate practical implications. For example, the seasonal modulation delays the normal M_2 tide at Immingham in the Humber estuary by four minutes in September: at mid-tide this is equivalent to a level difference of 0.07 m.

Scientifically, it is generally assumed that the seasonal modulation is due to tide−surge interaction. As surges are bigger in winter, the winter amplitudes should be less than the amplitudes obtained by the analysis of data collected in the quieter summer months. The observations are generally consistent with this expected behaviour, but presenting evidence for a systematic link is not straightforward.

In this chapter we review previous analyses, including an analysis undertaken by Cath Allen after she graduated from Liverpool University in the summer of 1974, and present new data from a regional analysis of the Celtic Sea. Alternative mechanisms are considered, as are alternative methods of representing the modulations. Finally, we consider how predictable

Mixing and Transport in the Environment. Edited by K.J. Beven, P.C. Chatwin and J.H. Millbank
©1994 John Wiley & Sons Ltd

the modulations are in space and in time, and to what extent they can be incorporated into tidal predictions.

13.2 PREVIOUS ANALYSES

Corkan (1934) is generally attributed with the first systematic analysis of the seasonal M_2 modulation. He carried out harmonic analyses of five years of hourly height data from Liverpool, and a year of hourly height data from eight other British ports and from five other locations world-wide. He calculated that the modulation, which was consistent from year to year, existed generally in British waters and also throughout the world at large. He noted that the observed tidal variations were less than those predicted in the winter months, and greater than those predicted in the summer months. At Liverpool the discrepancy in the range was of the order of 2%. Corkan's analyses represented the modulations in terms of two harmonic constituents $MA_2(\sigma = 28.9430°/\text{hour})$ and $Ma_2(\sigma = 29.0252°/\text{hour})$, whose angular speeds are separated from the principal lunar semidiurnal constituent M_2 ($\sigma = 28.9841°/\text{hour}$) by the equivalent of 365 days. He also expressed his results more directly in the form of an annual perturbation in the amplitude of M_2

$$R\cos(\alpha + \frac{d}{365} \cdot 360)$$

where R is the amplitude of the perturbation, α is the phase lag of the perturbation on 1 January and d is the number of days elapsed from 1 January.

Geographically, he found that the phase lag of MA_2 steadily increases from north to south along the British coast of the North Sea, reaches a maximum at Immingham and then decreases slightly at Harwich and Southend. The amplitude of MA_2 is generally greater than that of Ma_2 and the phase lag of the latter is more variable. The value of R was approximately 2% of the M_2 amplitude, except at Immingham and Aberdeen where it was approximately 1%. Corkan looked for explanations in terms of meteorological effects and river discharges, but his limited data led to inconclusive results. Notably, he made no distinction between astronomical and non-astronomical components of the seasonal variations.

A much fuller analysis, which makes this distinction between astronomical and non-astronomical components by applying the response analysis technique was published by Cartwright (1968). He identified the influence of solar parallax on the moon's orbit as the astronomical factor which gives an annual modulation to the M_2 tide. In the response analysis smoothed amplitude and phase responses are calculated across the semidiurnal species; these responses may be used to calculate the harmonic equivalents from a harmonic analysis of the gravity potential (Cartwright and Taylor, 1971); these show that the astronomical amplitudes of the two terms are 0.345 and 0.305% of the M_2 amplitude, respectively. This method allows a separation of the astronomical forcing of the M_2 modulation from that due to other non-astronomical effects by vector subtraction. In his paper, Cartwright makes an explicit distinction between *annual* modulation, due to solar parallax, and *seasonal* modulation, which he assumes depends on the solar declination, i.e. the seasonal solar heating cycle. In practice the two effects are indistinguishable in analysis, as they relate to the anomalistic year of 365.2596 days and the tropical year of 365.2422 days, but the terminology is useful. It identifies gravitational modulations of the M_2 tide as *annual* and non-gravitational modulations as *seasonal*.

Analyses of three years of data (1 January 1959−2 January 1962) were made for six stations:

Stornoway, Lerwick, Aberdeen, Immingham, Southend and Newlyn. Cartwright represented the (non-astronomical) modulation

$$[1 + \delta\cos(\sigma t + \alpha)]H_2\exp[i\omega t + i\epsilon\cos(\sigma t + \beta)]$$

where $H_2 e^{i\omega t}$ is the undisturbed tide, δ and α are parameters which represent the amplitude and timing of the amplitude modulation and ϵ and β similarly represent the phase modulation. The frequencies ω and σ are for the semidiurnal lunar tide and the seasonal modulation, respectively. The phase lead α is equivalent to Corkan's α and the amplitude δ to Corkan's R. Maximum amplitude modulations (δ) occur at Aberdeen (1.25%) and at Newlyn (0.91%), and tend to give maximum M_2 amplitudes in the period March to June. There are possible geographical relationships with Stornoway, Lerwick and Aberdeen forming one group, Immingham and Southend another, with Newlyn showing its own separate behaviour.

Cartwright suggested that Immingham and Southend, as estuary ports, may be more influenced by seasonal changes in temperature and rainfall, and noted that the seasonal modulation in phase is significantly greater (1.32° and 1.43°, respectively) for these two ports than for the other four ports. In general, the results tend to confirm the earlier, less refined, analyses of Corkan (1934).

Cartwright also noted that dynamic interactions of semidiurnal tides with seasonal changes in mean sea level are far too small to account for the modulations; however, it would be very difficult to identify which of the many seasonally varying ocean quantities such as currents, densities, depths or ice boundaries could have an influence on the tides.

Subsequently, Cath Allen analysed one year (1972) of simultaneous hourly data, for 11 stations from Lerwick to Southend along the British coast of the North Sea (Pugh and Vassie, 1976). This paper designated the two seasonal modulating harmonic constituents at speeds 28.94304° and 29.02517°/hour, as H_1 and H_2, respectively. H_1 was the larger constituent, except at Buckie. It reached an amplitude of 2% of M_2 at Immingham, Harwich and Southend, but also at Aberdeen; the phase lag increases steadily from north to south. Similarly, although slightly smaller in amplitude, the H_2 phases increase progressively from north to south in a way which also closely follows that of the M_2 harmonic tidal constituent. Figure 13.1 shows these close relationships in more detail. Also plotted are the percentage astronomical amplitudes of the two terms which contribute to the M_2 modulations; these are both much smaller than the non-astronomical contributions.

In their paper, Pugh and Vassie (1976) also considered the possible explanations for the modulations. Phase modulations could be due to a varying depth of the North Sea: the M_2 tide propagating in shallow water with a speed (water depth × gravitational acceleration)$^{1/2}$ will arrive later if the water depth is lower. For the North Sea, where M_2 propagation speeds are consistent with a mean depth of 30 m, the annual mean sea level variation from +0.06 m in November to −0.06 m in May is equivalent to 0.8 minutes, or 0.4° in phase, with early arrival in November and late arrival in May. The observed modulations are a delay of three minutes in July/August and an advance of three minutes in January/February, which is more than can be accounted for by mean sea level changes. At ports further north the mean sea level effect on the timing of the M_2 tide would be even less. Evidently, sea-level changes can account for only a small part of the observed M_2 seasonal modulations.

Pugh and Vassie (1976) argued for local effects being a primary factor in the modulations on two grounds. Firstly, the seasonal modulations of offshore currents at the Inner Dowing were weak, and the ellipses and directions of progressive wave propagation were not consistent with the normal observed pattern for the semidiurnal band. Secondly, the amplitudes of the

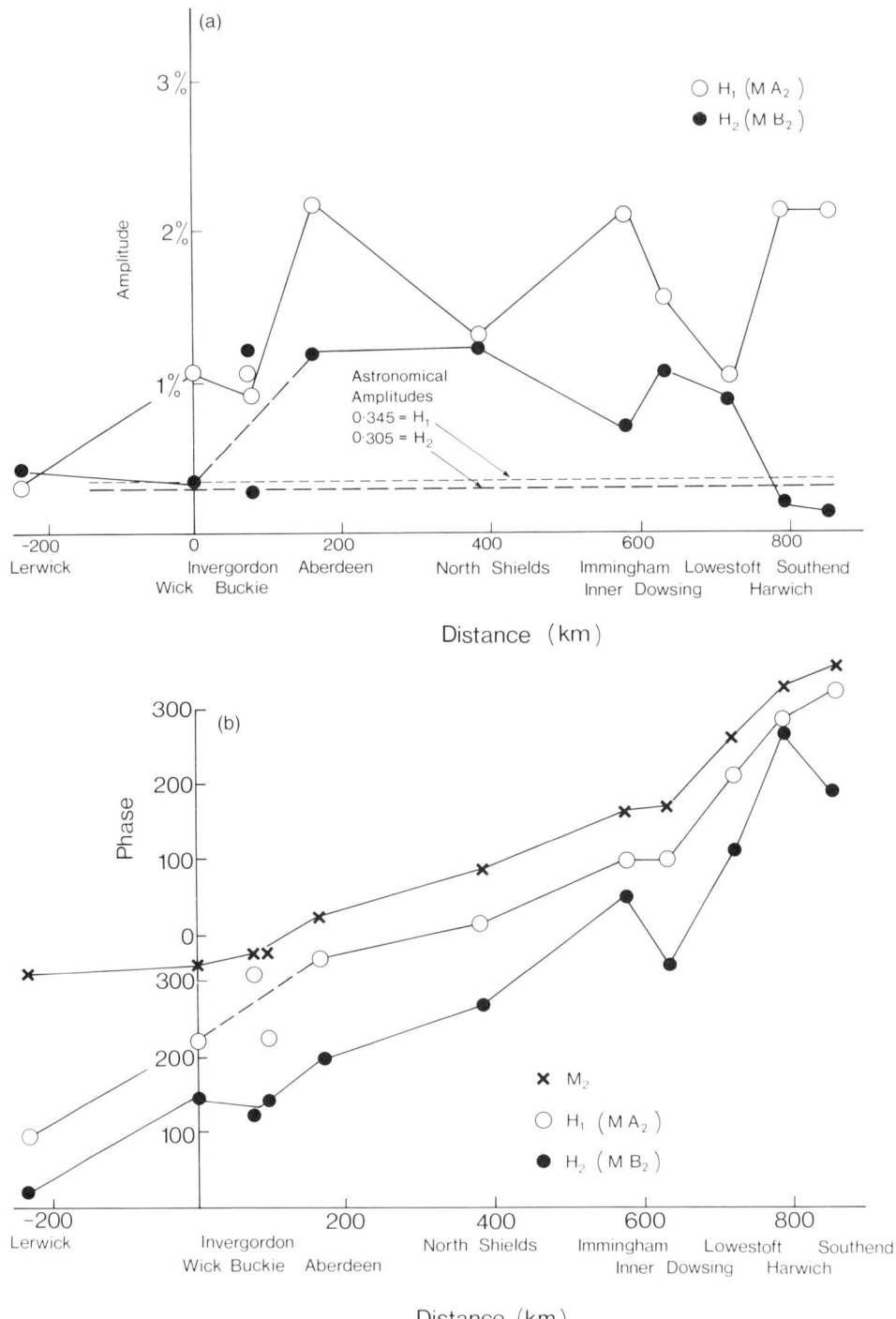

Figure 13.1 MA$_2$ and MB$_2$ variations in (a) amplitude and (b) phase along the British coast of the North Sea

modulations are variable in a non-systematic local way along the coast. Although there is consistency of phase propagation, they considered that this consistency could be due to the modulations being related to the phase of M_2 by non-linear local interactions.

Subsequent to this study the annual modulations of M_2 have been incorporated into the analysis and prediction of tides prepared for operational purposes by the Proudman Oceanographic Laboratory. However, the terminology MA_2 and MB_2 has now replaced earlier conventions for the two modulating harmonics (see Table 13.1): the suffix '2' is essential to designate the tidal species and the 'B' was chosen in favour of Corkan's lower case 'a' to make computer printout easier (Cartwright, personal communication). We will use this standardized convention for the remainder of this chapter.

Amin (1982) analysed long-period tidal records from eight ports on the west coast of Great Britain. He noted large variations in both MA_2 and MB_2 from year to year and suggested a correlation between the annual mean sea level change S_a and the constituent MA_2 at Liverpool. Nevertheless, the evidence is not conclusive, as it is based on only five years of analyses; firm correlations would require further investigation with much longer periods of data. He also noted an annual perturbation of the semidiurnal solar tide S_a, but this is more difficult to recognize because the angular speeds of the modulating terms are the same as those of the astronomical constituents T_2 and R_2. Amin (1983) has also made detailed analyses of Southend tidal constituents, including MA_2 and MB_2, for the period 1929−79, to which we shall return later. Baker and Alcock (1983) have analysed sea level and earth tilt data to show modulations in M_2, S_2 and K_1, which they related to non-linear interactions and atmospheric tides.

Guoy (1989) has shown similar modulations in equatorial regions, with approximately 1% modulation amplitude in the Malacca Straits, but values 3% or higher in the South China Sea. Most recently, Woodworth, Shaw and Blackman (1991) have examined the trends in mean tidal range around the British Isles, and along the adjacent European coastline. Seasonal changes in M_2, and, by implication, tidal range, are represented by the terms MA_2 and MB_2. They conclude that small changes of local water depth may be responsible for non-linear interactions which affect the local tidal ranges. These ideas increase the importance of a better understanding of the seasonal modulation effect. They imply that such an understanding may shed new light on the relationship between secular sea-level increases, anticipated in 'greenhouse' global warming scenarios, and the secular changes in high and low water levels. Woodworth, Shaw and Blackman (1991) show that trends in the latter are in general equal neither to each other, nor to the trend in mean sea level itself.

13.3 CELTIC SEA SEASONAL MODULATIONS

Data from a study of subtital variations of sea level around the Celtic Sea have been reanalysed to investigate further the question of the regional coherence of seasonal sea level modulations.

Table 13.1 Various nomenclature used to designate the seasonal modulating tidal harmonics of M_2

Angular speed (degrees/hour)	Corkan (1934)	Cartwright (1968)	Pugh and Vassie (1976)	Recommended nomenclature
28.94304	MA_2	(2 0 −1)	H_1	MA_2
29.02517	Ma_2	(2 0 1)	H_2	MB_2

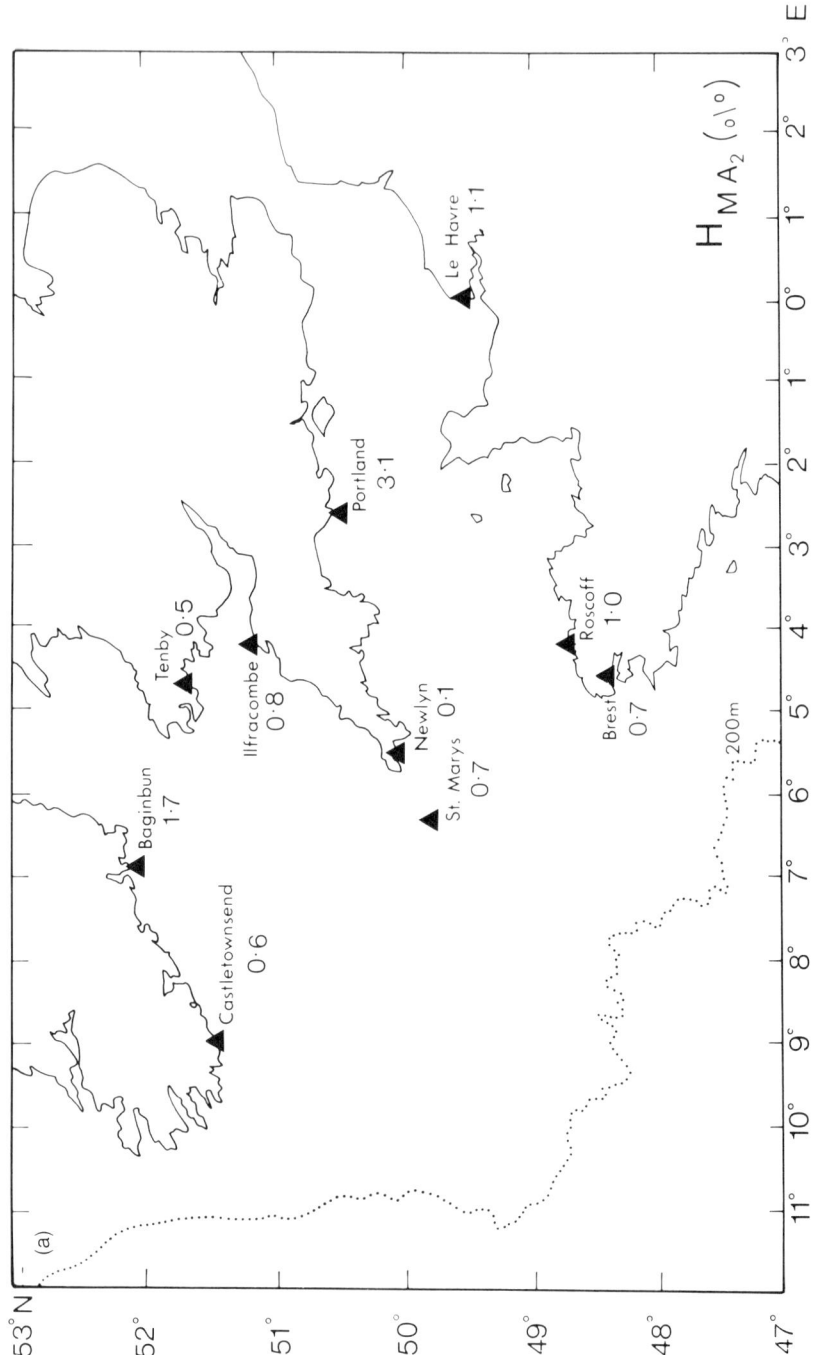

$H_M A_2 (o\o)$

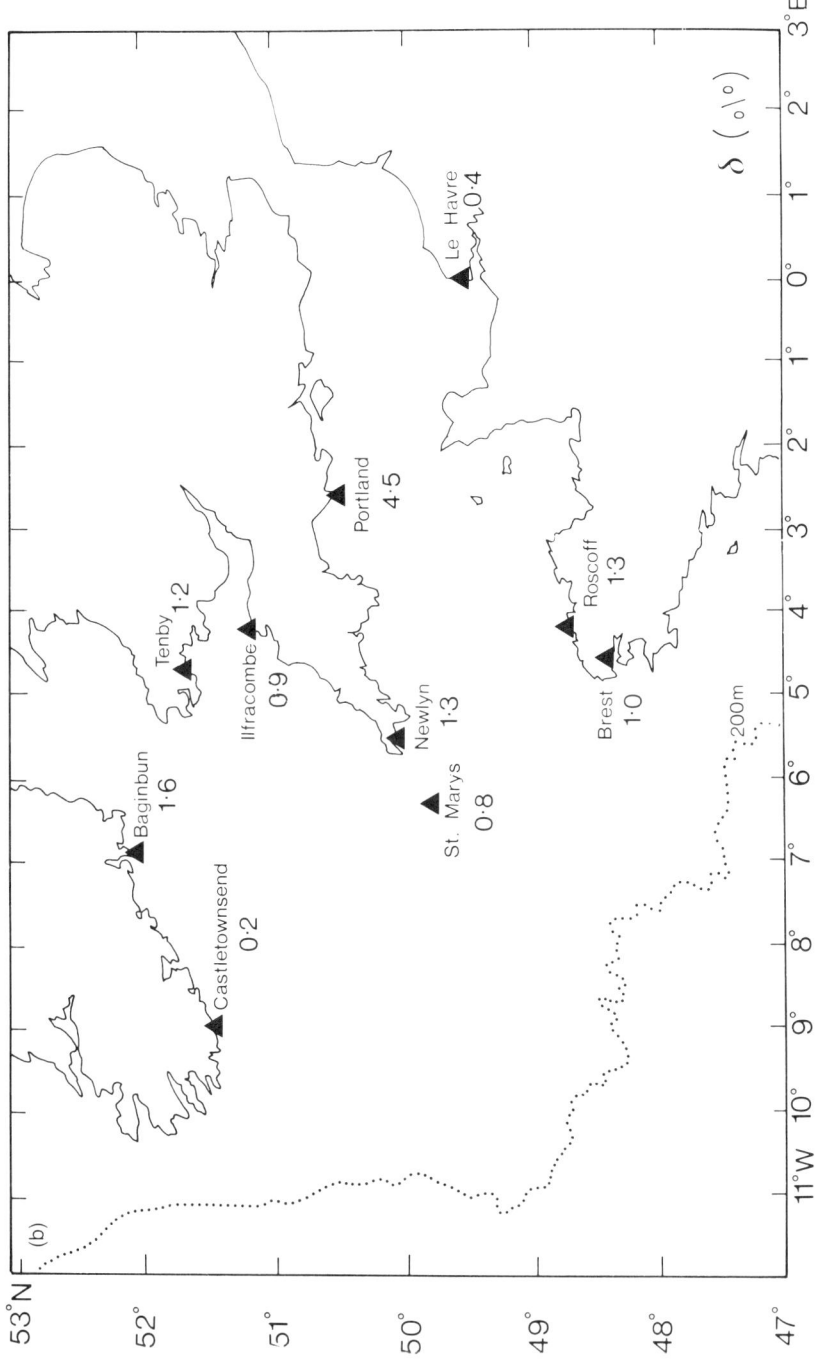

Figure 13.2 (a) Network of one year sea-level stations in the Celtic Sea and the English Channel used to investigate annual tidal modulations. The amplitude of the MA_2 harmonic is shown for each station as a percentage of M_2. (b) Amplitude of the total non-astronomical seasonal modulation, also as a percentage of M_2

Full details of the region and the sea-level stations are given in Pugh and Thompson (1986), where it is also shown that there is a high degree of regional coherence in the tidal and subtidal sea-level variations.

Coastal sea-level measurements were made over a common one-year period from July 1977 to June 1978. Bubbler pressure gauges were installed for the experiment at Castletownsend and Baginbun on the Irish coast and at Tenby at the entrance to the Bristol Channel. Data from permanent gauges at Ilfracombe, St Mary's, Newlyn and Portland were also available, as were data from the French ports of Brest, Roscoff and Le Havre (see Figure 13.2).

The year of data was harmonically analysed at each port for terms which included MA_2 and MB_2. The results of these standard analyses are shown in Table 13.2, together with the corresponding M_2 amplitudes and phases. Also shown are the ratios of MA_2 and MB_2 amplitudes (H) to M_2, and the phase lags (g). The modulating amplitudes of the larger harmonic term MA_2 are also shown in Figure 13.2a as a percentage of the local M_2 amplitude.

Table 13.3 shows these MA_2 and MB_2 harmonics recast as total amplitude and phase modulations according to the same formulation as Cartwright, but without complex number notation (see Appendix)

$$H_{M_2}\{1+\delta\cos(\nu t - \alpha)\}\cos\{(\omega_{M_2}t - g_{M_2}) + \epsilon\cos(\nu t - \beta)\}$$

Here, the modulating terms are capable of more direct interpretation: δ is the annual modulation in the M_2 amplitude H_{M_2}, α is the phase lag of the maximum value of the amplitude modulation from the vernal equinox (in days this is $\alpha(365/360)$ from the equinox, which is close to 21 March; Table 13.3 shows the day of maximum M_2 amplitude, counting from 1 January — this definition of α is slightly different from that used in the Corkan and Cartwright formulations given earlier), ϵ is the amplitude of the phase modulation, and is positive if the tide is earlier and β defines the time of maximum phase lag — the greatest phase lag occurs when $\cos(\nu t - \beta)$ is -1. Table 13.3 gives the day of maximum tidal phase lag, counting from 1 January.

The frequencies ω_{M_2} and ν are for the semidiurnal lunar tide and the seasonal modulation respectively; g_{M_2} is a phase lag.

Finally, the gravitational modulations were removed from MA_2 and MB_2, assuming the same amplitude and phase responses to astronomical forcing as for M_2 itself. The δ, α, ϵ and β terms have been computed for this non-astronomical modulation and are shown in Table 13.3, using the same procedure as outlined above. The δ values are also shown in Figure 13.2b.

Table 13.2 shows that for this region also, in the total modulation, the amplitude of MA_2 is larger than that of MB_2, and the phases are slightly more stable. The large MA_2 amplitude at Portland tends to distort the percentage averages, because the M_2 amplitude at Portland is small. At Newlyn the MA_2 amplitude is very small (0.1%), but the MB_2 amplitude is 1.06% of M_2; this contrasts with the MA_2 amplitude at nearby St Mary's, which is 0.74%. Apart from Portland and Newlyn, the MA_2 amplitudes lie in the range 0.5–1.65% of the M_2 amplitudes. The mean MA_2 amplitude, $1.0 \pm 0.8\%$ of H_{M_2}, is similar to that for the North Sea shown in Figure 13.1a ($1.5 \pm 0.5\%$, excluidng Lerwick); for MB_2 the corresponding amplitudes are $0.8 \pm 0.4\%$ of H_{M_2} for the Celtic Sea, and $0.7 \pm 0.4\%$ for the North Sea, again very similar.

Turning to the modulations of the combined constituents, shown in Table 13.3, the δ amplitude has a mean value of $1.5 \pm 1.2\%$ of H_{M_2}, with a maximum amplitude on day 128 ± 35 (in May), which is a few months earlier than in the North Sea. The total phase modulation

Table 13.2 Total MA$_2$ and MB$_2$ amplitudes and phases for the Celtic Sea and English Channel ports. $H(\frac{2}{1})$ and $H(\frac{3}{1})$ are the amplitude ratios for MA$_2$ and MB$_2$ compared with M$_2$. $g(2\text{-}1)$ and $g(3\text{-}1)$ are the corresponding phase differences

Location	M$_2$		MA$_2$		MB$_2$		$H(\frac{2}{1})$	$g(2\text{-}1)$	$H(\frac{3}{1})$	$g(3\text{-}1)$
	H	g	H	g	H	g				
Castletownsend	1.178	139.2	0.007	162.4	0.003	223.6	0.0059	23.2	0.0025	84.4
Baginbun	1.275	150.2	0.021	149.0	0.015	231.7	0.0165	-1.2	0.0118	81.5
Tenby	2.719	171.5	0.014	165.4	0.031	269.5	0.0051	-6.2	0.0114	104.1
Ilfracombe	3.070	162.2	0.023	95.5	0.006	221.7	0.0075	-66.7	0.0020	58.5
St Marys	1.754	131.8	0.013	234.0	0.020	213.7	0.0074	102.2	0.0114	81.9
Newlyn	1.700	135.5	0.002	125.2	0.018	238.4	0.0012	-10.3	0.0106	102.9
Portland	0.607	191.2	0.019	200.0	0.008	183.5	0.0313	8.8	0.0132	-7.7
Le Havre	2.609	286.1	0.030	298.6	0.019	37.0	0.0115	12.5	0.0073	110.9
Roscoff	2.678	142.2	0.27	191.3	0.016	95.9	0.0101	49.1	0.0060	-46.3
Brest	2.063	109.4	0.015	102.1	0.019	178.5	0.0073	-7.3	0.0092	69.1
Mean							0.0104	10.4	0.0085	63.9
Standard deviation							0.0084	43.7	0.0040	51.3

Table 13.3 Amplitude and phase modulations for the composite semidiurnal lunar M_2 tides including gravitational modulations and with gravitational modulations removed. Days are counted from 1 January

	$\delta(\%)$	Day of maximum M_2	ϵ	Day of maximum M_2 lag
Annual modulations: total				
Castletownsend	0.6	82	0.4	128
Baginbun	2.2	114	1.1	133
Tenby	1.2	154	0.7	112
Ilfracombe	0.9	146	0.3	241
St Marys	0.4	171	1.1	70
Newlyn	1.1	178	0.6	99
Portland	4.5	71	1.0	162
Le Havre	1.0	107	1.0	137
Roscoff	1.6	129	0.2	84
Brest	1.4	128	0.5	107
Mean	1.5	128	0.7	127
Standard deviation	1.2	35	0.3	48
Annual modulations: non-astronomical				
Castletownsend	0.2	256	0.4	125
Baginbun	1.6	129	1.1	130
Tenby	1.2	188	0.7	109
Ilfracombe	0.9	192	0.3	244
St Marys	0.8	232	1.1	69
Newlyn	1.3	210	0.6	96
Portland	4.5	70	0.7	154
Le Havre	0.4	156	1.0	134
Roscoff	1.3	8	0.2	106
Brest	1.0	148	0.5	103
Mean	1.3	159	0.7	117
Standard deviation	0.45	75	0.3	25

ϵ has an amplitude of $0.7 \pm 0.3°$, equivalent to 1.5 minutes in the time of M_2 high and low waters. The greatest lag occurs around day 127 ± 48, again in early May, and also a few months ahead of the maximum lag in the North Sea.

For prediction purposes it is necessary to consider the total modulation as represented by MA_2 and MB_2, determined by harmonic analysis. To investigate the non-astronomical effects it is necessary to remove the astronomical modulations, as in Table 13.3. As in the other tabulations, Portland appears to be anomalous. The δ modulation has a mean value of $1.3 \pm 1.2\%$ of H_{M_2} if Portland is included, and $1.0 \pm 0.45\%$ if Portland is omitted from the calculations. The maximum modulated amplitude due to non-astronomical effects is on day 159 ± 75, that is in June; in this instance both Portland and Roscoff are anomalous, and if they are excluded the day of maximum modulated amplitude is 189 ± 43, in early July. Non-astronomical modulations of phase are $0.7 \pm 0.3°$, equivalent as previously to 1.5 minutes, reaching a maximum lag on day 117 ± 25 — that is, in late April. Note, however, that for Newlyn at least, analysis (Section 13.4) of a very long period of data shows that the 1977–8 phase modulation was not typical of the mean behaviour.

In summary, the amplitudes of the total seasonal modulations are similar and close to 1.5% in both the Celtic Sea region and the North Sea, but the time of maximum amplitudes is

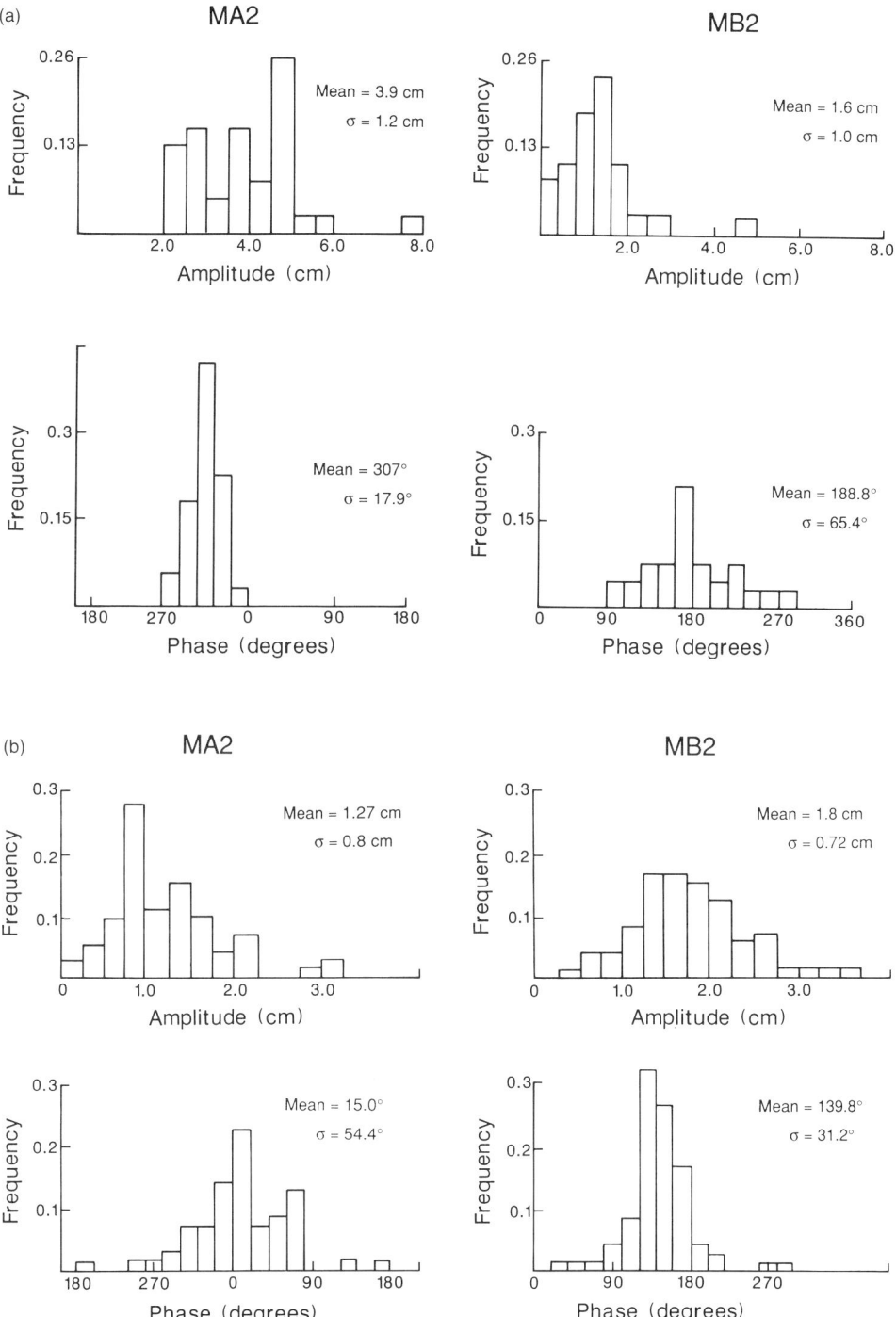

Figure 13.3 (a) Histograms of MA$_2$ and MB$_2$ derived from 30 years of Southend data. (b) Histograms of MA$_2$ and MB$_2$ derived from 70 years of Newlyn data

earlier in the Celtic Sea region by about 60 days. It is worth noting that the smallest modulations were at Castletownsend and Lerwick, both stations close to the open Atlantic Ocean. Anomalously large amplitudes were found at Portland in the Celtic Sea area and at Aberdeen, Immingham, Harwich and Southend in the North Sea.

13.4 INTERANNUAL STABILITY OF SEASONAL MODULATIONS

The regional comparisons in the preceding section were for different years, and it is necesary to consider how variable the modulation is from year to year. Amin (1983) has prepared a detailed analysis of the Southend tidal constituents, which we have used to study the interannual variability. Data from 1929 to 1979, analysed in four blocks, shows a high degree of stability (Table 13.4) from period to period. Clearly, amplitudes and phases of MA_2 are more stable than those of the weaker MB_2, but this is to be expected. Comparison of these long-term means with the year of the Pugh and Vassie (1972) North Sea analyses shows this to have been a typical year within the long-term averages. The results of Corkan (1934), although only for a single year, are consistent with the other analyses.

The question of the reliability of a single year within these long-term averages requires further investigaton. Although the analysis of Amin (1983) shows stability in MA_2 and MB_2 over time spans of 19 years, the histograms in Figure 13.3a and the statistics in Table 13.5 for Southend show that individual years can be fairly variable. These statistics were derived from 30 years of Southend hourly height data by tidal analysis and in general confirm the findings of earlier researchers. As found by Corkan (1934), the phases of MB_2 are more variable and MA_2 is usually larger than MB_2. However, it is interesting to note that the amplitude of MB_2 varies from almost zero to a magnitude 50% greater than the mean value of MA_2.

For comparison, Figure 13.3b and Table 13.5 contain the equivalent statistics for Newlyn. These were produced from 70 years of hourly height data from 1915 onwards. Here the dominance of MA_2 is not obvious; the variation in both constituents is approximately equal and the largest phase variation is in MA_2. As with Southend, the value from any individual year can vary considerably from the mean.

Some of the variability could be due to instrumental effects, but there is no way of checking this. Most of the year to year variability is undoubtedly oceanographic, and again suggests that M_2 modulations are driven by influences which have a strong but irregular seasonal component.

Table 13.4 Long-term stability of MA_2 and MB_2 at Southend (after Amin, 1983)

	MA_2		MB_2	
	H(mm)	g(degrees)	H(mm)	g(degrees)
1929–47	39	305.2	11	169.7
1941–59	40	299.2	6	204.1
1951–69	39	304.4	11	210.7
1961–79	38	310.8	11	193.3
Mean	39	304.9	9.8	194.5
Standard Deviation	1.0	4.7	2.5	18.0
Pugh and Vassie (1976): 1972	44	314	3	186.2
Corkan (1934): 1925	55	330	23	182

13.5 RELATIONSHIPS BETWEEN M_2 MODULATIONS AND SURGE ACTIVITY

To test the conventional idea that modulations in M_2 are caused by storm surge interactions with the tides, variations in M_2 amplitude can be tabulated or plotted against variations in the surge residuals. Determination on a daily basis is not possible because of the astronomical variation in the semidiurnal tide. We have analysed in two month (59 day) blocks for the M_2 constituent and computed the corresponding surge parameter as the root mean square (RMS) amplitude of the non-tidal residual from the analysis.

Table 13.6 shows results from this analysis grouped according to season. There are six bands through the year, each representing a two month period. The mean and standard deviation of the RMS surge height and of the amplitude and phase of M_2 are shown.

Figure 13.4 shows the same data for the amplitudes in graphical form. Each individual 59 day analysis is plotted according to season so the scatter in each can be identified. The upper trace is the amplitude of M_2 and the lower trace is the RMS surge height (both with an arbitrary datum shift to enable comparison). There is an obvious seasonal modulation in

Table 13.5 Statistics for MA_2 and MB_2 from yearly analyses: 30 years of Southend data since 1930 and 70 years of continuous Newlyn data from 1915. Amplitudes are in millimetres and phases in degrees

	MA_2			MB_2		
	Mean ± SD	Minimum	Maximum	Mean ± SD	Minimum	Maximum
Southend						
H	39 ± 14	21	80	16 ± 10	3	53
G	308 ± 18	266	341	189 ± 65	80	356
Newlyn						
H	13 ± 8	0	55	18 ± 7	5	46
G	15 ± 54	− 164	161	140 ± 31	36	219

Table 13.6 Surge amplitudes and M_2 modulations by season for Southend and Newlyn. Surge and M_2 amplitudes in millimetres and phases in degrees

	Mean ± SD RMS surge amplitude (mm)	Mean ± SD M_2 amplitude (mm)	Mean ± SD M_2 phase (degrees)
Southend			
Jan−Feb	266 ± 45	2037 ± 23	352 ± 1
Mar−Apr	204 ± 35	2036 ± 20	353 ± 1
May−Jun	135 ± 17	2087 ± 31	354 ± 1
Jul−Aug	145 ± 18	2058 ± 25	355 ± 1
Sep−Oct	209 ± 26	1999 ± 21	354 ± 1
Nov−Dec	273 ± 48	2036 ± 31	352 ± 1
Newlyn			
Jan−Feb	141 ± 28	1707 ± 12	135 ± 1
Mar−Apr	122 ± 22	1710 ± 9	135 ± 1
May−Jun	90 ± 15	1721 ± 12	134 ± 1
Jul−Aug	87 ± 27	1707 ± 13	134 ± 1
Sep−Oct	104 ± 19	1687 ± 10	135 ± 1
Nov−Dec	144 ± 48	1696 ± 14	135 ± 1

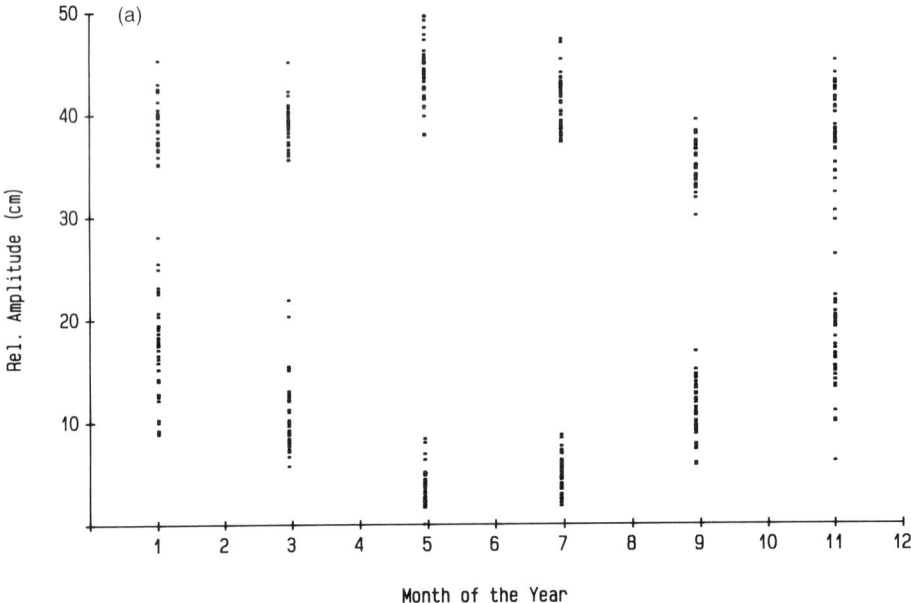

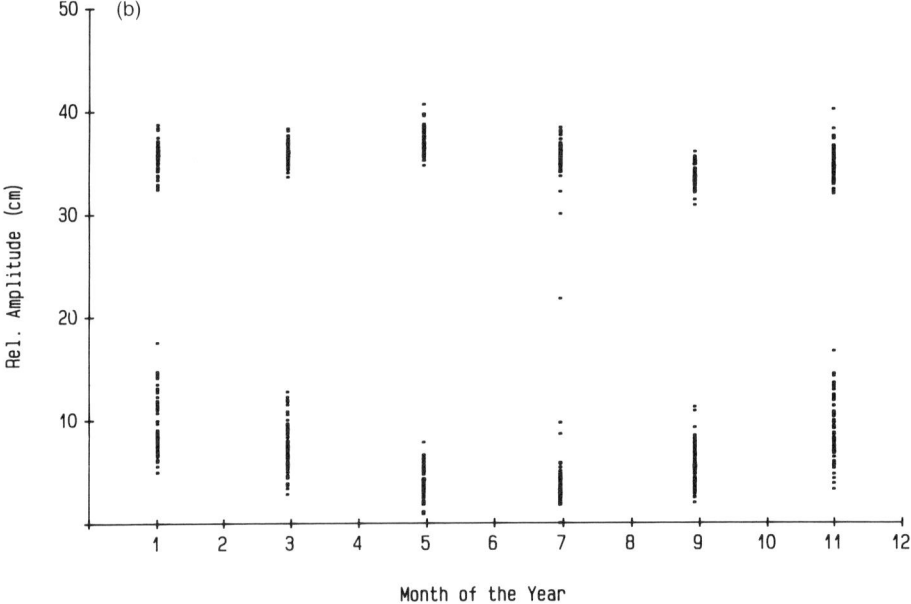

Figure 13.4 Relative amplitude of M_2 (upper trace) and RMS surge height (lower trace) derived from 59 day analyses. The results are regrouped into six seasonal blocks and offset for clarity. (a) 30 years of Southend data; (b) 70 years of Newlyn data

both the tides and in the surge activity. The amplitude of M_2 is largest in the summer when there is least surge activity.

This is a detailed statistical exposition of the evidence generally cited for the accepted hpyothesis that tide−surge interaction is responsible for seasonal modulations in the M_2 tide. However, even for the detailed summary in Table 13.6 the seasonal links between tide and surge anomalies are not entirely convincing. At Newlyn the largest M_2 amplitudes occur in May−June, whereas the smallest surges occur in the period July−August; at Southend the May−June surge amplitudes have their smallest variability (17 mm), whereas the variability in the M_2 amplitude during the May−June period has a maximum value (31 mm).

Correlation between two variables is not conclusive proof of a cause and effect linkage. It is necesary to consider the possibility that both surge amplitudes and M_2 amplitudes have seasonal signals which are both related to the summer/winter seasonal cycle, but not directly linked causally to each other. A more critical test is to remove the seasonal signal from the two variables and determine whether there is a correlation between the deseasonalized anomalies in both variables.

The results of this further analysis are shown in Figure 13.5. The seasonal 'linear' regression relationship from the data given in Table 13.6 and the least-squares fit to the plotted, deseasonalized data, are also shown. The correlation between residual surge activity and M_2 tidal amplitude and phase is very weak (error bands are for one standard deviation). This unexpected result suggests that the residual, and probably the seasonal, tidal modulations are affected by surge activity in only a limited way.

13.6 DISCUSSION

We have shown that there is no clear link between the level of surge activity and the modified M_2 amplitude for Newlyn and Southend. When seasonal variations in M_2 and surges are removed, larger surges in the residuals do not reduce the amplitude of M_2 consistently. This result is unexpected and important; it reopens the question of other influences on tidal amplitude and phase. A detailed approach in terms of numerical modelling experiments, with controlled surge and tide inputs, would be the most effective way to proceed.

However, it is also useful to develop a simple theoretical model to estimate the effects of surges on tides through loss of energy due to the bottom friction, which increases non-linearly as the current speed increases. This energy loss is generally found to be related to the cube of the speed, and on this assumption the following analysis shows that the third power frictional effects of the enhanced currents, due to surges and tides acting together, cannot alone account for the observed energy losses. The following simple model of tidal propagation and attenuation in a channel allows an estimate of the upper limit on energy losses due to this process.

Consider a tidal wave propagating in a channel of constant width and depth in the positive x direction. Averaged over a tidal cycle the energy flux through the section at x_1 will be

$$\phi_y H_{M_2}^2$$

where ϕ_y is a constant in terms of water depth, density, channel width and gravitational acceleration (Pugh, 1981).

The difference between the flux through x_1 and $x_1 + \Delta x$ is

$$2\phi_y H_{M_2} \left(\frac{dH_{M_2}}{dx} \right) \Delta x$$

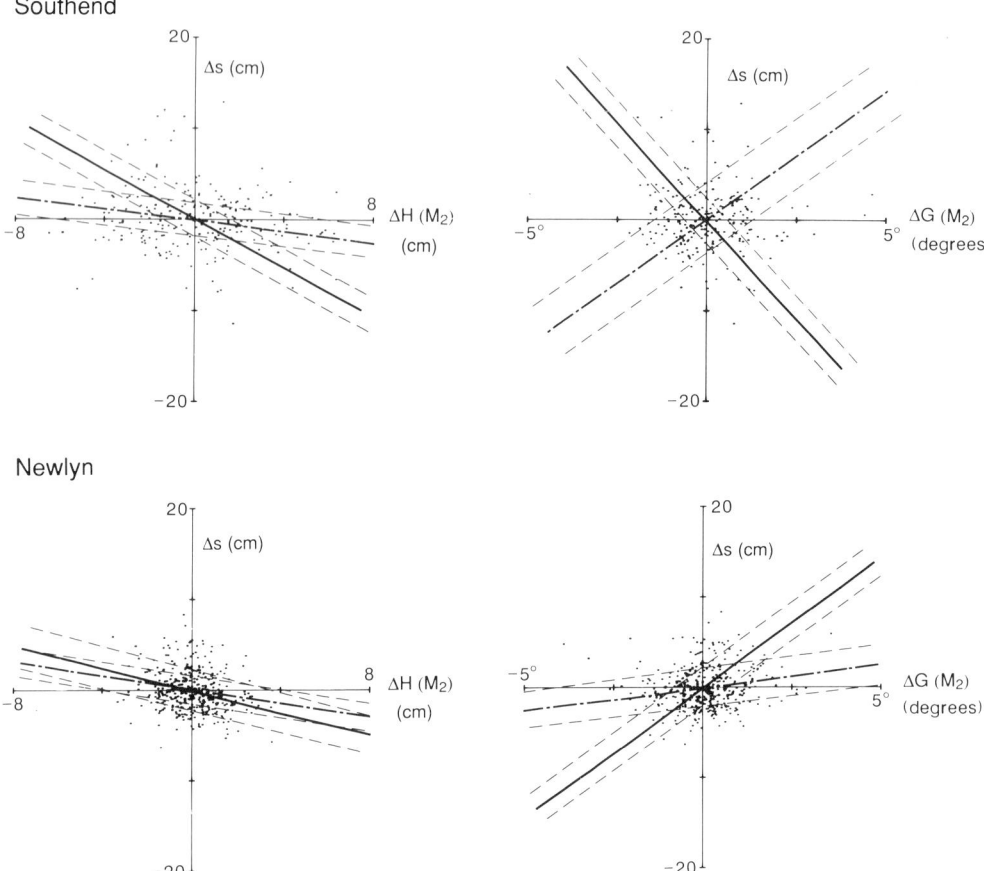

Figure 13.5 Scatter diagrams of residual surge amplitudes and residuals for M_2 amplitudes and phases at Southend and Newlyn. The broken lines show the regression fit to the plotted data. The solid lines show the regression fit to the seasonally averaged data in Table 13.6. The lack of any evident correlation between the two gradients shows that the seasonal modulations of M_2 and the seasonal variations in surge activity are not strongly linked in a causal sense

whereas the energy lost due to dissipation is

$$\lambda H^3_{M_2} \Delta x$$

where λ is a constant which accounts for the channel width, the water depth and the frictional drag of the channel bed. For a steady-state energy balance condition

$$2\phi_y H_{M_2}\left(\frac{dH_{M_2}}{dx}\right)+\lambda H^3_{M_2} = 0$$

Integrating between x_1 and x_2 where H_{M_2} at x_1 is reduced to αH_{M_2} at x_2 gives

$$\alpha = \left[1 + \frac{(x_2 - x_1)\lambda}{2\phi_y}H_{M_2}\right]^{-1}$$

$$\equiv [1 + KH_{M_2}]^{-1} \tag{13.1}$$

where K is a constant for a particular location.

For low energy loss and small attenuation ($\alpha \approx 1$), the attentuation of the tidal wave arriving at a fixed location x_2 in the channel, as the incoming tidal forcing H_M changes, may be written as

$$\frac{\mathrm{d}\alpha}{\mathrm{d}H_{M_2}} = -K \tag{13.2}$$

The value of K is not available in practice, but may be approximated at a particular location by separately identifying the direct and non-linear components of M_2 at the site. [Another method would be to examine the 18.6 year nodal modulations of M_2 and their non-linear attenuation (Amin, 1983) but we shall not do so here].

The response analysis technique gives a total gravitational M_2 and a non-linear component (technically I^{2+2-2}). The actual M_2 tide, given directly in harmonic analysis, is computed in response analyses as the vector addition of the gravitational and non-linear components. In a sense the semidiurnal tide is self-limiting, as the expression for ($\mathrm{d}\alpha/\mathrm{d}H_{M_2}$) shows.

At Newlyn the non-linear/gravitational amplitude ratio for M_2 is $0.179/1.87 = 0.096$, hence $\alpha = 0.904$ when $H_M = 1.87$ m (1977−8 data, as used in Pugh and Thompson, 1986). Substituting in Equation 13.1

$$K = 0.057/\text{metre } H_{M_2}$$

Suppose that at Newlyn the typical seasonal surge changes are ± 0.05 m (on a mean value of 0.15 m). The equivalent variation in α, if these 0.05 m values are directly added to the M_2 amplitudes, is from 0.901 to 0.907, a total modulation of $\pm 0.3\%$. This is substantially less than the 1% actually observed. When we also remember that surge currents in the region are generally of longer period (of the order of two days; Pugh and Thompson, 1986) and consequently the surge currents have a considerably lower speed for a given amplitude, it is clear that the effects of enhanced bottom friction due to a combination of tidal and surge currents would be relatively small.

Further detailed investigation of other mechanisms for enhanced energy losses is beyond the scope of this chapter, but it seems probable that the waves generated during the time of surge-generating storms may also dissipate energy, and wave activity may not be related closely to surge amplitudes. The combination of wave and tidal currents enhances bottom stress and energy losses substantially (Christoffersen and Jonsson, 1985; Davies, 1990). Numerical model studies in which tides, surges and waves are included in the simulation could be used to test the hypothesis.

13.7 CONCLUSIONS

Seasonal modulations of the principal semidiurnal lunar tide are evident in analyses from all areas around the British Isles. In the North Sea, the Celtic Sea and the English Channel they typically exeed 1% of the M_2 amplitude. Modulations are systematically related to the phases of the M_2 tide.

The harmonic constituents MA_2 and MB_2 are consistent from year to year, and averages taken over 19 year periods are very stable. Even values based on only a single year of analysis are worth including in the prediction of future tides.

The energy removed from the M_2 tide is not clearly related to the level of surge activity

Table 13.7 Amplitudes and phases of MA_2 and MB_2 at oceanic sites in the South Atlantic and Indian Ocean

Location	Year	MA_2		MB_2		M_2	
		H(mm)	g(degrees)	H(mm)	g(degrees)	H(mm)	g(degrees)
Ascension Island	1983–5	1	298	2	284	330	178
Ascension Island	1988	14	245	12	354	330	178
Ascension Island	1989	1	162	2	293	330	178
Ascension Island	1990	2	179	3	257	330	178
Tristan da Cunha	1988	1	54	1	348	330	178
Port Stanley	1990	18	123	11	269	444	280
Signy Island	1990	1	98	1	279	—	—
Amsterdam Island	1988	2	147	1	178	420	229
Kerguelen Island	1988	6	258	5	288	410	230
Scotia Sea (53.5°S 57.0°W)	1989	2	132	1	135	360	281
Scotia Sea (56.6°S 52.5°W)	1989	5	255	4	232	399	274
Scotia Sea (56.6°S 52.5°W)	1990	1	120	1	160	399	274
Scotia Sea (60.0°S 47.0°W)	1990	4	278	5	70	422	268

once seasonal trends are removed. At times of high surge activity, for example during the winter months, the M_2 amplitudes are lower than average; during the summer months they are greater than the annual average. However, this link may be due to other seasonal factors affecting the two parameters. It appears that frictional losses due to a combination of tide and surge current are not the dominant process which accounts for the energy loss. Another mechanism worth investigation is the possibility that friction and energy losses are enhanced during storms by wave−tide interaction. Numerical modelling simulations will be necessary to confirm this. Another influence could be the seasonal variations in stratification, with the development of seasonal thermoclines along the tidal transmission path, and their non-linear effect on tidal wave propagation.

In addition to regional modulations, there are clearly local effects and some evidence that ports in or near estuaries show the greatest variation from regional patterns of modulation (Harwich, Southend, Tenby, Immingham, Aberdeen). Two local dynamic effects which could account for this estuarine enhancement can be identified. The first is the influence of extensive low lying ground, which limits high combined tide plus surge levels by flooding at high water. The second is the generation of waves; the amplitude of these waves could be enhanced in estuaries, and the local wave set-up of water levels at tide gauge sites could be influenced by the extensive local shallow water. In addition, in the case of salt marshes, the response to flooding at high tide could vary because of seasonal effects on the vegetation.

It is noticeable that the smallest seasonal modulations were at Castletownsend and Lerwick, both ports close to the open Atlantic Ocean. It will be interesting to extend these analyses to other regions, including open ocean islands. However, even for the open ocean sites there could be seasonal modulations due to local wave set up related to storm activity.

Table 13.7 summarizes the seasonal modulations at selected oceanic islands and deep ocean sites. The Scotia Sea data were obtained by bottom pressure gauges. In general, the amplitudes of the modulations are relatively small, except at Port Stanley. At Ascension Island the analysis for 1988 is anomalous, and probably due to errors in the observations. A full analysis would require a better distribution of locations with several years of data from each site. A systematic and comprehensive analysis of seasonal modulations on a global scale is required.

REFERENCES

Amin, M. (1982) On analysis and prediction of tides on the west coast of Great Britain. *Geophys. J. R. Astron. Soc.* **68**, 57−78.

Amin, M. (1983) On perturbations of harmonic constants in the Thames Estuary. *Geophys. J. R. Astron. Soc.* **73**, 587−603.

Baker, T.F. and Alcock, G.A. (1983) Time variation of ocean tides. In: *Proceedings of the Ninth International Symposium on Earth Tides*. Schweizerbart'sche Verlagsbuchhandlung, Stuttgart, 341−350.

Cartwright, D.E. (1968) A unified analysis of tides and surges round north and east Britain. *Phil. Trans. R. Soc. London A* **263**, 1−55.

Cartwright, D.E. and Taylor, R.J. (1971) Raw computations of the tide-generating potential. *Geophys. J. R. Astron. Soc.* **23**, 45−74.

Christofferson, J.B. and Jonsson, I.G. (1985) Bed friction and dissipation in a combined current and wave motion. *Ocean Engin.* **12**, 387−423.

Corkan, R.H. (1934) An annual perturbation in the range of tide. *Proc. R. Soc. London A* **144**, 537−559.

Davies, A.G. (1990) A model of the vertical structure of the wave and current bottom boundary layer. In: *Modelling Marine Systems* (Ed. A.M. Davies). Vol. 2. CRC Press, Boca Raton, 263−297.

Guoy, T.K. (1989) Tides and tidal phenomena of the ASEAN region. *MSc Thesis*. Flinders University of South Australia.

Lamb, H. (1932) *Hydrodynamics*. 6th edn. Cambridge University Press, Cambridge, 738pp.

Pugh, D.T. (1981) Tidal amphidrome movement and energy dissipation in the Irish Sea. *Geophys. J. R. Astron. Soc.* **67**, 515–527.

Pugh, D.T. and Thompson, K.R. (1986) The sub-tidal behaviour of the Celtic Sea — 1. Sea levels and bottom pressures. *Cont. Shelf Res.* **5**, 293–319.

Pugh, D.T. and Vassie, J.M. (1976) Tide and surge propagation off-shore in the Dowsing region of the North Sea. *Dtsch. Hydrogr. Z.* **29**, 163–213.

Woodworth, P.L., Shaw, S.M. and Blackman, D.L. (1991) Secular trends in mean tidal range around the British Isles and along the adjacent European coastline. *Geophys. J. Int.* **104**, 593–609.

APPENDIX

Where a dominant constituent in a tidal band or species is modulated by smaller constitutents, and where all the separate constituents are determined from a harmonic analysis, it is usually difficult to appreciate how the combination of the several terms affects the time variation of the dominant constituent.

Provided that the modulating constituent amplitudes (a_1 and a_{-1}) are small compared with the amplitude a_0 of the major constituent, by making approximations it is possible to express the combinations as amplitude and phase modulations of the dominant constituent. The theory of combining M_2 and S_2 is given in Lamb (1932). For annual modulations of M_2 by MA_2 and MB_2 we have

$$H(t) = a_0\cos(\omega t - g_0) + a_1\cos[(\omega+\nu)t - g_1] + a_{-1}\cos[(\omega-\nu)t - g_{-1}]$$

$$\text{M}_2 \qquad\qquad\qquad \text{MB}_2 \qquad\qquad\qquad \text{MA}_2$$

where H is tidal amplitude, ω and ν are the semidiurnal (M_2) and seasonal modulation frequences, and g_0, g_1 and g_{-1} are phase lags.

The combined amplitude may be written for small MA_2 and MB_2 as

$$a_0\left\{1 + \left(\frac{a_1}{a_0}\right)\cos[\nu t - (g_1-g_0)] + \left(\frac{a_{-1}}{a_0}\right)\cos[-\nu t - (g_{-1}-g_0)]\right\}$$

or, by further manipulations, as

$$a_0\{1 + D\cos\nu t + E\sin\nu t\}$$
$$= a_0\{1 + \delta\cos(\nu t - \alpha)\}$$

where

$$D = \left(\frac{a_1}{a_0}\right)\cos(g_1-g_0) + \left(\frac{a_{-1}}{a_0}\right)\cos(g_{-1}-g_0)$$

$$E = \left(\frac{a_1}{a_0}\right)\sin(g_1-g_0) - \left(\frac{a_{-1}}{a_0}\right)\sin(g_{-1}-g_0)$$

and

$$\delta = (D^2 + E^2)^{1/2}$$

$$\alpha = \arctan(E/D)$$

Similarly, the phase of the combined constituents becomes ($g_0 + \Delta g$) where

$$\Delta g = \left(\frac{a_1}{a_0}\right)\sin[\nu t - (g_1-g_0)] + \left(\frac{a_{-1}}{a_0}\right)\sin[-\nu t - (g_{-1}-g_0)]$$

which, by further manipulation, becomes

$$F\cos\nu t + G\sin\nu t$$

$$\equiv \epsilon\cos(\nu t - \beta)$$

where

$$F = \left(\frac{a_1}{a_0}\right)\sin(g_1 - g_0) + \left(\frac{a_{-1}}{a_0}\right)\sin(g_{-1} - g_0)$$

$$G = \left(\frac{a_1}{a_0}\right)\cos(g_1 - g_0) - \left(\frac{a_{-1}}{a_0}\right)\cos(g_{-1} - g_0)$$

and

$$\epsilon = (F^2 + G^2)^{1/2}$$

$$\beta = \arctan(G/F)$$

Hence

$$H(t) = a_0[1 + \delta\cos(\nu t - \alpha)]\cos[(\omega t - g_0) + \epsilon\cos(\nu t - \beta)]$$

This is the formulation given in Section 13.3 with a_0, ω and g_0 replaced for clarity by H_{M_2}, ω_{M_2} and g_{M_2}, respectively.

In harmonic analyses the phases of M_2, MA_2 and MB_2 are measured from the vernal equinox, approximately 21 March, when the sun's mean ecliptic longitude is zero.

14 Study of Shear Dispersion in Tidal Waters by Applying Discrete Particle Techniques

G.C. VAN DAM

Rijkswaterstaat, Tidal Waters Division, The Hague, The Netherlands

14.1 INTRODUCTION

In her 1982 paper, Allen applied a particle technique to simulate dispersion in a horizontal oscillating current in a vertical plane, caused by current shear and vertical mixing. Although she first intended to take into account a height dependency of vertical mixing intensity, she restricted herself, for reasons of computing time, to vertical displacements fixed in size, but random in direction. For modelling the essential features of the shear dispersion mechanism this is not a severe limitation. Therefore, in the present paper, a similar simplification is used. However, instead of giving all the transverse displacements the same absolute value, their average size is kept constant (with a top hat distribution of the individual values) so that, for point sources, the particles do not only occupy positions at a finite number of discrete levels. The latter point was not important for Allen as she confined her study to line sources.

Allen chose the size of the vertical random displacements in her calculations on the basis of laboratory experiments as one-tenth of the water depth, apparently as a kind of effective mixing length. This physical point of view makes it unnecessary to consider accuracy in the sense of how good the results are as an approximation to a solution of mathematical equations. However, the step size chosen by Allen is small enough to make the results a fair numerical approximation of the exact solution of the advection−diffusion equation in the vertical plane for the given vertical velocity distribution, with a transverse diffusion coefficient independent of place and time and a longitudinal diffusion coefficient $K_x = 0$ (not to be confused with the effective longitudinal dispersion coefficient caused by the velocity shear and the transverse mixing).

Allen points out in the second section of her paper that she wanted to obtain a method that is more precise than the commonly applied one-dimensional advection−diffusion equation. In the light of the preceding paragraph, we can say that her results, in fact, show that a two-dimensional approach is much better than a one-dimensional approach, and, second, that a particle technique is suitable for treating the two-dimensional problem. In a convenient way, the periodic and long-term behaviour of longitudinal dispersion can be explained and various details of the early stages after release can be analysed, including skewed distributions, initial delay effects and the influence of the tidal phase at the time of release.

In a book of research papers in Cath Allen's memory, it seems appropriate to begin with the case of dispersion of dissolved matter in a longitudinal system with simplified geometry

Mixing and Transport in the Environment. Edited by K.J. Beven, P.C. Chatwin and J.H. Millbank
©1994 John Wiley & Sons Ltd

as dealt with in Allen's 1982 paper, and then to present some further extensions of particle tracking techniques which were indicated in Allen's paper, but not taken further there.

14.2 MODELLING THE SPREAD OF DISSOLVED MATTER IN LONGITUDINAL SYSTEMS WITH SIMPLE GEOMETRY

14.2.1 Method and definitions

In this section, the depth and width are kept uniform and constant, with, in particular, neglect of the systematic water level variations caused by tides.

Two-dimensional spreads in both a vertical and a horizontal plane are simulated by the same method, but the simulations termed 2DH (two-dimensional horizontal) use the symmetrical distributions of longitudinal velocity shown in Figure 14.1a; simulations termed 2DV (two-dimensional vertical) use asymmetrical velocity distributions such as those in Figure 14.1b. Three-dimensional simulations can be represented by a velocity distribution F in the cross-section, which is the product of the distributions f and g of the 2D cases:

$$F(y', z') = f(y')g(z') \tag{14.1}$$

with $y' (= y/b)$ and $z' (= z/d)$ being the relative positions from the shore and bottom, with b and d being the width and depth, respectively. This supposes that the cross-section is a rectangle.

The models discussed in this section are unbounded in the longitudinal direction x and all velocities are longitudinal and independent of x. Diffusion is assumed to be of the classical or Fickian type (no scale effects), independent of place and time, but not isotropic, and the only non-zero entries in the diffusion tensor are K_{xx}, K_{yy} and K_{zz}. It is approximated, for all directions, by an ordinary random walk. In the present computations $K_{xx} = K_x$ has been taken zero. For a sufficiently small time step Δt and a sufficiently large number of particles, the history of the particle distribution is close to the exact solution of the advection–diffusion equation, which in theory is reached only for zero time step and an infinite number of particles (Einstein, 1905). Acceptable approximations can be obtained with a finite time step and a finite number of particles.

To obtain results independent of depth and width, cross-exchange coefficients K_c ($= K_y$ or K_z) are replaced by cross-sectional mixing times T_c ($= T_y$ or T_z) which are proportional to s_c^2/K_c (s_c = width b or depth d). The choice of the coefficient of proportionality between T_c and s_c^2/K_c is somewhat arbitrary. In this paper we define

$$T_c = \tfrac{1}{2} \frac{s_c^2}{K_c} \tag{14.2}$$

To avoid (for point sources) particles only occupying specific levels, the random walk is not performed by steps of constant absolute value, but by steps of constant *average* size. The diffusion coefficient K (in a particular direction) generated by random steps depends on the character of the distribution ϕ of the steps Δ by the relation (Einstein, 1905)

$$K = \frac{1}{\Delta t} \int_{-\infty}^{\infty} \frac{\Delta^2}{2} \phi(\Delta) d\Delta \tag{14.3}$$

The only requirements for ϕ are

$$\int_{-\infty}^{\infty} \phi(\Delta) d\Delta = 1 \quad \text{and} \quad \phi(\Delta) = \phi(-\Delta) \tag{14.4}$$

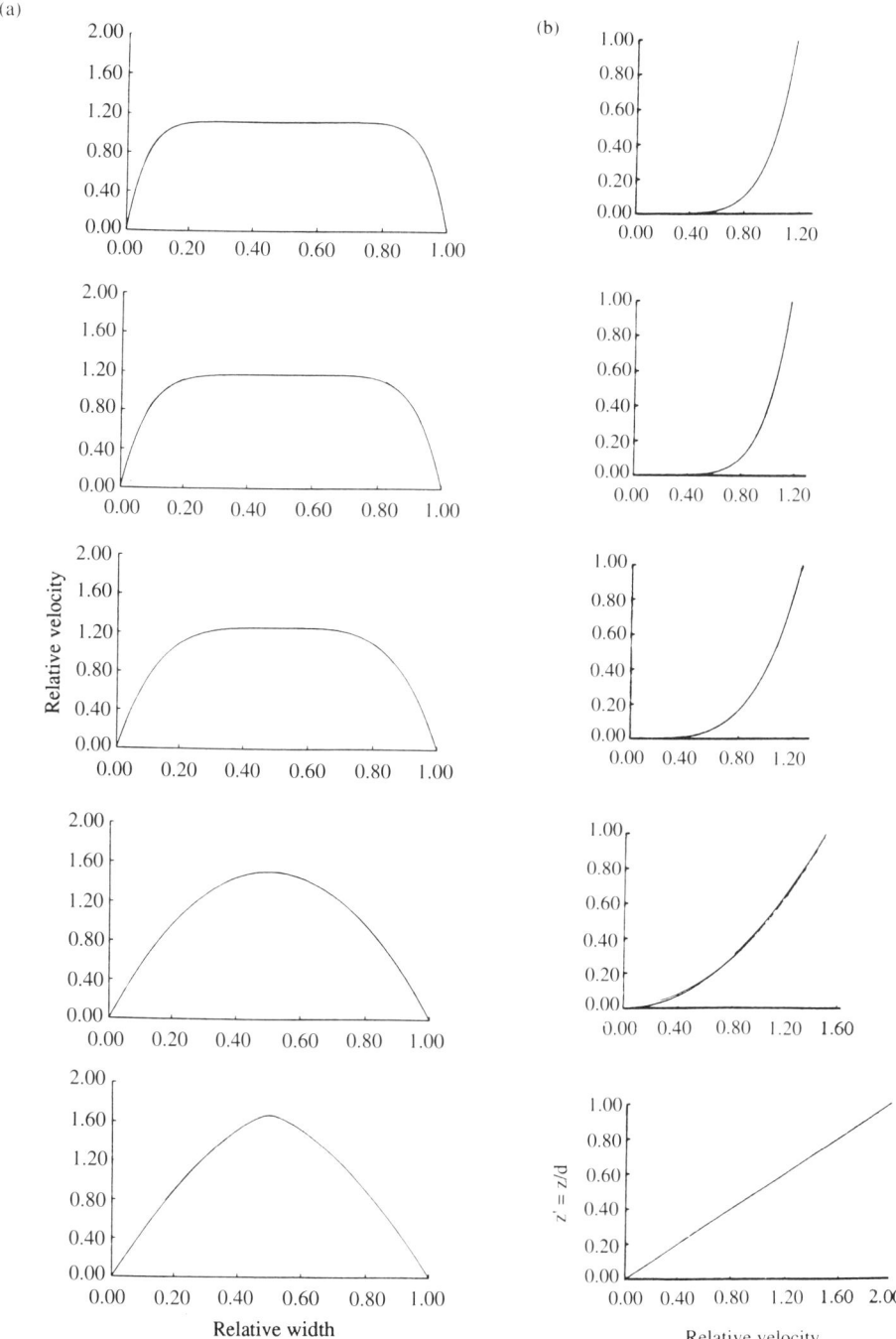

Figure 14.1 (a) Symmetrical velocity distributions in the horizontal plane as calculated by Equation 14.16 (ordinary and higher parabolae), drawn for (top to bottom) $r = 8, 6, 4, 2$ and 1.5. (b) Velocity distributions in vertical plane as calculated by Equation 14.12 (Van Veen profiles), drawn for $q = 8, 6, 4, 2$ and 1, respectively

In the present example the distribution $\phi(\Delta)$ is a 'top hat distribution'. It follows then that

$$K = \frac{2}{3} \frac{(\overline{|\Delta|})^2}{\Delta t} \tag{14.5}$$

and

$$\Delta = R(6K\Delta t)^{1/2} \qquad [R \text{ random in } (-1,1)] \tag{14.6}$$

As in this study the use of T_y and T_z is preferred, Equation 14.6 is rewritten as

$$(\Delta y)_r = bR(3\Delta t/T_y)^{1/2} \tag{14.7}$$

or, in dimensionless units,

$$(\Delta y')_r = R(3\Delta t/T_y)^{1/2} \tag{14.8}$$

[and analogously for $(\Delta z)_r$ and $(\Delta z')_r$].

All sources are taken in the $x = 0$ plane and can be point sources, horizontal or vertical line sources and (in the three-dimensional simulations) plane sources (uniformly distributed in the cross-section). In this section, as in Allen's 1982 paper, only instantaneous sources are considered and the particle distributions after the instantaneous release are only considered in the x-direction (longitudinal dispersion) and are characterized by moments about the mean

$$\mu_k = \frac{\sum\limits^{n} (x_n - \bar{x})^k}{N} \tag{14.9}$$

and properties derived therefrom (skewness, excess kurtosis).

A longitudinal dispersion coefficient

$$D_x = \frac{1}{2} \frac{d\sigma_x^2}{dt} \qquad (\sigma_x^2 = \mu_2) \tag{14.10}$$

is introduced by definition, analogous to the diffusion coefficient K_x in classical, gradient-type diffusion, where Equation 14.10 holds as a property of K_x (after instantaneous point release in a one-dimensional system).

The velocity distributions represented in Figure 14.1 are

$$f(y') = \frac{(r + 1)}{r} \{1 - (2|y' - 0.5|)^r\} \tag{14.11}$$

for the horizontal transverse direction and

$$g(z') = \frac{(q + 1)}{q} (z')^{1/q} \tag{14.12}$$

for the vertical direction. The latter are the so-called Van Veen distributions (Van Veen, 1936; 1938). They approach logarithmic distributions fairly closely but are more convenient to use and are common in physical oceanography. For well mixed systems, $q = 5, 6, 7$ are common values (larger values for greater depths of water). In stratified situations, the best possible approximations are obtained with lower q values (down to $q = 1$ or even less), but usually the degree of approximation in these instances is poor, just as for any other function not particularly adapted to the relevant vertical density distribution.

Horizontal transverse distributions in nature are variable, but may well agree with functions of Equation 14.11 with values of r in the range $2-6$ (or even 8), larger values corresponding to wide rivers; values of $r < 1$ are unrealistic.

For some special purposes the program contains two other functions to be added to the velocities generated by the use of Equation 14.11 or 14.12. They are

$$u(y') = a_y \sin(2k\pi y') \tag{14.13}$$

for the horizontal transverse direction and

$$u(z') = u_{ds}\left(\frac{1+\gamma}{\gamma-\delta}z'^\gamma + \frac{1+\delta}{\delta-\gamma}z'^\delta\right) \tag{14.14}$$

for the vertical transverse direction.

In Equation 14.13 $u(y')$ can also be used as an exchange profile (e.g. $k = 1$), but also to obtain special profiles by combining it with Equation 14.11, e.g. profiles containing a representation of 'dead zones' (see Section 14.2.3). $u(z')$ is an exchange current such as can be caused by wind or by a density gradient; it does not change the profile-averaged velocity.

Equations 14.11 and 14.12 are used for steady as well as time dependent (tidal) currents. In the latter, the deformation of the velocity profile by inertia, especially important around the turn of the flow, is completely neglected. Such effects, just like the precise shape of the velocity profiles, are unimportant in the present context.

14.2.2 Effective component of the velocity profile

The effective component of the total velocity profile for longitudinal dispersion is $u' = u - \bar{u}$, where $\bar{u}$ is the average of u over the cross-section.

It is well known (Taylor, 1953; 1954; Aris, 1956; Fischer et al. 1979) that the coefficient D_x, as defined by Equation 14.10, tends to a constant equilibrium value $(D_x)_{eq}$ for large times. For a steady current, this equilibrium is fully determined by u' and the transverse exchange rate characterized by K_c or T_c. Fischer et al. (1979) derived a general expression for $(D_x)_{eq}$ in terms of u' and K_c for the two-dimensional case; in the present notation this result is

$$(D_x)_{eq} = -2T_c \int_0^1 u' \int_0^{s_2'} \int_0^{s_1'} u' \, ds' \, ds_1' \, ds_2' \tag{14.15}$$

with $s' = y'$ or z' and $u' = u'(s') = u'(s_2')$.

Except when $\bar{u} = 0$ (velocity profiles such as Equation 14.14) we can write $u = \bar{u}F$, where F is as in Equation 14.1, so that $u' = (F-1)\bar{u}$ and it is then easily seen that, for steady flow

$$(D_x)_{eq} = c_d T_c \bar{u}^2 \tag{14.16}$$

in which the coefficient c_d depends on F in a complicated way as illustrated by Equation 14.15 for the two-dimensional case. With the present model, c_d can be found from the numerical simulations.

For a periodic flow with period T, a relation such as Equation 14.16 will hold approximately with instantaneous values as long as u varies slowly ($T \gg T_c$). Maintaining the suffix $_{eq}$ (as there will be some initial influence of the tidal phase at the moment of release), we expect for periodic (tidal) averages

$$(\tilde{D}_x)_{eq} = c_d' T_c \tilde{\bar{u}}^2 \tag{14.17}$$

with $\sim$ for tidal averaging and a c'_d that differs only slightly from c_d. For $T_c/T \to 0$, Equation 14.17 should become exact and $c'_c \to c_d$.

If, instead, T_c/T gradually increases, the well known counteractive effect of the oscillatory motion begins to suppress the longitudinal dispersion (Fischer *et al.*, 1979 and references cited therein; Van Dam and Louwersheimer, 1992). This effect becomes dominant when T_c becomes larger than T, so that the maximum longitudinal dispersion occurs when $T_c \approx T$. When there are no other dispersive mechanisms (e.g. a steady current component), the longitudinal dispersion will tend to zero for $T_c/T \to \infty$. Figure 14.2 illustrates these effects in the 2DV case by showing the development of an instantaneous line source of particles for $T_c = T_z = T$ for $T_z = 6T$, and for $T_z = \frac{1}{6}T$.

Apparently, for arbitrary T_c, the expression on the right-hand side of Equation 14.17 includes a function Φ of T_c/T (contained in c'_d) which tends to unity for $T_c/T \to 0$ and to zero for $T_c/T \to \infty$; as $c'_d \to c_d$ for $T_c/T \to 0$, we obtain

$$(\tilde{D}_x)_{eq} = c_d T_c \tilde{\bar{u}}^2 \Phi(T_c/T) \tag{14.18}$$

in which, for a simple harmonic flow, $\tilde{\bar{u}}^2$ can be replaced by $\frac{1}{2}u_a^2$ (u_a = profile-averaged amplitude).

Because of the linear character of the mechanisms (see also Fischer, 1972), we finally expect for a mixture of a single harmonic and a steady current (u_0 is $\bar{u}$ for the steady flow component)

$$(\tilde{D}_x)_{eq} = c_d T_c [u_0^2 + \frac{1}{2}u_a^2 \Phi(T_c/T)] \tag{14.19}$$

This relation is confirmed by numerical simulations with the present model (Section 14.2.3).

14.2.3 Numerical simulations

The numerical simulations with the simple geometry model of Section 14.2 can be seen as a direct extension of the work presented in Allen's 1982 paper and were facilitated by the great increase in computer power since 1982. They can now be obtained using a standard PC.

Oscillatory flows in this subsection are all simplified to a single sinusoidal component (as assumed in Equation 14.19). To relate the results to experimental data, the period T is chosen to be 45 000 s (only 0.67% more than the period of the M2 tide). Figure 14.3 shows $\sigma_x^2(t)$ for such an oscillating current, without the steady component. All results are for the 2DV case with various $T_c = T_z$, but a fixed velocity distribution: Equation 14.12 with $q = 6$. This is done to clearly show the influence of T_z. In nature, the best q will generally be lower as T_z is larger.

All the graphs in Figure 14.3 have the same scale so that it is easily seen that the average dispersion rate initially grows with increasing T_z as in steady flow and that for further increases in T_z the trend reverses due to the compensating effect of the flow reversal described in Section 14.2.2 and Figure 14.2. The same effect is the cause of the increase with T_z of the degree of modulation of the $\sigma_x^2(t)$ curves. In the limit $T_z \to \infty$ the modulation tends to 100% maximum temporal growth within each tidal period, but no net growth in longer periods. Instantaneous values of D_x as defined by Equation 14.10 oscillate but remain positive for moderate values of T_z. At a certain T_z/T the compensating reaction after the turn of the flow becomes strong enough to make σ_x^2 shrink after each reversal of the current; consequently D_x becomes periodically negative. It is clear that the instantaneous values of

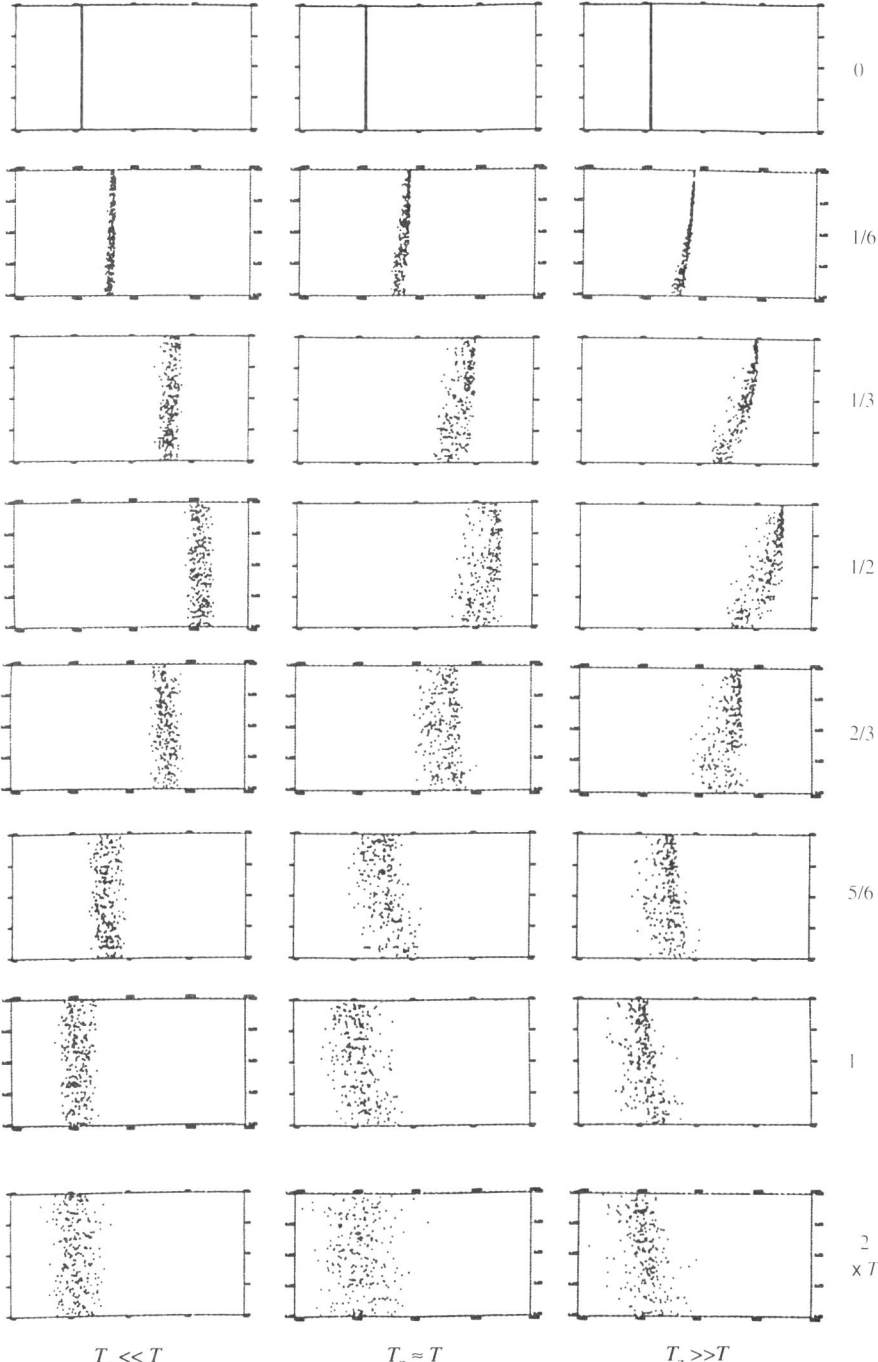

Figure 14.2 Development of vertical line source in two tidal periods for $T_z \ll T$, $T_z \approx T$ and $T_z \gg T$ (net current = zero)

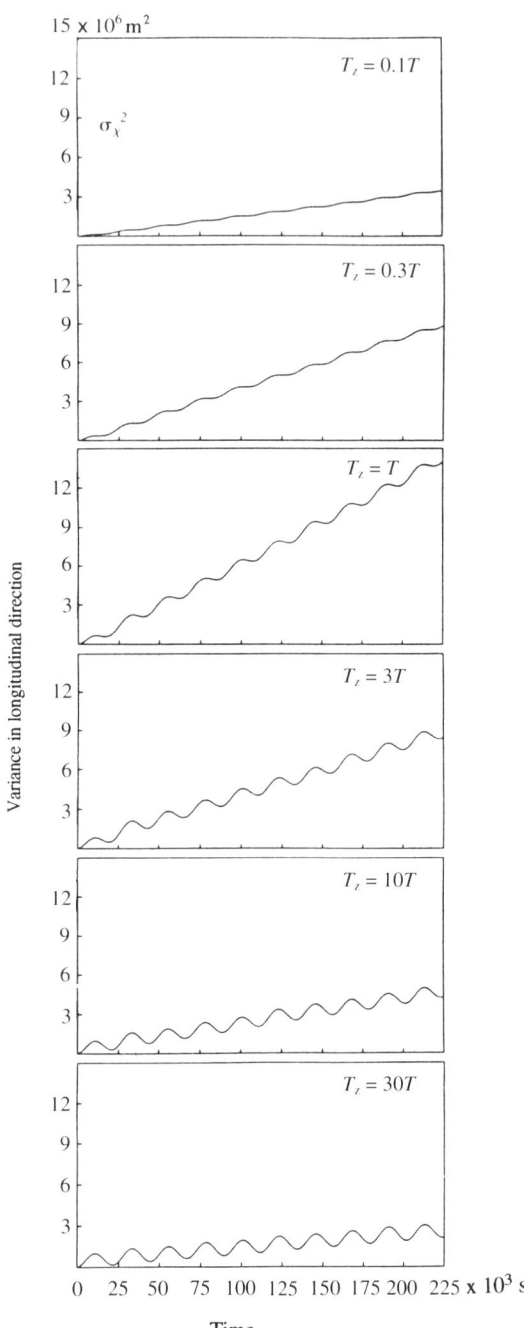

Figure 14.3 Variance in longitudinal direction as function of time for oscillatory flow for various T_z values (2DV case, profile 12, $q = 6$). Source: vertical line over full depth

this D_x cannot be equal to a gradient-type diffusion coefficient; if so, there would always be a positive growth of the cloud.

Figure 14.3 also illustrates that after some time (which increases with T_z) the periodic sections of the $\sigma_x^2(t)$ curve become identical in shape, so that the tidal average $\tilde{D}_x$ reaches an equilibrium value $(\tilde{D}_x)_{eq}$. In Figure 14.4 this $(\tilde{D}_x)_{eq}$ is plotted as a function of T_z/T, showing behaviour of the expected type (Equation 14.18).

The computations were performed with a profile averaged amplitude $u_a = 1\,m\,s^{-1}$, so $\bar{u}^2 = \frac{1}{2}u_a^2 = \frac{1}{2}m^2\,s^{-2}$. When $(\tilde{D}_x)_{eq}$ is also computed for the corresponding case of steady flow (same kinetic energy, so $u_0 = \frac{1}{2}2^{1/2}\,m\,s^{-1}$), it is indeed found that the two cases agree as $T_z \to 0$ (Figure 14.5). The value of the dimensionless coefficient c_d found for the chosen velocity distribution (Equation 14.12 with $q = 6$) is 3.56×10^{-3}. The ratio Φ has the expected behaviour as a function of T_z/T; it tends to unity for $T_z/T \to 0$ and to zero for $T_z/T \to \infty$. The decay at large T_z/T is as $(T_z/T)^{-p}$ with $p \approx 1.65$.

Holley, Harleman and Fischer (1970) derived an analytical expression for Φ in the case of a linear velocity profile ($q = 1$ in terms of Equation 14.12 in the present paper); this Φ behaves as $(T_z/T)^{-2}$ for large T_z/T. Simulation with the present model in the case $q = 1$ gives excellent agreement with their result (Figure 14.6a). However, the weaker decay for larger q values found in the present simulation is significant (Figure 14.6b). It can be qualitatively understood by noting that the thickness of the effective shear layer decreases with increasing q (Figure 14.1b). This smaller effective thickness can be seen as a thinner effective mixing zone, resulting in shorter effective mixing times than those that apply for full depth.

The graphs of Φ for the various values of q shown in Figure 14.6b can be fairly well approximated by the formula

$$\frac{1}{1 + a(T_z/T)^p} \tag{14.20}$$

with p decreasing and a increasing with increasing q. A similar relation (Equation 14.20 with T_y instead of T_z) will hold in a horizontal plane.

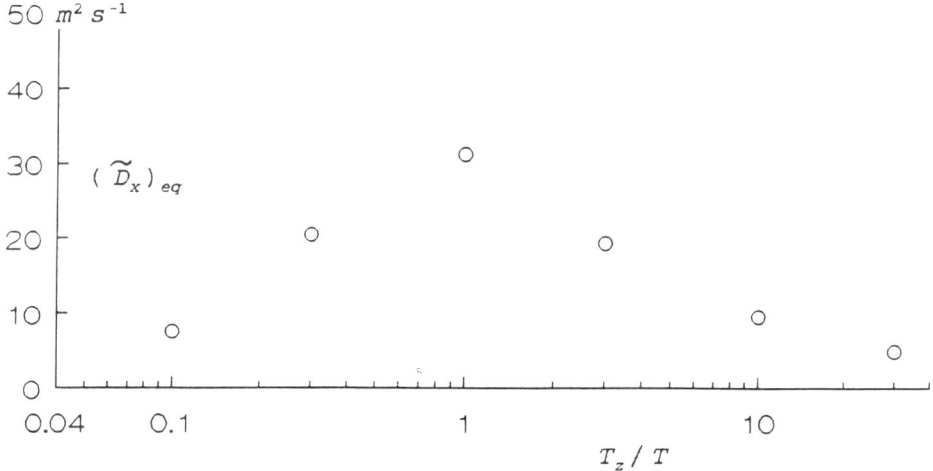

Figure 14.4 Equilibrium values of tidally averaged longitudinal dispersion coefficient as from 2DV particle simulations for various T_z/T values

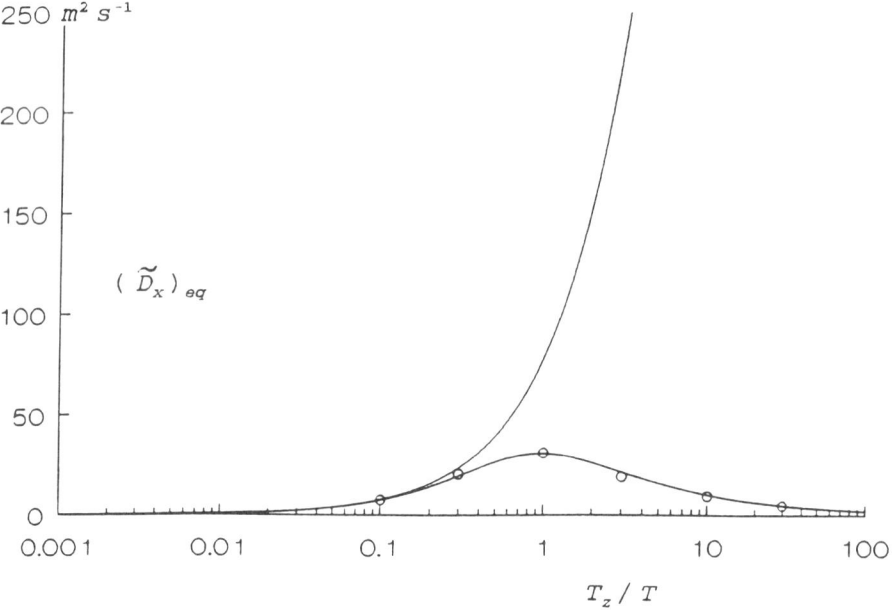

Figure 14.5 Convergence for $T_c \rightarrow 0$ of longitudinal dispersion for 100% steady and 100% oscillatory flow with equal kinetic energy

Figure 14.7 shows results for a number of intermediate cases between the 100% steady and 100% oscillatory flow represented in Figure 14.5, again for the velocity distribution of Equation 14.12 with $q = 6$. The curves were obtained using Equations 14.19 and 14.20. For each curve, one point was computed by direct calculation with the particle model; these points are shown in the figure.

Next, some aspects of the above 2DV results and some 2DH results of the model will be discussed with reference to some experimental evidence from rivers and estuaries.

In Figure 14.7 it can be seen that as long as the steady component of the flow is small (as it is in most estuaries) and certainly if $T_z < T$ (as in well mixed estuaries), the magnitude of longitudinal dispersion coefficients due to vertical shear is low and does not explain the much larger total values observed in real estuaries. Simulations with the 2DH mode of the model show that in these instances a dominant role of shear dispersion in the horizontal plane can be made plausible if the lower values of r in the velocity distributions of Equation 14.11 are combined with mixing times of the order of the tidal period T or longer. Such transverse mixing times T_y are fairly common in estuaries of medium width. For extremely wide estuaries T_y may become much longer and, as far as tidal flow alone is concerned, dispersion by horizontal shear will decrease due to Φ tending to zero as $T_y \rightarrow \infty$, similar to the behaviour noted earlier for large values of T_z.

For both the 2DV and 2DH contribution to the total longitudinal dispersion, a relatively small steady flow component due to river discharge can play a dominant part for large mixing times (see Figure 14.7 for T_z) but in wide estuaries the steady component u_0 is so small that this effect will usually not be important.

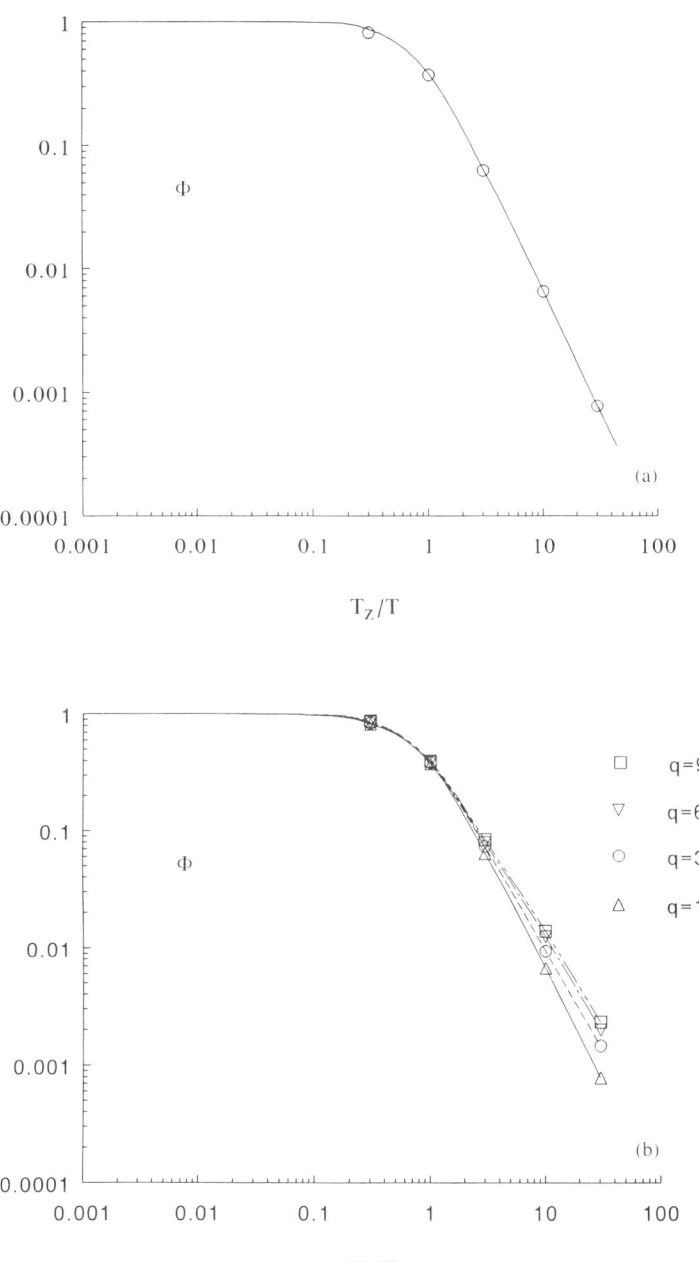

Figure 14.6 Ratio Φ between dispersion coefficients for oscillatory and steady flow for various T_z/T values. (a) ($\circ$) computed values for linear velocity profile ($q = 1$); (——) analytical solution by Holley, Harleman and Fischer (1970). (b) Computation for various (vertical) profiles ($q = 1,3,6,9$)

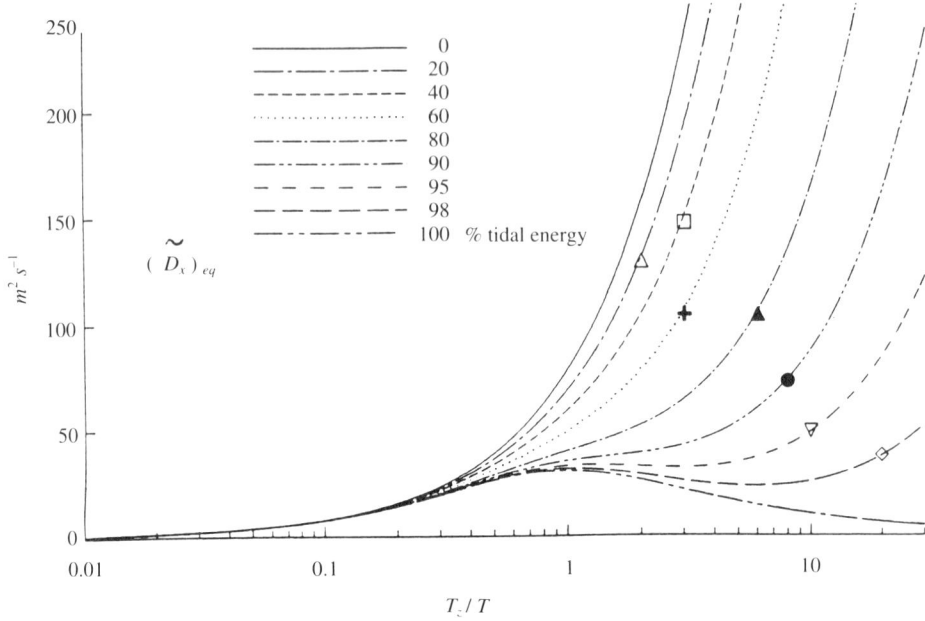

Figure 14.7 $(D_x)_{eq}$ values computed by Equations 14.19 and 14.20 for 100, 98, 95, 90, 80, 60, 40, 20 and 0% tidal energy; total energy of flow the same in all instances. Individual points $\diamond$, $\triangledown$, $\bullet$, $\blacktriangle$, $+$, $\square$ and $\triangle$ from explicit particle simulations

Some enhancement of the dispersion due to vertical shear occurs when the vertical velocity distribution becomes less uniform due to suppressed vertical mixing (stratification). For example, if $q = 3$ instead of 6, the value of $(\tilde{D}_x)_{eq}$ found for $T_z = T$ in a 2DV simulation is 3.5 times higher. However, much higher $\tilde{D}_x$ values are reached when strong longitudinal density gradients cause considerable density currents, as for example in the Rotterdam Waterway. Such exchange currents, which are of the type described by Equation 14.14, are so effective because they do not change sign like the tidal current. In the Rotterdam Waterway the surface value u_{ds} can be as high as 0.5 m s^{-1}, sometimes even higher. As a result, local values of $\tilde{D}_x$ as high as $1500 \text{ m}^2 \text{ s}^{-1}$ are observed. Such a value was also obtained with the model using the profile of Equation 14.14 with $u_{ds} = 0.6 \text{ m s}^{-1}$, $\gamma = 3$, $\delta = 1/6$ (comparable with $q = 6$ in Equation 14.12) and a T_z of one day. The same order of magnitude would be obtained with a higher T_z value and a smaller u_{ds}. The largest values of u_{ds} are observed in the prototype with high river discharges. The strong density currents in the Rotterdam Waterway are enhanced by the relatively narrow (artificial) cross-section. The strong inward dispersion of salt balances the strong advective seaward transport that would otherwise push the salt wedge out of the channel into the open sea.

In wide well-mixed estuaries with moderate river discharges, pronounced density effects do not occur and the longitudinal dispersion coefficients are lower (typical values of $100-300 \text{ m}^2 \text{ s}^{-1}$) with the main contribution coming from the horizontal transverse shear and mixing. Dye experiments in this type of estuary have shown that horizontal cross-sectional mixing times can be of the order of several days to a week. If u_a is taken as 1 m s^{-1} with a mixing time of five days and a velocity profile of type Equation 14.11 with $r = 2$, the model simulation gives $(\tilde{D}_x)_{eq} = 219 \text{ m}^2 \text{ s}^{-1}$. This order of magnitude cannot be obtained with the same mixing time and a much higher value of r. The simple profile with $r = 2$ should not be seen as an accurate description of the actual profile but rather as a simplification to the much more complex real profile. For a full picture of the main agents of longitudinal dispersion in relatively wide and well mixed estuaries, one more contribution has to be mentioned. The slow cross-sectional mixing (in the above case with T_y = five days) is due to several mechanisms but probably mainly to horizontal eddies, the latter causing a similar, possibly even stronger mixing in the longitudinal direction. Suppose the estuary width is 7000 m; if T_y = five days, $K_y = \frac{1}{2} \times 7000^2/(5 \times 86\,400) = 57\,\mathrm{m^2\,s^{-1}}$. If K_x is of the same order, its contribution is not entirely negligible compared with the $219\,\mathrm{m^2\,s^{-1}}$ above.

It became evident above that the periodic current reversal has a moderating effect on longitudinal dispersion and that therefore in an estuary very high dispersion coefficients only occur when a significant non-reversing flow component is present. In particular a strong density current, with its pronounced profile can generate dispersion coefficients which are an order of magnitude larger than usual.

In a non-tidal river the current is persistent and large longitudinal dispersion coefficients are to be expected. Indeed, in a river such as the Rhine, dispersion coefficients of 1500 to $2000\,\mathrm{m^2\,s^{-1}}$ (Van Mazijk *et al.*, 1991) are found in certain river sections with profile-averaged velocities of about $1\,\mathrm{m\,s^{-1}}$. A calculation with the 2DV mode of the model gives values of $(D_x)_{eq}$ of about $3\,\mathrm{m^2\,s^{-1}}$ (well mixed water column). It is clear, that as in the case of well mixed estuaries, the main agent of the dispersion is the horizontal transverse shear and the accompanying mixing. The order of magnitude of the cross-sectional mixing time found in dye experiments and monitoring effluent plumes (Van Mazijk, 1986) is one day. When this is combined with a profile of type 11, the large values mentioned cannot be fully explained, unless very unrealistic values of the parameter $r(<1)$ are used in the simulation. It is known, however, that in the relevant regions the 'dead zones' along one or both banks are agents for the enhancement of the longitudinal dispersion (Van Mazijk *et al.*, 1991). If a velocity profile more typical of such a situation is taken, much higher dispersion coefficients easily emerge. The profile shown in Figure 14.8 (obtained by a combination of Equations 14.11 and 14.13, with $k = 1.5$) generates a $(D_x)_{eq}$ of $3100\,\mathrm{m^2\,s^{-1}}$ for $u = 1\,\mathrm{ms^{-1}}$. This D_x is higher than the observed values but the dead zones may often be narrower and, additionally, the dead zones have a smaller depth than the middle of the river. These effects can be taken into account in a more refined model.

In the above simulations of tidal systems, a line source was used (as in Allen, 1982) to reduce the length of the adjustment period before reaching the periodic equilibrium regime with constant $\tilde{D}_x$. In practice, small sources are more common. The influence of source size and position is unimportant with rapid transverse mixing. But, as shown in Figure 14.9 for both tidal and steady flow, even for a T_c of 2.5 days (2DH case, $T_c = T_y$) the effect is still not very large. One can see that the greater shear near the bank gives a greater initial dispersion for a source at the bank than for one in the middle of the channel. The difference is smaller than the effect of the initial transverse spread obtained for a line source (lower part of the figure). It should be realized that σ_x^2 (or σ_x) is not a measure of dilution. In this respect the line source is more effective than the figure suggests.

Another factor influencing the early stages of the dispersion is the phase of the tide at the time of release. Figure 14.10 illustrates the convenience of the model in computing and depicting the effects. Some other examples (with a vertical line source in 2DV) have already been given by Allen (1982); the topics of discharge timing and position have also been studied in numerous papers by Smith (e.g. Smith, 1985).

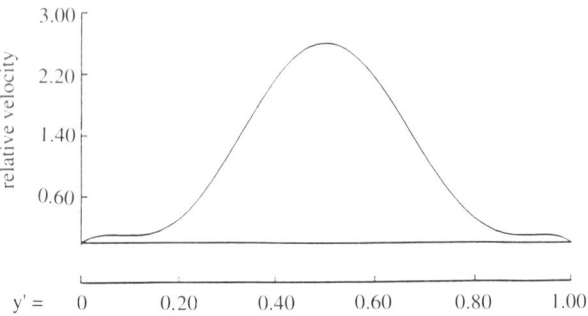

Figure 14.8 Velocity profile for simulating 'dead zones' along both banks. Composed by using Equations 14.11 plus 14.18 (with $k = 1.5$)

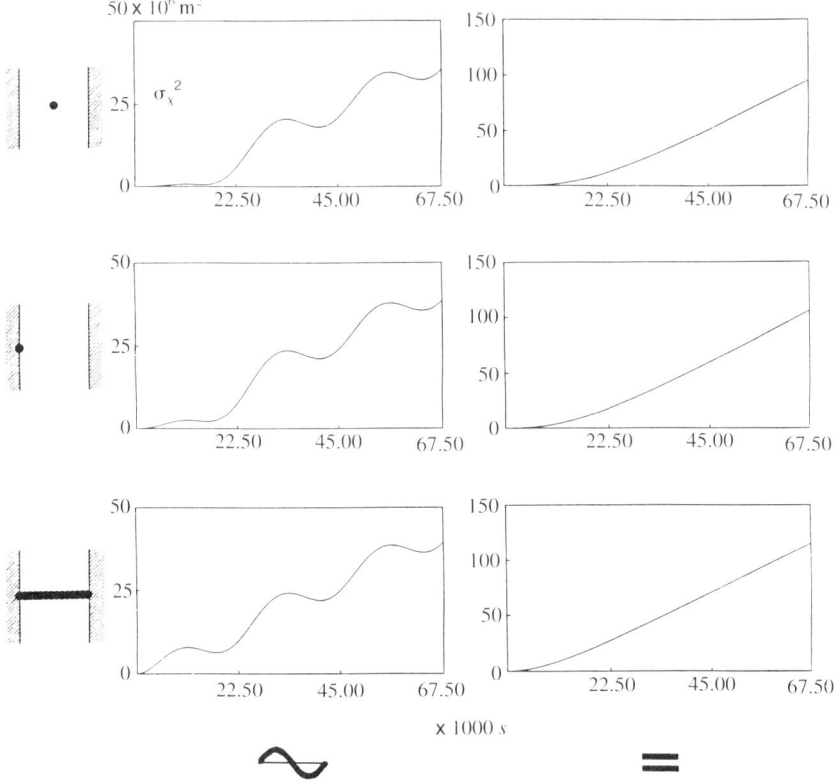

Figure 14.9 Comparison of $\sigma_x^2(t)$ for line source and two point sources in horizontal plane. Velocity profile of Equation 14.11 with $r = 2$; tidal and steady flow; $T_v = 2.5$ days

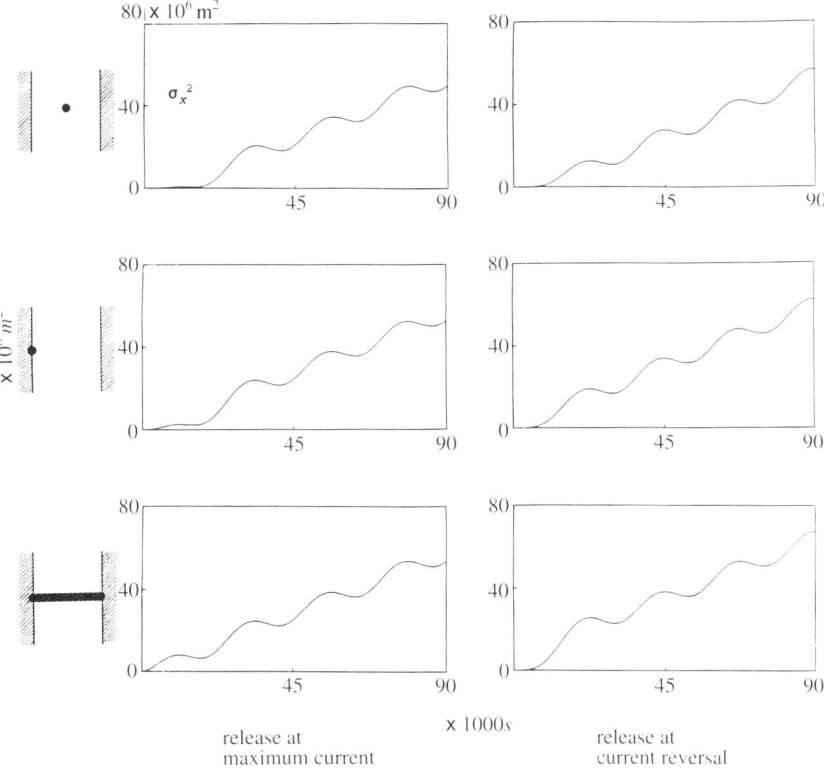

Figure 14.10 Influence on $\sigma_x^2(t)$ of tidal phase at time of release; various examples

A specific detail must be noted here. In the lower left part of Figure 14.10, the period of σ_x^2 is half that of the (tidal) oscillation of the flow. This is due to the symmetry of the two directions of the current and the corresponding phase at the time of release: the material returns (slightly spread) after half a period and then makes a similar move in the opposite direction. If the release is at the turn of the flow, the material moves initially in one direction twice as far and only 'returns' after a full period of the flow; therefore σ_x^2 in the lower right part of Figure 14.10 reaches a much higher first maximum, but in twice the time; the period in σ_x^2 would remain a full period of the flow if there was no longitudinal spread due to the vertical mixing. However, the longitudinal spread blurs the position of the cloud and the initial condition is quickly forgotten. Only with rather unrealistically long mixing times is an initial doubling of the period of σ_x^2 (and so also of D_x) maintained during several cycles (Figure 14.11). Thus in general the periods of σ_x^2 and D_x are half that of the flow.

For instantaneous point sources regarded only from the limited viewpoint of longitudinal spreading, there is little point in giving results from the three-dimensional mode of the model. The processes in the horizontal (2DH mode) and vertical (2DV mode) planes are essentially mutually independent as soon as the particles have spread in the cross-section, so that the resulting equilibrium values of D_x simply add. The simulations confirm this.

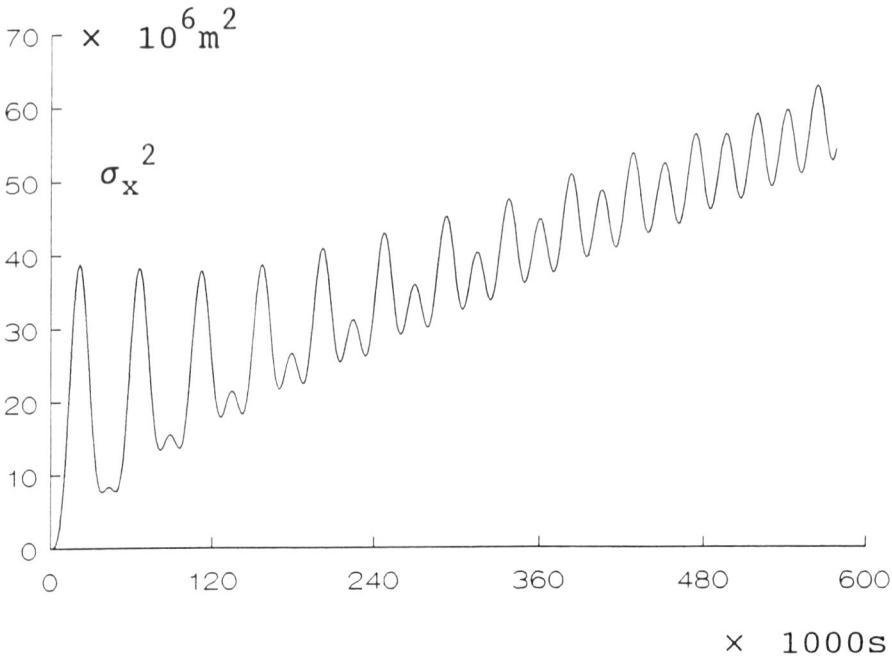

Figure 14.11 A line source with release at the time of current reversal shows a period in σ_x^2 equal to flow period T that remains during various flow cycles if T_c is very long, being gradually replaced by $\frac{1}{2}T$. Present example: $T_c = T_y = 24T$

14.2.4 Character of shear dispersion process; accuracy of computations

The requirements for accuracy depend on the purpose of a computation. In the case of shear dispersion the following two aims can be distinguished. One is that an approximation to the exact solution of the two-dimensional (possibly three-dimensional) advection-diffusion equation is sought. This can be obtained to any desired degree of accuracy by taking a sufficiently large number of particles and sufficiently small time steps. The other aim is the physical viewpoint mentioned in the introduction and applied by Allen (1982). Here, random steps of a magnitude representing a distinct exchange distance of water parcels (mixing length) are chosen. Such an approach could also be considered for the 2DH case. The result is still a certain approximation to the exact solution of an advection-diffusion equation. As long as the velocity profile is smooth, with a dominating component with a wavelength of the order of one or two times the width or depth, the result with (average) steps of one tenth of the width or depth (Allen's choice) is fairly close to the exact solution of the advection-diffusion equation, provided the number of particles is sufficiently high, except for a short initial phase (when only a small number of steps has been taken). The variations in σ_x^2 (after the initial phase) are fractionally about $N^{1/2}/N$ (where N is the number of particles in the simulation) as found by Allen (1982). When higher wavenumbers in the velocity distribution function become dominant, step lengths have to be reduced to obtain the same degree of sampling as before.

The relative errors in the time derivative of σ_x^2 (so in D_x) and in higher moments (so also in the skewness and kurtosis) are larger than those in σ_x^2. This is only relevant when these quantities are needed, for example in an investigation of how the shape of the particle distributions depends on the nature of the dispersion process.

For steady flows shear dispersion in tubes and channels has been studied extensively (Aris, 1956; Taylor, 1953; 1954; Fischer *et al.* 1979 and references cited therein) and it is well known that after an initial phase, depending on T_c, the dispersion coefficient defined by Equation 14.10 becomes constant and the longitudinal distribution tends to normality. In other words, regarded as a one-dimensional process in the x-direction the dispersion becomes equivalent to a classical diffusion process with D_x as the diffusion coefficient. This can be understood; the shear dispersion mechanism amounts to a longitudinal exchange process over a finite distance of the order of $\overline{|u'|} \times T_c$, with an exchange velocity of the order $\overline{|u'|}$. As $\overline{|u'|}$, for a given profile, is proportional to $\bar{u}$, this simple reasoning also leads to Equation 14.16. After some time, the constant exchange distance becomes small compared with σ_x and this is the condition for a classical, gradient-type diffusion.

The reasoning does not change much in the case of a reversing current. As long as $T_c \ll T$, the situation is essentially unchanged. For larger values of T_c the advective shift by the shearing velocity is partially compensated by the opposite shift (before the material has had enough time to spread in the relevant transverse direction) due to the reversal of the u- and therefore the u'-profile. The net result (after each flow cycle) has the same features as a longitudinal exchange over a finite distance, but with a reduced efficiency (represented by Φ in the equations).

Figure 14.12 shows the gradual decrease of skewness α_3 and excess kurtosis λ_4 [$\alpha_3 = (\mu_3/\mu_2)^{3/2}$; $\lambda_4 = \alpha_4 - 3$ with $\alpha_4 = (\mu_4/\mu_2)^2$ where $\alpha_4 = 3$ for a Gaussian distribution] for 100% steady and 100% oscillatory flow for a particular value of T_c and a particular velocity distribution, together with the time behaviour of σ_x^2 and D_x in the model simulation. The initial part of the α_3 and λ_4 curves has only schematic significance because of limited accuracy due to the initially small number of time steps (see earlier). Taking smaller time steps increases the accuracy, leading to higher values of $|\alpha_3|$ and $|\lambda_4|$ in the initial phase. Furthermore, the accuracy in α_3 and λ_4 is smaller than that in σ_x^2 because they involve higher moments. The accuracy in σ_x^2 is such that in the upper right graph of Figure 14.12 the curve seems straight after the initial period of about 45 000 s. The corresponding D_x curve shows the variations in the time derivative due to the much smaller oscillations in σ_x^2 itself.

14.3 EXTENSIONS

In this section some further applications of particle modelling are given as an illustration of how the more sophisticated applications foreseen by Allen (1982) have been realized to some extent since then.

14.3.1 Open sea: simplified model

In the open sea, the horizontal velocity shear is diverse in direction and organized on numerous scales. In an estuary, the width bounds the size of the eddies; in the sea the bounds of the basin are so wide that they are often beyond either the domain of interest or modelling and consequently the size of the eddies is virtually unbounded. In a simplified dispersion model, comparable with the longitudinal model of the preceding section (e.g. with constant uniform depth), the best approximation to the horizontal velocity field is a uniform tidal current field with a superimposed field of eddies of all sizes. Two ways exist to account for the latter:

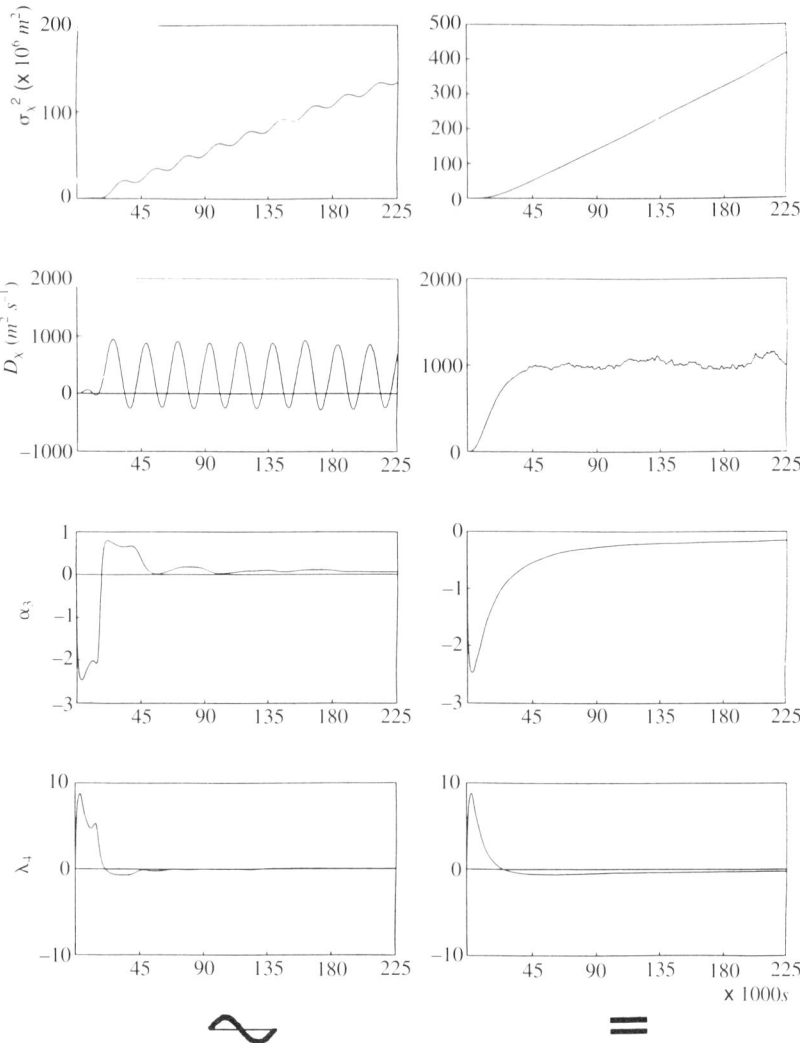

Figure 14.12 Behaviour of σ_x^2, D_x skewness α_3 and excess kurtosis λ_4 for 100% oscillatory and 100% steady flow. $T_c = T_y$ (2DH case) = 2.5 days; velocity profile of Equation 14.11 with $r = 2$; point source at $y' = 0.5$. Although the σ_x^2 curve looks straight for steady flow, its derivative D_x shows its irregularities (10 000 particles)

a 'scaling' in the random step procedure instead of the ordinary random walk used in the longitudinal model of Section 14.2, or an explicit simulation of the field itself by a finite but relatively large number of superimposed 'waves' in the form of sinusoidal velocity components with a discrete but broad spectrum of wavenumbers and, in principle, random directions. Each of these components has essentially the form of Equation 14.13, but randomly directed and with a random phase (banks at large or virtually infinite distance).

The two techniques, the scaled random walk and the synthetic eddy field approach, are described in more detail elsewhere (Van Dam, 1991; Van Dam and Louwersheimer, 1992).

In Van Dam 1991, it is shown that as far as the horizontal dispersion of small instantaneous sources [measured either by maximum concentration or size $(2\sigma_\xi \sigma_\eta)^{1/2}$] is concerned, both methods are fully equivalent, except that the scaled random walk produces smooth dilution curves, almost identical in repeated simulations (just as in the longitudinal model of the preceding section), whereas the (randomized) synthetic eddy field delivers fluctuating realizations, which require ensemble averaging to obtain the same smooth curves found in the corresponding scaled random walk simulation. The eddy approach thus seems more realistic, but its single realizations are not necessarily representations of real individual cases, the degree of fluctuation depends on the applied number of eddy sizes ('density or spectral lines') in the artificial, discretized spectrum. One could tune the spectral density on prototype data to obtain a high degree of agreement between the model and nature in which the peaks and dips in the natural (continuous) spectrum would relate to the 'lines' in the discrete artificial spectrum.

Figures 14.13a and 14.13b compare the two methods. In Figure 14.13b only two of the curves of Figure 14.13a are approximated with the eddy simulation method, and show the scatter between the various realizations and the agreement of the average values with the curves obtained in the scaled random walk simulation (Figure 14.13a). Figure 14.13a also shows a number of intermediate cases; the smoother results make the scaled random walk more suitable for investigating the influence of parameters. The parameter concerned is the vertical mixing time T_z, which varies considerably in a shallow, locally and temporarily stratified sea such as the North Sea, which the simulation pertains to.

The upper curve (a straight line due to the choice of a simple kinetic energy spectrum) represents the dilution due only to the horizontal structure of the velocity field. In nature this limiting case is closely approximated by the dilution curves in regions with fast vertical mixing. The bundle of curves is also bounded on the lower side; just as in the longitudinal case (Section 14.2) the dispersion has a maximum for mixing times of the order of the period of the oscillating flow (Figure 14.4). The maximum dispersion in the simulations of Figure 14.13a is reached at a somewhat larger T_z than $T_z \approx T$, due to the circumstance that slower mixing of matter is accompanied by a suppressed mixing of momentum, so that the vertical velocity distribution becomes more pronounced (lower q in terms of Equation 14.12) and the dispersion for larger T_z is further enhanced. This additional enhancement of the dispersion due to vertical shear has been taken into account globally in Figure 14.13a (not in Figures 14.4 to 14.7, where no attempt was made to cover a set of realistic examples).

The adequacy of the three-dimensional model is illustrated by Figure 14.13c and 14.13d, where the two bounds (minimum and maximum effect of vertical shear dispersion) are plotted in one graph with the compiled data from dye experiments in the North Sea (each dot represents one patch in a particular phase of the dilution process) performed at scattered locations in regions of diverse character. In Figure 14.13c the simplest possible kinetic energy spectrum is used; in Figure 14.13d the spectrum is more adapted to the field data. Some points lie outside the envelope in the upper left part of the graphs because the initial size of the dye patches in the relevant experiments was too large.

14.3.2 Complex geometries

In practical applications, often located in coastal regions and estuaries, the positions of patches and plumes and the effects due to local bottom topography and shoreline geometry are often of such importance that there is a need for models that account for these factors. In such cases, the water movement above a certain length scale can be simulated by two- or even

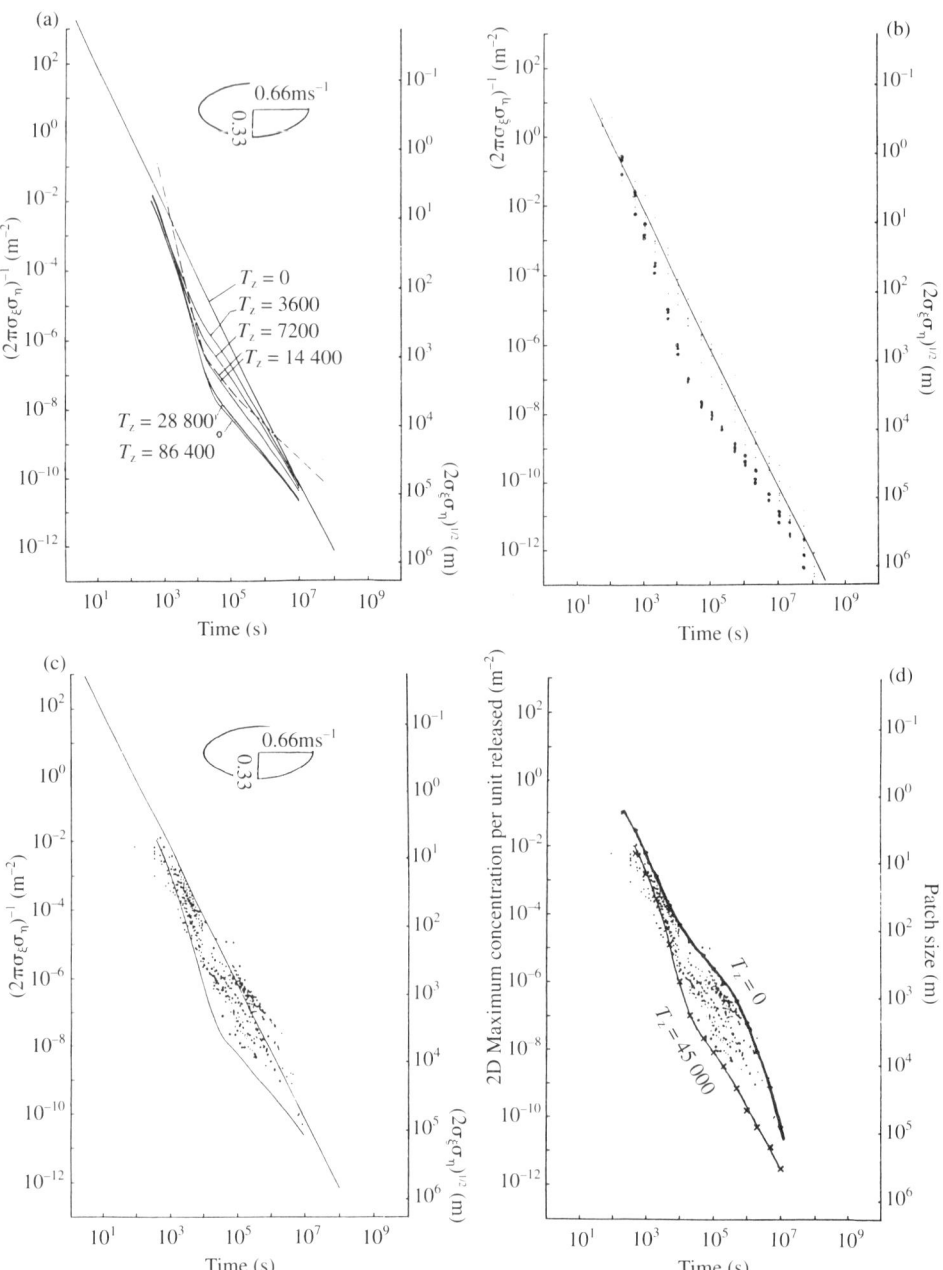

Figure 14.13 (a) Behaviour of maximum concentration of instantaneous point source at sea for various vertical mixing rates and simple (k^{-1}) spectrum of horizontal kinetic energy density, simulated by scaled random walk technique. $(—\cdot—\cdot)$ $T_2 = 14\,440$ s, shear dispersion alone (no eddies). (b) Same behaviour (for two mixing rates only), simulated by synthetic eddy velocity field (several realizations). (c) Upper and lower curve of Figure 14.13a combined with compiled North Sea data from tracer experiments (mainly dye release). (d) Same North Sea data with curves analogous to Figure 14.13c but calculated with adapted energy spectrum. Obtained with synthetic eddy technique (averages of five realizations for each of the two curves)

three-dimensional hydrodynamic numerical flow models on some kind of grid (rectangular, finite elements or curvilinear). For water quality and other transport problems, a transport model is needed in addition. Often this is a finite difference or finite element model on the same grid as the flow model. This approach limits the spatial resolution and encounters certain numerical problems leading to spatial oscillations (including negative concentrations) and/or numerical diffusion. In the present paper, this approach will not be considered further, but some attention will be paid to the alternative or particle simulation.

Three-dimensional particle simulation on the basis of a 2DH flow model can be realized in much the same way as the simpler models of the preceding section. The simplified velocity field is replaced by the time- and place-dependent field generated by the numerical flow model; its three-dimensional extension can be obtained with analytical distributions as in models with uniform depth. Van Dam Louwersheimer (1992) described a model of this type in more detail, with some results. In the present context it suffices to mention that, because of the structure of the velocity field produced in the two-dimensional flow model at greater length scales (several times the mesh size), the scaling of the random displacement or, alternatively, the use of a synthetic eddy field, is needed only in a limited scale range. This implies a (gradual) cutoff of the spectrum beyond a certain wave length, or a (gradual) reduction in the growth of random step length when the size of the spreading cloud has reached the scale at which the flow model begins to produce the required structure of the velocity field. Also here the eddy field approach is the most elegant and physically the most adequate.

Maier-Reimer (1973) was the first to propose particle simulation combined with numerically computed flow fields in large and complex configurations. He was also the first to propose a scaling (by growth in time) of random displacements to obtain better agreement with tracer experiments. The examples in his thesis were restricted to relatively small scales using a flow model on a very coarse grid (37 km). Therefore he did not need the cutoff of the step growth first proposed by Van Dam (1982).

14.3.3 Suspended matter with non-neutral buoyancy

Particle models of various types like those described in the preceding sections can be adapted to simulate the transport of suspended materials with positive or negative buoyancy. Suspensions of neutrally buoyant particles are transported like solutes.

Buoyancy is introduced by a positive or negative fall velocity, either the same for all particles or following a prescribed distribution. In the case of positive buoyancy, no other provisions are needed. Negative buoyancy, such as with fine sediments, is usually accompanied by the possibility of deposition and erosion. The conditions for these are expressed in current velocities near the bottom, or bottom shear stress. Further, erosion can be dependent on other conditions such as the local amount of deposited material (here represented by model particles) per square metre and the degree of consolidation, both factors contributing to a 'scour lag'.

Some results from a (still simple) model, without scour lag, are presented in Van Dam and Louwersheimer (1990). The flow model is of the complex, 'geographic' type, 2DH with extension to three dimensions by analytical vertical velocity distributions. However, certainly at the present stage, models of simple geometry such as those of Section 14.2 can be very useful. In Figure 14.14 a result is shown, obtained with a uniform depth−infinite length 2DV model including fall velocity, deposition and resuspension criteria, and also a zone with a density current and accompanying suppression of vertical exchange, representing the salt wedge in a partially stratified estuary. Maximum distortion of the vertical velocity distribution and

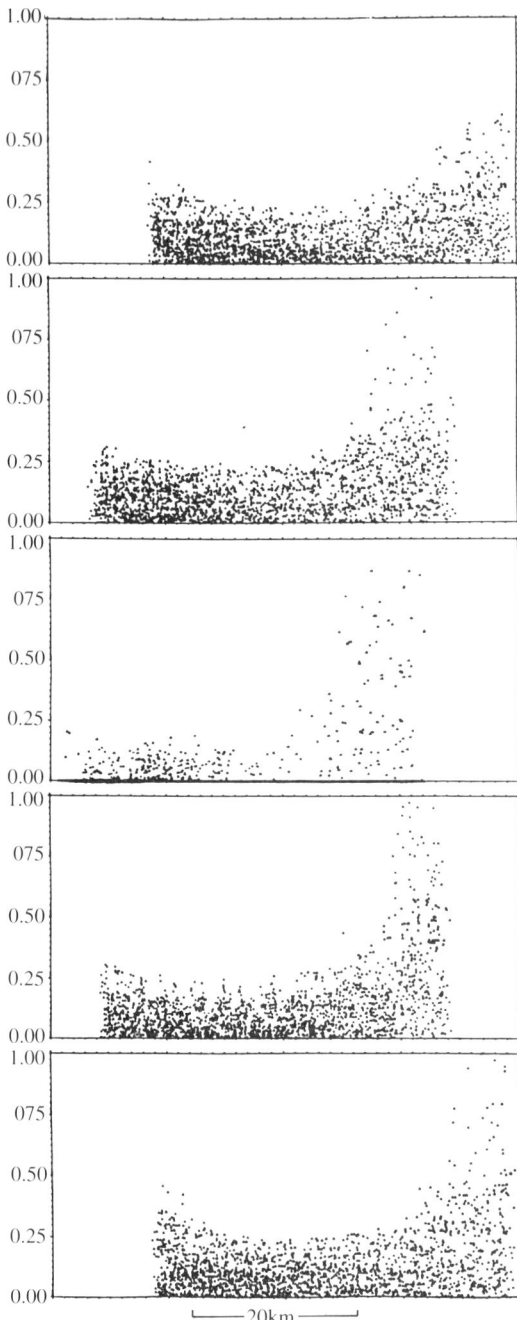

Figure 14.14 Distributions in vertical plane at five (equidistant) instants in M2 tidal period obtained with 2*DV* model accounting for brackish zone with density current and suppressed vertical mixing. Several tides after initial homogeneous distribution. Middle panel: high water slack, most particles lying on the bottom

maximum suppressed vertical mixing are about in the middle of the range represented in the figure (the thinnest layer of particles); both are gradually decaying to the left (river side) and to the right (sea side). A small net velocity from left to right represents a river discharge, responsible for the density current and the density stratification. An effect showing clearly in the simulation is the development of a turbidity maximum (in the figure on the left) some time after starting with a uniform particle distribution in the vertical plane represented in the figure. No particles are left in the region upstream of the turbidity maximum. At the sea side particles are slowly leaking away in the seaward direction. Particles in the middle region and the turbidity maximum may be permanently trapped; they reside most of the time in the lower part of the water column where the net flow is inward because of the density current. Figure 14.14 typifies the complexity of the problem of sediment transport in an estuary. Although the model is rather complicated, it still does not include the horizontal transverse dimension with its horizontal eddies, transverse density currents, wind effects and complex topography. However, even within the constraints of the simplified 2DV model simulation, Figure 14.14 is just an example; the schematized regime will be sensitive to parameter choices representing the various conditions and processes found in the diverse types of real estuaries. The author intends to investigate this problem further.

14.4   DISCUSSION AND CONCLUSIONS

Simulation of transport in surface waters by particle models is a convenient technique to deal with diffusion/dispersion and advection problems; in essence, particle methods are a way to approximate solutions of the advection-diffusion equation in a broad sense, i.e. including the advection by 'subgrid' or other small scale water movements. The numerical analysis is usually simpler and more straightforward than in other solution methods and in computing time the approach is competitive — in several instances more economic. Especially attractive in particle techniques is the flexible 'subgrid' character with a potentially unlimited degree of spatial resolution with respect to the statistical approach to small-scale dispersion and mixing as well as the small-scale structures induced by the time dependence of deterministic velocities such as in tidal movements.

In longitudinal systems, a simple shear dispersion model appears to be sufficient to explain the orders of magnitude of the observed longitudinal dispersion and cross-sectional mixing without necessarily including details of the smaller (eddy) structures of the velocity field (Section 14.2).

For applications where more detail is required, such as in near field problems at sea and in estuaries, the introduction of the influence of the spectral structure of the velocity field can very conveniently be established in a particle approach, while at the same time the far field can remain included in the modelling.

By its direct and visualizing character, a particle approach is often helpful in analysing the essential physical mechanisms and in obtaining general, sometimes analytical, descriptions or explanatory presentations.

Shear dispersion plays a dominant part in spreading dissolved and suspended matter in surface waters. In longitudinal systems such as rivers and estuaries the systematic velocity shear in transverse directions, combined with the mixing, explains the major part of observed longitudinal dispersion. In certain areas extra vertical shear caused by density currents and the suppressed vertical mixing by density stratification has to be taken into account to explain the sometimes strongly enhanced longitudinal dispersion in estuarine regions. In some cases these density effects cause vertical shear and mixing to be dominant; normally the horizontal transverse shear and mixing dominate longitudinal dispersion in rivers and estuaries. In non-tidal rivers the (horizontal) transverse shear may bring about very large longitudinal dispersion coefficients as high as $2000 \, m^2 \, s^{-1}$ and more; these values are never induced by horizontal shear in tidal areas because of the suppressing action of the periodic current reversal. The same orders of magnitude are only reached in some estuaries because of the aforementioned enhancement of the vertical shear mechanism by density effects. The main component of the flow responsible for the strong enhancement is not the tidal but the density current, which does not reverse and is steady in magnitude.

In the open sea the horizontal velocity shear is not only diverse in direction but also relatively weak. Therefore, at sea there is an important range, up to time-scales of several days, in which dispersion by vertical shear and mixing plays a dominant part. For large (time and length) scales, horizontal velocity gradients gradually take over and become dominant in the end, due to the spectral composition of the horizontal velocity field that contains 'eddies' at all scales up to the size of the entire basin. The wide range of vertical mixing rates with locally or temporarily important stratification effects causes a considerable divergence in dilution curves, followed later by a convergence at large scales (Section 14.3.1 and Figure 14.13).

The particle approach has largely contributed to understanding the complex combined action of the various mechanisms.

The application of particle modelling to the transport of suspended matter, including deposition and erosion, seems adequate and promising but has not yet been used extensively and has certainly not been verified or calibrated by observations. The models also need further refinement. Verification and calibration will probably be much more difficult than for dissolved matter.

ACKNOWLEDGEMENTS

Thanks are due to Aqua Systems International (Poeldijk, The Netherlands) for the development and construction of programs and for performing the computations without charge.

REFERENCES

Allen, C.M. (1982) Numerical simulation of contaminant dispersion in estuary flows. *Proc. Soc. London A* **381**, 179–194.

Aris, R. (1956) On the dispersion of a solute in a fluid flowing through a tube. *Proc. R. Soc. London A* **235**, 67–77.

Chatwin, P.C. and Allen, C.M. (1985) Mathematical models of dispersion in rivers and estuaries. *Annu. Rev. Fluid Mech.* **17**, 119–149.

Einstein, A. (1905) Über die von der molekularkinetischen Theorie der Wärme geforderte Bewegung von in ruhenden Flüssigkeiten suspendierte Teilchen. *Ann. Phys.* **4**, 17.

Fischer, H.B. (1972) Mass transport mechanisms in partially stratified estuaries. *J. Fluid Mec.* **53**, 671–687.

Fischer, H.B., List, E.J., Koh, R.C.Y., Imberger, J. and Brooks, N.H. (1979) *Mixing in Inland and Coastal Waters*. Academic Press, New York and London.

Holley, E.R., Harlemann, D.R.F. and Fischer, H.B. (1970) Dispersion in homogeneous estuary flow. *J. Hydr. Div. ASCE* **96** (HY 8), 1691–1709.

Maier-Reimer, E. (1973) Hydrodynamisch-numerische Untersuchungen zu horizontalen Ausbreitungs und Transportvorgangen in der Nordsee. *PhD Thesis*. Mitt. Institut für Meereskunde der Universität Hamburg, XXI.

Smith, R. (1982) Contaminant dispersion in oscillatory flows. *J. Fluid Mech* **114**, 379–398.

Smith, R. (1985) When and where to put a discharge in an oscillatory flow. *J. Fluid Mech* **153**, 479–499.

Taylor, G.I. (1953) Dispersion of soluble matter in solvent flowing slowly through a tube. *Proc. R. Soc. London A* **219**, 186–203.

Taylor, G.I. (1954) The dispersion of matter in turbulent flow through a pipe. *Proc. R. Soc. London A* **223**, 446–468.

Van Dam, G.C. (1982) Distinct-particle simulations. In: *Pollutant Transfer and Transport in the Sea.* Vol. I, pp. 131–135. (Ed. G. Kullenberg). CRC Press, Boca Raton.

Van Dam, G.C. (1990) Two- and three-dimensional modeling of dispersion and mixing in shallow tidal waters. *IUTAM Symposium on Fluid Mechanics of Stirring and Mixing, La Jolla, California* [Abstract (1991): *Phys. Fluids A* **3**, 1450].

Van Dam, G.C. and Louwersheimer, R.A. (1992) A three-dimensional transport model for dissolved and suspended matter in estuaries and coastal seas. In: *Dynamics and Exchanges in Estuaries and the Coastal Zone* (Ed. D. Prandle). *Coastal and Estuarine Studies.* Vol. 40 AGU, Washington, DC, 481–506.

Van Mazijk, A. (1986) Representativiteit van de monstername-punten Lobith en Kleve-Binnen. Working Group Hydrology, Co-operating Rhine- and Meuse-Water Boards (RIWA). *Monograph (August 1986)/Summary (September 1986).*

Van Mazijk, A., Verwoerdt, P., van Mierlo, J., Bremicker, M. and Wiesner, H. (1991) Rheinalarm-modell Version 2.0, Kalibrierung und Verifikation (Alarm model 'Rhine', version 2.0. Calibration and verification). *Report No. II-4 of the CHR/KHR (International Commission for the Hydrology of the Rhine Basin).*

Van Veen, J. (1936) *Onderzoekingen in de Hoofden in verband met de gesteldheid der Nederlandse kust.* Alg. Landsdrukkerij, The Hague, 258pp.

Van Veen, J. (1938) Water movements in the Straits of Dover. *J. Cons. Int. Explor. Mer* **13**, 7–36.

15 Coastal Surface Current and Wave Measurement with High Frequency Radar: Limitations and Prospects

LUCY R. WYATT

School of Mathematics and Statistics, University of Sheffield, Sheffield, UK

15.1 INTRODUCTION

High frequency (HF) radar systems have been used with great success to provide high resolution surveys of surface currents in coastal waters. In the UK the OSCR radar has been used for both industrial and scientific measurement programmes (Prandle, 1991), showing a measurement accuracy of about $3-4$ cm/s with a spatial resolution of up to 1 km^2 and a maximum range of about 40 km (see Table 15.1). Comparable systems are also in operation in the USA (Lipa and Barrick, 1990), Canada (Howell and Walsh, 1989), Australia (Heron, 1985) and Europe (Forget *et al.*, 1981; Gurgel, Essen and Shirmer, 1986).

High frequency radars operate by transmitting a radio wave which is scattered from any surfaces with which it interacts. This scatter is received at the radar which measures the time and phase differences between the transmit and receive waves. Time differences measure the range to the scatterer and scatter from different ranges can be identified at a range resolution which depends on the parameters of the radar transmission. Spatial resolution depends on the range and azimuth resolution of the system. The latter is determined by the beam width of the receiving antenna.

Backscatter from moving surfaces is shifted in frequency from that of the transmitted radio frequency due to the Doppler effect. The power spectrum of the backscatter, often referred to as the Doppler spectrum, provides information about the velocities of scatterers and their relative amplitudes. The power spectrum of the backscattered signal at HF is generally characterized by two large peaks (see Figure 15.1), which in the absence of a surface current are symmetrically located around zero Doppler frequency at frequency shifts of $\pm (2gk_r)^{1/2}$ in deep water. They thus represent moving targets with frequency shifts that clearly identify them as ocean surface waves with twice the radio wavenumber (k_r) moving directly towards (positive Doppler shift) and away (negative) from the radar. The strong backscatter from these waves is a Bragg resonance effect. A surface current with a component, v, towards (or away from) the radar beam imposes an additional positive (negative) Doppler shift (vk_r/π) on both peaks. This provides the simple basis for HF radar ocean current measurements. Two radars are required to measure the full current vector.

The two peaks referred to above are a first-order effect. The mathematical problem of analysing the interaction of radio waves with ocean waves has been carried out to second

Mixing and Transport in the Environment. Edited by K.J. Beven, P.C. Chatwin and J.H. Millbank
© 1994 John Wiley & Sons Ltd

Table 15.1 Oceanographic parameters measured by high frequency radar systems in the UK and comparisons with conventional instruments. The ranges of significant wave height (in the wave band of comparison) for which measurements have been compared are shown

Oceanographic parameter	System	Instrument comparison	Bias	Standard deviation	Maximum range (km)
Surface current	OSCR	Current meters		3–4 cm/s	40
Surface current	PISCES	Current meters	'good agreement'		>200
Significant wave height	PISCES	WAVEC	−1%	13.8%	150
Mean period	PISCES	WAVEC	−1%	11.2%	150
Long waves (>10 s)	PISCES	WAVEC			150
Amplitude (1.5–8 m)			5%	24.6%	
Direction			−7.2°		
Period			−2%	4.8%	
Waves 10–5 s	PISCES	WAVEC			150
Amplitude (1–5 m)			−3%	13.8%	
Direction			−4.7°		
Period			−1%	4.3%	
Short waves (<5 s)	PISCES	WAVEC			150
Amplitude (0.5–2 m)			10%	32.2%	
Direction			−3.2°		
Period			−3%	3.7%	

order (Barrick, 1972) and successfully describes the continuum in the Doppler spectrum surrounding the first-order peaks and marked as B in Figure 15.1. This is also a Bragg resonance phenomenon which is achieved from (1) non-linear ocean waves of the correct wavelength (i.e. π/k_r) but which do not satisfy the linear dispersion relationship and hence propagate with different speeds giving rise to different Doppler shifts and (2) double electromagnetic scattering processes which also give different Doppler shifts. This continuum

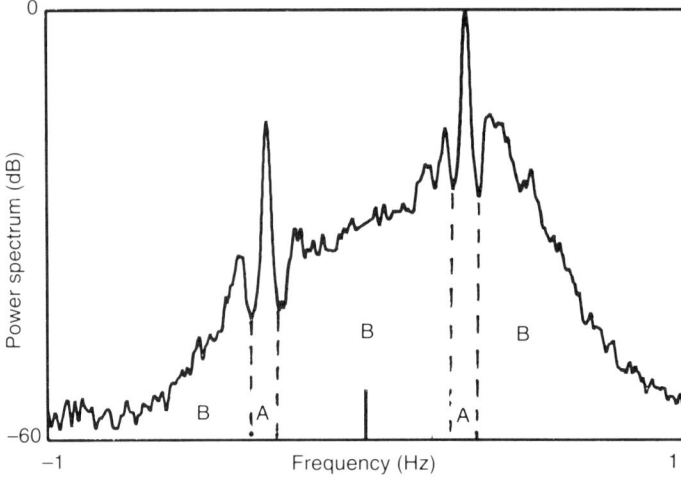

Figure 15.1 Doppler (power) spectrum from one of the PISCES radars during a NURWEC2 storm, showing (A) first- and (B) second-order contributions. The power levels shown are on a decibel scale relative to the peak. The measurement was taken at 0600 h on 28 March 1987

can, in principle, be inverted to provide measurements of the ocean wave directional spectrum with the same spatial resolution as the current measurements. Once again, two radars are required to resolve directional and amplitude ambiguities (Wyatt, 1987). Thus HF radar systems have the potential to provide simultaneous wave and current measurements.

The relationship between the Doppler spectrum of the backscattered signal and the ocean wave directional spectrum involves a non-linear first-kind Fredholm integral equation. The solution of this equation is not straightforward and standard matrix inversion methods cannot be used because of the ill-conditioning inherent in such problems. Wave measurements from HF radar that have been published fall into three categories: (1) parameters determined from semiempirical relationships; (2) model-fitting to find parameters of the directional spectrum; and (3) estimates of the directional spectrum using integral inversion methods that attempt to account for the ill-conditioning. Semiempirical methods are discussed by Wyatt (1988), who showed that significant wave height could be successfully measured in this way. Differences between radar and wave-buoy measurements were of the order of 13% (see Table 15.1). Mean period estimates with good accuracy have also been obtained (Wyatt *et al.*, 1986). Model-fitting methods have been used by Maresca and Georges (1980), who fitted the JONSWAP wind–wave model to skywave radar data, and by Lipa, Barrick and Maresca (1981) and Wyatt *et al.* (1986) and Wyatt (1991), who fitted a frequency dependent cardioid directional model to obtain the long-wave part of the spectrum.

Integral inversion methods were first used by Lipa (1977), although at this stage the directional distribution was assumed to be uniform with frequency. The method has been developed further and now allows the measurement of five frequency dependent Fourier coefficients (Lipa, Barrick and Maresca, 1981). The Fourier decomposition model has also been used by Howell and Walsh (1989) in implementing a singular value decomposition method. Wyatt (1990a; 1990b) developed an iterative inversion method which provides a model-independent measurement of the ocean wave directional spectrum. All these integral inversion methods have an upper limit on the ocean wave frequency that can be measured which is dependent on the radar operating frequency. Howell and Walsh (1989) quote an upper limit of 0.215 Hz at 25.4 MHz with a lower limit as the radar frequency decreases. Wyatt's method has an upper limit of about 0.2 Hz at 7 MHz and 0.3 Hz at 27 MHz and uses a wind–wave model with radar derived parameters to estimate the HF spectrum. This method has shown an ability to measure wind, swell and mixed seas with good accuracy (see Table 15.1). It also appears to have the intriguing ability to measure bimodal directional distributions, i.e. waves of the same frequency travelling in very different directions. This is not at the present time a standard feature of wave-buoy measurements, although maximum entropy analysis methods show some promise (Lygre and Krogstad, 1986).

Much of the work on wave measurement has been carried out with the PISCES radar system (Shearman and Moorhead, 1988; Table 15.1), which is capable of such measurements to ranges of up to 200 km. Current measurements beyond 200 km have also been achieved with this system (Venn *et al.*, 1988). Wave measurements have been obtained with good accuracy on most occasions to ranges of 100 km (Wyatt and Holden, 1992). Longer range measurements have been obtained but have not been verified against wave-buoys. At these long ranges range resolution is not usually required to be as good as would be necessary in coastal waters.

There has occasionally been evidence of within-cell current structure, seen as multiple first-order peaks in the averaged spectrum, which made inversion for wave measurement impossible. This was sometimes known to be associated with a significant change in current amplitude during the half-hour data collection period required for wave measurement. Such data can only be identified if the signal is constantly monitored throughout the half-hour. In a quasi-operational mode, when only the final average spectrum is seen, it is not easy to distinguish between a lack of temporal stationarity and a lack of spatial homogeneity.

This chapter will describe some work with the OSCR system off the North Yorkshire coast, where there is strong evidence in the radar data of current shears on scales of less than 1 km during periods of strong tidal currents. The problems these cause in the measurement of currents and waves will be discussed. It has been suggested that the multiple peak structure that is seen in the radar data under these circumstances could be used to estimate some measure of the associated mixing processes. This idea will be explored.

15.2 PROBLEMS IN CURRENT AND WAVE MEASUREMENTS

The most accurate current measurements are obtained from good quality Doppler spectra of the type shown in Figure 15.1. The two peaks are clearly identified and their frequencies (either point or centroid) can be measured to the resolution of the spectral analysis. (Note that there are minor differences in the signal processing strategy required for wave and current measurements). Such a spectrum also provides good quality wave measurements because the first- and second-order parts of the spectrum are clearly identified and separable. The first-order peaks provide a good estimate of the short-wave direction and can be used with confidence for normalizing the second-order spectrum for further analysis.

To obtain such a measurement requires: (1) a stationary (over the measurement period) and homogeneous (over the radar resolution cell) underlying wave spectrum and surface current field (note that the surface current effect is compensated for before wave measurement is attempted); (2) the use of sea-state dependent radar frequency (low HF in high sea-states); (3) low radio interference levels; and (4) a lack of ship clutter at Doppler shifts of interest. The last of these will not be dealt with further here except to say that signal processing strategies need to be developed to deal with such clutter.

The use of radio frequencies at the upper end of the HF band has developed for the current measurement systems to minimize problems with interference. This is fine for high resolution coastal work where the range limitation (40 km for currents and 15 km for waves for the OSCR system) is not a problem and where the primary interest is not in storm measurement. The second-order theory that provides the theoretical framework for wave measurement breaks down in high sea-states. Barrick (1986) has suggested an upper limit of 4 m significant wave height at 25 MHz. Unfortunately, the lower frequencies used in HF radar are in heavily occupied communications bands and automatic frequency and band width selection and reselection is constantly required. However, these frequencies provide a much longer range and have provided successful wave measurements in storm conditions (up to 10 m significant wave height has been measured).

Potentially the most difficult problem to overcome is the first. Both current and wave measurements are obtained from Doppler spectra which are averaged from a number of independent spectra. Current measurements only require the peak in the spectrum to have a stable frequency and this can be achieved with very little averaging. Wave measurements require amplitude stability and hence many averages and long data collection times to achieve the required frequency resolution with DFT (Discrete Fourier Transform) processing. (From the point of view of simultaneous wave and current measurements it should be noted that, in addition to the different time frames, different processing is required to bring out the second-order features, which can be 20−30 dB down on the first-order peak.)

Non-stationarity is therefore likely to be a more serious problem for wave measurements. There are likely to be two different effects. One involves a changing wave spectrum, particularly at the high frequencies which generate the first-order peaks. This will lead to errors in wave measurements which may or may not be more serious than those that will also arise in wave-buoy measurements in non-stationary wave conditions as the time for data collection is similar. The second effect involves non-stationary currents which lead to a smearing of the first-order peaks and can prevent the separation of first- and second-order which is required for inversion. As has already been mentioned, such effects can be monitored visually during data collection, but automatic monitoring and compensation is not yet possible in an operational context.

To minimize the problem of non-stationarity shorter data collection times are required. Signal processing strategies that produce spectra with good frequency resolution and amplitude stability with short data sets are needed. The use of maximum entropy spectral analysis methods in HF sea-state radar was discussed by Georges (1980), who indicated a number of problems associated with poor amplitude estimates. There have been many advances in autoregressive spectral analysis methods (see e.g. Marple, 1987) and preliminary results (Wyatt, unpublished data) indicate a future for such methods in HF analysis.

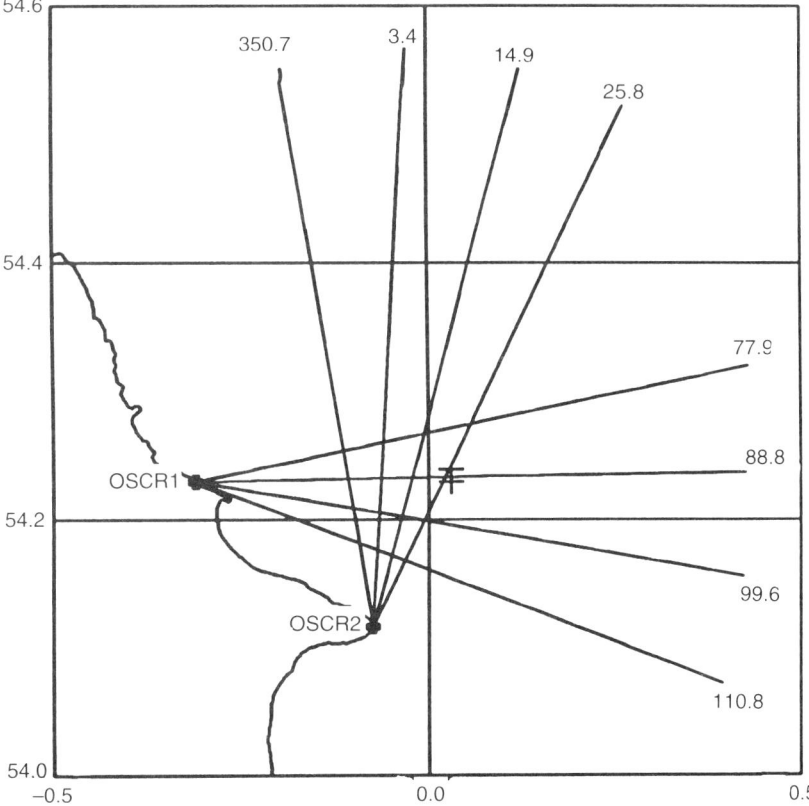

Figure 15.2 Area of the OSCR measurements showing the two sites on the Yorkshire coast, the beams used and the location of the WAVEC buoy. The data described in the text were obtained at the intersection point of the 3.4° and 88.8° beams. The digital coastline data were prepared by the Defence Mapping Agency of the USA and supplied by BODC

Spatial inhomogeneity can perhaps be dealt with by improving the range and azimuthal resolution. There are technical limitations here. Improved range resolution requires wider band widths and hence a greater susceptibility to interference. Azimuthal resolution depends on beam widths and might be improved by using digital methods. Reductions in spatial scales, even if technically feasible, are limited by the requirement to have a sufficient number of waves for Bragg resonance. Spatial inhomogeneity in currents leads to multiple first-order peaks which, if the range of currents is sufficiently wide, can obscure the separation between the first- and second-order structure and hence render inversion impossible. Heron (1985) has discussed this phenomenon and shown that the within-cell structure observed using current meters is of the correct order to explain the observed Doppler line broadening.

15.3 OSCR CURRENT MEASUREMENTS AND PROBLEMS

The measurements that are used here to illustrate the points made in the preceding sections and to raise further questions were obtained using the OSCR system off the north-east coast of England (see Figure 15.2). The data set was collected over an eight and a half day period in March 1986 to provide a data set for exploring the wave measurement potential of OSCR. Wave measurements have been made with this data set and are briefly described. First, however, some of the problems and interesting structures in the measurements will be discussed. These are only preliminary and qualitative comments. When the full data set is analysed it is hoped that a more coherent and satisfactory explanation for the phenomenon observed will emerge.

The analysis concentrated on the intersection point of the OSCR2 beam directed towards 3.4°N from Flamborough Head and the OSCR1 88.8° beam from Grisethorpe. The WAVEC buoy, deployed to provide sea truth for the radar wave measurements, was located about 7.5 km further east, but unfortunately the signal to noise ratio was not sufficient for radar wave measurement at this range.

15.3.1 Current measurements

It was obvious that there was a large tidal signal in the 3.4° Doppler spectrum. This is seen in Figure 15.3, which shows the time series of currents measured using only the larger of the first-order returns and the corresponding power spectrum. The current measurements used only the larger peak because it was obvious that the lower first-order return was difficult to interpret; it had multiple peaks, the largest of which did not necessarily correspond to the upper peak because the frequency differences were not the same. Although there is still one clearly identifiable peak in the spectrum and hence an estimate of surface current can be obtained, the existence of the additional peaks means that it is not so clear how to interpret this current estimate. This phenomenon was particularly marked at the time of the strongest tidal currents. Once the tidal current had been removed it became clear that the width of the first-order return (i.e. the envelope of the multiple peaks), both for positive and negative frequencies also varied smoothly with a tidal cycle. This width was larger for the weaker return. This is seen in Figure 15.4, which shows Doppler spectra down to 30 dB aligned so that the largest peak is located at the Bragg frequency, for the first part of the data set. If the largest peak is in fact reflecting the tidal current, this figure shows that the additional currents are either moving more slowly in the same direction or in the opposite direction and at speeds up to that of the tide itself.

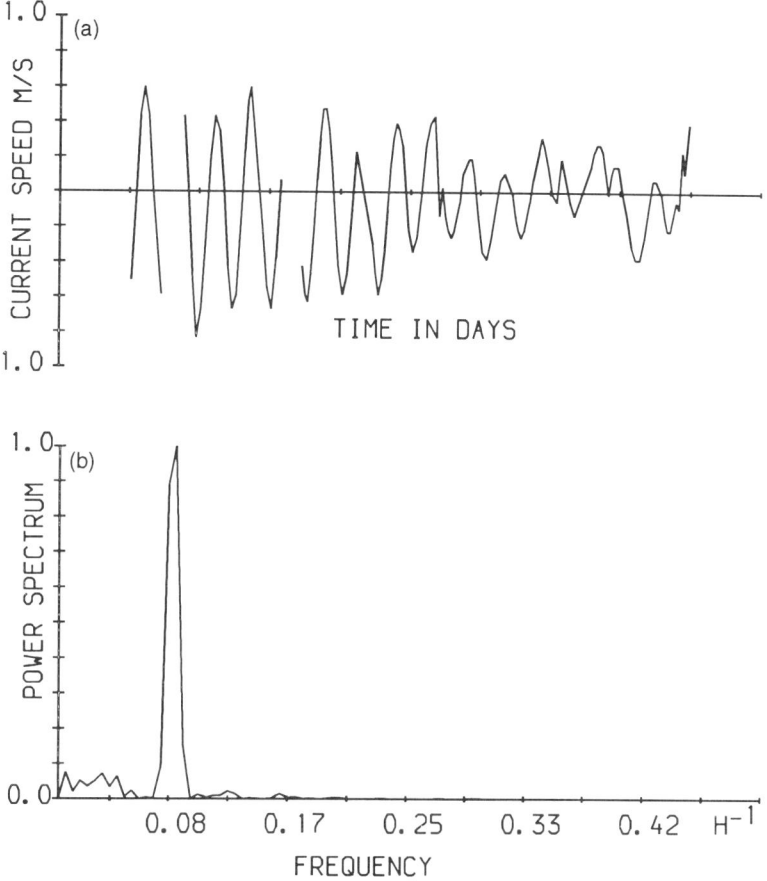

Figure 15.3 (a) Current component along the 3.4° beam. A positive current is directed away from the radar. The horizontal axis extends from 12 March 1986 to 21 March 1986 inclusive and is marked in one day intervals. (b) Power spectrum (on a linear scale normalized to the peak) of current component. The peak is located at a frequency of $\approx 1/12\ h^{-1}$

The widths have been measured using a 25 dB cutoff either side of the maximum and are plotted (measured as the frequency bin separation of the cutoff from the first-order zero-current Bragg bin) as a function of time in Figure 15.5, together with the corresponding power spectrum. The tidal current shown in Figure 15.3 was removed before this calculation. The widths themselves show a strong tidal signal. Note that the method used to identify the 25 dB cutoff loses some of the width at the tidal current maxima. No attempt has been made to select a measure of width that avoids this problem because the effect is not sufficient to lose the tidal signal in the spectrum. The clear separation between the peaks contributing to the first-order return suggests that the within-cell current field is made up of several distinct streams. This interpretation, of course, presumes that the effect of the current within these different streams on the wave phase speed is still the simple addition assumed in HF radar current measurements. It is the wave phase speed at a fixed wavenumber that is being measured. The implications of any wave refraction effects have not yet been explored.

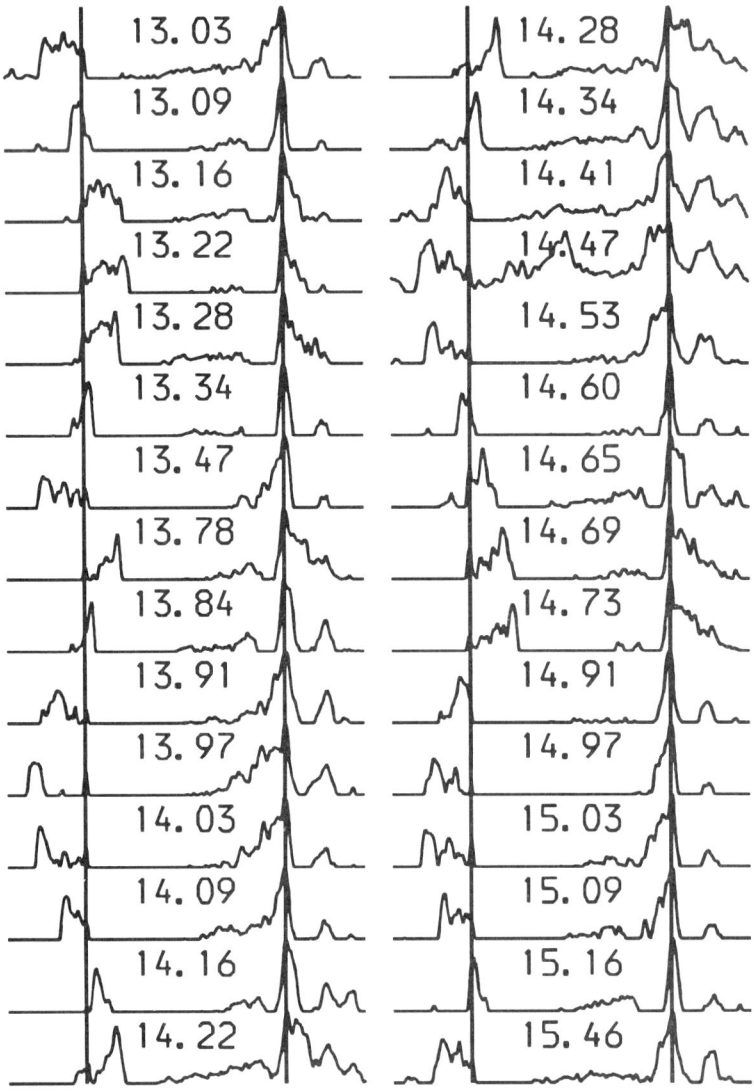

Figure 15.4 Doppler spectra plotted with a 30 dB vertical scale normalized to the peak. The time of each measurement is shown in days (from 13 March 1986). Vertical lines are drawn at the frequencies of zero-current first-order peaks. The main peak in each spectrum is aligned to these lines

15.3.2 Current differences

So far the multiple current structure has been discussed as a problem in that, when present, it limits both wave and current measurements. The possibility that measurements of this type can have a useful practical application was discussed by Heron (1985). He was considering the problem of line broadening due to surface current distributions on the radar cell scale associated with turbulence. His argument suggests that the mean energy dissipation rate within the radar cell could be estimated from the current range inferred from the broadened first-order Bragg line.

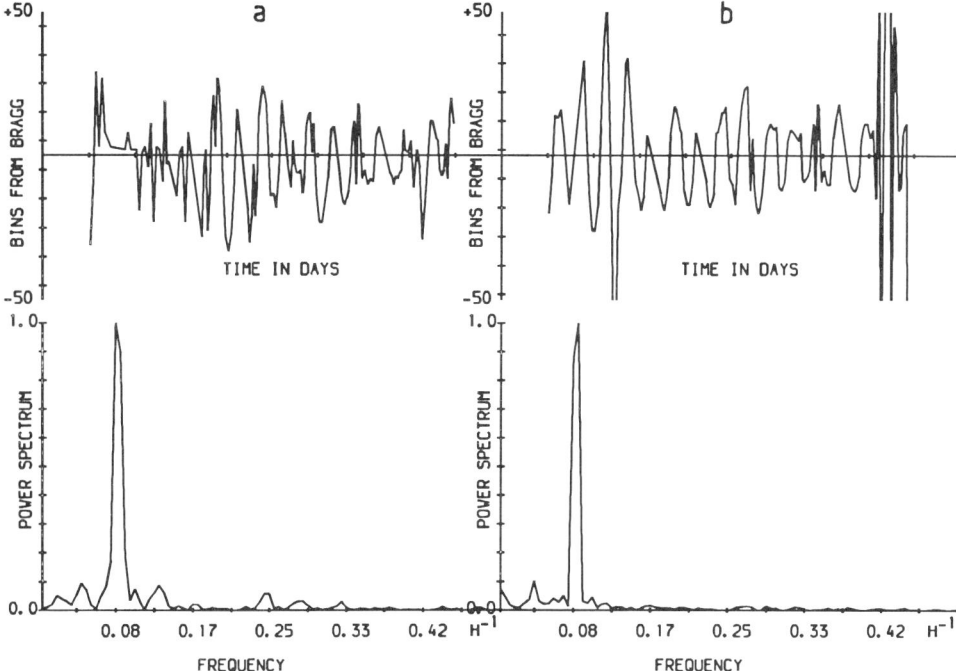

Figure 15.5 Time series and power spectrum (on a linear scale normalized to the peak) of the width of the envelope of multiple first-order peaks. Fifty bins correspond to a speed difference of ≈ 2 m/s. (a) Lower first-order return; (b) upper first-order return

The along-beam current speeds shown in Figure 15.3 range between ± 0.8 m/s, decreasing in range during the later part of the measurement period. The widths of the envelope of the multiple peak structure correspond to current differences of up to 1.5 m/s. With a cell scale size of 2400 m, Heron's argument, which assumes that the horizontal current shears that are being measured by the radar are driving the turbulence, implies an energy dissipation rate 1.4×10^{-3} m^2 s^{-3} at the time of maximum tidal current. Whether these data can actually be interpreted in this way and what the implications of such a large dissipation rate would be, requires further study.

Piau *et al.* (1984, unpublished data) also considered the possibility of using HF radar for turbulence measurements. Their measurements off the north-west coast of France were also in a region of strong tidal currents, but the turbulent velocity scale that they determined was substantially less than that found here.

15.4 OSCR WAVE MEASUREMENTS

The problems already discussed made wave measurement particularly difficult. All data go through a preliminary automatic quality assessment before inversion. Much of the data with multiple first-order peaks were rejected at this stage. However, for some of the data the multiple peak structure was misclassified as second order and therefore inverted. This generated very large, very long wave components in the wave spectrum and contributed to anomalous significant wave height and mean period estimates. In many such instances, however, the inversion at higher frequencies appeared to be consistent with measurements made in the absence of multiple peak structure. It is likely that the apparently anomalous measurement of long wave peaks is less serious for this data set than it would have been in the Celtic Sea area where the PISCES data were obtained. Swell with periods of 10 seconds or more driven by Atlantic depressions is endemic in this region and is a feature of a large proportion of PISCES measurements. The OSCR data set referred to here has indicated a very different wave climatology in the North Sea. Although the range of meteorological conditions was not large, the existence of land rather than sea at the western boundary inevitably results in predominantly fetch-limited waves with peaks of low amplitude and high frequency. For most of this data set the spurious peaks are well separated from the measured wave fields, which peak at about 0.2 Hz.

Figure 15.6 shows amplitude (significant wave height over the band) and direction measurements in four wave bands. The effect of the anomalous peaks is still evident in the direction estimates for the longer waves but, apart from that, the variation with time is fairly smooth and on the whole consistent with the prevailing meteorological conditions. Intercomparisons with wave-buoy data are not presented here because the buoy data were not available. The gap from about midday on 19 March 1986 until early on 20 March 1986 reflects the poor data quality at that time. It was reported that there was a storm developing suggesting that the data were running into the higher order non-linearity problem referred to in Section 15.2.

It is worth noting that the same data have been used here to obtain both wave and current measurements. Data sets of this type are likely to be valuable for the study of wave—current interaction.

15.5 CONCLUDING REMARKS

This chapter has raised questions about the interpretation of HF radar data. Two particular problems have been highlighted: the limitation of radar operation at the high frequencies of currently operational systems and the interpretation and analysis of data where there is inhomogeneity in the surface currents on scales less than the radar resolution in both space and time. The second of these in particular emphasizes the need for care in interpeting wave and current measurements obtained with such a system. The quality of the data must be closely monitored.

In spite of these difficulties, HF radar has proved to be, and will continue to be, of great value in oceanographic research and operations where surface data, and particularly the spatial variability of such data, are required. Further work is required to establish whether the current shear problems can be turned to advantage by providing measurements of turbulence scales.

The use of such systems for combined wave and current measurements in all weather conditions requires flexibility in radar frequency and band width. Alternatively, new theories must be developed that allow for the strong non-linearities in the scattering process at high frequencies in high sea-states.

ACKNOWLEDGMENTS

The support of SERC (grant ref. GR/F25860) is gratefully acknowledged. The OSCR wave analysis was carried out under contract with the Department of Energy, as part of its offshore research programme, now transferred to the Health and Safety Executive. The PISCES system is being developed by Neptune Radar Ltd. The OSCR system was originally developed at

a   AMPLITUDE IN METRES

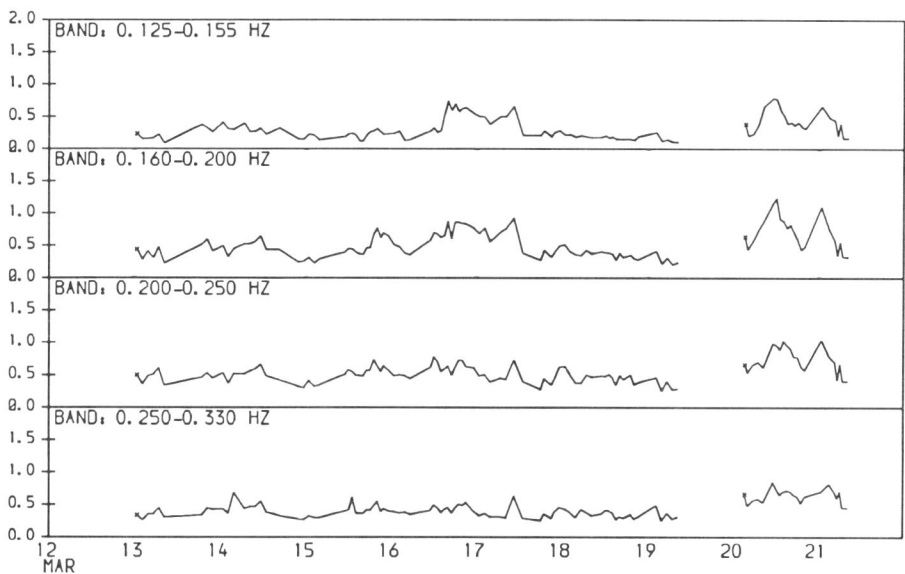

b   DIRECTION

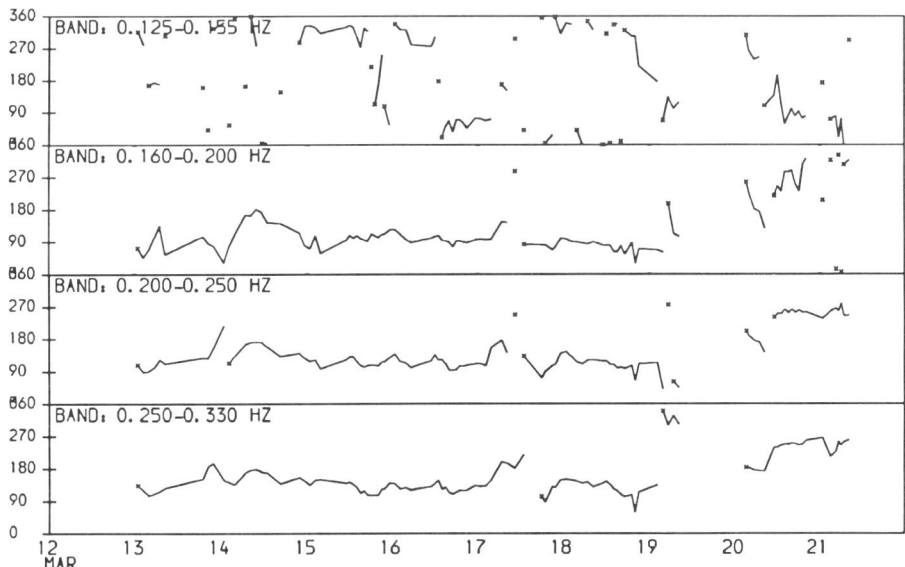

Figure 15.6 Wave measurements from OSCR. (a) Amplitude i.e. significant wave height measured over the frequency bands shown. (b) Mean direction; this is scattered in the longer wave band because some of the directional characteristics of the spurious wave peaks associated with the first-order multiple peak structure are included in the average

the Rutherford and Appleton Laboratories. It is being further developed and marketed by Marex Ltd.

REFERENCES

Barrick, D.E. (1972) Remote sensing of sea state by radar. In: *Remote Sensing of the Troposphere* (Ed. V.E Derr). US Government Printing Office, Washington DC, Ch. 12.

Barrick, D.E. (1986) The role of the gravity wave dispersion relation in HF radar measurements of the sea surface. *IEEE J. Ocean. Engin.* **OE-11**, 286−292.

Forget, P., Broche, P., McMaistre, J.C. and Fontanel, A. (1981) Sea-state frequency features observed by ground wave HF Doppler radar. *Radio Sci.* **16**, 917.

Georges, T.M. (1980) Progress toward a practical skywave sea-state radar. *IEEE Trans. Antennas Propagation AP-28*, 751−761.

Gurgel, K.-W., Essen, H.-H. and Shirmer, F. (1986) CODAR in Germany — a status report valid November 1985. *IEEE J. Ocean. Engin.* **OE-11**, 251−257.

Heron, M.L. (1985) Line broadening on HF ocean surface radar backscatter spectra. *IEEE J. Ocean. Engin.* **OE-10**, 397−410.

Howell, R. and Walsh, J. (1989) Measurement of ocean wave spectra using narrow-beam HF radar. *International Geoscience and Remote Sensing Symposium IGARSS'89 Digest*, IGARSS, 2969−2972.

Lipa, B.J. (1977) Derivation of directional ocean-wave spectra by inversion of second order radar echoes. *Radio Sci.* **12**, 425−434.

Lipa, B.J. and Barrick, D.E. (1990) Codar wave measurements from a North Sea semisubmersible. *IEEE J. Ocean. Engin.* **15**, 119−125.

Lipa, B.J., Barrick, D.E. and Maresca, J.W. (1981) HF radar measurements of long ocean waves. *J. Geophys. Res.* **86**, 4089−4102.

Lygre, A. and Krogstad, H.E. (1986) Maximum entropy estimation of the directional distribution in ocean wave spectra. *J. Phys. Oceanog.* **16**, 2052−2060.

Maresca, J.W. and Georges, T.M. (1980) Measuring rms waveheight and the scalar ocean wave spectrum with HF skywave radar. *J. Geophys. Res.* **85**, 2759−2771.

Marple, S.L. (1987) *Digital Spectral Analysis with Applications*. Prentice-Hall, Englewood Cliffs.

Prandle, D. (1991) A new view of near-shore dynamics based on observation from HF radar. *Prog. Oceanogr.* **27**, 403−438.

Shearman, E.D.R. and Moorhead, M.D. (1988) Pisces: a coastal groundwave radar for current, wind and wave mapping to 200 km ranges. *Proc. Int. Geoscience and Remote Sensing Symposium IGARSS'88* (Eds. T.D. Guyenne and J.J. Hunt), ESA Publications, 773−776.

Venn, J.F., Mardell, G.T., Howarth, J. and Holmes, C.G. (1988) Current measurement by long range HF ground-wave radar. *Proc. Int. Geoscience and Remote Sensing Symposium IGARSS'88* (Eds. T.D. Guyene and J.J. Hunt), ESA Publications, 783−786.

Wyatt, L.R. (1986) The measurement of the ocean wave directional spectrum from HF radar Doppler spectra. *Radio Sci.* **12**, 473−485.

Wyatt, L.R. (1987) Ocean wave measurement using a dual-radar system: a simulation study. *Int. J. Remote Sensing* **8**, 881−891.

Wyatt, L.R. (1988) Significant waveheight measurement with HF radar. *Int. J. Remote Sensing* **9**, 1087−1095.

Wyatt, L.R. (1990a) A relaxation method for integral inversion applied to HF radar measurement of the ocean wave directional spectrum. *Int. J. Remote Sensing* **11**, 1481−1494.

Wyatt, L.R. (1990b) Progress in the interpretation of HF sea echo: HF radar as a remote sensing tool. *IEE Proc.* **137**, 139−147.

Wyatt, L.R. (1991) HF radar measurements of the ocean wave directional spectrum. *IEEE J. Ocean. Engin.* **16**, 163−169.

Wyatt, L.R. and Holden, G. (1992) Developments in ocean wave measurement by HF radar. *IEE Proc. Part F* **139**, 170−174.

Wyatt, L.R., Venn, J., Burrows, G.D., Ponsford, A.M., Moorhead, M.D. and van Heteren, J. (1986) HF radar measurements of ocean wave parameters during NURWEC. *IEEE J. Ocean. Engin.* **11**, 219−234.

16 Cohesive Sediment Transport in Estuaries

J.R. WEST

School of Civil Engineering, The University of Birmingham, Birmingham, UK

16.1 INTRODUCTION

Estuaries represent only a tiny proportion of the world's surface waters, but they are of considerable interest to mankind due to their ability to provide a combination of food supply, fresh water availability and transport facilities within a small geographical area. Early small settlements were often located on the sheltered headwaters of estuaries where fresh water is plentiful for most of the time. More recently the improvement in water supply technology and the need for deeper water port facilities led to the development of large communities on the banks of the lower reaches of estuaries. These communities exerted changes in the hydrodynamics and quality of estuarine waters due to local coastal defence works, dredging and effluent discharges. Changes may also be induced by engineering works located far away from estuaries in the upper parts of catchments. For example, changes in river flow and water quality entering an estuary can be caused by the construction of reservoirs to prevent flooding, to generate power, or to alleviate the effects of low river flow conditions during summer periods.

Estuarine systems are extremely complex due to the interaction of physical, chemical and biological processes. Water motion is influenced by tides, river flow, wind and the buoyancy effects of fresh water and meteorologically induced temperature fluctuations. Chemical processes are influenced by the changing salinity, pH and dissolved oxygen concentration conditions, whereas the biology can be strongly modulated by the hydrodynamics, chemistry and interactions between species.

A further influential factor in the control of many estuarine systems is the presence of considerable amounts of small clay and silt sized particles, either in suspension or in the bed sediments. The presence of this fine sediment is thought to be due to a combination of physical, chemical and biological phenomena. For example, buoyancy-induced gravitational circulation patterns can transport marine sediments into estuaries (Festa and Hansen, 1978), salinity-induced flocculation of suspensions in river discharges aids sedimentation in the upper reaches of estuaries (Krone, 1986) and the excretion of mucopolysaccharides by diatoms can help to stabilize sediments on muddy intertidal zones (Paterson, 1989). The fine sediment can modify the flow of water through density and sedimentation effects, the transport of pollutants through adsorption phenomena and the ecology of estuarine systems through a number of mechanisms, for example light extinction.

With the demands made by humans on the estuarine environment showing little sign of abating, and with the development of a better public awareness of the desirability of

Mixing and Transport in the Environment. Edited by K.J. Beven, P.C. Chatwin and J.H. Millbank
©1994 John Wiley & Sons Ltd

environmental protection, there is a need to understand the effects of current estuary management strategies and to be able to predict the effects of any proposed changes in those practices. The changes may be required to improve the quality of the environment or to improve the quality of life for a group of individuals.

This chapter examines the importance of cohesive sediments with respect to the management of estuaries and describes the processes that can influence sediment transpport. A review is undertaken of some of the recent advances in the understanding of estuarine fine sediment transport processes in some British estuaries. Some thoughts are offered, where appropriate, on where further research effort may be directed to improve the practice of estuarine environmental management.

16.2 INFLUENCE OF COHESIVE SEDIMENTS ON ESTUARIES

Fine sediment, usually considered to have a characteristic size of less than 0.063 mm, tends to occur in estuaries predominantly in the zone known as the turbidity maximum. This zone is often located in the region of the limit of saline intrusion: the fresh–salt water interface or FSI. The term turbidity maximum results from the higher than average concentrations of suspended solids occurring in this region compared with the rest of the estuary. However, the term is in some ways misleading as during periods of low hydrodynamic activity the cohesive nature of fine sediment leads to much of the material being trapped on the bed, either in main channels or on muddy intertidal zones. It is only during periods of highly active hydrodynamic conditions caused by spring tides, floods or storms that a significant proportion of this sediment may be taken into suspension and moved up and down an estuary. The interchange of particles between the suspension and the bed occurs over a range of time-scales varying from a few seconds to centuries.

The discharge of effluents to rivers leads to the conveyance of soluble pollutants to the shelf seas and ultimately to the deep ocean. Some initial dilution takes place in the river flow, but this is small compared with the potential dilution available in the oceans. The presence of fine sediment in the upper reaches of estuaries can lead to the adsorption of some heavy metals and persistent organic chemicals and to their build-up in the bed sediments to concentrations many times higher than those in the overlying water. The resuspension of these sediments, through natural flood and storm events or through changes to the hydrodynamics or salinity distributions in an estuary, can lead to the transport of pollutants to the shelf seas many years after their initial release into a water body. If anoxic conditions occur within the fine-grained bed sediments this can lead to the release of adsorbed chemicals from particle surfaces into the interstitial pore water and the slow diffusion of chemicals into the overlying water.

The release of persistent pollutants to coastal seas from coastal outfalls or by dumping from ships can lead to soluble and particle adsorbed species entering estuaries as a constituent of the sea water. Hydrodynamic effects such as gravitational circulation can lead to transport inland to the FSI. Thus the sediments in the upper reaches of estuaries have the unique potential for building up inventories of persistent chemicals resulting from industrial and agricultural activity from within a catchment, and from coastal and marine disposal practices. The particles may be locked into the bed sediments of an estuary for many years until channel migration or a particularly severe storm or flood lead to erosion.

The ability of tides, floods and wind-generated waves to erode and redistribute fine sediments can influence the ecology of an estuary in two main ways. Firstly, the presence of suspended sediment increases the attenuation of light in the water column. This leads to an inhibition of photosynthetic activity and hence, a reduction in primary production in the water column. This in turn affects the higher organisms which depend on the primary production. Fluvial and marine phytoplankton will be found in the upper and lower reaches of estuaries, but those reaches that have rapid changes of salinity and that are light starved to a greater or lesser extent will usually exhibit much lower algal activity.

Secondly, the presence of mud deposits composed of fine particles that have adsorbed organic matter from the water column before sedimentation leads to the availability of a good nutrient source and hence some of the most productive areas in estuaries. However, the periodic occurrence of erosion events can severely inhibit the population development of many species. Species that can colonize rapidly and bury themselves during inclement conditions can survive in profusion if the erosion events are not too frequent or severe. Deposition events can also influence the biota on a mud flat as frequent and heavy deposition will severely limit the range of species that can survive in those conditions. The availability of prey in the water column and in the intertidal mud flats will control the distributions of their predators such as birds and fish.

Muddy intertidal zones and the associated salt marshes provide a first defence against flooding. They have the ability to absorb wave energy and hence to protect the flood banks which are normally constructed at the edge of salt marshes. Severe storms may lead to considerable erosion, but the following spring tides will usually lead to new accretion and the replenishment of the protective layers of sediment, provided nothing has taken place to disrupt a natural equilibrium.

Thus it may be concluded that the management of estuaries requires a satisfactory knowledge of the transport of fine sediments. Such knowledge needs an understanding of the water flow and of the settling, consolidation and erosion of fine sediment. The state of some aspects of our current understanding is examined in the following sections.

16.3   ESTUARINE HYDRODYNAMICS

Early work on estuarine flow mechanisms considered the propagation of tidal waves along estuaries and hence only area mean velocities were considered (Ippen, 1966). Interest in pollutant and sediment transport, particularly on the east coast of the USA, led to the concept of gravitational circulation, in either a turbulent mean or a tidal mean sense (Officer, 1976). The study of gravitational circulation requires the consideration of longitudinal and vertical (u, w) components of velocity in the $x-z$ plane. In vertically well mixed conditions, vertical stratification effects can be ignored in some circumstances and two-dimensional (u, v) in plan $(x-y)$ representation can be used as aids to management decision-making (McDowell and O'Connor, 1977). In the last decade a small number of studies have investigated the existence of transverse gradients in estuaries and have revealed the presence of fronts and secondary circulations (Nunes and Simpson, 1985; Simpson and Turrell, 1985; Guymer and West, 1991).

Recent advances in computer hardware and numerical techniques have led to three-dimensional time dependent models being available. their application is very expensive due to the computing and input data requirements. Their predictive capability is constrained by limitations in our knowledge of turbulence. A sound knowledge of turbulence is essential as the generation and decay of turbulence in estuarine flows leads to the transport of momentum, salt and thermal energy. The momentum transport determines the turbulent mean velocity distributions and hence in part the rate of turbulence generation. The turbulent transport of salt is often the determining factor influencing the vertical density structure in an estuary, which in turn affects the momentum transport, velocity distribution and turbulence generation. In well mixed flows ($d\rho/dz = 0$, where ρ is the fluid density and z the vertical axis) turbulence generation is mainly in the region of the bed and hence strongly influenced by the changing bedform and roughness.

In stratified flows (here taken to mean continuous stratification $d\rho/dz < 0$; z is positive upwards) the bed-generated turbulence is inhibited and internal waves are found within the flow (Turner, 1973). Their generation can be by a number of mechanisms induced by channel irregularities or by internal fluid shear (Dyer, 1982). These waves have been observed to break or their breaking has been inferred (Partch and Smith, 1978; Geyer and Smith, 1987; New, Dyer and Lewis, 1987). The breaking of internal waves leads to turbulence generation and the production of a packet of a locally well mixed zone of water within the flow. This zone is influenced by shear and gravity effects and probably spreads out in a plane appropriate to its density within the water column (Darbyshire, 1990).

The vertical stratification of the water column is usually strongly time dependent during a tidal cycle (Figure 16.1), particularly in shallow coastal plain estuaries that experience meso- and macrotidal ranges. Figure 16.1 shows the temporal variation of depth mean values and the distribution over the water column of turbulent mean values of velocity, salinity and suspended solids concentration at Cotehele in the upper Tamar estuary. The vertical distributions are plotted as deviations from the depth mean value, each ordinate being placed on the abscissa at the appropriate time of observation. The scales relating to the deviations are given for each of the three parameters. The salinity profiles show significant stratification over a period of five hours. The temporal dependence of the salinity gradients results from the temporal change in velocity and flow depth, and the differing effects for flood and ebb tides of the interaction between longitudinal current-induced fluid shear and longitudinal salinity-induced density gradients. The potential for turbulence generation increases with flow velocity. The effects of turbulence generation on the structure of the water column increase with a decrease of flow depth. The effect of flow direction is such that on flood tides the shear in the flow interacts with the longitudinal density gradient leading to gravitationally unstable conditions. As the velocity gradients tend to be greatest near to the bed, any vertical density structure remaining from the ebb tide tends to be eroded from below during the flood tide. On ebb tides the combination of shear and longitudinal density gradient lead to the production of stable vertical density gradients (West, Knight and Shiono, 1984).

The difference in the vertical density structure on flood and ebb tides leads to better vertical momentum transfer on the flood and hence to smaller vertical gradients of the longitudinal component of velocity (du/dz). During ebb tides the combination of velocity gradients and salinity gradients leads in effect to denser saltier water being left behind the depth mean flow, thus causing a net landwards dispersion of salt. This is needed to counteract the seawards advection of salt by the fluvial discharge if the saline intrusion is to be maintained. There are, however, further influences.

If an estuary is fairly narrow (< 200 m), then the drag of the banks can lead to sufficient transverse fluid shear to cause transverse salinity gradients in the presence of a longitudinal density gradient of sufficient magnitude. On a flood tide this results in denser water in the middle of a channel and less dense water in the shallow bank side regions. Figure 16.2a shows salinity values measured in the Conwy estuary during a flood tide, which are a good example of transverse density gradients (Guymer and West, 1991). Such a flow structure leads to the sinking of the denser water in the centre of the channel as it moves landwards. This allows

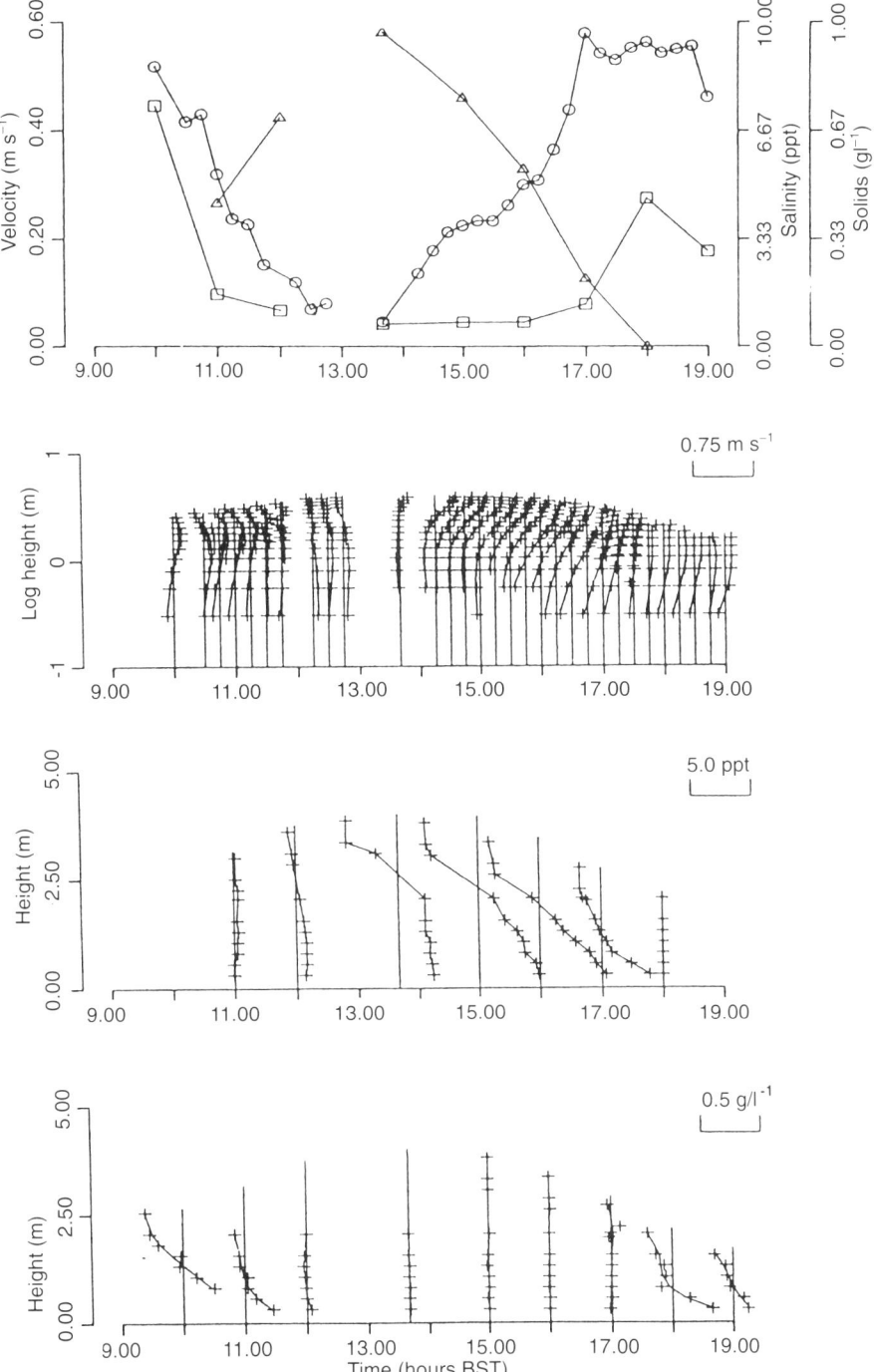

Figure 16.1 (a) Vertical profiles and depth mean values of (b) velocity, (c) salinity and (d) suspended solids at Cotehele, River Tamar, 8 July 1988. High water 1315 h. (o) velocity; (△) salinity; (□) solids

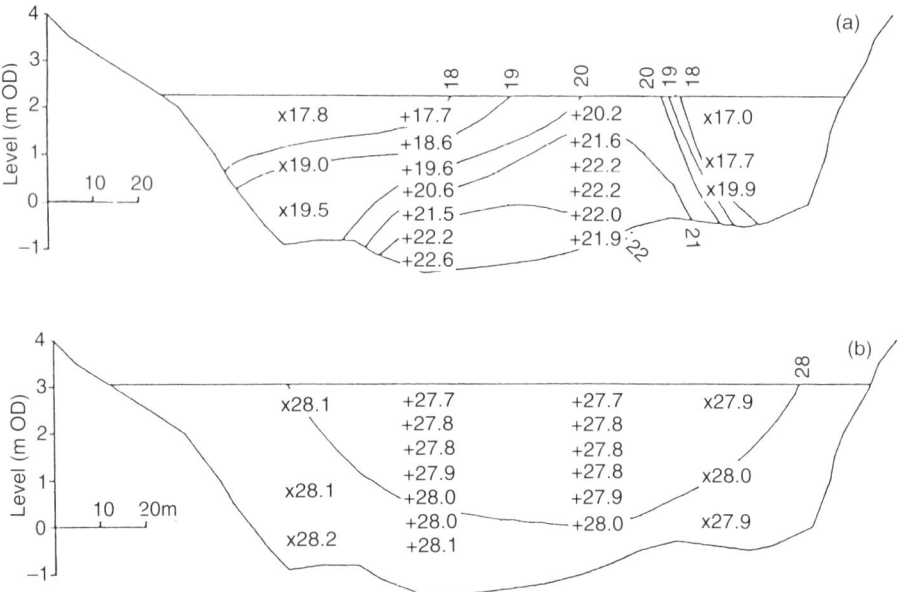

Figure 16.2 Salinity distributions on a cross-section at Tal-y-Cafn, River Conwy, 15 July 1983. (a) Flood, 89 minutes before high water. (b) Ebb, 68 minutes after high water (Guymer and West, 1991; reproduced by permission of Academic Press Ltd.). S_A: (a) 19.8; (b) 27.9 ppt

the formation of a secondary flow structure with an axial convergence zone (Nunes and Simpson, 1985), which is usually marked by a well defined line of scum and floating detritus. Bends can cause distortion to these patterns due to the faster flows in the deeper water near to the outer banks (West and Mangat, 1986).

On the ebb tide the denser water occurs near to the banks and a secondary flow in the opposite sense may be anticipated. However, the stratification may be expected to reduce the influence of bed friction and to inhibit secondary flow. Visual evidence suggests that a weak secondary flow can occur on the ebb tide as the author has observed the remains of the flood tide scum line to be deposited on the banks during the early ebb. Figure 16.2b shows another example of cross-sectional salinity distribution from the Conwy estuary. The data are interesting in that a weak transverse salinity gradient is observed, as expected, but the vertical gradients are rather less than those observed in the Tamar (Figure 16.1). The smaller gradients are probably partly due to mixing caused by a sharp bend upstream of the Conwy study site, highlighting the longitudinal variability of density structure due to topographic features.

In wider estuaries the influence of broad intertidal zones, and sand banks within channels, can lead to sufficient transverse shear to create transverse salinity gradients and longitudinal fronts. These can occur in the presence of fairly small longitudinal salinity gradients and thus may be found in the lower reaches of estuaries. The heating or cooling of shallow intertidal zones can also create density effects and hence influence the nearshore processes. The transverse effects induced by salinity are most significant in the surface layers in the presence of deep water in the main channels. This is because the presence of longitudinal salinity gradients leads to stable vertical gradients for a substantial part of the tidal cycle and thus the differential effects of bed friction drag between the shallower better mixed water and the deeper stratified water are accentuated.

The effects of turbulence may also be demonstrated in another way by considering the area mean flow of water in an estuary and assuming, to a first approximation, that the river flow is sensibly constant and that successive tidal ranges are similar. Over a tidal cycle the cross-sectional average flow should show a net flow seawards equivalent to the river discharge. Within a tidal cycle, the temporal distribution of flow through a cross-section can show considerable asymmetry between flood and ebb tides. This is due to a temporally varying combination of surface slope, water depth and bed friction characteristics. The asymmetry of the bed shear stress, depth and velocity functions lead to the probability of area mean flux terms involving solute and suspended solids showing net upstream or downstream residuals when integrated over a tidal cycle (West *et al.*, 1991). If a two-dimensional view of the flow field is adopted then the temporal variation in transverse or vertical vector and scalar fields leads to tidal mean residual fluxes that vary with respect to the transverse or vertical directions respectively (Uncles, Elliott and Weston, 1985).

Over recent years field studies of estuarine hydrodynamics have helped to develop a more detailed picture of the complexities of the interaction of fluvial, marine, atmospheric and topographical influences on estuarine flows. Most field exercises only provide a limited view in time and space and there is still a dearth of good observations of turbulent and internal wave processes. The continuing development of better transducers and data logging systems should lead to a further expansion of our knowledge in the near future. There is a particular need to study the production of fine-scale turbulence under stratified conditions as this controls the ebb tide salinity dispersion processes and the inhibition of sediment transport. Such studies should then allow further progress in the improvement of the predictive capability of mathematical models of estuarine water motion.

16.4 COHESIVE SEDIMENTS: LABORATORY STUDIES

The recent study of estuarine cohesive sediment transport may be arbitrarily split into laboratory and field studies. The former have yielded knowledge of the basic properties of the sediments. The latter have helped to show how the interaction of sediment particles with water and air affects their transport in the natural environment. Attention is given here to the settling, consolidation and erosion properties of cohesive sediments.

A good pragmatic description of cohesive sediment transport properties has been given by Delo (1988). Settling velocity, to a first approximation, is dependent on suspension concentration and may be given by

$$w_{50} = 0.001C \qquad (0.05 < C < 2.0) \tag{16.1}$$

where w_{50} is the median settling velocity of flocs (m/s) and C is the suspended sediment concentration (kg/m^3). Equation 16.1 holds up to the point where the concentration is high enough for hindered settling effects to become significant. This happens at concentrations of the order of $5-10\,\text{kg/m}^3$. The rate of deposition to the bed is given by

$$\frac{dm}{dt} = -Cw_{50} \tag{16.2}$$

where dm/dt = mass flux deposited/unit area (kg/m^2/s). Such calculations should be treated as first estimates as in practice the suspension will be composed of primary particles and aggregates which almost certainly have temporally and spatially varying size and density distributions as the salinity and fluid shear fields vary with time. A further caution is necessary as most of the fall velocity data on which Equation 16.1 is based have been obtained in the quiescent conditions of the Owen tube. These data show a considerable variation between estuaries (Delo, 1988) and at a point in an estuary with time (Burt, 1986; Barton et al., 1991). Figure 16.3 shows an example of the variation of fall velocity data obtained on six occasions from the upper Tamar estuary. The variation is thought to be due to the migration of the tubidity maximum leading to low concentrations of coarse sediment being present in October 1987 and high concentrations of fine sediment being present in July 1986. The other observations represent intermediate conditions.

The settling of flocculating suspensions has been the subject of many theoretical and laboratory investigations (for example Krone, 1986; Hunt, 1986). This work has been slow to find extensive practical application in estuary management models due to the idealized conditions of the investigations not being closely satisfied in the natural environment. One of the recurring difficulties is the presence of a wide variety of organic coatings on the mineral particles found in estuaries that have been acquired either within an estuary, or in a river or sea before entering an estuary.

The settled sediment on the bed will dewater at rates depending on the floc structure of the original suspension and the drainage path availability, which will in turn depend on the rate of sedimentation. Delo (1988) suggests that a reasonable estimate of initial bed dry density is $80 \, \text{kg/m}^3$. The subsequent rate of increase in density will depend strongly on the above variables. Delo (1988) gives an example of approximately $225 \, \text{kg/m}^3$ being reached in about two days. Sills and Elder (1986) give laboratory data showing similar results for two days, and $300-400 \, \text{kg/m}^3$ for the order of a month for Combwich mud. Combwich is located near to the mouth of the Parrett estuary which discharges into Bridgwater Bay, on the southern side of the Bristol Channel.

In practice a bed will dewater fastest at the bottom due to the effects of loading. Hence there will be a vertical variation in sediment density, with density increasing with depth below the water—sediment interface. The bed density $[\rho_b(z)]$ varies in a complex manner, but to a first approximation the ratio ρ_b/ρ_{bav} (ρ_b = bed density, ρ_{bav} = depth mean bed density) varied between 0.5 and 1.5 over the middle 80% of the bed depth for an example of laboratory conditions given by Delo (1988).

The erosion of a bed that has undergone consolidation has been found to be a difficult process to study accurately in the laboratory. The main difficulty has been the production of spatially uniform bed shear stress distributions under laboratory conditions. A straight rectangular flume shows boundary layer development effects over a substantial length before longitudinally uniform conditions are achieved, and even then there will still be a transverse shear stress distribution due to wall and secondary flow effects. Surface driven circular flumes have the advantage of continuity of flow, but can lead to centrifugally induced secondary flow effects. In general, as secondary flows increase with velocity it is easier to study the erosion of poorly consolidated beds than beds of greater density.

A bed erodes until a level is reached where the density and cohesion of the bed have developed to a level where they are able to resist the applied bed shear stress. The rate of erosion may be given by

$$\frac{dm}{dt} = -E(\tau_\beta - \tau_\epsilon) \tag{16.3}$$

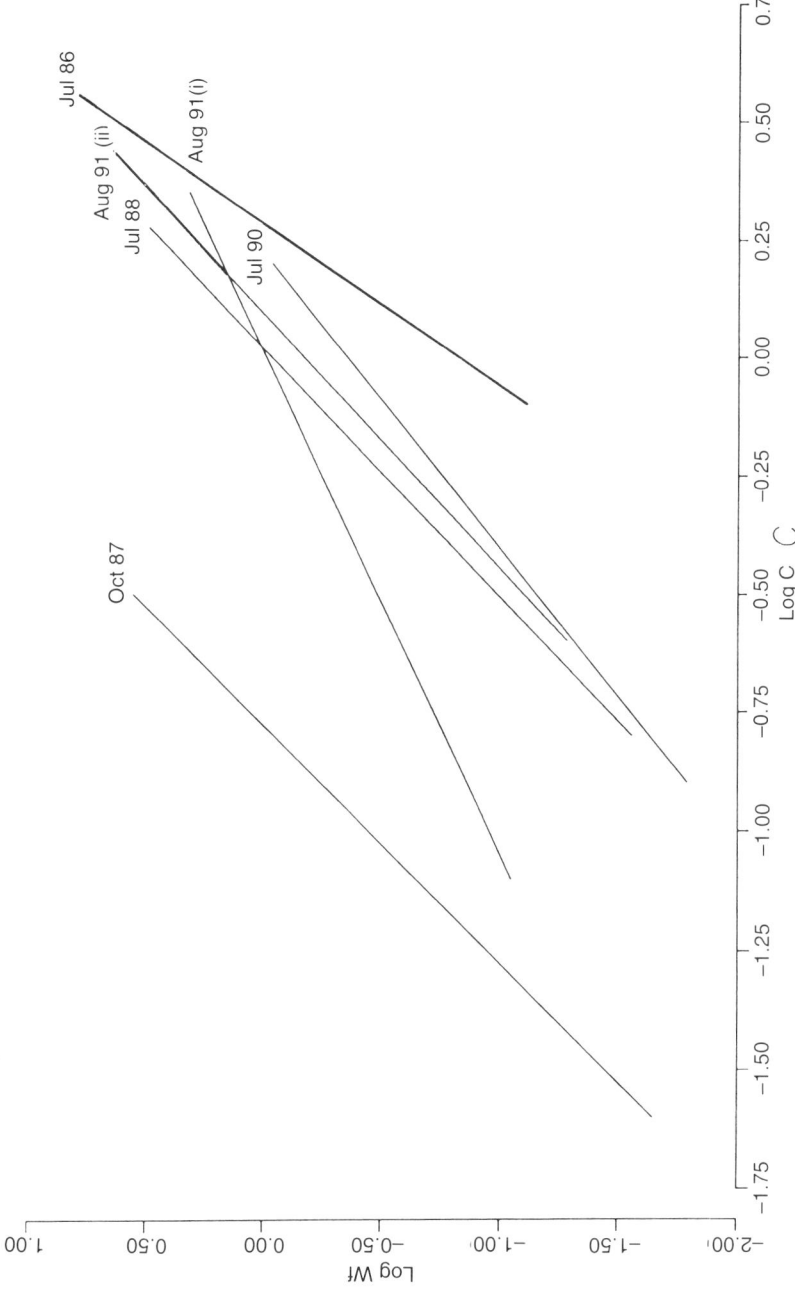

Figure 16.3 Fall velocity variation with suspension concentration, upper estuary of River Tamar 1986–91

where E is erosion constant (kg/N/s), τ_β = bed shear stress (N/m^2) and τ_ϵ = critical erosion shear strength of mud (N/m^2). A first approximation for the erosion constant and critical erosion shear strength has been given (Delo, 1988) as $E = 0.002$ kg/N/s and

$$\tau_\epsilon = 0.0012\rho_b^{1/2} \tag{16.4}$$

In practice it is again essential to undertake laboratory measurements on a bed prepared by settling a dilute suspension of a slurry to simulate a bed that is as close as is practicable to prototype conditions.

Within a tidal cycle there is often insufficient time over slack water for the newly formed bed to achieve a significant erosional strength and the bed may be considered as fluid mud. If the bed has formed from a fairly high concentration of suspended solids (5–10 kg/m^3) then it may have sufficient thickness to move as a layer under the shear stress applied at the water–fluid mud interface. The further application of shear stress may result in the breaking of internal waves at the density interface or the complete disruption of the mud layer and the rapid transport of the sediment into suspension. Mud layers can also move under the effects of gravity on sloping surfaces.

Waves can provide a further complicating influence on the processes of sedimentation, consolidation and erosion. Waves generate an additional shear stress, particularly in the wave boundary layer at the bed, and also additional pressure fluctuations which can induce sufficient strain in some beds to cause a weakening of the cohesive strength (Suhayda, 1986). These effects can induce fluidization in weak beds and surface erosion in stronger beds. The maximum peak wave-induced bed shear stress may be estimated (Delo, 1988) from semiempirical equations for the wave boundary layer. Caution is needed when considering intertidal zones as the most intensive turbulent activity takes place where waves are breaking.

From the above discussion it is apparent that first estimates of the deposition, consolidation and erosion properties of estuarine cohesive sediments may be obtained from generalized empirical formulae. However, careful laboratory measurements are essential if site-specific values are to be determined. Owing to the scale effect differences between laboratory and natural flows, and the idealized nature of most laboratory flow sytems, the laboratory results need to be treated with caution. Thus laboratory derived values should be considered as useful precursors to the modelling of cohesive sediment transport, as in practice it is necessary to integrate the *in situ* hydrodynamic characteristics of a flow with the *in situ* suspension/sediment characteristics.

16.5 COHESIVE SEDIMENTS: SOME EXAMPLES OF FIELD STUDIES

A simplistic view of estuarine cohesive sediment transport is to consider a body of water having a high suspended solids concentration of partially flocculated cohesive sediment traversing the upper reaches of an estuary on each flood and ebb tide. The body of water has a background concentration of discrete particles and flocs that have small settling velocities, such that a high percentage of the particles do not settle out over slack water periods. Sediment is then added to the water column by bed erosion during the acceleration period of each tide and settles out of suspension during the succeeding deceleration to the next slack water. The turbidity maximum thus has higher concentrations during the higher bed shear stress periods of spring tides compared with those of neap tides. The location of the reaches traversed by the turbidity maximum changes with river flow, with floods pushing the sediment seawards followed by a slow return of the sediment upstream to the reach occupied during low flow periods.

A simple, and in most instances essential, addition to this view is that much of the fine sediment in an estuary is not in the main channel at any given time. It is stored on the bed of the intertidal zones of the reaches traversed by the turbidity maximum at various stages of the annual cycle of fluvial discharge, or is lying in sheltered embayments or creeks. The sediment dynamics of both the turbidity maximum and of the intertidal zones need to be understood if reasonable models of cohesive sediment transport are to be produced.

The existence of the turbidity maximum is thought to be due to several mechanisms, which vary in their importance in time and space in a given estuary. In small tidal range estuaries where a salt wedge is well established throughout a tidal cycle gravitational circulation will probably be important. Fluvial sediment settles out of the upper seaward flowing fresh water layer into the lower sea water layer. The lower layer moves landwards to replace water entrained into the upper layer by breaking internal wave-induced transport processes. In medium to high tidal range estuaries where the water column becomes partially to well mixed for some of the tidal cycle in the upper reaches of the estuary, tidal pumping has been shown to produce a net movement of sediment upstream (Uncles, Elliott and Weston, 1985). A consideration of the interaction of the temporally varying values of cross-sectional area of flow, cross-sectional area mean velocity and cross-sectional area mean suspended solids concentration shows that as each term departs fairly significantly from a simple harmonic function then the net tidal average residuals usually result in a landwards transport of sediment (Uncles, Elliott and Weston, 1985).

There are several reasons for the component terms departing from simple harmonic functions. In the upper reaches of estuaries the flooding tidal wave may be considered as a positive surge and thus has steep longitudinal surface gradients, particularly during spring tides in high tidal range estuaries. These gradients produce a rapid rise in water level. On the ebb tide, the retreating tidal wave is like a negative surge and is dissipative in nature. The water drains out of the upper estuary under gravity, thus leading to a rate of change of water level that is generally slower than during the flood tide. This behaviour results in the velocity distribution showing a short, faster flowing flood tide and a longer, slower ebb tide. Not all estuaries behave in this manner and topographical effects can lead to some having faster ebb tides than flood tides, for example the Hamble and Forth. The suspended solids concentration will tend to respond to the bed shear stresses associated with the velocity distribution. However, this variation in the suspended solids is affected by several factors. The stability of the water column during the first part of the ebb tide will result in the suppression of vertical turbulent transport processes (West, Oduyemi and Shiono, 1991). Thus much of the sediment deposited on the bed during high slack water will remain there until the vertical density gradients are reduced. These gradients may be reduced by either fresh water replacing the salinity as the FSI is advected downstream, or by the flow becoming shallow and fast enough for the combination of bed and internally generated turbulence to overcome the gradient producing effect caused by the interaction of fluid shear and the longitudinal salinity gradient. This reduction of stratification may take place after one or two hours of ebb flow (see Figure 16.1), thus the particles will have effectively been left behind by the body of water from which they fell during the latter stages of the flood tide. In Figure 16.1 the suspended solids that settled out before high water were not resuspended in any substantial amount until 1800 hours. Such behaviour results in further asymmetry in the temporal variation of the area mean suspended solids concentration and a tendency for a net upstream transport to occur.

As sediment is deposited towards the end of the ebb tide in the region of the seaward end of the tidal excursion of the FSI, a layer of virtually unconsolidated sediment is formed. As the flood tide accelerates, this layer is rapidly eroded and thereafter the concentration of fine sediment in the water column is limited by the lack of availability of further sediment to be eroded. This causes additional asymmetry in the temporal variation of the area mean suspended solids concentration. The net effect of these various mechanisms is for sediment to be transported upstream in the region traversed by the FSI. The location of the FSI is also influenced considerably by the early ebb tide water column stability in many estuaries as this is the main period within a tidal cycle in which upstream dispersion of salt takes place to counteract the downstream flushing of salt induced by the fluvial flow.

Well upsteam of the FSI and below the tidal limit, the flood tide currents are weak as the upstream tidal volume is small. The ebb currents are of long duration and often stronger than the flood currents, particularly in valleys that have slopes that are steeper than those of the surface of the tidal wave. The net result is a tidally averaged transport of sediment seawards. Thus the combination of the tidal and buoyancy effects aids the formation of turbidity maxima in most estuaries. The magnitude of the maximum concentration and the longitudinal extent vary according to the availability of kinetic energy and sediment. Suitable sediment sources and a high tidal range, such as found in the Severn estuary−Bristol Channel system, result in maximum concentrations of greater than 100 mg/l extending over more than 100 km of channel (Figure 16.4). These data show the maximum and minimum values of suspended solids concentration observed during a spring and neap tidal cycle at eight stations along the estuary. A well defined maximum occurs in the Sharpness region for the observed conditions, with the spring maximum concentration being about five times the neap maximum concentration.

A good understanding of cohesive sediment transport also requires an understanding of the size distribution of the primary particles and of the aggregates. The development of an *in situ* particle size analyser based on a Malvern laser system has allowed some advance for moderate concentrations of suspended solids (Bale and Morris, 1987). The size and density of suspensions in the Tamar estuary have been shown to vary through a tidal cycle. There is a need to extend this work to higher concentrations and to a wider range of low conditions in several estuaries.

Much of the previous field work in estuaries has not given full recognition to the transverse variation of bed sediment properties. In general, the sediments on the beds of main channels are coarser (sand and gravel) than those of the banks (mud). There is a need to undertake studies of the main channel processes that focus on the interaction of sand and mud transport processes. Studies of the dynamics of the turbidity maximum usually ignore the influence of the intertidal zones as sinks and sources of sediment.

During flood tides water enters the intertidal zone from generally faster moving main channel flows and thus sedimentation may be anticipated. During ebb tides there is a very good chance that sediment will be retained on an intertidal zone as the ebb tide shear stresses are generally less than the flood tide values and the critical erosion bed shear stress is greater than the depositional stress. In addition, during the ebb tide the water level may have fallen far enough to expose the sediment before the critical erosion bed shear stresses are reached. Over the period when the sediment is uncovered and exposed to the air, wind and solar radiation lead to rapid dewatering and strengthening. This helps to armour the surface layer against erosion but slows down the up flow of water from the underlying layers, hence inhibiting their compaction. Cores taken from intertidal zones confirm this layering. The cores also showed that lenses of sand can occur which may be able to aid dewatering under some circumstances

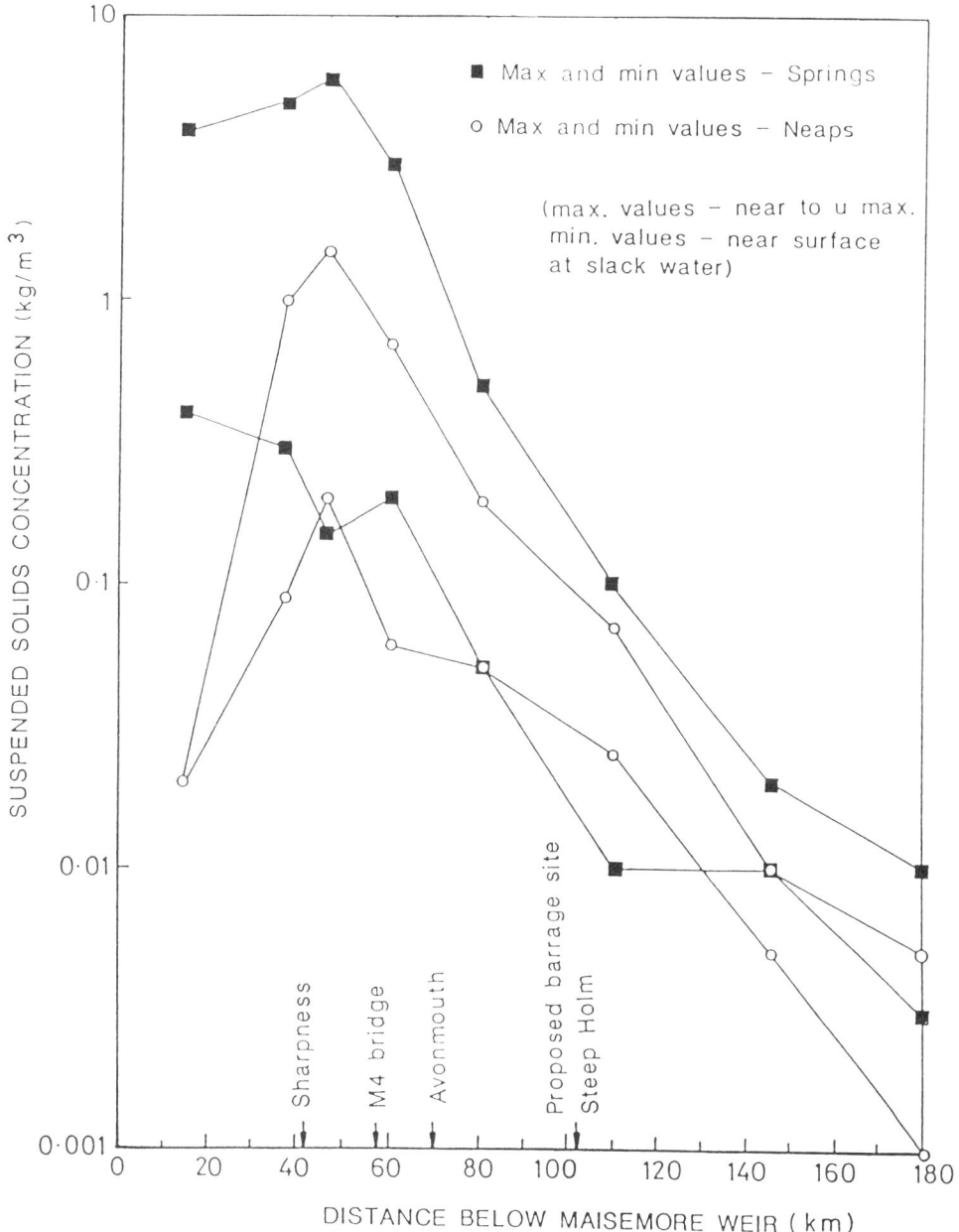

Figure 16.4 Maximum and minimum values of suspended solids for spring and neap tides, River Severn, Bristol Channel 1980

(West, 1991). The combination of conditions of hydrodynamic activity and sediment supply that lead to sand being deposited, or the fine particles being preferentially removed from a mixed deposit, need to be researched.

The effect of the net removal of sediment from suspension and its storage on the intertidal zones is to reduce the stock of fine sediment available for transport in the main channel. This appears to be a significant factor during low flows as the sedimentation builds up over a period of weeks, leaving relatively low concentrations in the turbidity maximum compared with those found immediately subsequent to a flood event that has remobilized bank sediments. Visual observation by the author suggests that in some reaches of upper estuaries the banks can achieve a condition close to an equilibrium where they become very steep and only a small wavelet is sufficient to erode the surface layer, thus mainting an equilibrium between the deposition and erosion processes.

The cyclical erosion of the fine sediment from the intertidal zones tends to be dominated by fluvial floods in the upper reaches of estuaries and by wind wave-induced erosion in the wider lower reaches. The remobilization of bank sediments in the upper reaches of estuaries can be caused by a number of mechanisms. Erosion can take place from the surface, or scouring of the toe of a bank can lead to slips. The latter is usually the most important mechanism as the main force of a flood is felt at low water when the opposing effects of fluvial and tidal action are not present. The complexity of the bank structure, which is usually composed of many layers of sediment which have reached varying stages of dewatering and cohesion, leads to the prediction of erosion rates being very difficult. The lack of field observations is another difficulty, which is surprising as these failures are of some considerable importance to flood protection works in low lying areas such as East Anglia and the Somerset levels.

In the lower reaches of estuaries the supply of sediment is greatest when fluvial floods flush the suspended, and some of the bed sediments, from the upper estuary. This cycling has been reported in the Tamar estuary by Bale, Morris and Howland (1985). Fluvial floods have little influence on the intertidal zone velocity distributions, but the greater channel width allows the wind to generate bigger surface waves. The seasonal nature of the wave climate leads to the build-up of sediment during calm conditions, particularly over spring tides, and then to erosion during the succeeding stormy periods. Studies in the Severn estuary demonstrated this cycle with the observed conditions showing sedimentation tending to occur during the summer and erosion during the winter gales (West and West, 1991). Figure 16.5 shows the annual variation of accretion and erosion averaged across two intertidal zones in the Severn estuary. Day 0 during the study was 25 October 1988. Erosion rates may be anticipated to be a complicated function of time as they may be assumed to depend on the direction of the wind, the direction of the tide which can attenuate wave energy transmission, the position of the water level when the storm is at its height, and the antecedent consolidation and erosion of the bed. There is a need for observations of these events.

There is evidence of longer term variations of sediment cycling on and off intertidal zones. These variations have time-scales of a few years to centuries. The variation of sea levels associated with the ice ages have led to estuaries cutting down through old sediments and forming new areas of accretion. In the Severn estuary it has been postulated that several periods of intertidal zone accretion and erosion have taken place in the last 2000 years, probably reflecting changes in the intensity of storm activity (Allen and Rae, 1987). The recent exposure of artefacts in the Severn estuary provide evidence that substantial areas of the intertidal zone of the estuary are undergoing a period of erosion. The relative importance of meteorological trends, sea-level changes and the control of flooding by reservoir and sea wall construction is at present unclear. Observations of heavy metal and radionuclide distributions in sediments of the Ribble estuary by the author suggest that accretion rates in excess of 4 cm/year have occured in the last 15 years. This is thought to be due in part to the reduction in navigational dredging and to land reclamation schemes. Again, no firm evidence exists to allow the determination of the relative importance of the various contributory mechanisms.

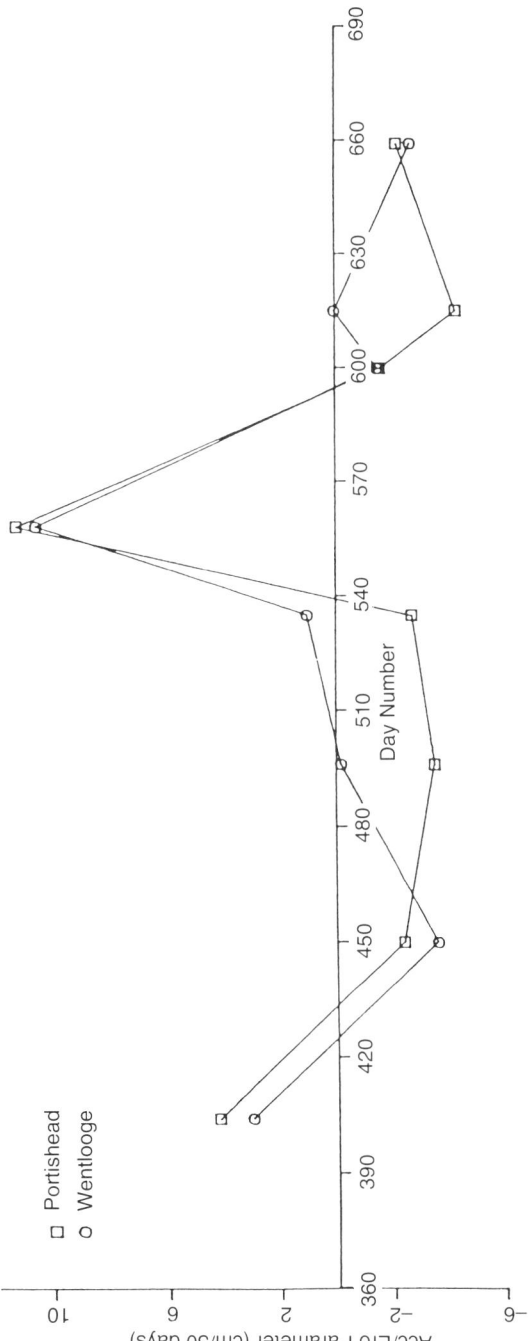

Figure 16.5 Variation of erosion and accretion rates at Portishead and Wentlooge, River Severn, 1987–8 (West and West, 1991; reproduced by permission of Olsen & Olsen)

16.6 CONCLUSIONS

A review of some of the recent work on cohesive sediment transport in British estuaries has shown that considerable progress in deepening our understanding has been made by undertaking field studies. The work has shown that a broad understanding of cohesive sediment transport is essential for the prediction of chemical and biological processes needed for the management of the estuarine environment to meet the conflicting demands of exploitation and conservation of natural resources. The transport of suspended sediment and the storage of sediment on intertidal zones has been shown to depend on time-scales that relate to turbulence events lasting a few seconds through to interglacial periods. There is a need for further studies, particularly in the field, and preferably by teams so that the skills available to different scientific disciplines can be brought together to further elucidate this intriguing topic.

REFERENCES

Allen, J.R.L. and Rae, J.E. (1987) Late Flandrian shoreline oscillations in the Severn estuary: a geomorphological and stratigraphical reconnaissance. *Trans. R. Soc. London* **B315**, 185−230.

Bale, A.J. and Morris, A.W. (1987) *In situ* measurement of particle size in estuarine waters. *Est. Coast. Shelf Sci.* **24**, 253−263.

Bale, A.J., Morris, A.W. and Howland, R.J.M. (1985) Seasonal sediment movement in the Tamar estuary. *Oceanol. Acta* **8**, 1−6.

Barton, M.L., Stephens, J.A., Uncles, R.J. and West, J.R. (1991) Particle fall velocities and related variables in the Tamar estuary. In: *Estuaries and Coasts* (Eds. M. Elliott and J.P. Ducrotoy). Oslen & Alsen, Fredensborg, 31−36.

Burt, T.N. (1986) Field settling velocities of estuary muds. In: *Estuarine Cohesive Sediment Dynamics* (Ed. A.J. Mehta). Springer Verlag, New York, 126−150.

Darbyshire, E.J. (1990) Vertical turbulent transport in the Tamar estuary. *PhD Thesis*, University of Birmingham.

Delo, E.A. (1988) *Estuarine Muds Manual. Rep. No. SR 164.* Hydraulics Research, Wallingford.

Dyer, K.R. (1982) Mixing caused by lateral internal seiching within a partiallmixed estuary. *Est. Coast. Shelf Sic.* **15**, 443−457.

Festa, J.F. and Hansen, D.V. (1978) Turbidity maxima in partially mixed estuaries: a two-dimensional numerical model. *Est. Coast. Mar. Sci.* **7**, 347−359.

Geyer, W.G. and Smith, J.D. (1987) Shear instability in a highly stratified estuary. *J. Phys. Oceanogr.* **17**, 1668−1679.

Guymer, I. and West, J.R. (1991) Field studies of the flow structure in a straight reach of the Conwy estuary. *Est. Coast. Shelf Sci.* **32**, 581−592.

Hunt, J.R. (1986) Particle aggregate breakup by fluid shear. In: *Estuarine Cohesive Sediment Dynamics*. (Ed. A.J. Mehta). Springer Verlag, New York, 85−109.

Ippen, A.T. (1966) *Estuary and Coastline Hydrodynamics*. McGraw-Hill, New York.

Krone, R.B. (1986) The significance of aggregate properties to transport processes. In: *Estuarine Cohesive Sediment Dynamics*. (Ed. A.J. Mehta). Springer Verlag, New York, 66−84.

McDowell, D.M. and O'Connor, B.A. (1977) *Hydraulic Behaviour of Estuaries*. Macmillan Press, London.

New, A.L., Dyer, K.R. and Lewis, R.E. (1987) Internal waves and intense mixing periods in a partially stratified estuary. *Est. Coast. Shelf Sci.* **24**, 15−33.

Nunes, R.A. and Simpson, J.H. (1985) Axial convergence in a well mixed estuary. *Est. Coast. Shelf Sci.* **20**, 637−649.

Officer, C.B. (1976) *Physical Oceanography of Estuaries*. Wiley, London.

Partch, E.N. and Smith, J.D. (1978) Time dependent mixing in a salt wedge estuary. *Est. Coast. Mar. Sci.* **6**, 3−19.

Paterson, D.M. (1989) Short term changes in the erodibility of intertidal cohesive sediments related to the migratory behaviour of epipelic diatoms. *Limnol. Oceanogr.* **34**, 223−234.

Sills, G.C. and Elder, D.M. (1986) The transition from sediment suspension to settling bed. In: *Estuarine Cohesive Sediment Dynamics*. (Ed. A.J. Mehta). Springer Verlag, New York, 192−205.

Simpson, J.H. and Turrell, W.R. (1985) Convergent fronts in the circulation of tidal estuaries. In: *Estuarine Variability*. (Ed. D.A. Wolfe). Academic Press, New York, 139−152.

Suhayda, J.N. (1986) Interaction between surface waves and muddy bottom sediments. In: *Estuarine Cohesive Sediment Dynamics*. (Ed. A.J. Mehta). Springer-Verlag, New York, 401−428.

Turner, J.S. (1973) *Buoyancy Effects in Fluids*. Cambridge University Press, Cambridge.

Uncles, R.J., Elliott, R.C.A. and Weston, S.A. (1985) Dispersion of salt and suspended sediment in a partially mixed estuary. *Estuaries* **8**, 256−269.

West, M.S. (1991) Aspects of inter-tidal sediment dynamics in the Severn estuary. *PhD thesis*, University of Birmingham.

West, J.R. and Mangat, J.S. (1986) The determination and prediction of longitudinal dispersion coefficients in an arrow, shallow estuary. *Est. Coast. Shelf Sci.* **22**, 161−181.

West, M.S. and West, J.R. (1991) Spatial and temporal variations in inter-tidal zone sediment properties in the Severn estuary. In: *Estuaries and Coasts* (Eds. M. Elliott and J.P. Ducrotoy). Olsen & Olsen, Fredensborg, 25−30.

West, J.R., Knight, D.W. and Shiono, K. (1984) A note on flow structure in the Great Ouse estuary. *Est. Coast. Shelf Sci.* **19**, 271−290.

West, J.R., Oduyemi, K.O.K. and Shiono, K. (1991) Some observations on the effect of vertical density gradients on estuarine turbulent transport processes. *Est. Coast. Shelf Sci.* **32**, 365−383.

West, J.R., Uncles, R.J., Stephens, J.A. and Shiono, K. (1991) Longitudinal dispersion processes in the upper Tamar estuary. *Estuaries* **13**, 119−124.

17 Numerical Modelling of Tidal Eddies in Coastal Basins with Narrow Entrances Using the $k-\epsilon$ Turbulence Model

R.A. FALCONER

Department of Civil Engineering, University of Bradford, Bradford, UK

and

LI GUIYI

Southern Science Ltd, Worthing, UK

17.1 INTRODUCTION

The occurrence of tidal eddies in coastal basins with narrow entrances, in the lee of headlands and in coastal bays and estuaries is important to engieners and scientists in terms of sediment deposition, pollutant mixing and the siting of sea outfalls, the erosion of coastal headlands and promontories and navigation problems. Such time dependent tidal eddies are complex hydrodynamic flow processes which are not easy to model numerically and laboratory model scale predictions can often lead to erroneous results, particularly when Froudian law distorted models are used.

In simultaneous numerical model studies undertaken by Falconer (1980; 1982; 1985) to predict the tide-induced circulation in laboratory-scale rectangular harbours with narrow entrances and in Port Talbot Harbour, South Wales, results showed that the circulation strength of the flood tide eddies, observed within the laboratory models and Port Talbot Harbour, were underpredicted using a simple turbulence model governed by bed-generated turbulence only. Furthermore, in the laboratory models of rectangular harbours the tracked pathlines and the measured depth mean velocities confirmed the existence of two or more counter-rotating tidal eddies for an aspect ratio (or length to breadth ratio) greater than two, whereas the corresponding numerical model predictions showed only one eddy for all aspect ratios (Falconer, 1980). In both these studies the flood tide eddy (or eddies) arose as a result of the separation streamline and the corresponding tidal jet occurring downstream of the harbour entrance and beyond the entrance breakwater tip. Hence the exclusion of free shear-generated turbulence from the numerical model was partly assumed to account for the lower flood tide eddy strength with the idealized model harbours, particularly as the simple mixing length turbulence stress representation included in the numerical model did not take account of the advection of turbulence by the mean flow.

In view of the common occurrence of large-scale tidal eddies in coastal basins with narrow

Mixing and Transport in the Environment. Edited by K.J. Beven, P.C. Chatwin and J.H. Millbank
© 1994 John Wiley & Sons Ltd

entrances and the importance of such complex hydraulic phenomena, a study was undertaken to improve and refine an existing two-dimensional depth-integrated numerical model to reproduce, with improved accuracy, the tidal eddies that often occur in coastal basins with narrow entrances, estuaries and the lee of headlands. The main improvements considered in the model simulations were the inclusion of an empirically based zero-equation turbulence model, giving rise to a vorticity flux source at separation points (see Falconer *et al.*, 1986) and the more refined unsteady two-equation $k-\epsilon$ turbulence model (Rodi, 1984).

It is worth noting at this stage that the benefits of an improved turbulence model of tidal eddy simulation, with the inclusion of free shear turbulence, are potentially not just applicable to hydrodynamic modelling, but are also particularly relevant to modelling solute transport diffusion. At present most two- or three-dimensional combined hydrodynamic and water quality models assume bed-generated turbulent diffusion only, based on a logarithmic velocity distribution (Elder, 1959) or field data generally acquired for nearly isotropic−homogeneous turbulent flow (Fischer, 1973). In including a refined $k-\epsilon$ turbulence model, the depth mean eddy diffusivity field can be calculated directly and, for a non-stratified flow, can be equated directly with the turbulent diffusion coefficients of the advective-diffusion equation, with these numerically predicted coefficients including both free shear and bed-generated turbulence. This inclusion of free shear turbulent mixing in water quality models is of particular importance in the mixing associated with tidal eddies.

17.2 MODEL DIFFERENTIAL EQUATIONS

The differential hydrodynamic equations included in the numerical model were based on the depth-integrated form of the Reynolds equations, with the advective accelerations being expressed in their pure differential form, thereby enabling momentum conservation in the finite difference model. The corresponding continuity and momentum equations in the x, y directions were written in the following form

$$\frac{\partial \zeta}{\partial t} + \frac{\partial UH}{\partial x} + \frac{\partial VH}{\partial y} = 0 \tag{17.1}$$

$$\frac{\partial UH}{\partial t} + \beta \left[\frac{\partial U^2 H}{\partial x} + \frac{\partial UVH}{\partial y} \right] - fVH + gH\frac{\partial \zeta}{\partial x} + \frac{\tau_{bx}}{\rho} - \nu_t H \left[\frac{\partial^2 U}{\partial x^2} + \frac{\partial^2 U}{\partial y^2} \right] = 0 \tag{17.2}$$

$$\frac{\partial VH}{\partial t} + \beta \left[\frac{\partial UVH}{\partial x} + \frac{\partial V^2 H}{\partial y} \right] + fUH + gH\frac{\partial \zeta}{\partial y} + \frac{\tau_{by}}{\rho} - \nu_t H \left[\frac{\partial^2 V}{\partial x^2} + \frac{\partial^2 V}{\partial y^2} \right] = 0 \tag{17.3}$$

where ζ = water surface elevation above (positive) datum, t = time, U, V = depth average velocity components in x, y directions, H = total depth of flow, β = momentum correction factor for non-uniform vertical velocity profile, f = Coriolis parameter, g = gravitational acceleration, τ_{bx}, τ_{by} = bed shear stress components in x, y directions, ρ = fluid density and ν_t = depth average turbulent eddy viscosity.

For the momentum correction factor, β, this was evaluated assuming a logarithmic vertical velocity profile, giving

$$\beta = \left[1 + \frac{g}{\kappa^2 C^2}\right] \qquad (17.4)$$

where κ = von Karman's constant (= 0.41) and C = Chezy roughness coefficient, obtained from the Colebrook—White equation for transitional and fully developed rough turbulent flow and given as

$$C = -18\log_{10}\left[\frac{k_s}{12H} + \frac{5C}{18Re}\right] \qquad (17.5)$$

where k_s = Nikuradse equivalent roughness and Re = flow Reynolds number (= $4HU_s/\nu_\ell$ where U_s = depth average fluid speed and ν_ℓ = laminar viscosity). In the case of practical applications where wind effects are significant, then the value of β and the advective accelerations can be modified for an assumed second-order parabolic vertical velocity profile (Falconer and Chen, 1991). Similarly, for the bed shear stress components a quadratic friction law was assumed giving, for the x-direction

$$\tau_{bx} = \rho g U U_s C^{-1} \qquad (17.6)$$

with a similar expression for the y-direction.

For the zero-equation turbulence model considered in the early part of the study, the lateral shear stress terms were evaluated using an empirically based mixing zone velocity profile to give the free shear layer turbulence, with the bed-generated turbulence being represented by Taylorian diffusion. This representation of the depth-integrated lateral shear stresses was similar to that given by Lean and Weare (1979) and extended in Falconer et al. (1986) and Falconer and Mardapitta-Hadjipandeli (1987).

For the second and more significant part of the study a pure differential form of the depth integrated two-equation $k-\epsilon$ turbulence model was used, with the equations being expressed in the following form

$$\frac{\partial kH}{\partial t} + \frac{\partial kUH}{\partial x} + \frac{\partial kVH}{\partial y} = \frac{\partial}{\partial x}\left[\frac{\nu_t H}{\sigma_k}\frac{\partial k}{\partial x}\right] + \frac{\partial}{\partial y}\left[\frac{\nu_t H}{\sigma_k}\frac{\partial k}{\partial y}\right]$$
$$+ \nu_t H\left[2\left(\frac{\partial U}{\partial x}\right)^2 + 2\left(\frac{\partial V}{\partial y}\right)^2 + \left(\frac{\partial U}{\partial y} + \frac{\partial V}{\partial x}\right)^2\right]$$
$$+ C_f^{3/2} C_k U_s^{3/2} - \epsilon H \qquad (17.7)$$

$$\frac{\partial \epsilon H}{\partial t} + \frac{\partial \epsilon UH}{\partial x} + \frac{\partial \epsilon VH}{\partial y} = \frac{\partial}{\partial x}\left[\frac{\nu_t H}{\sigma_\epsilon}\frac{\partial \epsilon}{\partial x}\right] + \frac{\partial}{\partial y}\left[\frac{\nu_t H}{\sigma_\epsilon}\frac{\partial \epsilon}{\partial y}\right]$$
$$+ C_{1\epsilon} C_\mu kH\left[2\left(\frac{\partial U}{\partial x}\right)^2 + 2\left(\frac{\partial V}{\partial y}\right)^2 + \left(\frac{\partial U}{\partial y} + \frac{\partial V}{\partial x}\right)^2\right]$$
$$+ \frac{C_\epsilon C_f^2 U_s^2}{H} - C_{2\epsilon}\frac{\epsilon^2}{k} \qquad (17.8)$$

where k = turbulent kinetic energy, σ_k, σ_ϵ, C_μ, $C_{1\epsilon}$, $C_{2\epsilon}$ = constant coefficients, ϵ = dissipation rate of turbulent kinetic energy, C_f = friction coefficient (= gC^{-2}), $C_k = C_f^{-1/2}$ and $C_\epsilon = 3.6 C_{2\epsilon} C_\mu^{1/2} C_f^{-3/4}$. In the event that no detailed laboratory measurements have been undertaken to date to evaluate precisely the turbulence parameter coefficients for the $k-\epsilon$

model, standard values for the constants were used in the model simulations (Rodi, 1984), giving $\sigma_k = 1.0$, $\sigma_\epsilon = 1.22$, $C_\mu = 0.09$, $C_{1\epsilon} = 1.44$ and $C_{2\epsilon} = 1.92$. From these equations, the time and spatial variation of the depth average eddy viscosity was evaluated from the equation

$$\nu_t = C_\mu k^2 \epsilon^{-1} \tag{17.9}$$

For the hydrodynamic boundary conditions relating to the application of the $k-\epsilon$ turbulence model, and the governing momentum equations 17.2 and 17.3, various boundary representations were considered and compared. The best results were obtained using the conditions outlined in Falconer and Li (1992) and given here for completeness

(i)   Along a solid wall normal to the x-direction

$$U = 0 \quad \text{and} \quad \frac{\partial U}{\partial x} = 0$$

$$\left.\begin{array}{l} \dfrac{\partial V}{\partial x} = \dfrac{2V_{adj}}{\Delta x} \text{ for a lower boundary} \\[2em] \dfrac{\partial V}{\partial x} = \dfrac{-2V_{adj}}{\Delta x} \text{ for an upper boundary} \end{array}\right\} \quad \text{No-slip boundary condition}$$

$$\frac{\partial k}{\partial x} = 0$$

$$\frac{\partial \epsilon}{\partial x} = 0$$

where Δx = grid size and V_{adj} = velocity parallel to the wall and $\frac{1}{2}\Delta x$ away from the wall. Similar expressions were obtained for a wall normal to the y-direction.

(ii)   For a point adjacent to a wall, the law of the wall was used to calculate k and ϵ (Rodi, 1984) giving

$$k = U_*^2 C_\mu^{-1/2}$$

$$\epsilon = U_*^3 \kappa^{-1} (0.5\Delta x)^{-1}$$

where U_* = depth average velocity.

(iii)   For an open seaward boundary in the x-direction, an inflow boundary condition was prescribed as suggested by Rodi (1984)

$$k = \epsilon = 0$$

$$\frac{\partial V}{\partial y} = 0$$

with the lateral zero velocity gradient giving rise to a Coriolis slope along the boundary. However, for an outflow boundary condition, k, ϵ and the velocity gradient along the boundary were all obtained by linear extrapolation from the corresponding values computed just within the domain.

17.3 MODEL DIFFERENCE EQUATIONS

The governing differential equations of mass and momentum conservation were expressed in an alternating direction implicit finite difference form, with the difference equations being second-order accurate in space and time. The advective accelerations and the turbulent diffusion terms were centred in time by iteration, with the resulting scheme being theoretically unconditionally stable, but requiring a maximum Courant number (i.e. $\Delta t (gH)^{1/2} \Delta x^{-1}$ where Δt = time step) of about eight. If the Courant number exceeds this approximate value, then the accuracy of the scheme decreases rapidly with increasing Courant number as the numerical wave increasingly propagates more slowly than the physical wave. A space-staggered grid scheme was used, with the depths below datum in the x and y directions being specified at the centre of the corresponding grid sides. This representation differs from the conventional staggered grid notation, but has two advantages. Firstly, more depth data are included in the model for increased accuracy in representing features such as dredged channels and, secondly, the depths in the x and y directions are prescribed at exactly the same locations as for the U and V velocity components, respectively. The advective accelerations were expressed in a finite difference form similar to the marker and cell technique (Williams and Holmes, 1974), with the derivatives being fully centred in space except for the velocity component normal to the direction of motion being considered. The revised scheme is basically second-order accurate and has no advection stability constraints.

For the finite difference representation of the $k-\epsilon$ turbulence equations, the time dependent form of these equations was found to exhibit severe spurious wave characteristics — particularly when applied to the current study. The numerical results were unacceptable for the fully centred difference scheme and were caused by the relatively large kinetic energy gradients occurring for adjacent grid cells and the inability of first- and second-order accurate schemes to represent such high gradients with sufficient accuracy. In the case of the first-order upwind difference scheme, although numerically smooth and stable results were obtained, the degree of artificial diffusion (of magnitude $0.5U_\ell \Delta x$, where U_ℓ = local depth mean velocity) was unacceptable with peak values being of the order of $1000\,\text{mm}^2\,\text{s}^{-1}$. The values were often significantly greater than the physical turbulent diffusion occurring over much of the flow field. On the other hand, although the central difference scheme showed considerably less diffusion, the large spurious negative k values were unacceptable and were progressive and unstable in time.

In overcoming the considerable numerical difficulties associated with solving numerically the unsteady $k-\epsilon$ equations, various higher order accurate difference schemes were considered, including the QUICK scheme (Falconer and Liu, 1988) third-order upwind differencing and the EXQUISITE scheme (Leonard, 1981). The main difference between these schemes lay in the interpolation method for the turbulent advection terms. In the case of central differencing, linear interpolation is used with only three grid points included in the difference representation. The first refined scheme considered, namely the QUICK scheme, includes the use of quadratic upstream interpolation. This scheme is at least second-order accurate and involves up to five grid points in the interpolation procedure. The EXQUISITE scheme consists of a combination of quadratic, exponential and linear interpolation with the scheme being designed to make the best use of each interpolation procedure depending on the flow direction and the form of the spatial variation of the property (namely k and ϵ in the current study). Finally, the third-order upwind difference scheme is based on the principle of weighting the grid points considered in the derivative in favour of those points upsteam of the gradient location, with the scheme being third-order accurate and again including up to five grid points. When all three schemes were coded up and compared with the laboratory data for the current study it was found that the EXQUISITE scheme gave the closest agreement with the laboratory data, although this scheme also required more computational effort than the others. Full details of the finite difference formulation for the $k-\epsilon$ equations using the EXQUISITE scheme are given in Falconer and Li (1992).

17.4 MODEL TESTS AND APPLICATION

17.4.1 Physical model tests

The main application and testing of the $k-\epsilon$ turbulence model was undertaken for a series of laboratory experiments to determine the tidal current structure and tidal exchange characteristics for a range of idealized rectangular harbours with varying geometric, kinematic and hydrodynamic conditions. The hydraulic models were treated as vertically distorted models of idealized prototype rectangular harbours, which had dimensions and experienced tidal conditions similar to many small harbours and marinas in Puget Sound, Washington, USA (Nece, Smith and Richey, 1980). The prototype harbour was assumed to have a planform area of 18.7 ha, an asymmetrical entrance of width 48 m and a horizontal bed of mean depth 6 m. For the current study the model harbour was set to be square, with prototype dimensions of 432 m × 432 m and was assumed to experience sinusoidal repeated semidiurnal tides of period 12.4 hours and range 4 m. The horizontal scale ratio was varied from 1:200 to 1:400 and with the vertical scale varying from 1:12.0 to 1:3.125. An illustration of the prototype model dimensions and coordinate system is shown in Figure 17.1, with the corresponding physical model being designed using a conventional Froude law scaling relationship. As for previous studies, the dimensions and geometrical features of the laboratory model were governed by the objective of fitting several different grid sizes perfectly into the model harbour shape and across the asymmetrical entrance, with the numerical model tests being undertaken using three different grid sizes of 12, 6 and 4 cm.

The model harbours were constructed of plywood and Perspex and were located in a tide tank having a working area of overall length 5.32 m and width 3.80 m and a working depth of 0.62 m, as shown in Figure 17.2. Tides were produced by a tide generator in the form of a variable elevation waste weir fed by a constant rate water supply entering the tank through a perforated manifold. Dye observations made in the tank showed that flows approaching

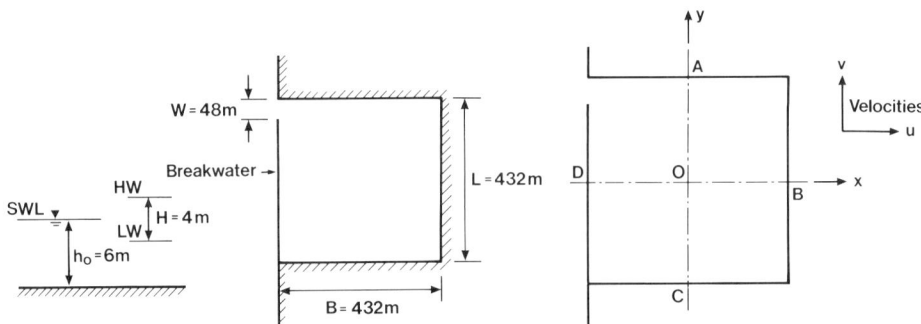

Figure 17.1 Harbour dimensions and coordinate system

the harbour on the flood tide were uniform as they passed the outer end of the sloping beach shown in Figure 17.2, and there were no measurable cross-currents.

The weir was moved vertically by screw jacks driven by a variable speed DC motor, controlled by a data logger and control unit and driven by a computer. Water surface levels in the tank were measured using a water level recorder, calibrated in place and located near one wall of the tank as shown in Figure 17.2. Tidal data were recorded at two second intervals and verified that the actual tide curves could be considered as sinusoidal, with the tidal wave lagging the weir motion by approximately 6°.

Velocities in the harbour were determined from plotted pathlines of floats, this being consistent with the numerical model obejctive of obtaining depth average velocities. The floats used were 5 mm diameter plastic fishing bobbers, weighted with lead shot so that they maintained a vertical orientation and with lengths selected so that they penetrated most of the water column. Float penetration depths were 150, 130 and 85 m for the corresponding distortion ratios of 1:12, 1:10 and 1:6.25 and with the mean water depths being 180, 150.2 and 93.1 mm, respectively. Floats were tracked visually, with their positions being marked by hand at two-second increments on a Perspex overlay sheet. An axial rod projecting from

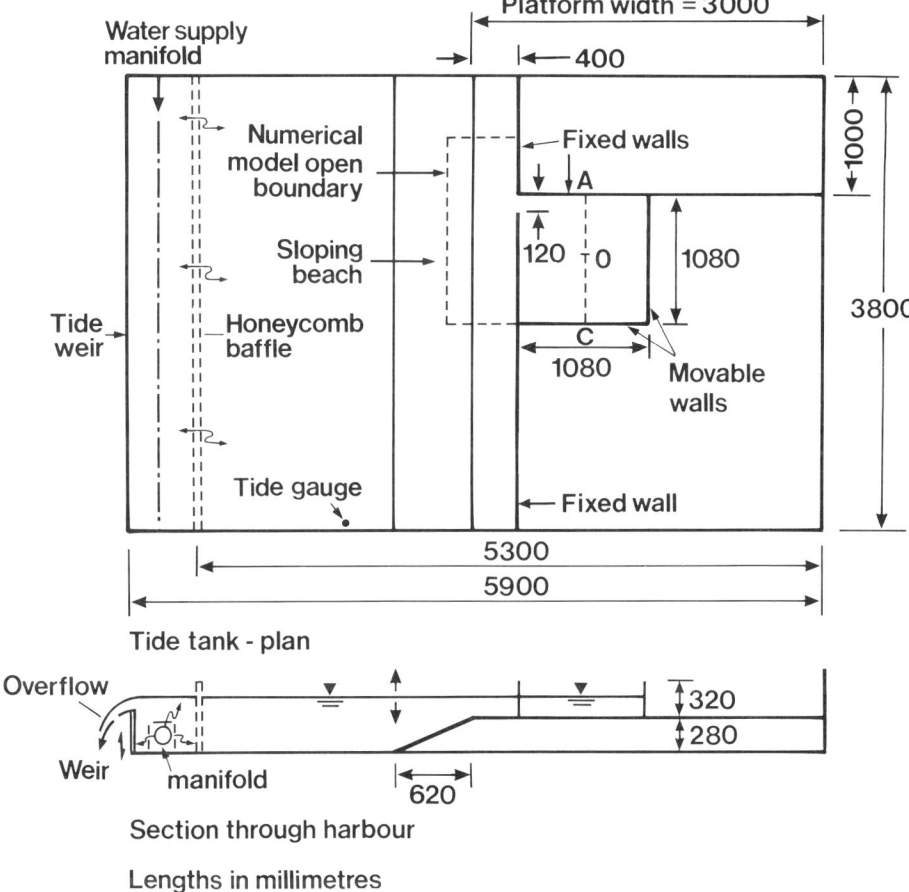

Figure 17.2 Schematic illustration of laboratory tidal basin and harbour configuration

the top of the float allowed an observer positioned over the harbour to align the tip of a marking pen with the vertical float axis so that parallax errors were minimized. Measurements of the float coordinates marked on the overlay sheet allowed velocities to be calculated by $u = dx/dt$ and $v = dy/dt$.

For simplicity of operation, the timing of the float measurements was keyed to the sinusoidal motion of the weir, with the objective being to measure current velocities along the harbour axes (i.e. AOC and BOD in Figure 17.1) near to the mean water level flood and ebb tide. For example, when flood tide data were being taken, floats were released at least 10 seconds before mean water level, with the objective being to position the release point so that the float would cross the axis under consideration at the time corresponding to the mean water level. All measured velocities reported herein were for axis crossing times of within ± 4 s of the desired time, and with most crossing times being within ± 2 s of mean water level. Although a uniform spatial distribution of velocity measurements was not obtained, sufficient data were obtained for the velocity profiles to be defined along the habour axes. Further details of the laboratory measuring procedure are given in Nece and Falconer (1989) and Falconer and Yu (1991).

17.4.2 Numerical model tests

In applying the zero-equation and the $k-\epsilon$ turbulence numerical models to the physical model illustrated in Figures 17.1 and 17.2, the model domain consisted of a mesh of 29×29 grid squares covering the model harbour and the region just beyond the harbour entrance and contained within the model open boundary illustrated in Figure 17.2. The grid spacing for the simulations reported herein was 60 mm and the time step differed from 0.36 to 0.5 s for the various distortion ratios to ensure that the Courant number remained at approximately eight. Based on the results of calibration tests, the bed roughness k_s was generally assumed to be 0.15 mm for all distortion ratios — a requirement governed by the physical model comparisons wherein the bed roughness was not varied with the distortion ratio. The de Chezy roughness coefficient was recalculated every 6 s and the eddy viscosity distribution determined every half time step.

For the open boundary conditions of the numerical model a sinusoidal tide of period 646, 709 or 894 s was specified at the open boundary parallel to the plane of the harbour entrance for distortion ratios of 1:12, 1:10 and 1:6.25, respectively. Similarly, for the two open boundaries normal to the harbour entrance plane the boundaries were assumed to be free streamline boundaries with the velocity component normal to the boundary set to zero. For the closed boundaries, the no-slip boundary condition was applied, together with the corresponding boundary conditions for the $k-\epsilon$ model as given previously. As for the physical model studies the tidal range and the mean harbour depth for each distortion ratio were 0.12 and 0.18 m (for 1:12 distortion), 0.10 and 0.15 m (for 1:10 distortion) and 0.062 and 0.094 m (for 1:6.25 distortion). The numerical model was always started from a state of rest at low tide and took approximately three hours per tidal cycle of simulation on an IBM PS/80 personal computer.

The numerical model was then run for a range of geometric, kinematic and dynamic boundary conditions, as considered in the laboratory model programme, using the $k-\epsilon$ turbulence model and the simple mixing length model, with the eddy viscosity coefficient (i.e. ν_t/U_*H) being obtained by calibration. Typical examples of the predicted velocity fields at mean water level ebb and flood tide are shown in Figures 17.3–17.5 for the simple mixing length model and the $k - \epsilon$ turbulence model using the QUICK and EXQUISITE schemes. A general comparison of these velocity field predictions — together with others produced for this study — indicated at first sight that the $k - \epsilon$ turbulence model did not appear to show a marked difference in the predicted velocity field from that produced using the simple mixing length model. The difference in velocity magnitude produced by the two kinds of models was typically found to be about 2.5%. This close agreement between the velocity field distributions obtained using the simple mixing length and the $k - \epsilon$ turbulence models was attributed to a number of factors including (i) the low Reynolds numbers of the flow, (ii) the reduced effects of bottom friction production as a result of the relatively large depth due to physical model distortion and (iii) the artificial diffusion introduced through the quasi-centred difference scheme used for the numerical representation of the advective accelerations. However, there were certain dissimilarities in the general flow structure, particularly in the vicinity of the harbour entrance, thereby suggesting the inadequacy of the simple mixing length model in representing the free shear layer turbulence accurately along the free streamline. In comparing the velocity profiles across the AOC axis for the simple mixing length (Figure 17.6a) and the $k - \epsilon$ (Figure 17.6b) turbulence models, it can be seen that the flood tide jet is broader for the simple mixing length model, with the two grid squares adjacent to the wall at A having similar velocity magnitudes for the simple mixing length model. Several higher order accurate schemes were also considered for modelling the turbulent advection process including, in particular, third-order upwind differencing, the QUICK scheme (Figure 17.4) and the EXQUISITE scheme (Figure 17.5). Of these schemes the EXQUISITE scheme gave the closest agreement with the laboratory data, although this scheme also required more computational effort. A typical example comparison of the numerically predicted and laboratory measured velocity profiles across the AOC and BOD harbour axes is shown in Figure 17.7.

17.4.3 Comparisons between physical and numerical model results

Although no experimental data were available for the eddy viscosity distributions at the time of this study, numerical model predictions were also obtained of the eddy viscosity distributions, with typical results at mean water flood tide being shown in Figures 17.8−17.10 for a simple mixing length model and the $k - \epsilon$ turbulence model using the EXQUISITE and first-order upwind difference schemes. In comparing these figures it can be seen that there is a marked difference in the eddy viscosity magnitudes and distributions. In comparison with the $k - \epsilon$ (EXQUISITE) turbulence model, the simple mixing length model has grossly underpredicted the eddy viscosity values, with the results differing by over two orders of magnitude in the vicinity of the harbour entrance. The reason for this significant difference in results is that the $k - \epsilon$ model takes into account both the turbulence time history effects and turbulence transport, whereas the simple mixing length model only equates the turbulence stresses to the local bed friction velocity and the assumption that the turbulence is dissipated locally. The region of high velocity and shear gradients along the separation streamline is clearly depicted using the $k - \epsilon$ model, together with the high shear gradients along the harbour wall. The advection into the harbour of the relatively high turbulence generated at the harbour entrance can be seen for the $k - \epsilon$ model (Figure 17.9), whereas this is not the case for the simple mixing length model, even though the velocity distributions in Figures 17.3 and 17.5 did not differ markedly. This large variation in the eddy viscosity distribution between the mixing length and $k - \epsilon$ turbulence models has since been included in the diffusion process of a water quality model (Falconer and Li, 1989), where the results have proved to be of

(a)

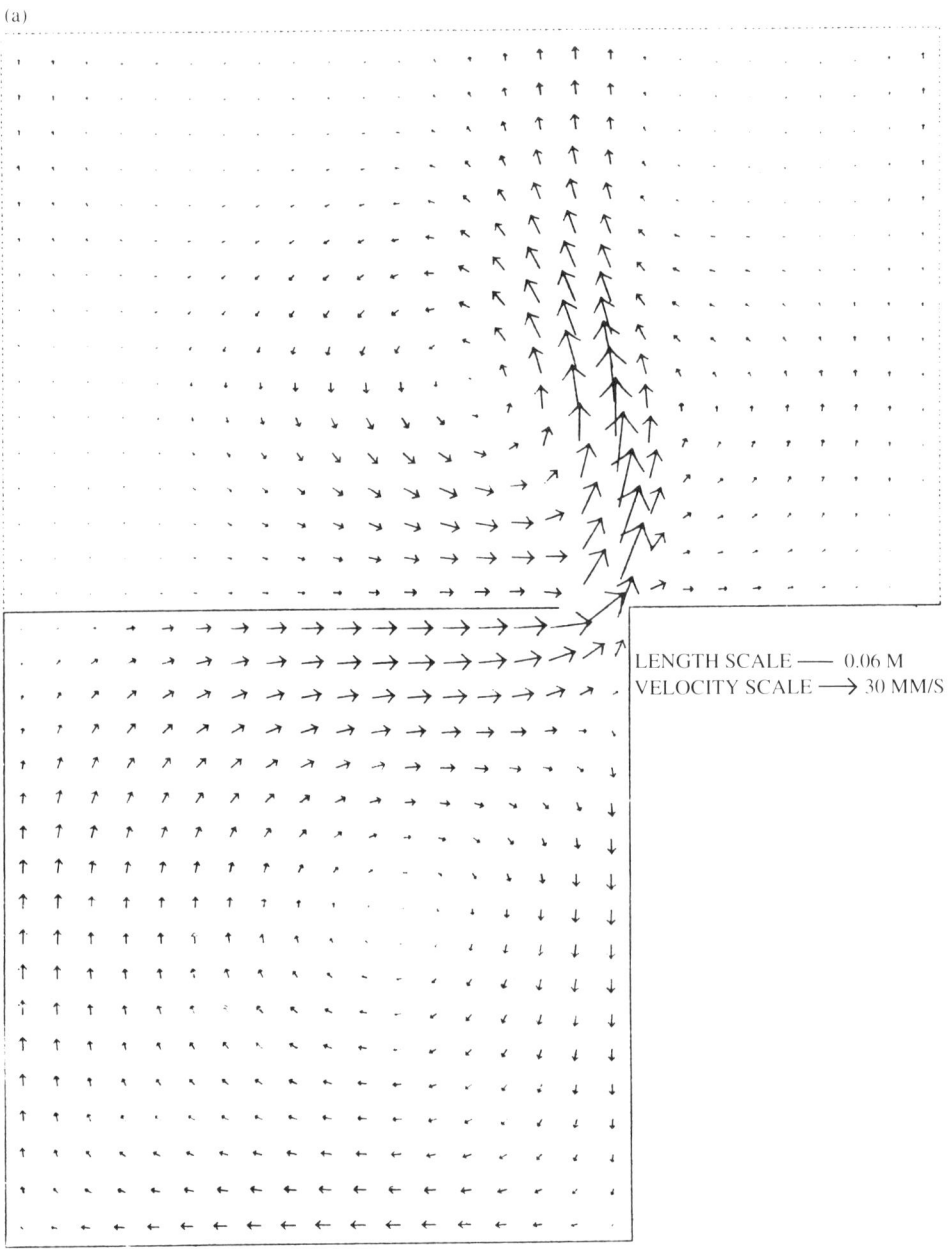

LENGTH SCALE —— 0.06 M
VELOCITY SCALE ⟶ 30 MM/S

Figure 17.3 Tidal currents in a rectangular harbour. Tidal height, 0.10 m; average depth, 0.15 m; tidal period, 707 s; roughness length, 0.15 mm. (a) Numerically predicted velocity field at mean water level ebb tide using the simple mixing length model. Time = 2651 s

(b)

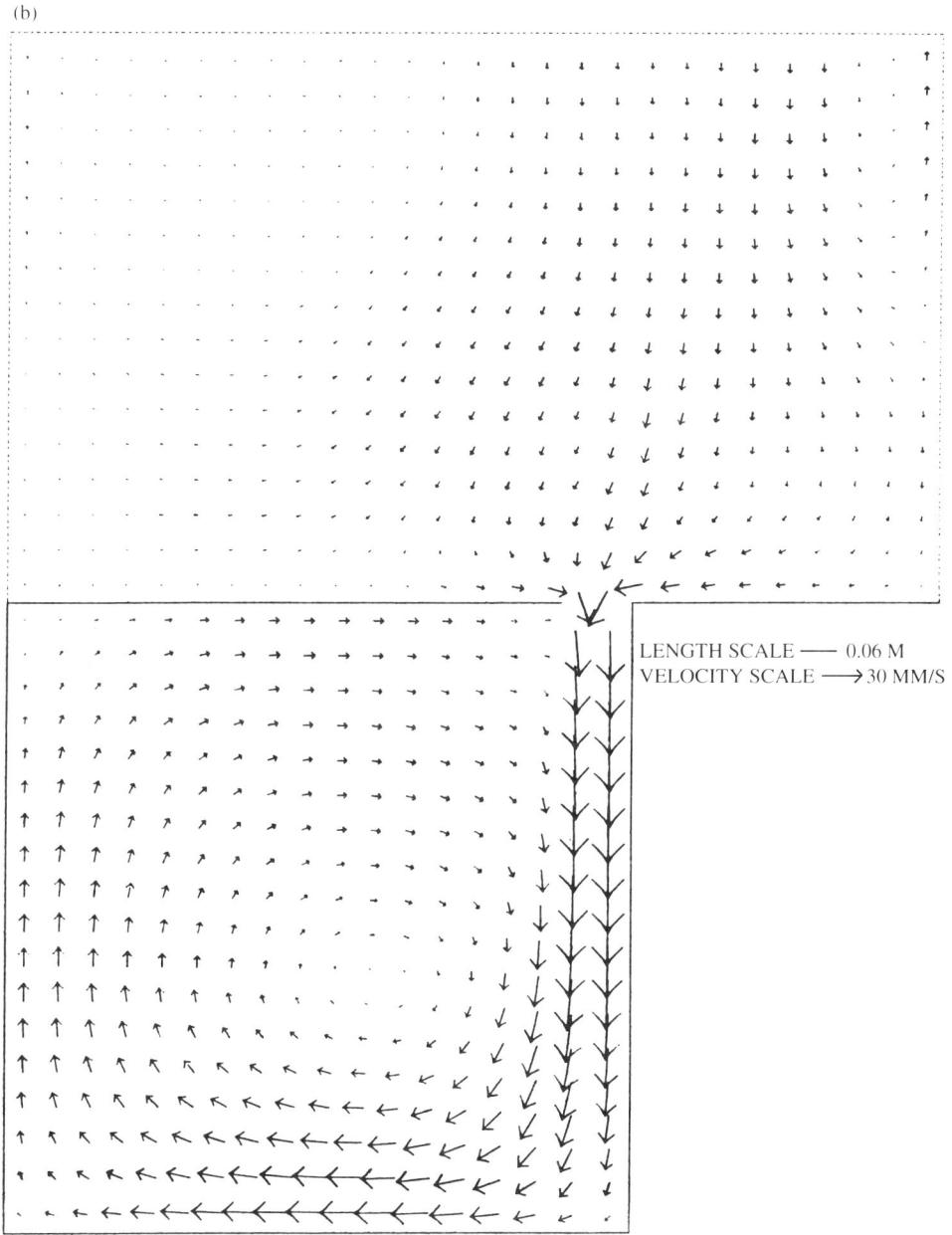

LENGTH SCALE —— 0.06 M
VELOCITY SCALE ——→ 30 MM/S

Figure 17.3 (*continued*) (b) Numerically predicted velocity field at mean water level flood tide using the simple mixing length model. Time = 2298 s

(a)

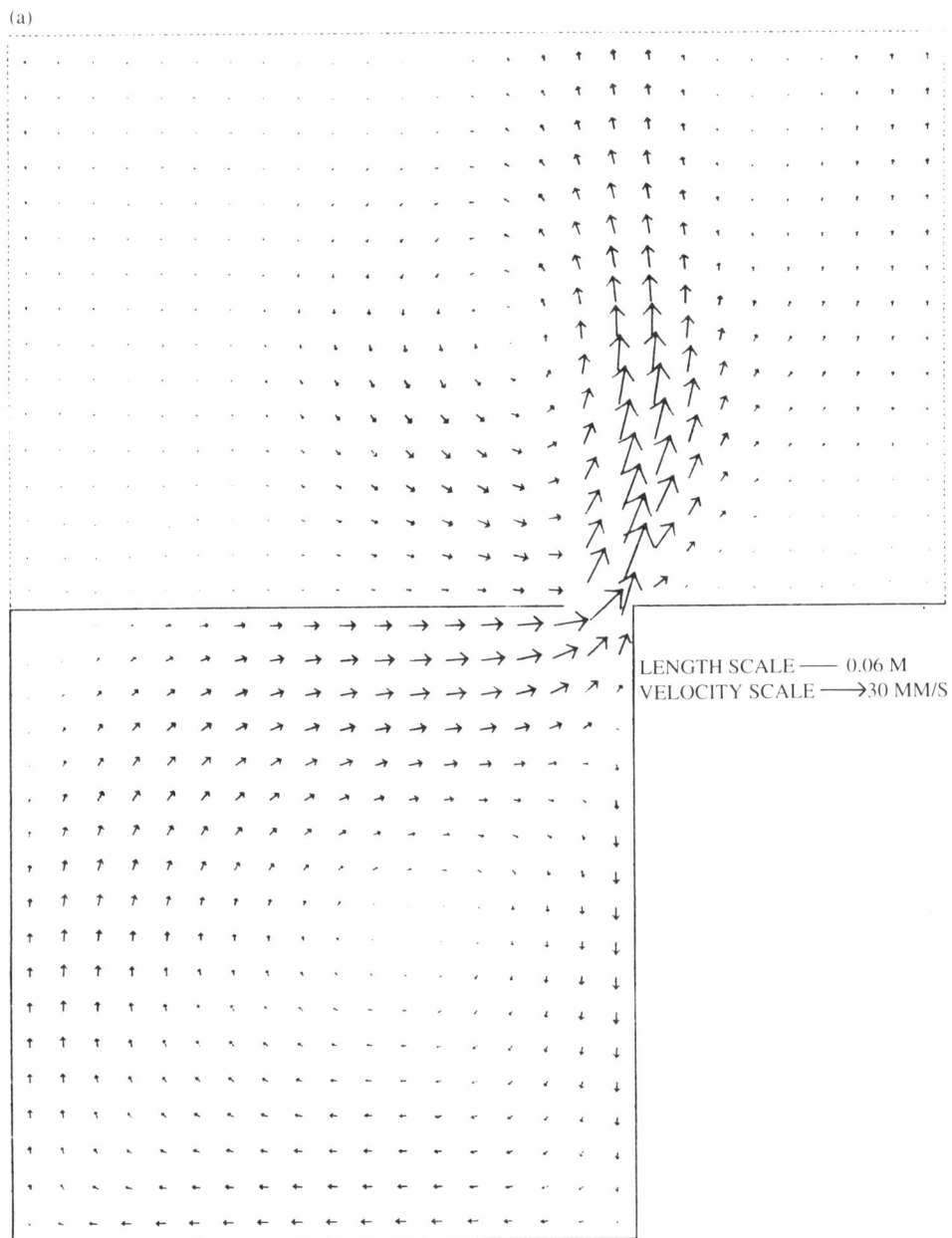

LENGTH SCALE —— 0.06 M
VELOCITY SCALE ⟶30 MM/S

Figure 17.4 Tidal currents in a rectangular harbour. Parameters as in Figure 17.3. (a) Numerically predicted velocity field at mean water level ebb tide using the QUICK scheme $k-\epsilon$ turbulence model

(b)

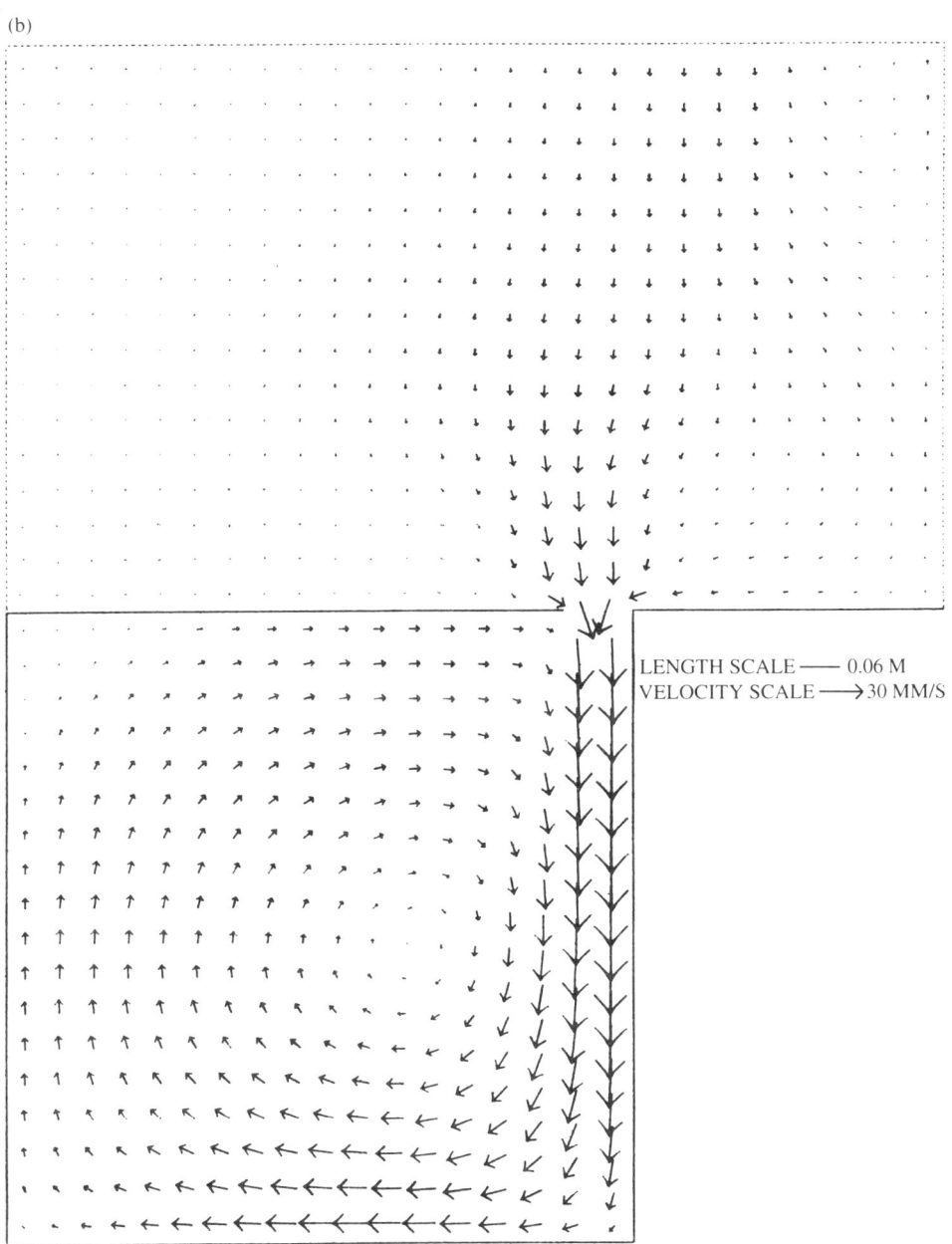

LENGTH SCALE ———— 0.06 M
VELOCITY SCALE ——→ 30 MM/S

Figure 17.4 (*continued*) (b) Numerically predicted velocity field at mean water level flood tide using the QUICK scheme k–ϵ turbulence model

(a)

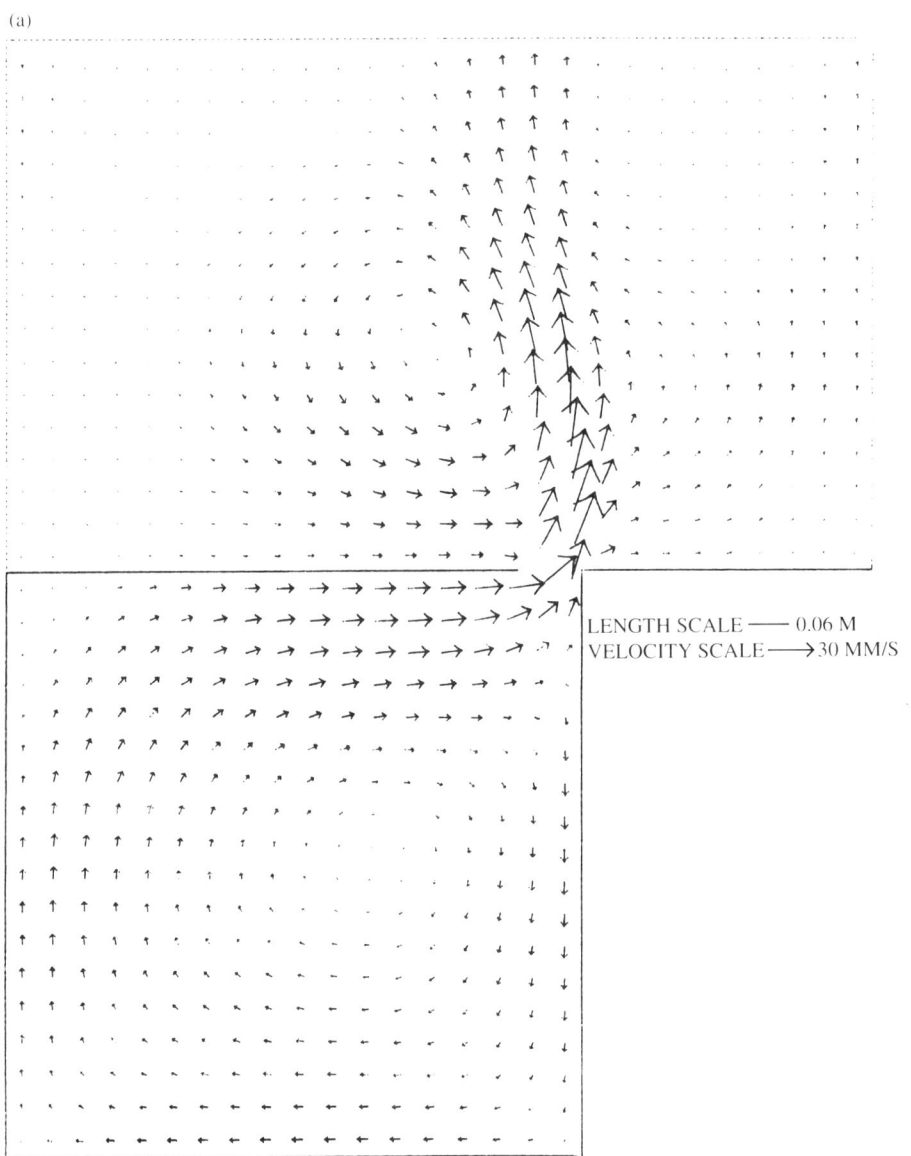

LENGTH SCALE ⸺ 0.06 M
VELOCITY SCALE ⸺→ 30 MM/S

Figure 17.5 Tidal currents in a rectangular harbour. Parameters as in Figure 17.3. (a) Numerically predicted velocity field at mean water level ebb tide using the EXQUISITE scheme $k-\epsilon$ turbulence model

(b)

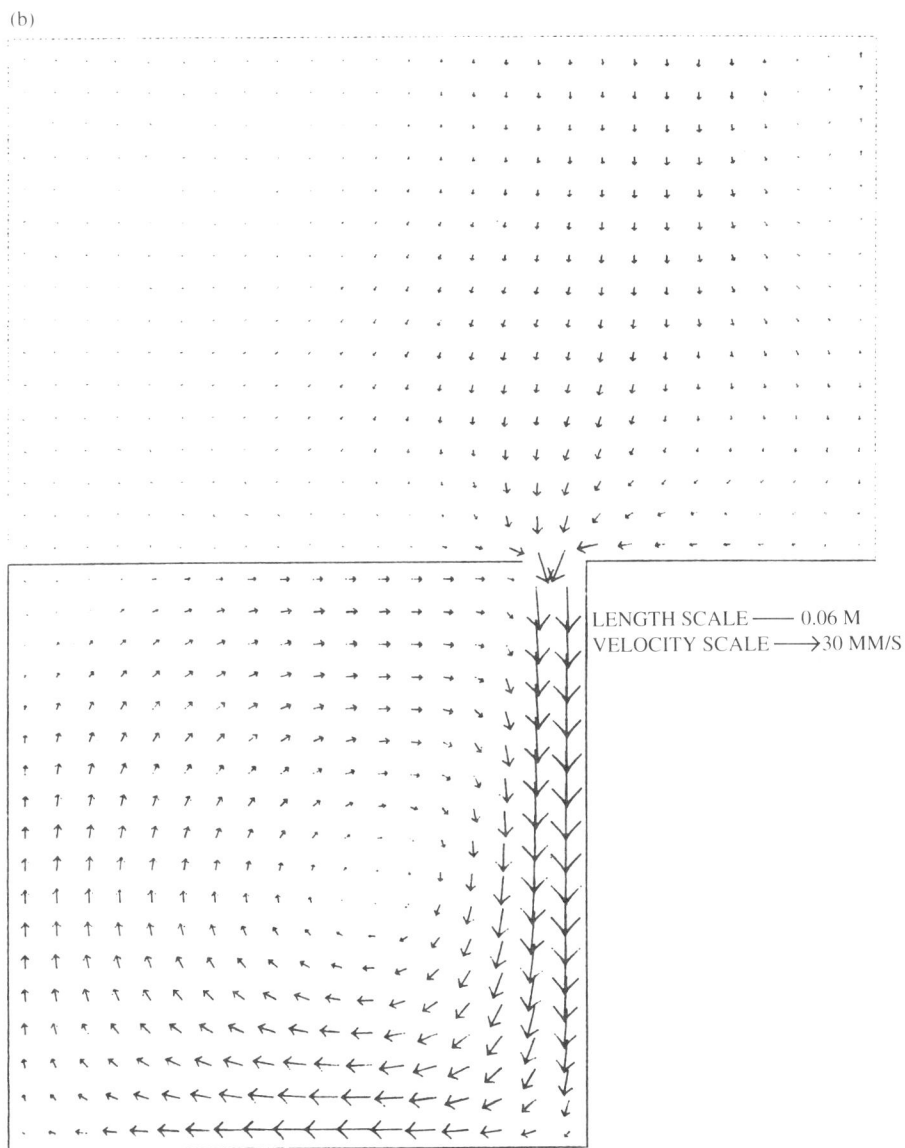

LENGTH SCALE —— 0.06 M
VELOCITY SCALE ——→30 MM/S

Figure 17.5 (*continued*) (b) Numerically predicted velocity fields at mean water level flood tide using the EXQUISITE scheme $k-\epsilon$ turbulence model

(a)

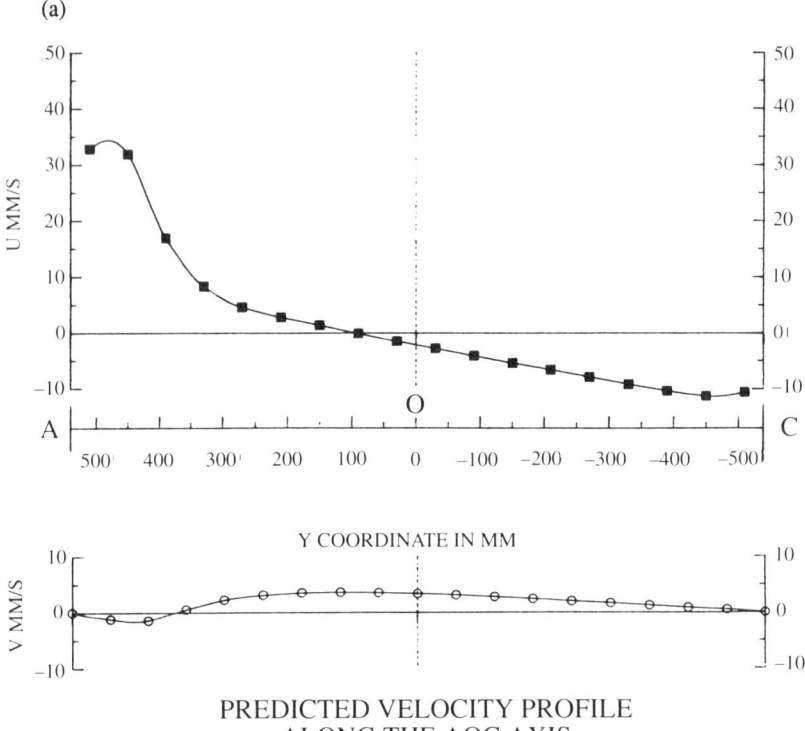

PREDICTED VELOCITY PROFILE
ALONG THE AOC AXIS

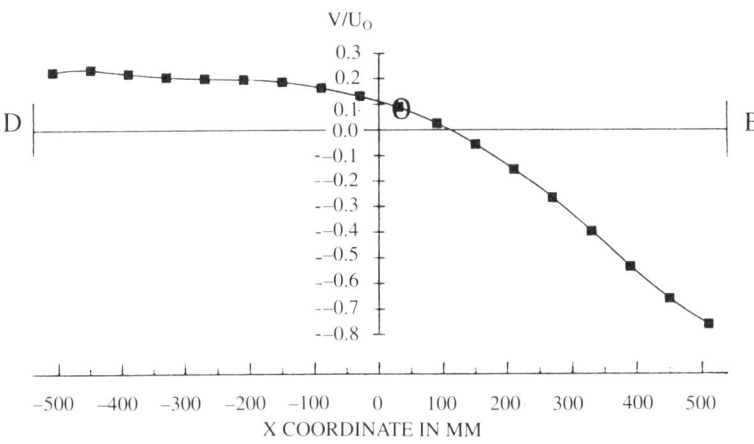

PREDICTED DIMENSIONLESS VELOCITY PROFILE
ALONG THE BOD AXIS

Figure 17.6 Numerically predicted velocity distribution across the central harbour axis at mean water flood tide using (a) the simple mixing length model

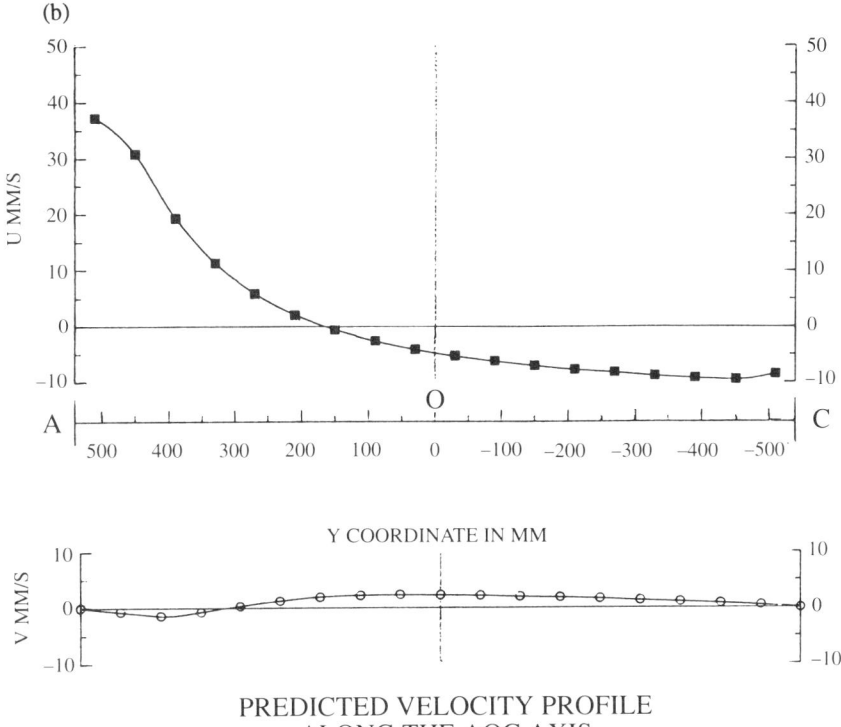

PREDICTED VELOCITY PROFILE
ALONG THE AOC AXIS

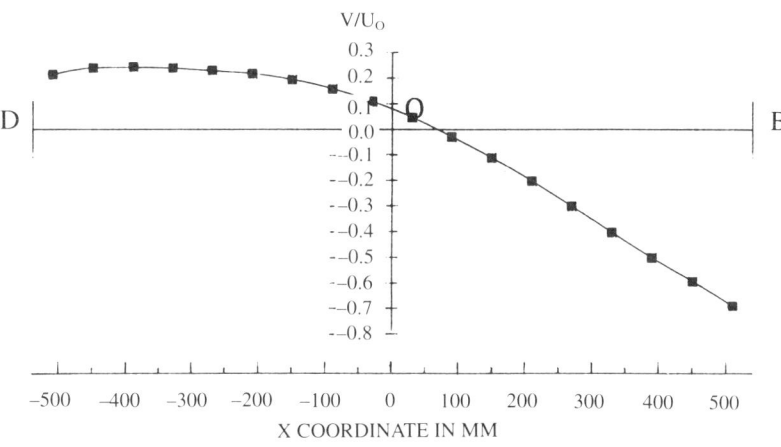

PREDICTED DIMENSIONLESS VELOCITY PROFILE
ALONG THE BOD AXIS

Figure 17.6 (*continued*) Numerically predicted velocity distribution across the central harbour axis at mean water flood tide using (b) the $k-\epsilon$ turbulence model. Time = 2298 s

much more significance in solute transport modelling. The results of these studies have therefore tended to indicate that, although horizontal turbulent momentum transport was not significant in modelling large water bodies using a relatively coarse finite difference grid, the representation of free shear turbulence and turbulent advection in the flow field using the $k-\epsilon$ model led to noticeable refinements in the accuracy of turbulent modelling of water quality consituents.

Research is currently being supervised by the first author to address some of the points highlighted earlier as to the reasons why the velocity field distributions are thought to be similar for significantly different eddy viscosity distributions. This research is focusing on the use of higher order accurate finite difference representations of the advective accelerations for larger Reynolds number flows and with increased bed friction effects. Finally, to illustrate the need to use a higher order accurate numerical scheme to solve the $k-\epsilon$ equations, comparisons between Figures 17.9 and 17.10 shows the highly diffusive effects of using the first-order upwind difference scheme which includes an unacceptably high level of numerical (or artificial) diffusion.

Further comparisons of the various predicted velocity and eddy viscosity distributions confirmed a number of other finds which can be summarized as follows:

(1) The lateral turbulent shear stresses and the advective accelerations were found to be extremely important in the numerical prediction of tidal eddies. When both of these terms were exluded no recirculation was predicted within the harbour as confirmed by previous studies by Falconer *et al.* (1986).

(2) The influence of the bathymetry and the mean depth on the structure and strength of tidal eddies was also found to be particularly important, hence confirming the advantages of using refined grid schemes to give increased resolution of the bathymetry and also highlighting the potential disadvantages of using bed generating packages for reproducing complex non-linear bathymetries on a regular grid (Falconer and Yu, 1991).

(3) The use of a quadratic friction law was shown to give rise to components of both tidal eddy generation and dissipation, as indicated by a detailed analysis of the vorticity transport equation (Falconer and Mardapitta-Hadijipandeli, 1987), although the scope of tidal eddy generation through this generally dissipative term is not always appreciated by engineers and scientists.

(4) Variations in the bed rougness length k_s resulted in a marked change in the bed boundary stresses, although this appeared to have little effect on the vortex strength of the corresponding tidal eddies in the range of roughness values considered. This result was in contrast to (2) above, where small variations in the bathymetry led to considerable influences on the tidal eddy structure.

(5) The choice of the grid size was found to be critical in accurately reproducing the tidal eddies and previous tests indicated that the eddies could only be reproduced to a reasonable level of accuracy if there were at least eight grid squares across an eddy (Falconer and Mardapitta-Hadjipandeli, 1987; Abbot, Larsen and Jianhua Tao, 1985).

In a more recent extension of the current study not yet published the first author has undertaken a range of tests to compare the accuracy of the predicted and experimentally measured velocity distributions across the central axes of the harbour for a range of grid sizes. The tests included discretizing the harbour with 9×9, 18×18, 27×27 and 36×36 grid square, with the corresponding grid sizes being 120, 60, 40 and 30 mm. The resulting predicted velocity distributions showed very little variation across the

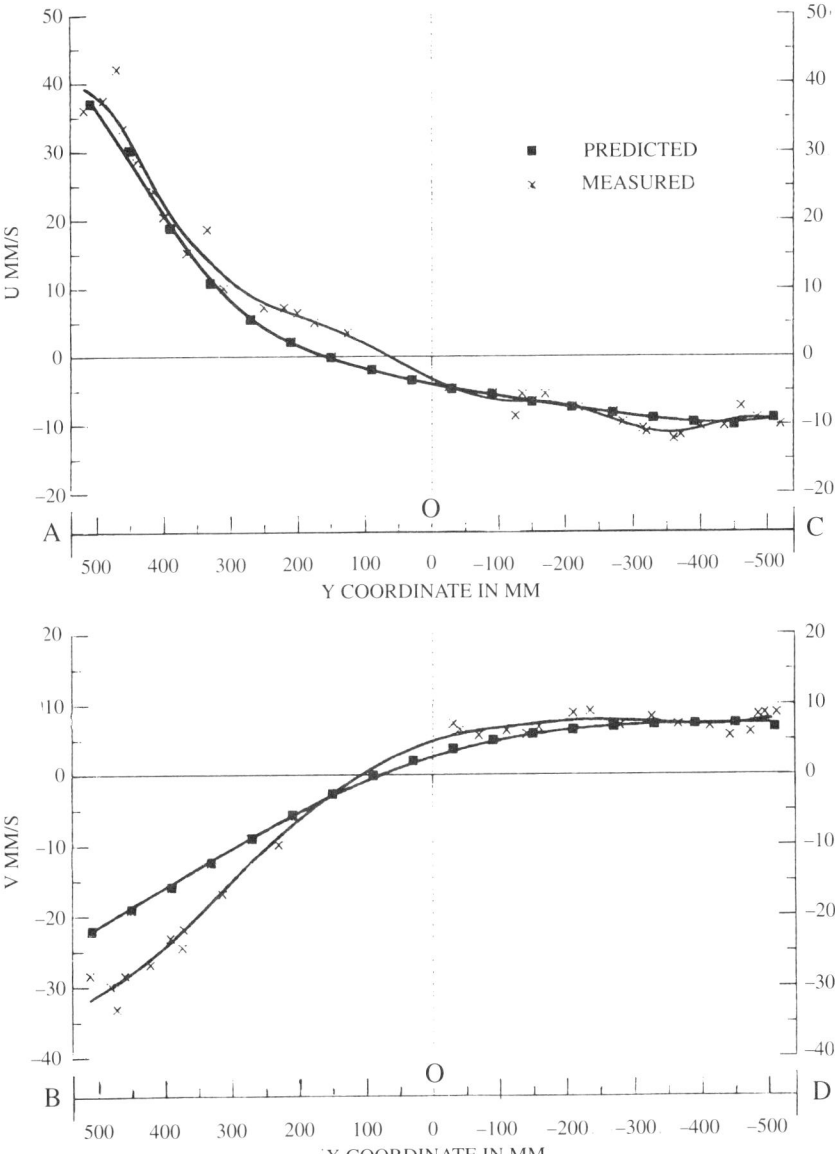

Figure 17.7 Measured and predicted velocity profiles along the harbour axes at mean water level flood tide using the EXQUISITE formulation $k-\epsilon$ turbulence model

harbour and were all in good agreement with the experimentally measured results. This observation has further indicated that, provided there are at least eight grid squares across the eddy, the current difference scheme introduces only limited numerical diffusion as a result of the grid resolution. This observation is in line with similar findings reproduced by Langendoen (1992).

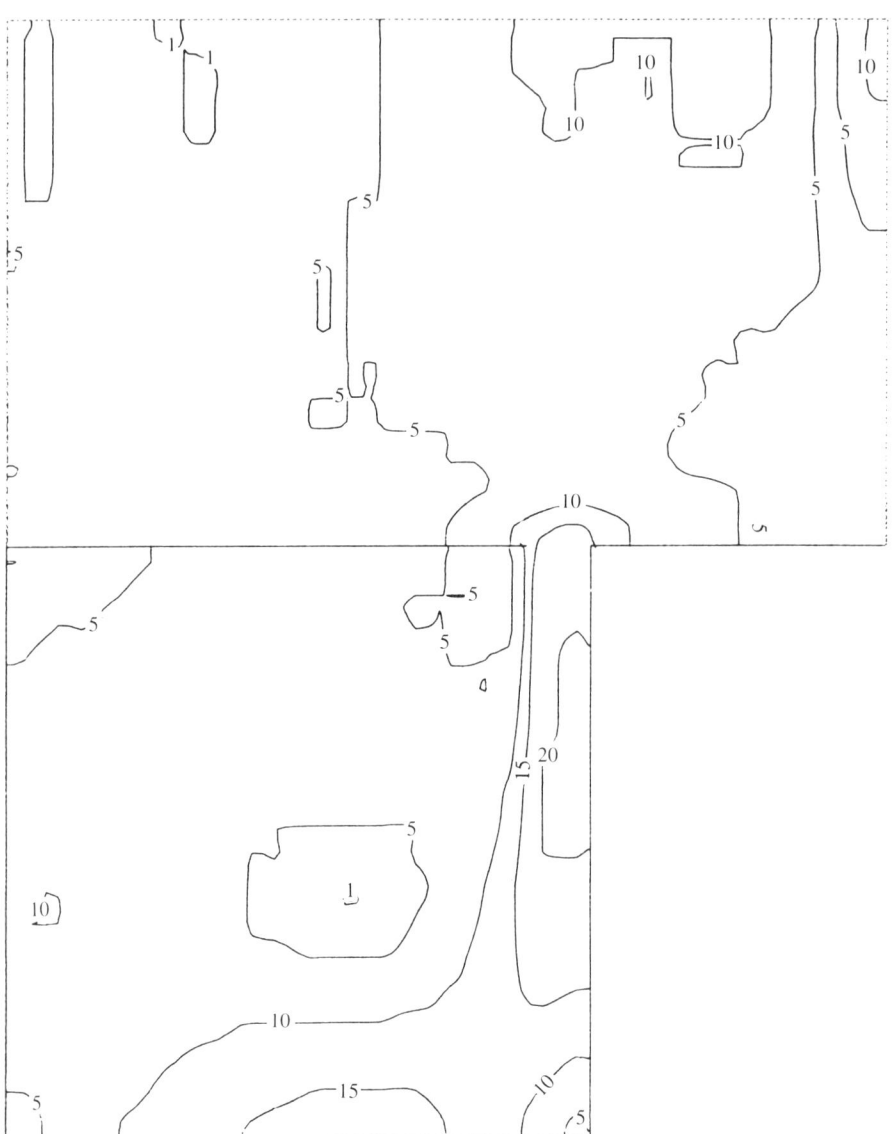

Figure 17.8 Numerically predicted eddy viscosity distribution at mean water level flood tide using the simple mixing length model. Units, mm²/s. Time = 1591 s

(6) In an analytical analysis of the influence of the various terms of the equations of motion on the hydrodynamics of tidal eddies, and comparisons of the numerical model results for simulations of the flow conditions within the equivalent prototype and laboratory model rectangular harbours, questions have been highlighted as to the suitability of using distorted physical models for tidal eddy simulation. The numerical model simulations confirmed the results of a previous analysis, based on non-dimensionalizing the equations of motion and the advective−diffusion processes (Mardapitta-Hadjipandeli and Falconer, 1990), showing that in general phycial models produce (i) an overestimation of momentum transfer by bed-generated turbulence, (ii) an underestimation of the effects of bed friction and (iii) an overestimation of dispersion−diffusion processes.

(7) Further tests on the influence of the various sold wall boundary conditions on the flow characteristics relating to the eddies predicted within the harbours showed that the turbulent shear stresses and the no-slip boundary condition had a marked effect on the tidal eddy structure near the wall, both in the prototype and physical model studies.

(8) In investigating the significance of the various terms of the $k-\epsilon$ equations on the tidal eddy simulations for the rectangular harbours, it was found that the bottom friction production term was more significant in the distorted models than the free shear production and dissipation. Tests also confirmed that great care had to be taken in the treatment of solid wall boundaries in the turbulence model equations.

(9) Finally, a sensitivity analysis of the k and ϵ open boundary values suggested that the assumed zero value did not noticeably affect the predicted values with the basin domain.

17.5 CONCLUSIONS

This chapter outlines a study of the tidal eddy characteristics within scaled laboratory rectangular harbours and, in particular, the numerical prediction of these eddies using the depth integrated $k-\epsilon$ turbulence model. The results between the numerically predicted and laboratory measured velocity distributions have been compared for a range of geometric, bathymetric, kinematic and dynamic boundary conditions, together with comparisons of the numerically predicted eddy viscosity distributions for various conditions. The main findings from these and other comparisons can be summarized as follows:

(1) A simple zero-equation turbulence model, including an empirical representation of free shear turbulence, was found to give velocity field predictions which were similar to those predicted using the depth integrated $k-\epsilon$ turbulence model. However, comparisons between the numerically predicted and laboratory measured profiles across the central axes of the model harbours showed that the $k-\epsilon$ turbulence model resulted in particularly close agreement with the measured data, whereas the simple mixing length model showed some slight discrepancies with the measured data, particularly in the region of the flood tide jet which was more diffused in the simple mixing length model.

(2) The $k-\epsilon$ turbulence model solutions for the tidal eddies in the rectangular harbours gave markedly different eddy viscosity distributions from those predicted using the simple mixing length model. In the vicinity of the harbour entrance the eddy viscosity predicted using the $k-\epsilon$ model was up to three orders of magnitude greater than that obtained from the simple mixing length model and there were particularly marked differences in both sets of results along the harbour walls.

CONTOUR MAP OF EDDY VISCOSITY
(MM **2/S)
TIME = 2298 S

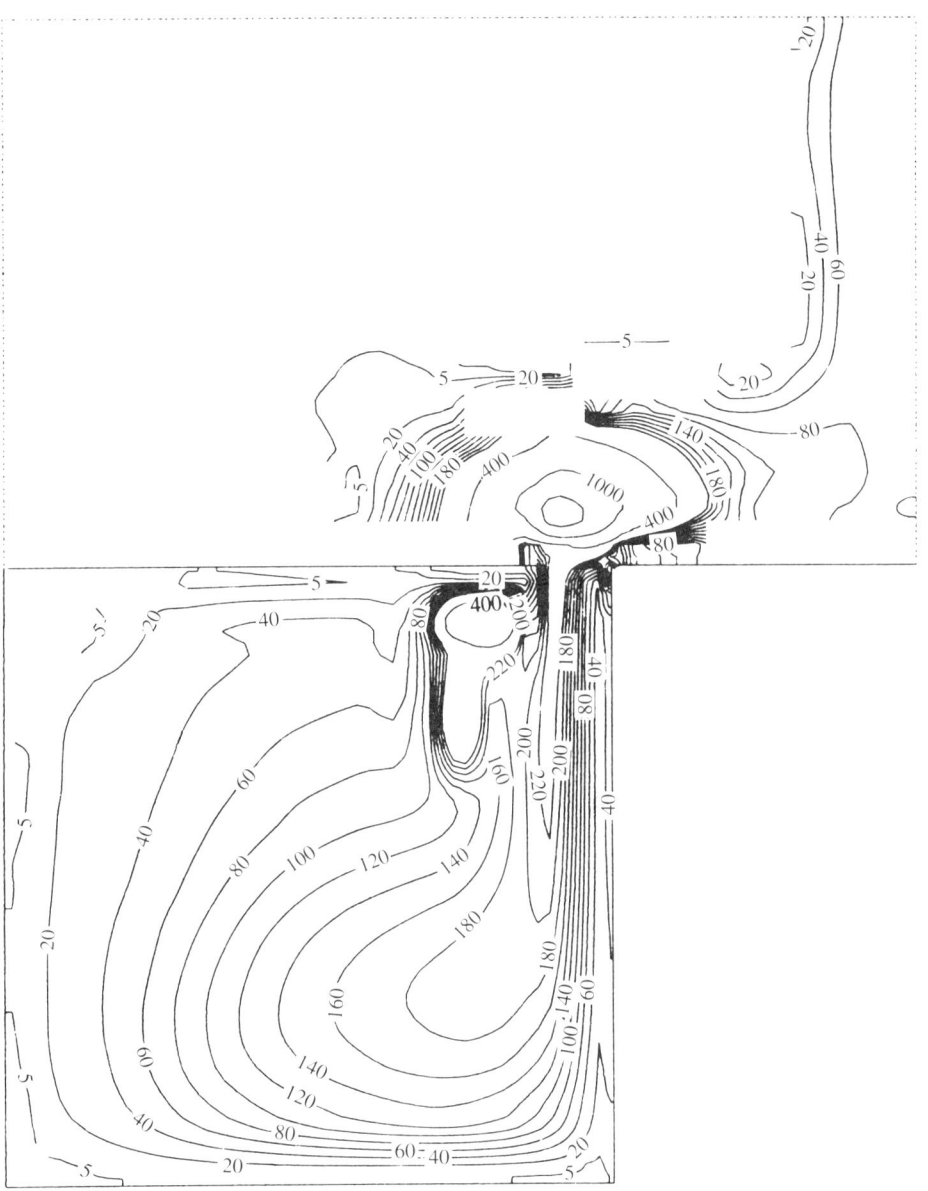

Figure 17.9 Numerically predicted eddy viscosity distribution at mean water level flood tide using the EXQUISITE scheme $k-\epsilon$ turbulence model. Units, mm^2/s. Time = 2298 s

CONTOUR MAP OF EDDY VISCOSITY
(MM **2/S)
TIME = 2298 S

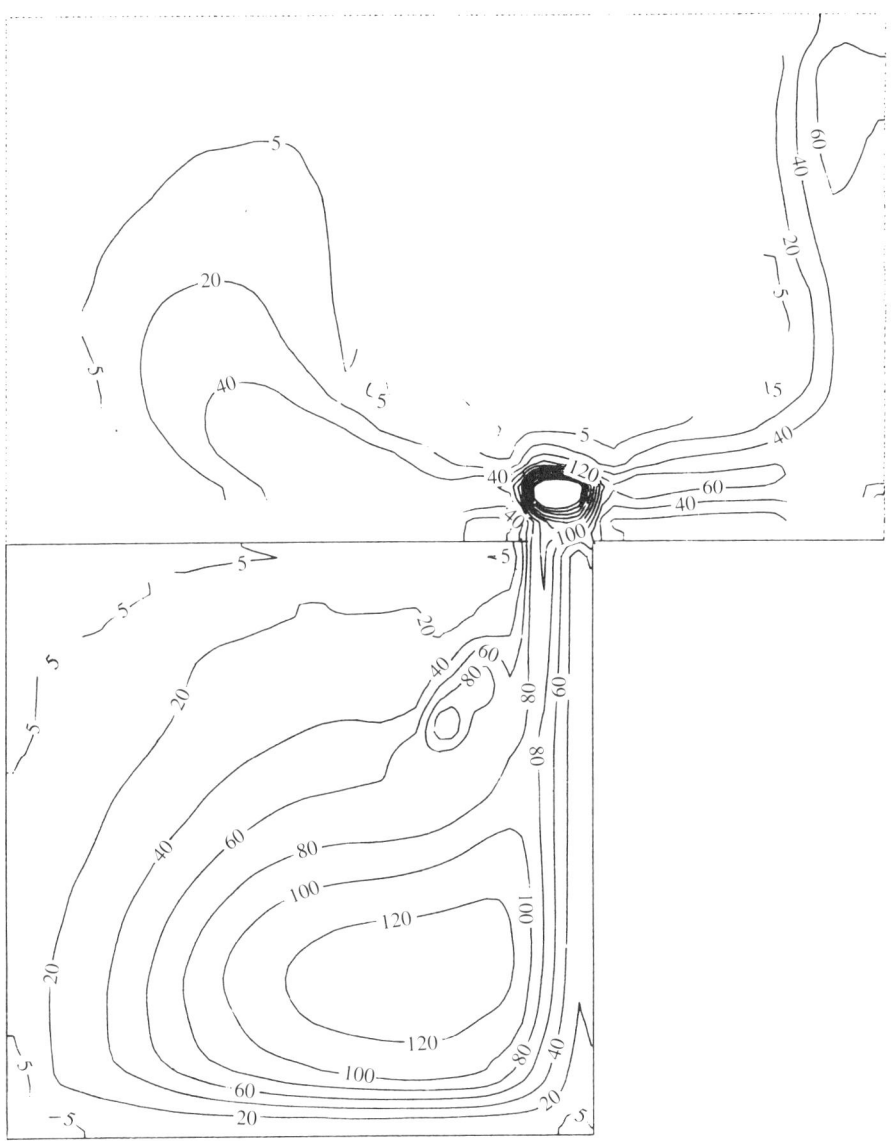

Figure 17.10 Numerically predicted eddy viscosity distribution at mean water level flood tide using the first-order upwind scheme $k-\epsilon$ turbulence model. Units, mm^2/s. Time = 2298 s

(3) A previous theoretical analysis of the depth mean vorticity transport equation indicated that an important mechanism of tidal eddy generation was for bathymetric gradients resulting from the decomposition of the quadratic friction term. This result was confirmed via numerical tests and highlighted the need for caution in using automatic grid generating bathymetric packages, particularly where there are rapid changes occurring in the bed level due to dredged or natural channels.

(4) Both the advective accelerations and the lateral turbulent shear stresses were found to be extremely important in numerically simulating tidal eddies, with numerical model tests by the authors and others confirming that a minimum of eight grid squares is required across an eddy for realistic predictions.

(5) The unsteady $k - \epsilon$ turbulence model was found to be numerically unstable using a second-order implicit difference scheme and, although stable for a first-order upwind, but diffusive, difference scheme, the level of artificial diffusion was unacceptably high and resulted in a markedly reduced prediction of the eddy viscosity magnitudes. Several high order accurate schemes were considered for modelling the $k - \epsilon$ turbulence model, with the EXQUISITE (exponential and quadratic interpolation) scheme giving the closest agreement with the laboratory results.

(6) Finally, the treatment of the wall boundary conditions was found to be critical in modelling the $k - \epsilon$ equations.

The main findings from this study therefore suggest that, although inclusion of the $k - \epsilon$ turbulence model is unlikely to have a significant effect on the numerically predicted velocity fields in enclosed coastal basins, the turbulent diffusion process in the advective−diffusion equation is likely to be much more accurately represented when using this turbulence model. The $k - \epsilon$ model allows for turbulence generated locally, such as at the harbour entrance, to be advected with the mean flow and with the diffusion process being particularly important in water quality modelling.

ACKNOWLEDGMENTS

The study reported herein was principally funded by the Science and Engineering Research Council (grant ref. GR/E/42655) using IBM PS/2 PCs provided by an IBM study grant. The research also benefitted considerably from the provision of a tidal basin by Bradford University and from laboratory data acquired by Professor Ronald E. Nece (University of Washington, USA) and Dr Yu Guoping (Tongji University), who were consecutively on sabbatical leave with the department.

REFERENCES

Abbott, M.B., Larsen, J. and Jianhua Tao (1985) Modelling circulation in depth averaged flows. Part 1: the accumulation of evidence. Part 2: a reconciliation. *J. Hydr. Res.* **23**, 309−326; 397−420.

Elder, J.W. (1959) The dispersion of marked fluid in turbulent shear flow. *J. Fluid Mech.* **5**, 544−560.

Falconer, R.A. (1982) Numerical modelling of flushing in narrow entranced bays and harbours. *Proceedings of the International Symposium on Refined Modelling of Flows, IAHR, Paris, 7−10 September.* IAHR, Delft, 499−508.

Falconer, R.A. (1985) Residual currents in Port Talbot harbour: a mathematical model study. *Proc. Inst. Civ. Engin. Part 2, Res. Theory* **79**, 33−35.

Falconer, R.A. (1988) Numerical modelling of tidal circulation in harbours. *J. Waterway, Port, Coastal Ocean Div. ASCE* **106**, 31−48.

Falconer, R.A. and Chen. Y. (1991) An improved representation of flooding and drying and wind stress effects in a two-dimensional tidal numerical model. *Proc. Inst. Civ. Engin. Part 2, Res. Theory* **9**, 659−678.

Falconer, R.A. and Li, G. (1989) Numberical modelling of tidal eddies in narrow entranced coastal basins and estuaries. *Univ. Bradford Intern. Rep. GR/E/42655*, 1−16.

Falconer, R.A. and Li, G. (1992) Modelling tidal flows in an island's wake using a two-equation turbulence model. *Proc. Inst. Civ. Engin. Wat. Maritime Energy* **96**, 43−53.

Falconer, R.A. and Liu, S.Q. (1988) Modelling solute transport using the QUICK scheme. *J. Environ. Engin. ASCE* **114**, 3−21.

Falconer, R.A. and Mardapitta-Hadjipandeli, L. (1987) Bathymetric and shear stress effects on an island's wake: a computational model study. *Coastal Engin.* **11**, 57−86.

Falconer, R.A. and Yu, G.P. (1991) Effects of depth, bed slope and scaling on tidal currents and exchange in a laboratory model harbour. *Proc. Inst. Civ. Engin. Part 2, Res. Theory* **91**, 561−576.

Falconer, R.A., Wolanski, E. and Mardapitta-Hadjipandeli, L. (1986) Modelling tidal circulation in an island's wake. *J. Waterway, Port, Coastal Ocean Engin. ASCE* **112**, 234−254.

Fischer, H.B. (1973) Longitudinal dispersion and turbulent mixing in open channel flow. *Annu. Rev. Fluid Mech.* **5**, 59−78.

Langendoen, E.J. (1992) Flow patterns and transport of dissolved matter in tidal harbours. *Commun. Hydr. Geotech. Engin. Dep. Civ. Eng. Delft Univ. Techno. Rep. 92-8*, 1−185.

Lean, G.H. and Weare, T.J. (1979) Modelling two-dimensional circulating flow. *J. Hydr. Div. ASCE* **105** (HY1), 17−26.

Leonard, B.P. (1981) A survey of finite difference schemes with upwinding for numerical modelling of the incompressible convective diffusion equation. In: *computational Techniques and Turbulent Flow*. Vol. 2 (Eds C. Taylor and K. Morgan). Pineridge Press, Swansea, 1−35.

Mardapitta-Hadjipandeli, L. and Falconer, R.A. (1990) Some observations on nested modelling of flow and solute transport in rectangular harbours. *Proc. Inst. Civ. Engin. Part 2, Res. Theory* **89**, 15−38.

Nece, R.E. and Falconer, R.A. (1989) Modelling of tide induced depth averaged velocity distributions in a square harbour. In: *Hydraulic and Environmental Modelling of Coastal, Estuarine and River Waters* (Eds R.A. Falconer *et al.*). Gower Technical, Aldershot, 56−66.

Nece, R.E., Smith, H.N. and Richey, E.P. (1980) Tidal circulation and flushing in five western Washington harbours. *Univ. Washington, Seattle, Charles W. Harris Hydr. Lab. Techn. Rep. No. 63*, 1−51.

Rodi, W. (1984) *Turbulence Models and their Application in Hydraulics*. International Association for Hydraulic Research, Delft, 1−104.

Williams, J.W. and Holmes, D.W. (1974) Marker-and-cell Technique, a computer programme for transient stratified flows with free surfaces. *Rep. No. 134*, Hydraulics Research Station, Wallingford, 1−54.

NOTATION

C	De Chezy roughness coefficient
C_f	Bed friction coefficient
C_k	Constant turbulence coefficient
C_ϵ	Constant turbulence coefficient
$C_{1\epsilon}$	Constant turbulence coefficient
$C_{2\epsilon}$	Constant turbulence coefficient
C_μ	Constant turbulence coefficient
f	Coriolis parameter
g	Gravitional acceleration
H	Total depth of flow
k	Turbulent kinetic energy
k_s	Nikuradse sand grain roughness size
Re	Flow Reynolds number

t Time

U, V Depth average velocity components

U_s Depth average fluid speed

U_* Depth average shear velocity

V_{adj} Velocity parallel to a wall

x, y Cartesian coordinate axes in the horizontal plane

β Momentum correction factor for non-uniform vertical velocity profile

Δx Grid size

Δt Time step

ϵ Dissipation rate of turbulent kinetic energy

ζ Water surface elevation above datum

κ Von Karman's constant ($=0.4$)

ν_ℓ Kinematic laminar viscosity

ν_t Kinematic eddy viscosity

ρ Fluid density

σ_k Constant turbulence coefficient

σ_ϵ Constant turbulence coefficient

τ_{bx}, τ_{by} Bed shear stress components in x,y directions

Part III

DEVELOPING THEORIES OF TRANSPORT AND DISPERSION

18 Quantitative Models for Environmental Pollution: a Review

P.C. CHATWIN

School of Mathematics and Statistics, University of Sheffield, Sheffield, UK

and

P.J. SULLIVAN

Department of Applied Mathematics, University of Western Ontario, London, Ontario, Canada

18.1 INTRODUCTION

There are, quite rightly, growing concerns world-wide about the dangers, both actual and potential, of environmental pollution. Public pressure is causing governments and their agencies, sometimes acting multinationally, to support or require the development of regulations to control and monitor the degree of such pollution. Such regulations often involve mathematical modelling. Unfortunately, turbulent diffusion, one of the basic scientific processes that controls environmental pollution, is still not understood, at least not to an extent that allows such models to be reliable or accurate. Therefore, even leaving aside scientific curiosity, there is — in the authors' opinion — an overwhelming practical case for increased public support of basic research in turbulent diffusion. This case is not at present being articulated with sufficient force by scientists nor, partly as a consequence, is it recognized by governments and their agencies. Also, what publicly funded research there is tends to have aims that are dominated by perceived market needs (e.g. to construct a mathematical model of the consequences of the accidental release of chlorine on a factory site with account taken of the buildings *or* of the effect of hot water release from a planned power station on the chemistry and biology in a complex estuarine system). The point of this comment is not that such aims are not worthwhile, but that the present state of knowledge means that they are misguided and/or unrealistic because they cannot be met except in a cosmetic sense.

In a world that is becoming 'greener', the authors believe that an extended discussion of the points made in the previous paragraph is overdue and this chapter is an attempt at that, albeit a partial and selective one. In view of Cath Allen's scientific career and philosophy, it seems particularly appropriate for such a paper to appear in a volume in her memory.

18.2 SCIENTIFIC AND MATHEMATICAL BACKGROUND

The discussion will be restricted to cases of a pollutant whose concentration $\Gamma(x,t)$ at position x and time t is determined by two processes: advection by a turbulent flow with velocity field $\Upsilon(x,t)$, and molecular diffusion with constant diffusivity κ. The equation governing Γ

Mixing and Transport in the Environment. Edited by K.J. Beven, P.C. Chatwin and J.H. Millbank

is therefore

$$\frac{\partial \Gamma}{\partial t} + \Upsilon . \nabla \Gamma = \kappa \nabla^2 \Gamma \tag{18.1}$$

Hence cases where chemical changes occur are not included, but many of the points developed in this chapter also apply in general terms then. However, given this restriction, it is important to note that Equation 18.1 applies whether or not the pollutant is passive. This is so, in particular, when buoyancy forces (arising, for example, when a heavy gas such as chlorine is released into the atmosphere, *or* when hot water used to cool a power station is discharged into a river) have a significant influence on $\Upsilon(x,t)$ through the body force term in the Navier–Stokes equations. In such cases, experimental evidence suggests that it is nearly always reasonable to assume that the flow is incompressible, i.e. that

$$\nabla . \Upsilon = 0 \tag{18.2}$$

It is important first to emphasize that Γ and Υ are the real concentration and velocity fields in any one realization, and that they would be the values measured by transducers in a 'perfect' world without instrument noise and smoothing. In particular Γ and Υ are not average values over time or space except, of course, of the type required by the continuum hypothesis. As, however, all the flows considered are turbulent, Υ, and hence Γ, are random fields. This means that the values of $\Upsilon(x,t)$ and $\Gamma(x,t)$ for any particular x and t are (i) unpredictable and (ii) vary from realization to realization.

The key theme of this chapter will be that, for both theoretical correctness and practical validity, these random fields must be described in terms of their probability structures. Attention will be focused on one probability measure, namely the probability density function (pdf) of $\Gamma(x,t)$. This will be denoted by $p(\theta;x,t)$ and is defined by

$$p(\theta;x,t) = \frac{d}{d\theta}[\text{prob}(\Gamma(x,t) \leq \theta)] \tag{18.3}$$

for all $\theta \geq 0$. The variable θ therefore ranges over all the possible values that $\Gamma(x,t)$ can take. The probability that $\Gamma(x,t) \leq \theta$, appearing on the right-hand side of Equation 18.3, is the proportion of the realizations forming the underlying ensemble for which $\Gamma(x,t) \leq \theta$. The central concept of the underlying ensemble is discussed, for example, in Chatwin and Allen (1985a). Here it suffices to recall that the choice of the ensemble is for the investigator to decide, but that this choice must be made since it defines those conditions which are permitted to vary from realization to realization and, hence, determine $p(\theta;x,t)$ and all other probability measures. For example, when assessing the safety of a fixed storage tank containing dangerous gas, it is likely to be appropriate to consider an ensemble of releases in which the weather and the time of release are allowed to vary from realization to realization, but in which the source geometry and topography are not. If people are not in the vicinity of the storage tank at night, it is possibly more appropriate to consider another ensemble (a subensemble of the first) in which night-time releases (and hence night-time weather conditions) are excluded.

It is now convenient to state some of the more important properties of $p(\theta;x,t)$.

(i)   If $\delta\theta$ is small and positive, $p(\theta;x,t)\delta\theta$ is to first order in $\delta\theta$ the probability that $\theta \leq \Gamma(x,t)$
      $< \theta + \delta\theta$.

(ii) As $p(\theta;x,t)$ is a pdf

$$\int_0^\infty p(\theta;x,t)\mathrm{d}\theta = 1 \qquad (18.4)$$

(iii) In any ensemble there is a maximum possible value, $\Theta(x,t)$ say, of $\Gamma(x,t)$ so that $p(\theta;x,t)$ is identically zero for $\theta > \Theta(x,t)$ and the upper limit in the integral in Equation 18.4 could be replaced by $\Theta(x,t)$. Note that advection acting alone cannot change the concentration of pollutant in any fluid particle so that $\Theta(x,t)$ is strictly less than the source concentration θ_0 (supposed uniform) because of molecular diffusion, the only process that dilutes the pollutant. However, $\Theta(x,t)$ is unknown and is impossible to calculate except for some simple ensembles (Chatwin and Sullivan, 1979). Thus Equation 18.4 is usually a convenient form to use.

(iv) The expected value, or ensemble mean, of any function f of $\Gamma(x,t)$ is defined in the standard way by

$$E\{f[\Gamma(x,t)]\} = \int_0^\infty f(\phi)p(\phi;x,t)\mathrm{d}\phi \qquad (18.5)$$

Three special choices of f in Equation 18.5 are especially relevant.

$$f[\Gamma(x,t)] = \delta[\Gamma(x,t)-\theta] \Rightarrow p(\theta;x,t) = E\{\delta[\Gamma(x,t)-\theta]\} \qquad (18.6a)$$

$$f[\Gamma(x,t)] = \Gamma(x,t) \Rightarrow \mu(x,t) = E\{\Gamma(x,t)\} = \int_0^\infty \phi p(\phi;x,t)\mathrm{d}\phi \qquad (18.6b)$$

$$f[\Gamma(x,t) = [\Gamma(x,t)-\mu(x,t)]^2 \Rightarrow \sigma^2(x,t) = E\{[\Gamma(x,t)-\mu(x,t)]^2\}$$
$$\qquad (18.6c)$$
$$= \int_0^\infty [\phi-\mu(x,t)]^2 p(\phi;x,t)\mathrm{d}\phi = \int_0^\infty \phi^2 p(\phi;x,t)\mathrm{d}\phi - \mu^2(x,t)$$

In Equation 18.6b $\mu(x,t)$ is the (ensemble) mean concentration and in Equation 18.6c $\sigma^2(x,t)$ is the concentration variance, so that $\sigma(x,t)$ is the standard deviation of the concentration. It is also common to use the term rms concentration fluctuation for $\sigma(x,t)$.

(v) Use of Equations 18.5 and 18.6a in Equation 18.1 gives the evolution equation for $p(\theta;x,t)$ in the form

$$\frac{\partial p}{\partial t} + \nabla.E\{\mathbf{\Upsilon}\delta[\Gamma-\theta]\} = \kappa\nabla^2 p - \kappa\frac{\partial^2}{\partial\theta^2}E\{(\nabla\Gamma)^2\delta[\Gamma-\theta]\} \qquad (18.7)$$

where, for brevity, the dependence of p on θ,x,t and of $\mathbf{\Upsilon},\Gamma$ on x,t has not been shown explicitly. This will often be done in the rest of this chapter where there is no risk of confusion — for example, Equations 18.8 and 18.9 below. Use of Equations 18.6b and 18.6c with 18.7 gives equations for $\mu(x,t)$ and $\sigma^2(x,t)$

$$\frac{\partial\mu}{\partial t} + \nabla.E\{\mathbf{\Upsilon}\Gamma\} = \kappa\nabla^2\mu \qquad (18.8)$$

$$\frac{\partial\sigma^2}{\partial t} + \nabla.\mathbf{E}\{\mathbf{\Upsilon}\Gamma^2\} - 2\mu\nabla.E\{\mathbf{\Upsilon}\Gamma\} = \kappa\nabla^2\sigma^2 - 2\kappa E\{(\nabla\Gamma-\nabla\mu)^2\} \quad (18.9)$$

See, for example, Chatwin (1990) for further details of the derivation of Equations 18.7–18.9. It is important to emphasize that Equation 18.7 and hence 18.8 and 18.9, are completely general provided Equations 18.1 and 18.2 apply throughout each realization of the underlying ensemble. The expected values in Equation 18.7 depend on the choice of this ensemble, partly through its influence on the initial and boundary conditions.

18.3 SOME COMMENTS ON TRADITIONAL APPROACHES

Neither the notation nor the methods used in Section 18.2 are yet standard in turbulent diffusion, so some explanation is perhaps in order. The symbols μ and σ^2 have been used because this is universal practice for mean and variance in statistics. More traditional notation in turbulent diffusion would be, for example, C for Γ, $\bar{C}$ for μ, c' for $(\Gamma - \mu)$ and $\overline{(c')^2}$ for σ^2. This notation is inelegant, but there is a more serious criticism. This is that overbars have a precise connotation in statistics, namely to denote arithmetic means (the same connotation is used elsewhere, e.g. for centres of mass). Suppose, for example, that a finite number n of realizations (repetitions) of an experiment are made and that $\Gamma^{(r)}(x,t)$ is the concentration at (x,t) in the rth realization ($r = 1,2,\ldots,n$). Then the standard definition of $\overline{\Gamma}_n(x,t)$ would be

$$\overline{\Gamma}_n(x,t) = \frac{1}{n}\left\{ \sum_{r=1}^{n} \Gamma^{(r)}(x,t) \right\} \tag{18.10}$$

If there is no risk of confusion, $\overline{\Gamma}(x,t)$ can be used instead of $\overline{\Gamma}_n(x,t)$.

The key point is that $\overline{\Gamma}_n(x,t)$ is itself a random variable, unlike $\mu(x,t)$. The mean and variance of $\overline{\Gamma}_n(x,t)$ are $\mu(x,t)$ and $\sigma^2(x,t)/n$ respectively, and $\overline{\Gamma}_n(x,t)$ is an estimate of $\mu(x,t)$. The probability that the difference between $\overline{\Gamma}_n$ and μ exceeds any given quantity tends to 0 as $n \to \infty$ and, in that sense

$$\mu(x,t) = \lim_{n \to \infty} \{\overline{\Gamma}_n(x,t)\} \tag{18.11}$$

A related point about means is more serious. Many textbooks and papers still assert (usually without comment) that the only means with which turbulence and turbulent diffusion deal are time-averages. This assertion is wrong in principle and potentially dangerous if used in practice. Suppose, for example, that $\overline{\Gamma}_T(x,t)$ is defined by

$$\overline{\Gamma}_T(x,t) = \frac{1}{T}\left\{ \int_{t-T/2}^{t+T/2} \Gamma(x,s)ds \right\} \tag{18.12}$$

The first point to note is that, like $\overline{\Gamma}_n$ in Equation 18.10, $\overline{\Gamma}_T$ is a random variable. The statistical properties of $\overline{\Gamma}_T$ obviously depend on T, but, unlike those of $\overline{\Gamma}_n$, cannot be expressed simply in terms of those of Γ; for one thing the values of $\Gamma(x,s)$ in the integrand of Equation 18.12 are not statistically independent. Figure 18.1 illustrates how significantly time-average estimates of one statistical property of $\Gamma(x,t)$ depend on T. It follows that measured values and mathematical models of $\overline{\Gamma}_T$, and regulations expressed in terms of time-average concentrations, that do not specify T are valueless. (Although incidental to the main thrust of this paper, it is worth commenting also that even when T is specified in regulations its choice is not always supported by convincing evidence that $\overline{\Gamma}_T$ for that value of T is the correct measure of, for example, toxic risk.)

There is only one situation where time-averages are useful and that is when the ensemble is statistically steady, i.e. when all statistical properties of Γ, including p (and therefore μ,σ^2), are independent of t. This requires the release of pollutant to be at a steady rate into a flow which is itself statistically steady. The first restriction excludes, for example, sudden releases such as those occurring in accidents, and the second restriction excludes tidal flows. In those rare situations when these conditions are met, the mean concentration $\mu(x)$ can be estimated from the record of one experiment because, by ergodicity, $\overline{\Gamma}_T(x,t)$ then tends to

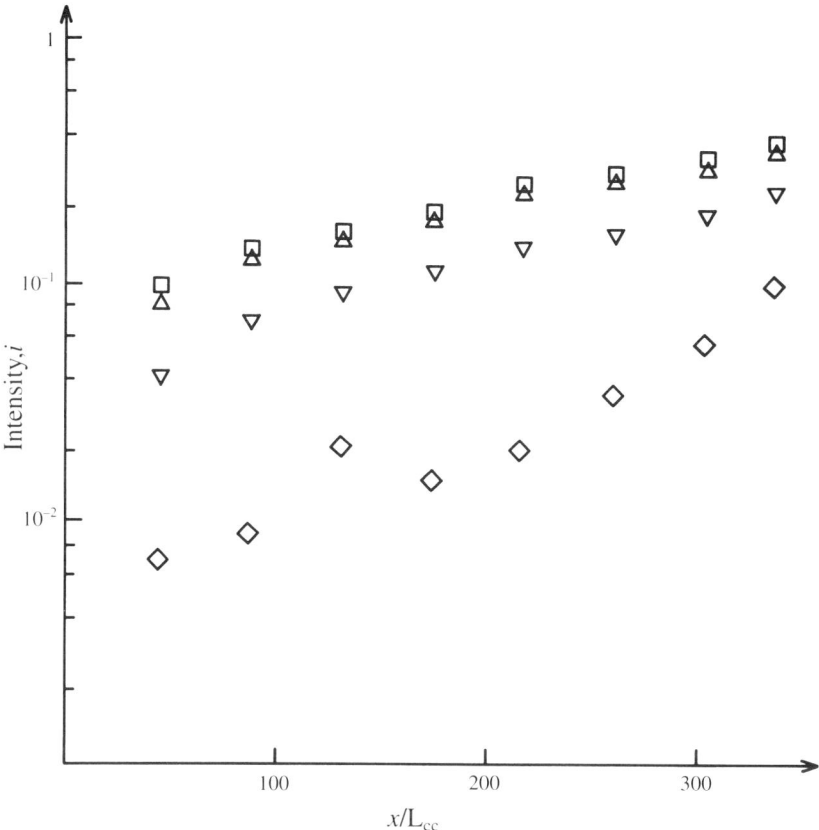

Figure 18.1 Dependence of estimate of intensity i on averaging time T for dispersion from a steady source, where $i = \sigma/\mu$, x is downwind distance and L_{cc} is a constant length scale. A running average of the time series over time T was made at each measuring station, and then μ and σ were estimated from the smoothed record (over identical segments). Symbols denote the following values of T in seconds: □ 0.07; △ 0.7; ▽ 7; and ◇ 70. Taken from Figure 14(b) of Davies (1990)

$\mu(x)$ as $T \to \infty$. When applicable, use of this result avoids costly repetitions. But, most turbulent diffusion problems, including nearly all those affecting the environment, are statistically unsteady. From the point of view of understanding, the use of time-averages is then at best an unnecessary complication and at worst a serious source of error and confusion; see Chatwin and Allen (1985b; 1985c).

Beginning in the early part of this century (Monin and Yaglom, 1971; pp. 606−676), much research on turbulent diffusion has focused on the behaviour of the mean concentration $\mu(x,t)$. The traditional approach followed that of Reynolds for the mean velocity field. If

$$U(x,t) = E\{\Upsilon(x,t)\} \tag{18.13}$$

is the mean velocity field and if $c(x,t)$ and $u(x,t)$ are the concentration and velocity fluctuations defined by

$$c(x,t) = \Gamma(x,t) - \mu(x,t), \ u(x,t) = \Upsilon(x,t) - U(x,t) \tag{18.14}$$

then Equation 18.8 can be rewritten in the more familiar form

$$\frac{\partial \mu}{\partial t} + (\boldsymbol{U}.\nabla)\mu = -\nabla.E\{\boldsymbol{u}c\} + \kappa\nabla^2\mu \tag{18.15}$$

In turbulent diffusion, the term involving $E\{\boldsymbol{u}c\}$ in Equation 18.15 cannot be small. However, it is not expressed in terms of μ and $\boldsymbol{U}$, and there is no known way of doing so that is mathematically correct or even physically sound. This is the simplest manifestation of the closure problem.

For the reason given in the next paragraph, little will be written here about the many different attempts made to circumvent the closure problem for μ. Understandably, these all involve empiricism and, also understandably, none has been fully justified scientifically, either theoretically from the fundamental Equation 18.1 or by satisfactory and comprehensive validation against data. Early methods such as Gaussian plume models or those involving eddy diffusivities are now becoming obsolete. Popular current methods, which extend in practice or principle to other statistical properties, include high-order closures (many available commercially despite the lack of proper validation) and random walks. It is interesting to recall that Cath Allen (Allen, 1982) was one of the first to apply the latter technique in an environmental context by developing the ideas of Sullivan (1971).

The mean concentration μ is the simplest statistical property of $\Gamma(\boldsymbol{x},t)$ and, indisputably, one of the most important. Unfortunately, too much research over too many years has proceeded on the assumption that it is the only one. The unfortunate result of such emphasis is that, until relatively recently, little attention has been paid to the question of whether other statistical properties are important, so that the essential randomness of turbulent diffusion has largely been ignored, i.e. it has been treated almost as a deterministic phenomenon.

18.4 PRACTICAL IMPORTANCE OF A STATISTICAL DESCRIPTION

Over the last 25 years or so, experimental evidence has accelerated to show that differences between the actual concentration $\Gamma(\boldsymbol{x},t)$ and the mean concentration $\mu(\boldsymbol{x},t)$ are not small. The simplest measure of such differences is $\sigma(\boldsymbol{x},t)$ defined in Equation 18.6c, and Figure 18.1 has already shown that experimental estimates of σ/μ, the intensity, are not small. Other examples from the many possible are given in Figures 18.2 and 18.3. It will be noted that values of the estimated intensity range from $\sim 10^{-1}$ to ~ 6, ignoring the even lower values in Figure 18.1 obtained by imposing artificially large averaging times T on the raw data. Some comments on the reasons for this variation will be made later. Figures 18.4 and 18.5 show measured values of $p(\theta;\boldsymbol{x},t)$ which, again, are chosen from many possible examples and confirm that turbulent diffusion is far from being deterministic. For, were the process deterministic, the only value possible for $\Gamma(\boldsymbol{x},t)$ would be $\mu(\boldsymbol{x},t)$ so that $p(\theta;\boldsymbol{x},t)$ would have the form

$$p(\theta;\boldsymbol{x},t) = \delta[\theta - \mu(\boldsymbol{x},t)] \tag{18.16}$$

where δ denotes the Dirac delta function; according to Equation 18.16 the graph of p versus θ would have a single infinite spike — very different from those shown in Figures 18.4 and 18.5.

Given the weight of this evidence, it would be surprising if it was not now recognized that accurate models of all types of environmental impact should take into account the statistical nature of $\Gamma(\boldsymbol{x},t)$. Summaries of practical examples are given by Chatwin (1990), Sullivan (1990), Weil, Sykes and Venkatram (1992) and by many others.

However, because of the traditional, but undue, emphasis on the mean concentration noted

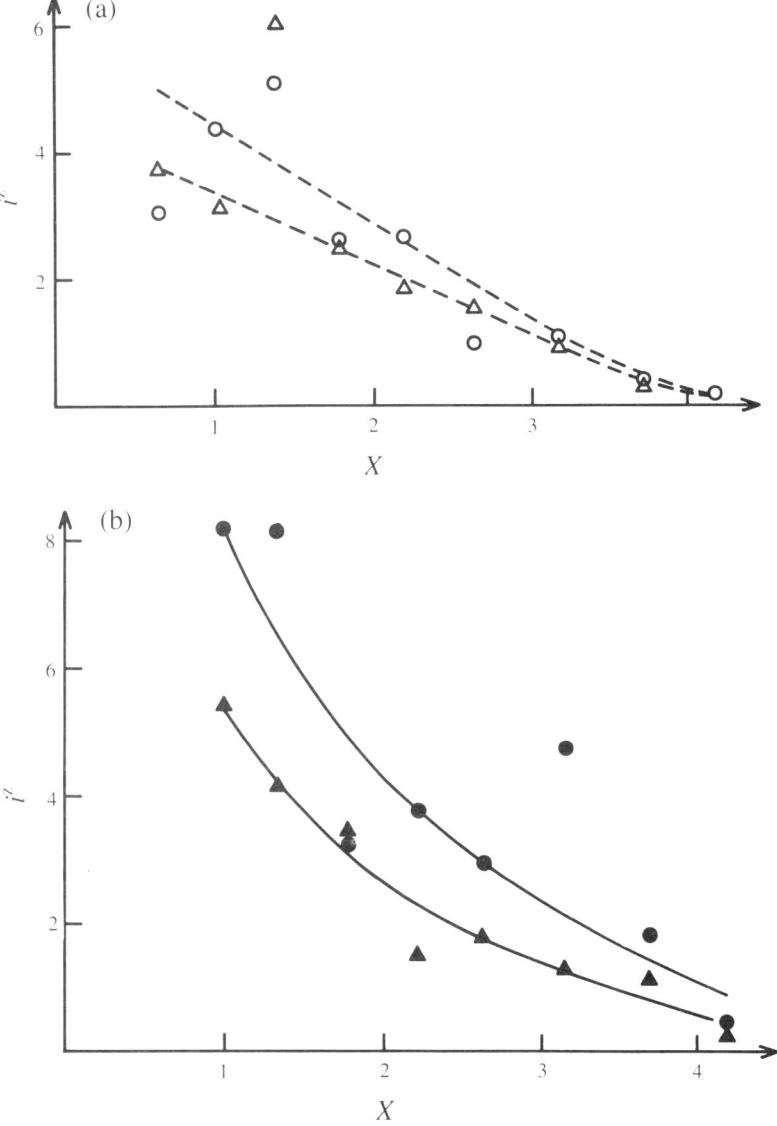

Figure 18.2 Estimates of i^2 (see caption to Figure 18.1) downstream of a steady (a) neutrally buoyant and (b) buoyant source from water tank experiments by Deardorff and Willis (1984). In each diagram the triangles and circles denote averages over the inner and outer parts of the plume, respectively. X is a dimensionless downstream distance measured from the source, as defined in the original paper (Reprinted from Deardoff and Willis, 1984, with kind permission from Pergamon Press Ltd, Headington Hall, Oxford OX3 0BW, UK)

above, most regulatory models now in use are based solely on its predicted behaviour. As turbulent diffusion is essentially a stochastic process (as evidenced by Figures 18.1–18.5), such models cannot (except perhaps for some particular types of environmental risk) ensure the protection that was presumably intended. In fairness, it should be noted that some such models attempt to recognize the existence of the large variations of actual concentrations from the mean by incorporating estimates of 'peak to mean ratio'. However, this measure is

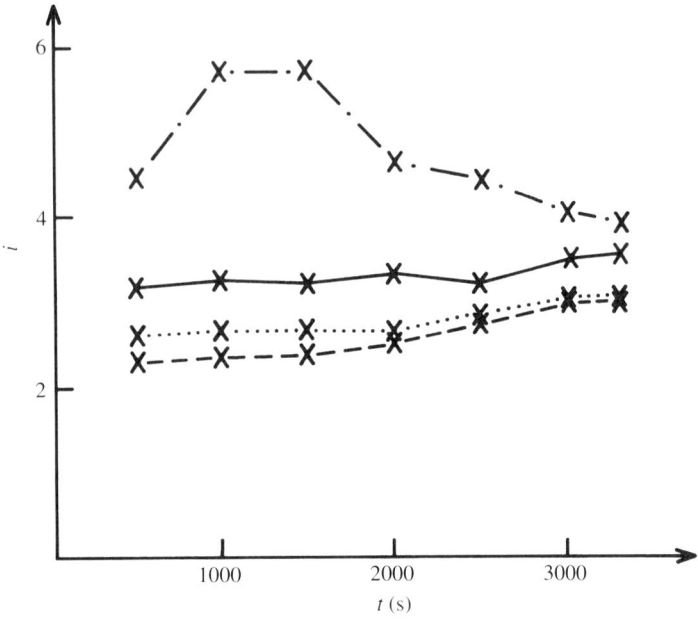

Figure 18.3 Estimates of i (see caption to Figure 18.1) for data from field trials conducted by Dr C.D. Jones under convective daytime conditions. The values of Γ were measured at four different cross-wind positions (channels) at 7.5 m downwind of the steady source and were digitized at 10 Hz. Estimates, denoted by crosses, were made for all four channels using all the data points up to the times shown to assess how statistical convergence of the estimates depended on n in Equation 18.10. The slight trends evident for the larger values of time probably indicate that the atmospheric conditions were only approximately statistically steady. Adapted from Figure C15 of Mole and Chatwin (1990)

scientifically unacceptable. Of course, there is a peak concentration Θ at each x and t as noted above in Section 18.2, but its value is unknown and very difficult, perhaps impossible, to calculate. The most serious objection to the use of the 'peak to mean ratio' is that it is almost impossible to measure Θ because its occurrence as an actual value of Γ is, by definition, an extremely rare event. This is especially true when, as is normally the case (see Figures 18.4 and 18.5), the graph of p versus θ has a long tail for the higher values of θ. It follows that it is extremely unlikely (and this conclusion can be quantified, in principle at least) that measured values of 'peak' concentration are at all close to the real value of Θ and may be orders of magnitude less. These measured values are, of course, themselves random variables, but random variables whose statistical properties depend mainly on quantities such as the length of data record and detailed characteristics of the instrumentation and data analysis. Even if Γ happens to be close to Θ for a particular value of x and t, it will be so only over a spatial region of dimensions of the order of the conduction cut-off length (Chatwin and Sullivan, 1979), typically of order 10^{-4} m in the atmosphere. This dependence of measured peak concentrations on the instruments and length of data record is well illustrated in Figure 15 of Koopman *et al.* (1982).

It is therefore clear that traditional regulatory models, based on the mean concentration μ (even when they predict its behaviour with x and t), are not satisfactory and must, eventually, be replaced by statistical models. Before discussing, in slightly more detail, the elements

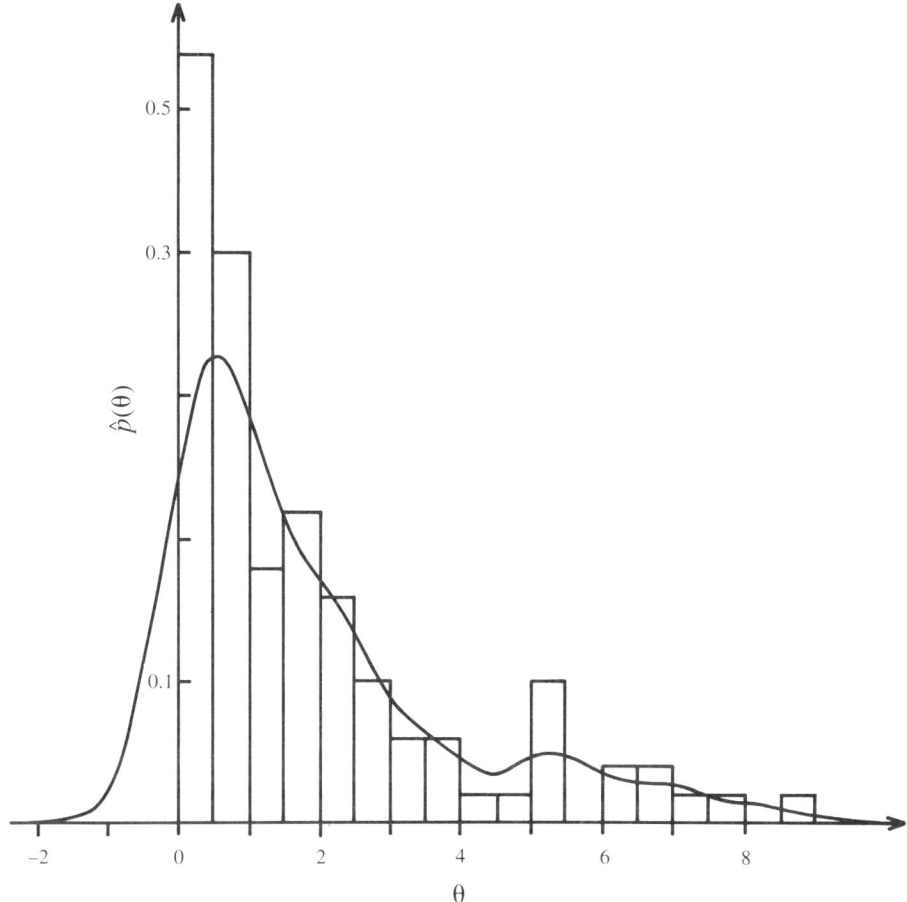

Figure 18.4 Histogram of concentrations from wind tunnel data obtained by Dr D.J. Hall, who carried out 100 repetitions of an experiment in which neutrally buoyant gas was instantaneously released. The curve shows an estimate $\hat{p}(\theta)$ of the pdf by G.W. Goodall found by a non-parametric technique. Notice the bimodality. See Chatwin and Goodall (1991) for further details of the experiments; this estimate is for the ground-level measuring station 0.70 m downwind of the source for a non-dimensional time τ after release of 9.80

of one such statistical model, it is important to reiterate a point made earlier in a different context. Any model for assessing, monitoring or regulating environmental risk, whether this arises through a potential accident or through persistent background concentrations (as is appropriate for most air and water quality control) must deal with the correct measure (or measures) of harm. Unfortunately, except perhaps for the case of local ignitability of a flammable gas, it is difficult to know what this correct measure is or, at least, whether any proposed measure (e.g. 'peak to mean ratio' again!) is supported by adequate evidence. It is understandable that this comment seems likely to be appropriate for most cases of toxicological risk.

One of the standard measures of toxicological risk is in terms of the dosage $D(x,t)$, where

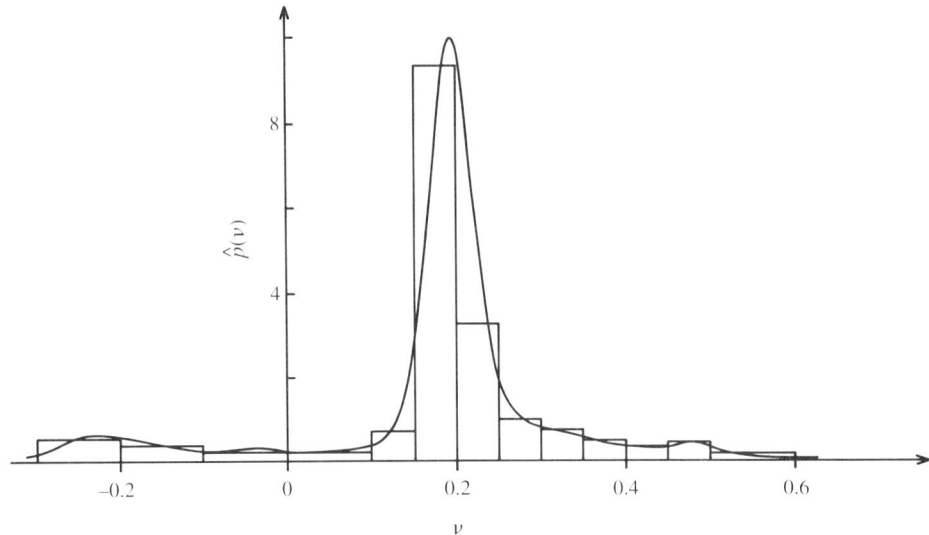

Figure 18.5 This figure has been derived by D.M. Lewis from data obtained by Dr C.D. Jones in field trials in 1991, and indicates some of the problems associated with noise. An attempt has been made to remove noise from the raw data, but about 12% of the data points still indicate negative voltages (concentrations). The curve is an estimate $\hat{p}(v)$ of the voltage pdf obtained using the same method as that in Figure 18.4; voltages are used pending final details on the calibration

$$D(x,t) = \int_{-\infty}^{t} \Gamma(x,s)ds \tag{18.17}$$

In Equation 18.17 the lower limit $t = -\infty$ allows for all possibilities but, in practice, it would be replaced by the time at which the risk began by, for example, the sudden accidental release of a toxic substance. (In view of comments above, it is interesting to note that an earlier 'definition' of dosage ignored any time variation of Γ, whether random or not, and postulated, wrongly of course, the existence of a constant concentration C and a time of exposure T, with D equal to CT.) Like $\Gamma(x,t)$, $D(x,t)$ is a random variable with pdf $p_D(\psi;x,t)$, where (see Equation 18.3)

$$p_D(\psi;x,t) = \frac{\mathrm{d}}{\mathrm{d}\psi}[\mathrm{prob}\{D(x,t) \leq \psi\}] \tag{18.18}$$

Figure 18.6 shows an example of p_D estimated from experimental data.

Provided D is the appropriate measure of toxic risk, and the only one, the mortality rate $M(x,t)$ of a population at x after time t (where M is a proportion so that $0 \leq M \leq 1$) is a random variable with statistical properties dependent on those of D, i.e. on p_D. The probability $P_M(m;x,t)$ that $M(x,t) \leq m$ for any m with $0 \leq m \leq 1$ depends on toxicological response and can be expressed precisely in terms of p_d and the conditional probability distribution of $M(x,t)$ for given dosage. This latter distribution has to be estimated by toxicologists. If p_D and it were to be known, $P_M(m;x,t)$ could be determined and used to take counter-measures. For example, choose a small value m_1 (e.g. $m_1 = 0.001$) and a large value P_1 (e.g. $P_1 = 0.999$) and insist that all the population is located so that

$$P_M(m_1;x,t) > P_1 \tag{18.19}$$

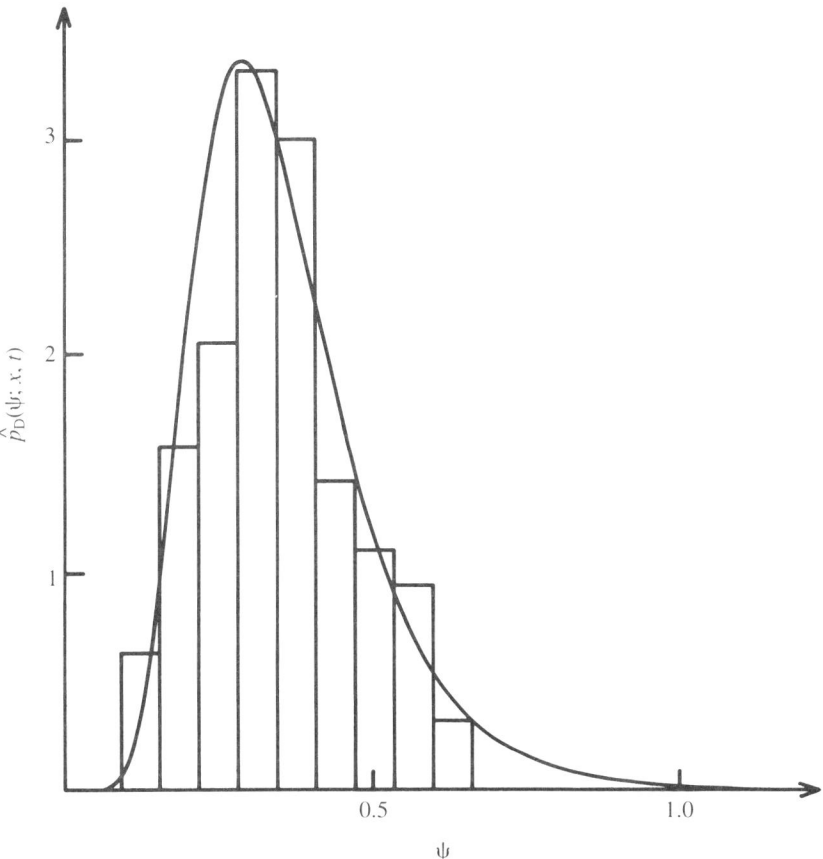

Figure 18.6 Histogram of dosages and estimate of dosage pdf $p_D(\psi;x,t)$ for Dr D.J. Hall's data. Details as in Figure 18.4 except that $\tau = 30.0$. The estimate, by W.L. Sweatman, uses a different technique from that used to obtain Figure 18.4; the curve is the log-normal pdf whose mean and variance are those obtained by maximum likelihood

With the given illustrative values for m_1 and P_1, Equation 18.19 ensures that there is a greater than 99.9% chance that less than 0.1% of the population will die; refer to Griffiths and Megson (1984), Ride (1984) and Griffiths and Harper (1985) for more detailed comments on toxicological risk.

It is clear that the approach to risk assessment just outlined can be applied in other cases, including those when Γ itself is the appropriate measure of the hazard. One such hazard is local ignitability of a mixture of air and a flammable gas such as CH_4. If the stoichiometric limits for the gas−air mixture are θ_1 and θ_2 (where $\theta_1 = 0.05$, $\theta_2 = 0.15$ for CH_4), the probability that the mixture is ignitable at x and t is determined very simply from $p(\theta;x,t)$ by

$$P_I(x,t) = \int_{\theta_1}^{\theta_2} p(\theta;x,t)d\theta \qquad (18.20)$$

Birch, Brown and Dodson (1980) presented convincing experimental evidence that Equation 18.20 is the correct quantification and, conversely, showed that their results could not be correlated with the mean concentration $\mu(x,t)$. In particular, there was no apparent connection between the level surfaces of P_I (i.e. surfaces on which the probability of local ignitability was constant) and those of μ; the two sets of surfaces did not even have similar shapes.

18.5 CURRENT RESEARCH NEEDS

18.5.1 Preamble

We contend that the above arguments show that the scientific case for using probabilistic/statistical methods to assess environmental pollution, rather than traditional deterministic methods, is overwhelming. There are three important factors which tend to inhibit progress towards the much more widespread adoption of such methods in practice.

First of all is the problem of public attitudes, in which it is convenient to include, for example, those of politicians, legislators and lawyers. In many countries, including Canada and the UK, the crucial concept of a probability is not something to which most people are ever exposed, let alone understand. (It could be argued that an exception needs to be made for gambling.) For example, when a public inquiry is set up following a serious incident, it is common to hear politicians use phrases like 'to ensure complete safety' in stating aims, especially in some of the most sensitive areas, such as food and nuclear reactors. It will be clear now that 'complete safety' is unachievable and that, instead, the public must be educated to understand, and to accept, that the only type of objective that can be achieved, in principle at least, is a large probability of harm less than a specified level occurring. For example, for the case of a toxic gas in which dosage is the appropriate measure of harm, the constants m_1, level of harm, and $(1-P_1)$, the probability of this level of harm not occurring, in Equation 18.19 are legitimate targets. The precise numerical values of these two measures have to be chosen by the public, bearing in mind cost — the smaller the probability and the degree of harm, the higher will be the cost.

The second factor has already been mentioned; it is the rapid acceleration in public pressure for regulations on environmental safety due to increased perceptions of its importance. As a result of this pressure many models based on invalid premises have been developed and sold, and are now being used. In detail, such models always involve assumptions, particularly for 'resolving' the closure problem, that are, at best, inadequately validated against data and, at worst, obviously erroneous. But, additionally and more fundamentally, they postulate a deterministic process that has no connection with reality. The existence of such models (and the willingness of some people to produce new ones) reduces the pressure on users, and potential users, to demand better ones. However (Hanna *et al.* 1992), there are a few encouraging signs of more realistic perceptions.

Finally, it is important to note that producing good statistical models of turbulent diffusion is difficult. On the one hand the underlying science is not adequately understood in the sense that the methods proposed for resolving the closure problem, inevitably present in any equation for a statistical property, such as Equations 18.7–18.9, cannot be asserted to have known limits of accuracy except, sometimes, in situations for which they have been directly validated against data. Too often, the results obtained in one set of circumstances are extrapolated without justification to other (usually more complicated) situations. For example, it has sometimes been claimed (or at least implied) that values of σ/μ measured in experiments on passive tracers on flat ground in neutral atmospheres with a steady continuous release can be applied to unsteady (including instantaneous) releases of heavy gases under all atmospheric conditions and on all terrains. The inevitable conclusion is that there must be more — much more — fundamental research. On the other hand, even if the fundamental scientific problems can be solved, the task of incorporating the new knowledge into practical models will be difficult. Particularly challenging is the need to have models that can respond to emergencies both quickly and with acceptable accuracy. Although the basic scientific problems will, in principle, be solved eventually (within one hundred years?) when computers are powerful enough for the statistical properties of Γ to be estimated from sufficiently many direct numerical solutions of Equation 18.1 and the Navier−Stokes equations (thereby avoiding the closure problem), it is almost certain that this technique will be far too slow for it to be used in, for example, a factory site in which dangerous gases are produced or stored.

18.5.2 Practical statistical models of the future

In view of the dilemma posed at the end of the last paragraph, it is proper to study the structure of statistical models that can be used in practice to assess actual or potential hazards. Such models ought to be simple enough to be applicable and useful, but not so simple that they are grossly inaccurate or misleading. Also, because of the large range of different hazards that exist (toxicity, flammability, malodour, etc.), each with its own measure, or measures, of harm (see Section 18.4), there is a need for many different models to be developed.

Given the unlikelihood, at least in the short to medium term, of being able to produce practical models (i.e. models yielding sufficiently accurate results quickly enough in real time) based on approximate solutions of equations such as 18.7−18.9 for $p(\theta;x,t)$, $\mu(x,t)$ and $\sigma^2(x,t)$, respectively, there are two promising but incipient lines of research. For illustrative purposes these will be discussed in terms of hazards for which $\Gamma(x,t)$ and $p(\theta;x,t)$ are appropriate measures; it will be clear that both approaches can be applied in other cases such as that when dosage, defined in Equation 18.7, is the relevant measure.

(1) If it is believed to be too ambitious to model p itself, the sensible goal is to predict only the magnitudes of μ and σ. As Γ cannot be negative, p cannot be the pdf of a normal distribution; see Figures 18.4 and 18.5. However, it is unlikely that the real distribution is so far from a normal distribution that probabilities obtained with the normal distribution with the same μ and σ will be dangerously different from their true values. This type of approximation can be especially useful for regulatory models expressed in terms of weekly, monthly or even annual averages, when use of the central limit theorem obviates the need for the precise form of $p(\theta;x,t)$. The principal goals in this type of research are therefore how μ and σ vary with x and t, and such factors as source geometry, flow characteristics and terrain. In some situations it may be more relevant or practical to use the same broad approach, but to replace μ and/or σ by other robust parameters such as the median concentration or cloud-average values (Sullivan, 1990; Chatwin and Sullivan, 1990b).
(2) An alternative and more ambitious line of work is to seek to model $p(\theta;x,t)$ itself in addition to μ and σ. Although successful practical use of Equation 18.7 is unlikely at the present time, the ready availability of data is now allowing the dependence of p on θ to be estimated experimentally. Figures 18.4, 18.5 and (for dosage) 18.6 were obtained in this way. The aim is to find simple two- or three-parameter model pdfs (e.g. log-normal, truncated normal, gamma, beta) that describe the data with adequate accuracy and then, on the basis of simple physical arguments and further validation, to predict how the form of the pdf varies with x,t and the other key factors. If this approach is to be successful,

the variation of the form of p will be given completely by the corresponding variations in the parameters. Usually these are chosen to be μ, σ and perhaps one other, often taken to be the intermittency factor. (The intermittency factor is conventionally defined to be the probability that the concentration is greater than zero. In practice this is an extremely useful concept but, from a scientific viewpoint, the definition is not viable (Chatwin and Sullivan, 1989) as the result of applying it depends significantly on extraneous factors such as instrument characteristics. Ride (1984) discusses an important use of the idea of the intermittency factor.)

The remainder of this section summarizes some of the more specific and more important research problems that crucially affect the successful achievement of the two programmes. It is convenient to consider these in two groups: those that largely relate to data and experiments, and those that are predominantly theoretical.

18.5.3 Research problems involving data

The mathematical problems associated with turbulent diffusion are so severe that experimental data must be used in any serious research. It is convenient to distinguish two separate roles for data.

The first is the use of data to test and, hopefully, validate existing models, however obtained. In view of the fact that recent studies (Hanna, Strimaitis and Change, 1991; Britter, 1991; Hanna, 1993; see also Hanna et al., 1992) have analysed the methodology of this process in some detail (in respect of hazardous gas dispersion models), little need be written here. However, three points that emerge from these studies are worth noting. The meaning of the term 'validation' is by no means clear-cut; consequently validation exercises require precise preparation and much subtlety. In general, the performance of a model is 'not related to its cost or complexity' and 'it is very difficult to demonstrate improved model performance as enhancements in model physics are added' (Hanna, 1993). Finally, the work of Hanna and co-workers uses certain measures to compare model predictions with data, and it would be worthwhile to investigate theoretically the statistical significance of the numerical values of these measures and, more generally, to assess similarly other measures of model performance.

The other role for data is in the development of models and, in particular, in research programmes like (1) and (2) above. It is first of all important to point out that the task of data analysis has changed its nature in the last 10 or 15 years principally because of the data acquisition systems that are now routinely used. Consequently, the amount of data available from nearly all experimental investigations has increased by orders of magnitude but, unfortunately, a lack of resources available to most analysts has caused many potentially valuable data sets to be inadequately analysed, or not analysed at all. Sadly, it is not always clear that this increase in the amount of data has been accompanied by a comparable rise in its quality; the studies of validation mentioned above all emphasize the magnitudes of data uncertainties. The authors believe that it is now essential that all experiments on environmental pollution be planned in conjunction with the end-users of the data and that their costs be realistically assessed. Only in this way will these mismatches be stopped.

In terms of the statistical models advocated in this review, particular attention needs to be given in advance to assessing the magnitudes of the standard deviations of experimental estimates of statistical properties such as μ, σ and p. This will lead to a choice, and costing, of the number of repetitions to be carried out or (in cases of statistically steady dispersion) the length of the experiment. Also, in view of the high cost of field trials involving many repetitions, it is inevitable that there will be even more use of wind and water tunnels in the future so that there must be more research into modelling complex environmental flows in the laboratory.

There is now a lot of evidence that the measured properties of $\Gamma(x,t)$ can be significantly dependent on the instruments used. In particular (see Figure 18.1), the magnitudes of the real concentration fluctuations, as measured, for example, by $\sigma(x,t)$, may be substantially underestimated; see, for example, Figures 5 and 7 of Chatwin and Goodall (1991). The principal cause of this is instrument smoothing which, though largely inevitable because of the minute length and time-scales present in the structure of dispersing pollutants, has arguably received inadequate attention in the past. There is a strong case for using at least two instrumentation systems of different types and characteristics in experiments so that the measurements of Γ can be cross-validated directly.

One consequence of instrument smoothing is to underestimate the magnitudes of the highest values of Γ and their frequency of occurrence; this may clearly have serious consequences of hazard assessment. In other instances, particularly those relating to overall air and water quality, the biggest contribution to dosage comes from the lowest values of Γ because they occur so often. It is important in such circumstances to be able to estimate the behaviour of $p(\theta;x,t)$ as $\theta \rightarrow 0+$. Unfortunately, and again inevitably, such estimation involves other important instrument effects, especially noise and, often, baseline drift. In the past such effects were too often dismissed in a cavalier fasion by applying arbitrary thresholds: fortunately, this practice appears to be dying rapidly. The proper treatment of noise requires its separate measurement and consequent deconvolution from the output signal; methods are available for this. The effects considered in this paragraph and the previous one have been considered by Mole (1990a; 1990b), but much remains to be done.

18.5.4 Research problems involving theory

Some theoretical problems have recently been discussed by us (Chatwin and Sullivan, 1993). These include: (i) extending the relationship between μ and σ proposed by Chatwin and Sullivan (1990a) to more general situations such as those near a steady continuous source; (ii) use of the representation of $p(\theta;x,t)$ in terms of the new definition of the intermittency factor (Chatwin and Sullivan, 1989); and (iii) investigation of the behaviour of $p(\theta;x,t)$ as $\theta \rightarrow 0+$ using both Equation 18.7 and the representation mentioned in (ii). Further details of these problems are contained in the cited references and will not be repeated here.

Two other classes of theoretical problem also merit brief mention.

One of these is the influence of source size and geometry on the statistical properties of Γ. It has been known for some time (Chatwin and Sullivan, 1979) that source size has a potentially profound effect on the size of σ (the smaller the source the larger is σ), but many intriguing questions remain. For example, it would be valuable to quantify the effects of molecular diffusion and source characteristics on σ and p, and to obtain a stronger theoretical basis for the robust correlation between μ and σ mentioned in (i) above. Some applications require models for multiple sources and work is in progress to extend the results obtained for single sources such as those cited above. This problem also has fundamental interest and data have been presented in papers by Warhaft (e.g. Warhaft, 1984) and others.

The second class of theoretical problem is essentially statistical rather than physical. Some such problems have been mentioned earlier, especially in the subsection on research problems involving data. Another interesting question is related to programme (1) discussed earlier, and that is to apply recent statistical research on extreme values, involving the Pareto distribution, to practical problems in which the appropriate measure of harm involves high values of concentration; see, for example, Smith (1989). Finally, there are interesting questions associated with the efficient estimation of $p(\theta;x,t)$ from data. Among these are the length of record needed to estimate statistical properties (such as μ and σ) to within acceptably small standard errors, the estimation of noise, the deconvolution of noise, and the criteria governing the use of parametric or non-parametric methods. Such problems are discussed by Mole and Jones (in press), where further references are given.

ACKNOWLEDGEMENTS

It is a pleasure to record our thanks to Gerald Goodall, David Lewis and Winston Sweatman for allowing us to show examples of their results in Figures 18.4, 18.5 and 18.6. We also thank Nils Mole. Our work is receiving financial support from the Natural Science and Engineering Research Council of Canada, from the Commission of the European Communities and from the Ministry of Defence of the UK.

REFERENCES

Allen, C.M. (1982) Numerical simulation of contaminant dispersion in estuary flows. *Proc. R. Soc. Lond. A* **381**, 179–194.

Birch, A.D., Brown, D.R. and Dodson, M.G. (1980) Ignition probabilities in turbulent mixing flows. *Rep. No. MRS E 374*, Midlands Research Station, British Gas, Solihull.

Britter, R.E. (1991) The evaluation of technical models used for major-accident hazard installations. *Report to Commission of the European Communities (DG XII)*.

Chatwin, P.C. (1990) Statistical methods for assessing hazards due to dispersing gases. *Environmetrics* **1**, 143–162.

Chatwin, P.C. and Allen, C.M. (1985a) Mathematical models of dispersion in rivers and estuaries. *Annu. Rev. Fluid Mech.* **17**, 119–149.

Chatwin, P.C. and Allen, C.M. (1985b) A note on time averages in turbulence with reference to geophysical applications. *Tellus* **37B**, 46–49.

Chatwin, P..C. and Allen, C.M. (1985c) Reply. *Tellus* **37B**, 315–316.

Chatwin, P.C. and Goodall, G.W. (1991) Research on continuous and instantaneous heavy gas clouds. *Report to Commission of the European Communities: Final Report on CEC Contract No. EV4T. 0025. UK(H)*. [Also Brunel University Department of Mathematics and Statistics Tech. Rep. TR/02/91.]

Chatwin, P.C. and Sullivan, P.J. (1979) The relative diffusion of a cloud of passive contaminant in incompressible turbulent flow. *J. Fluid Mech.* **91**, 337–355.

Chatwin, P.C. and Sullivan, P.J. (1989) The intermittency factor of scalars in turbulence. *Phys. Fluids A* **1**, 761–763.

Chatwin, P.C. and Sullivan, P.J. (1990a) A simple and unifying physical interpretation of scalar fluctuation measurements from many turbulent shear flows. *J. Fluid Mech.* **212**, 533–556.

Chatwin, P.C. and Sullivan, P.J. (1990b) Cloud-average concentration statistics. *Maths. & Computers in Simulation* **32**, 49–57.

Chatwin, P.C. and Sullivan, P.J. (1993) The structure and magnitude of concentration fluctuations. *Boundary-Layer Meteorol.* **62**, 269–280.

Davies, J.K.W. (1990) Research on continuous and instantaneous heavy gas clouds. *Report to Commission of the European Communities: Final Report on CEC Contract No. EV4T. 0005. UK(H)*.

Deardorff, J.W. and Willis, G.D. (1984) Groundlevel concentration fluctuations from a buoyant and a non-buoyant source within a laboratory convectively mixed layer. *Atmos. Environ.* **18**, 1297–1309.

Griffiths, R.F. and Harper, A.S. (1985) A speculation on the importance of concentration fluctuations in the estimation of toxic response to irritant gases. *J. Hazardous Mater.* **11**, 369–372.

Griffiths, R.F. and Megson, L.C. (1984) The effect of uncertainties in human toxic response on hazard range estimation for ammonia and chlorine. *Atmos. Environ.* **18**, 1195–1206.

Hanna, S.R. (1993) Uncertainties in air quality model predictions. *Boundary-Layer Meteorol.* **62**, 3–20.

Hanna, S.R., Chatwin, P.C., Chikhliwala, E., Londergan, R.J., Spicer, T.O. and Weil, J.C. (1992) Results from the model evaluation panel. *Plant/Operations Progr.* **11**, 2–5.

Hanna, S.R., Strimaitis, D.G. and Chang, J.C. (1991) Evaluation of 14 hazardous gas dispersion models with ammonia and hydrogen fluoride field data. *J. Hazardous Mater.* **26**, 127–158.

Koopman, R.P., Cederwall, R.T., Ermak, D.L., Goldwire, H.C. Jr., Hogan, W.J. McClure, J.W., McRae, T.G., Morgan, D.L., Rodean, H.C. and Shinn, J.H. (1982) Analysis of Burro series 40m^3 LNG spill experiments *J. Hazardous Mater.* **6**, 43–83.

Mole, N. (1990a) A model of instrument smoothing and thresholding in measurements of turbulent dispersion. *Atmos. Environ.* **24A**, 1313–1323.

Mole, N. (1990b) Some intersections between turbulent dispersion and statistics. *Environmetrics* **1**, 179–194.

Mole, N. and Chatwin, P.C. (1990) Fluctuations in atmospheric contaminants. *Final Report to Chemical Defence Establishment under Agreement No. 2006/71.* [Also Brunel University Department of Mathematics and Statistics Tech. Rep. TR/16/90.]

Mole, N. and Jones, C.D. Concentration fluctuation data from dispersion experiments carried out in stable and unstable conditions. *Boundary-Layer Meteorol.*, in press.

Monin, A.S. and Yaglom, A.M. (1971) *Statistical Fluid Mechanics: Mechanics of Turbulence.* Vol. 1 (Ed. J.L. Lumley). MIT Press, Cambridge, Mass.

Ride, D.J. (1984) An assessment of the effects of fluctuations on the severity of poisoning by toxic vapours. *J. Hazardous Mater.* **9**, 235–240.

Smith, R.L. (1989) Extreme value analysis of environmental time series: an application to trend detection in ground-level ozone. *Stat. Sci.* **4**, 367–393.

Sullivan, P.J. (1971) Longitudinal dispersion within a two-dimensional turbulent shear flow. *J. Fluid Mech.* **49**, 551–576.

Sullivan, P.J. (1990) Physical modeling of contaminant diffusion in environmental flows. *Environmetrics* **1**, 163–177.

Warhaft, Z. (1984) The interference of thermal fields from line sources in grid turbulence. *J. Fluid Mech.* **144**, 363–387.

Weil, J.C., Sykes, R.I. and Venkatram, A. (1992) Evaluating air quality models: review and outlook. *J. Appl. Meteorol.* **31**, 1121–1145.

19 Hydrodynamic Derivation of an Aggregated Dead Zone Model for Contaminant Dispersion in Estuary Flows

R. SMITH

Mathematical Sciences, Loughborough University, Loughborough, UK

19.1 INTRODUCTION

Scientific research is often portrayed as being a battleground between rival ideas or even between rival individuals. In the context of contaminant dispersion in bounded flows the high ground has long been occupied by Fickian diffusion (or Taylor dispersion) models based on hydrodynamics (Taylor, 1953). Threats to that dominance have come from random walk simulations (Allen, 1982)

> Principal advantages of such a method are that there are no restrictions due to conditions (i)–(iv) above; the technique models the physics of the situation much more realistically than the conventional method, and results show that it is very versatile: more so than finite difference models for example.

and from aggregated dead zone (ADZ) models (Beer and Young, 1983)

> There is little doubt that the models based upon identification and parameter estimation of the dead zone model of Eq.4 are bettter able to cope with real stream data than those based upon the Fickian diffusion equation.

The range of scientific approaches represented in this commenorative volume are testimony to Cath Allen's much more collaborative view of scientific research — as a shared quest for truth and understanding about nature. Different viewpoints give different perspectives which are most useful when they overlap. For steady, spatially uniform flows the hydrodynamic basis of ADZ or two-layer models is long established (Thacker, 1976). More recently, I have given a hydrodynamic derivation of a two-layer model for contaminant dispersion in river flows (Smith, 1987), where the flow is steady but non-uniform. Here a similar derivation is given for the dual problem of contaminant dispersion in estuary flows which are uniform but unsteady. Appropriately, it is Allen's (1982) numerical results that are used to test the efficacy of the resulting two-layer model.

Mixing and Transport in the Environment. Edited by K.J. Beven, P.C. Chatwin and J.H. Millbank
© 1994 John Wiley & Sons Ltd

19.2 VERTICAL DIFFUSION AND HORIZONTAL ADVECTION

The hydrodynamic equations governing contaminant dispersion are complicated. If that were not so, then there would be no need for simplified models and no controversy. The focus of concern for the present paper is time-dependent flows. We shall therefore make the following idealizations: the water depth, h, is taken to be constant; the flow, $u(z,t)$, is unidirectional in the x-direction; and the estuary is either so narrow or so wide that any u-dependence can be factored out (and therefore be ignored). The systematic effect of turbulence will be modelled in terms of an effective vertical eddy diffusivity $\kappa(z,t)$.

For a sudden discharge at position x_0 and time t_0 the subsequent evolution of the concentration $c(x, x, t)$ is assumed to satisfy the field equation and boundary conditions

$$\partial_t c + u \partial_x c - \partial_z(\kappa \partial_z c) = 0 \tag{19.1a}$$

with

$$\kappa \partial_z c = 0 \quad \text{on} \quad z = -h, 0 \tag{19.1b}$$

and

$$c = q(z)\,\delta(x - x_0) \quad \text{at} \quad t = t_0 \tag{19.1c}$$

The function $q(z)$ describes the vertical distribution of the discharge strength and the δ function shows that horizontally the discharge is confined to the single position $x = x_0$. The boundary conditions of Equation 19.1b correspond to there being no flux of contaminant across the bed ($z = -h$) or across the free surface ($z = 0$).

The objective of the Taylor dispersion, random walk, or ADZ models is to obtain simple approximations to the concentration $c(x, z, t)$, particularly at large enough times after discharge for the contaminant to have become vertically well mixed and carried back and forth at the vertically averaged velocity

$$\bar{u} = \frac{1}{h}\int_{-h}^{0} u\,dz \tag{19.2}$$

19.3 MOMENTS AND CENTROID DISPLACEMENT FUNCTIONS

A full solution of Equations 19.1a–19.1c would be formidably complicated. Aris (1956) showed that the x-derivatives could be eliminated if only the first few spatial moments were sought

$$c^{(p)}(z, t) = \int_{-\infty}^{\infty}\left(x - x_0 - \int_{t_0}^{t}\bar{u}(t')dt'\right)^p c(x, z, t)\,dt \tag{19.3}$$

The single equation 19.1a is replaced by the hierarchy of moment equations

$$\partial_t c^{(p)} - \partial_z(\kappa \partial_z c^{(p)}) = p(u - \bar{u})c^{(p-1)} \tag{19.4a}$$

with

$$\kappa \partial_z c^{(p)} = 0 \quad \text{on} \quad z = 0, -h, \tag{19.4b}$$

$$c^{(0)} = q(z), \quad 0 = c^{(1)} = c^{(2)} = \ldots \quad \text{at } t = t_0 \tag{19.4c,d}$$

Even these equations are too difficult to solve exactly unless the diffusivity κ and the velocity u have separable t and z-dependence

$$\kappa = \kappa_1(t)\,\kappa_2(z), \quad u = u_1(t)u_2(z) \tag{19.5a,b}$$

Fortunately, there are considerable simplifications if we restrict our attention to large times after discharge (Smith, 1985). The first few moments have the asymptotes

$$c^{(0)} (z, t) \sim \bar{q} \tag{19.6a}$$

$$c^{(1)} (z, t) \sim \bar{q} g_+ (z, t) + \overline{g_- q}|_{t_0} \tag{19.6b}$$

$$c^{(2)} \sim 2\bar{q} \left\{ \int_{t_0}^{t} \overline{u g_+} \, dt' - \overline{g_+ g_-}|_{t_0} + g_+^{(2)} (z, t) \right\}$$
$$+ 2\overline{g_- q}|_{t_0} g_+ (z, t) + 2\overline{g_-^{(2)} q}|_{t_0} \tag{19.6c}$$

$$c^{(3)} \sim 6\bar{q} \int_{t_0}^{t} \overline{(u - \bar{u}) g_+ g_-} \, dt' + \dots \tag{19.6d}$$

The forwards function g_+ gives the dependence of the centroid displacement $c^{(1)}/c^{(0)}$ on the observation time t and vertical position z, whereas the backwards function g_- gives the dependence on the discharge time t_0 and the vertical distribution $q/\bar{q}$. The equations satisfied by g_+ are the adjoint transverse diffusion equations

$$\pm \partial_t g_\pm - \partial_z (\kappa \partial_z g_\pm) = u - \bar{u} \tag{19.7a}$$

with

$$\kappa \partial_z g_\pm = 0 \quad \text{on} \quad z = -h, 0 \tag{19.7b}$$

and

$$\overline{g_\pm} = 0 \tag{19.7c}$$

For future reference we note that g_+, g_- satisfy the relationship

$$\partial_t \overline{g_+ g_-} = \overline{(u - \bar{u}) g_-} - \overline{(u - \bar{u}) g_+} \tag{19.8}$$

For the purposes of this chapter we shall not pursue the equations satisfied by the forwards and backwards variance functions $g_+^{(2)}$, $g_-^{(2)}$ (Smith, 1985. Equations 4.3a, 4.12a), nor pursue the observation (z, t) and discharge (q, t_0) contributions indicated by the dots in Equation 19.6d. What we shall glean from Equations 19.6a–19.6d is that if an approximate model equation can replicate the forwards and backwards centroid displacement functions g_+, g_-, then that model reproduces the exact large-time asymptotes for the lowest two moments $c^{(0)}$, $c^{(1)}$ and gives the dominant contribution for the next two moments $c^{(2)}$, $c^{(3)}$. By contrast, Taylor dispersion models do not extend to the skewness $c^{(3)}$ term (Chatwin, 1970).

19.4 EIGENMODE EXPANSION

If the flow velocity $u(z, t)$ and vertical diffusivity $\kappa(z, t)$ were of separable form (Equation 19.5a, b), then Equation 19.7a would be amenable to a series expansion in terms of separation of variables contributions. Smith (1987) was able to extend the separation of variables to encompass spatial non uniformity. Here we make the corresponding extension to encompass time dependence.

Eigenmodes $\psi_n^{(+)}(z, t)$ with associated decay rates $\lambda_n(t)$ and adjoint eigenmodes $\psi_n^{(-)}(z,t)$ are assumed to exist such that

$$\partial_z (\kappa \partial_z \psi_n^{(\pm)}) \, \lambda_n \psi_n^{(\pm)} = \pm \partial_t \psi_n^{(\pm)} \tag{19.9a}$$

with

$$\kappa \partial_z \psi_n^{(\pm)} = 0 \quad \text{on} \quad z = -h, 0 \tag{19.9b}$$

$$\overline{\psi_n^{(+)}\psi_n^{(-)}} = 1, \quad \overline{\psi_m^{(+)}\psi_n^{(-)}} = 0 \ (m \neq n) \tag{19.9c,d}$$

$$\overline{\psi_n^{(+)}\partial_t\psi_n^{(-)}} = \overline{\psi_n^{(-)}\partial_t\psi_n^{(+)}} = 0 \tag{19.9e}$$

We recall that overbars indicate vertical average values (Equation 19.2). It is the suppression (Equation 19.9e) of the t dependence that ensures that in steady flows the modes become independent of t.

The lowest mode is

$$\psi_0^{(+)} = \psi_0^{(-)} = 1 \quad \text{with } \lambda_0 = 0 \tag{19.10}$$

The higher modes are ordered in increasing values of the decay rates

$$0 < \lambda_1 < \lambda_2 \ldots \tag{19.11}$$

This ordering also corresponds to the modes becoming increasingly oscillatory with respect to z

$$\lambda_n = \overline{\kappa\partial_z\psi_n^{(+)}\partial_z\psi_n^{(-)}} \tag{19.12}$$

In terms of the eigenmodes the exact solutions for the centroid displacement functions g_+, g_- can be written in the remarkably neat form

$$g_+ = \sum_{n=1}^{\infty} \psi_n^{(+)} \int_{-\infty}^{t} \overline{(u-\bar{u})\psi_n^{(-)}} \exp\left(-\int_{t'}^{t} \lambda_n \, dt''\right) dt' \tag{19.13a}$$

$$g_- = \sum_{n=1}^{\infty} \psi_n^{(-)} \int_{t}^{\infty} \overline{(u-\bar{u})\psi_n^{(+)}} \exp\left(-\int_{t}^{t'} \lambda_n \, dt''\right) dt' \tag{19.13b}$$

If we use the eigenmodes to represent the concentration

$$c = \bar{c} + \sum_{n=1}^{\infty} c_n(x, t)\psi_n^{(+)}(z, t) \tag{19.14}$$

then instead of Equations 19.1a–19.1c we have the hierarchy of coupled equations

$$\partial_t\bar{c} + \bar{u}\partial_x\bar{c} = \sum_{n=1}^{\infty} \overline{(u-\bar{u})\psi_n^{(+)}}\partial_x c_n \tag{19.15a}$$

$$\partial_t c_n + \lambda_n c_n + [\bar{u} + \overline{(u-\bar{u})\psi_n^{(+)}\psi_n^{(-)}}]\partial_x c_n = \sum_{m \neq n} \overline{(u-\bar{u})\psi_m^{(+)}\psi_n^{(-)}}\partial_x c_m \tag{19.15b}$$

with

$$\bar{c} = \bar{q}\delta(x-x_0), \quad c_n = \overline{q\psi_n^{(-)}}\delta(x-x_0) \quad \text{at } t = t_0 \tag{19.15c,d}$$

In practice we could not expect to compute more than the first two non-trivial eigenmodes $\psi_1^{(\pm)}$. An alternative approach would be to compute the centroid displacement functions g_+, g_- and use these in imitating the structure of the exact solutions with a truncated model. If Ψ_+, Ψ_- denote approximations to $\psi_1^{(+)}$, $\psi_1^{(-)}$ and Λ the approximate decay rate, then a two equation model can be posed

$$c(x, z, t) = \bar{c}(x, t) + c_1(x, t)\Psi_+(z, t) \tag{19.16}$$

where the functions $\bar{c}$ and c_1 satisfy the equations

$$\partial_t\bar{c} + \bar{u}\partial_x\bar{c} = \overline{(u-\bar{u})\Psi_+}\partial_x c_1 \tag{19.17a}$$

$$\partial_t c_1 + \Lambda c_1 + [\bar{u} + \overline{(u-\bar{u})\Psi_+\Psi_-}]\partial_x c_1 = \overline{(u-\bar{u})\Psi_-}\partial_x \bar{c} \tag{19.17b}$$

with

$$\bar{c} = \bar{q}\delta(x-x_0), \quad c_1 = \overline{q\Psi_-}\delta(x-x_0) \quad \text{at } t = t_0 \tag{19.17c,d}$$

For the approximate model to replicate the forwards and backwards centroid displacement functions g_+, g_- we require that

$$g_+(z, t) = \Psi_+(z, t) \Big|_{-\infty}^t \overline{(u-\bar{u})\Psi_-} \exp\left(-\int_{t'}^t \Lambda \, dt''\right) dt' \tag{19.18a}$$

$$g_-(z, t) = \Psi_-(z, t) \Big|_t^\infty \overline{(u-\bar{u})\Psi_+} \exp\left(-\int_t^{t'} \Lambda \, dt''\right) dt' \tag{19.18b}$$

We remark that Equation 19.8 makes these formulae (19.18a,b) compatible with the ortho-normal property (Equation 19.9c) for the approximate modes.

19.5 TWO-LAYER MODEL

The physical arguments underlying the ADZ model are that the essential ingredients for contaminant dispersion are velocity differences, concentration differences and an exchange process between different parts of the flow. The simplest flow which has all these ingredients is a two-layer flow with a uniform velocity and concentration within each layer and with the exchange of contaminant between the layers. In the dead zone models there is the additional requirement that one layer is stationary. Here we shall allow movement in both layers.

The total depth h is split into two layers of depths h_a, h_b

$$h_a + h_b = h \tag{19.19}$$

The uniform velocities in each of these layers are denoted u_a, u_b with

$$u_a = \bar{u} - u'h_b/h, \quad u_b = \bar{u} + u'h_a/h \tag{19.20a,b}$$

The concentrations $c_a(x, t)$ and $c_b(x, t)$ in the two well mixed layers satisfy the coupled equations

$$\partial_t(h_a c_a) + h_a u_a \partial_x c_a + \mu\frac{h_a h_b}{h}(c_a - c_b) = \frac{1}{2}(\partial_t h_a + |\partial_t h_a|)c_b - \frac{1}{2}(\partial_t h_b + |\partial_t h_b|)c_a \tag{19.21a}$$

$$\partial_t(h_b c_b) + h_b u_b \partial_x c_b + \mu\frac{h_a h_b}{h}(c_b - c_a) = \frac{1}{2}(\partial_t h_b + |\partial_t h_b|)c_a - \frac{1}{2}(\partial_t h_a + |\partial_t h_a|)c_b \tag{19.21b}$$

with

$$c_a = q_a\delta(x - x_0), \quad c_b = q_b\delta(x - x_0) \quad \text{at } t = 0 \tag{19.21c,d}$$

The μ terms allow for diffusive equilibration of concentration between the layers, and the right-hand side terms allow for gain from or loss to the other layer as the depths h_a, h_b change. The effective source strengths in the respective layers are denoted q_a and q_b.

For the two-layer model (Equation 19.21a,b) we introduce the vertically averaged concentration $\bar{c}$ and the scaled concentration difference c'

$$h\bar{c} = h_a c_a + h_b c_b, \quad c' = \left[\frac{(h_a h_b)^{1/2}}{h}\right](c_a - c_b) \tag{19.22a,b}$$

The equations satisfied by $\bar{c}$ and c' are

$$\partial_t \bar{c} + \bar{u}\partial_x \bar{c} = u' \left[\frac{(h_a h_b)^{1/2}}{h} \partial_x c' \right] \tag{19.23a}$$

$$\partial_t c' + \left[\mu + \frac{h}{2h_a h_b} \partial_t |h_a| \right] c' + \left[\bar{u} + \frac{(h_a - h_b)}{h} u' \right] \partial_x c' = u' \left[\frac{(h_a h_b)^{1/2}}{h} \partial_x \bar{c} \right] \tag{19.23b}$$

$$\bar{c} = \bar{q}\delta(x-x_0), \quad c' = q'\delta(x-x_0) \quad \text{at } t = t_0 \tag{19.23c,d}$$

We remark that the representation

$$c_a = \bar{c} + \left(\frac{h_b}{h_a} \right)^{1/2} c', \quad c_b = \bar{c} - \left(\frac{h_a}{h_b} \right)^{1/2} c' \tag{19.24a,b}$$

is the two-layer equivalent of the eigenmode representation (Equation 19.14). So, c' can be identified with c_1. However, in general, we cannot make the reverse identification of a two-mode truncation as a two-layer flow because the right-hand-side coupling terms $\overline{(u-\bar{u})\Psi_+}$ and $\overline{(u-\bar{u})\Psi_-}$ in the two-equation model (Equation 19.17) are not necessarily equal.

19.6 MATCHING THE MODELS

Remarkably, there is a selection for the decay rate Λ and the approximate modes Ψ_+, Ψ_- which allows us to make the identification $c' = c_1$ and which replicates the forwards and backwards centroid displacement functions

$$2\Lambda = \frac{\overline{(u-\bar{u})g_+} + \overline{(u-\bar{u})g_-}}{\overline{g_+ g_-}} - \frac{\partial_t \overline{(u-\bar{u})g_+}}{\overline{(u-\bar{u})g_+}} + \frac{\partial_t \overline{(u-\bar{u})g_-}}{\overline{(u-\bar{u})g_-}} \tag{19.25a}$$

$$\Psi_+(z, t) = \frac{sgn(P)}{|P|^{1/2}} \overline{(u-\bar{u})g_-} g_+(z, t) \tag{19.25b}$$

$$\Psi_-(z, t) = \frac{sgn(P)}{|P|^{1/2}} \overline{(u-\bar{u})g_+} g_-(z, t) \tag{19.25c}$$

with

$$P = \overline{g_+ g_-} \; \overline{(u-\bar{u})g_+} \; \overline{(u-\bar{u})g_-} \tag{19.25d}$$

The principal awkwardness with these formulae are the simgularities which can arise if any of the vertical averages $\overline{g_+ g_-}$, $\overline{(u-\bar{u})g_+}$, $\overline{(u-\bar{u})g_-}$ change sign. (There is no singularity if the averages are zero in pairs.)

To define the equivalent two-layer flow it is convenient to introduce the skewness measure

$$s(t) = \frac{\overline{(u-\bar{u})g_+} \overline{g_-} sgn(P)}{2|P|^{1/2}} \tag{19.26}$$

The layer depths and velocities are

$$h_a = \frac{1}{2}h\left\{ 1 + \frac{s}{[1+s^2]^{1/2}} \right\}, \quad h_b = \frac{1}{2}h\left\{ 1 - \frac{s}{[1+s^2]^{1/2}} \right\} \tag{19.27a,b}$$

$$u_a = \bar{u} - \{[1 + s^2]^{1/2} - s\}\frac{sgn(P)\,|P|^{1/2}}{g_+g_-} \tag{19.27c}$$

$$u_b = \bar{u} + \{[1 + s^2]^{1/2} + s\}\frac{sgn(P)\,|P|^{1/2}}{g_+g_-} \tag{19.27d}$$

The 'diffusive' contribution μ to the exchange rate is given by

$$\mu = \Lambda - \frac{\partial_t\,|s|}{[1 + s^2]^{1/2}} \tag{19.28}$$

If the initial discharge is not vertically uniform, then the effective non-uniformity for the two-layer discharge is

$$q_a = \bar{q} + \left(\frac{h_b}{h_a}\right)^{1/2}q',\quad q_b = \bar{q} - \left(\frac{h_a}{h_b}\right)^{1/2}q' \tag{19.29a,b}$$

with

$$q' = \overline{q\Psi^{(-)}} \tag{19.29c}$$

19.7 ILLUSTRATIVE EXAMPLE

Bowden and Fairbairn (1982) found that in tidal flows the vertical distribution of velocity could accurately be represented by a parabolic profile

$$u(z,\ t) = \bar{u}(t) + u'(t)\left(\frac{1}{2} - \frac{3}{2}(z/h)^2\right) \tag{19.30}$$

Empirically, the velocity shear term $u'(t)$ is about one-third of the bulk velocity term $\bar{u}(t)$. (It turns out that u', as used in Equation 19.30, is precisely the same as the velocity difference u' as used in the two-layer model.)

A convenient model for the vertical eddy diffusivity which accounts for the decrease in eddy sizes towards the bed is

$$\kappa(z,\ t) = K_0(t)\frac{5}{2}[1 - (z/h)^2],\quad K_0 = \frac{3}{h^3}\int_{-h}^{0}z^2\kappa\,dz \tag{19.31a,b}$$

The time-dependence of κ and its consequences as regards the shear dispersion are discussed by Chatwin and Sullivan (1984). The vertical eigenmodes are Legendre polynomials

$$\phi_n^{(+)} = \phi_n^{(-)} = (4n + 1)^{1/2}P_{2n}(z/h) \tag{19.32a}$$

$$\lambda_n(t) = 5n(2n + 1)K_0(t)/h^2 \tag{19.32b}$$

The definition 19.31b for K_0 ensures that a small perturbation to κ does not change λ_1. The forwards and backwards centroid displacement functions g_+g_- involve just the $n = 1$ mode

$$g_+(z,\ t) = \left[\frac{1}{2} - \frac{3}{2}(z/h)^2\right]\int_{-\infty}^{t}u'(t')\exp\left(-\int_{t'}^{t}\lambda_1(t'')\,dt''\right)dt' \tag{19.33a}$$

$$g_-(z,\ t) = \left[\frac{1}{2} - \frac{3}{2}(z/h)^2\right]\int_{t}^{\infty}u'(t')\exp\left(-\int_{t}^{t'}\lambda_1(t'')\,dt''\right)dt' \tag{19.33b}$$

It is to achieve these simple expressions (Equation 19.33a, b) that we do *not* follow Allen (1982) and use a vertically uniform eddy diffusivity.

For the particular case of constant $\bar{\kappa}$ and of sinusoidal velocity

$$\lambda_1 = 15K_0/h^2 = \text{constant}, \quad u'(t) = U \sin \omega t \qquad (19.34\text{a,b})$$

the forwards and backwards centroid displacement functions are displaced sinusoids

$$g_+(z, t) = \left[\frac{1}{2} - \frac{3}{2}(z/h)^2 \right] U \frac{(\lambda_1 \sin \omega t - \omega \cos \omega t)}{\omega^2 + \lambda_1^2} \qquad (19.35\text{a})$$

$$g_-(z, t) = \left[\frac{1}{2} - \frac{3}{2}(z/h)^2 \right] U \frac{(\lambda_1 \sin \omega t + \omega \cos \omega t)}{\omega^2 + \lambda_1^2} \qquad (19.35\text{b})$$

The necessary integrals involving $u - \bar{u}$, g_+, g_- are now easy to evaluate

$$\overline{(u - \bar{u})g_+} = \frac{U^2}{5} \sin \omega t \left[\frac{(\lambda_1 \sin \omega t - \omega \cos \omega t)}{\omega^2 + \lambda_1^2} \right] \qquad (19.36\text{a})$$

$$\overline{(u - \bar{u})g_-} = \frac{U^2}{5} \sin \omega t \left[\frac{(\lambda_1 \sin \omega t + \omega \cos \omega t)}{\omega^2 + \lambda_1^2} \right] \qquad (19.36\text{b})$$

$$\overline{g_+ g_-} = \frac{U^2}{5} \left[\frac{(\lambda_1^2 \sin^2 \omega t - \omega^2 \cos^2 \omega t)}{(\omega^2 + \lambda_1^2)^2} \right] \qquad (19.36\text{c})$$

$$\overline{(u - \bar{u})g_+ g_-} = \frac{-2U^3}{35} \sin \omega t \left[\frac{(\lambda_1^2 \sin^2 \omega t - \omega^2 \cos^2 \omega t)}{(\omega^2 + \lambda_1^2)^2} \right] \qquad (19.36\text{d})$$

the apparent singularities in Equation 19.25a for Λ cancel out exactly leaving the neat (and obvious) result

$$\Lambda = \lambda_1 = 15K_0/h^2 \qquad (19.37)$$

The equivalent two-layer flow has constant value for the skewness measure

$$s = -\frac{5^{1/2}}{7} = -0.3194 \qquad (19.38)$$

Hence, the layer depths h_a, h_b are constant

$$h_a/h = 0.3479, \quad h_b/h = 0.6521 \qquad (19.39\text{a,b})$$

The layer velocities u_a, u_b are exactly in phase with the actual flow

$$u_a(t) = \bar{u}(t) - 0.6124u'(t), \quad u_b(t) = \bar{u}(t) + 0.3266u'(t) \qquad (19.40\text{a,b})$$

Thus, there is a thin lower layer of slow moving fluid and a thicker fast-moving layer (see Figure 19.1). (As remarked after Equation 19.30, the velocity difference $u_a - u_b$ is precisely u'.)

It deserves comment that the specification (Equations 19.37–19.40) of the two-layer flow does not involve the ratio ω/Λ of the tidal frequency and the mixing rate. For the complementary case of steady but non-uniform flows, Smith (1987) found similar robustness of the two-layer model for more complicated flows involving all the eigenmodes. In contrast, the shear dispersion process is sensitive to the ratio ω/Λ (Holley, Harleman and Fischer, 1970). For example, the tidally averaged shear dispersion coefficient (Equation 19.36a) has

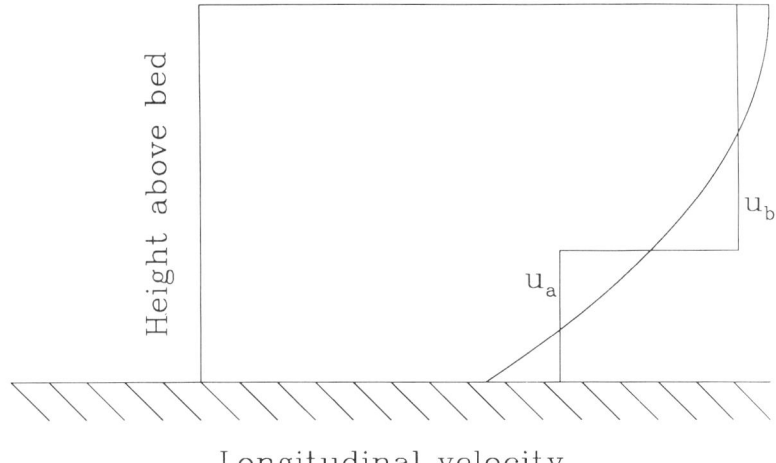

Longitudinal velocity

Figure 19.1 Two-layer approximation to the Bowden and Fairbairn (1952) model for the oscillatory velocity profile in an estuary

the value

$$\langle \overline{(u-\overline{u})g_+} \rangle = \left(\frac{U^2}{10\lambda_1} \right) \left\{ \frac{1}{[1 + (\omega/\Lambda)^2]} \right\} \tag{19.41}$$

Therefore a physical process which is time dependent and sensitive to the ratio ω/Λ has been modelled by layers whose thickness is independent both of time and of the ratio ω/Λ.

19.8 SKEWNESS

It is in replicating the asymptotic skewness that the ADZ type of model is superior to Fickian dispersion models. For rapidly mixed estuary flow with ω/Λ small, Allen (1982: Figures 6b and 7b) showed that the skewness at small times after discharge can be surprisingly large. In this section we shall evalue the skewness

$$\gamma(t, t_0) = \overline{c}^{(3)} (\overline{c}^{(0)})^{1/2} / (\overline{c}^{(2)})^{3/2} \tag{19.42}$$

for a uniform discharge in a rapidly mixed two-layer flow, and compare the results with those presented by Allen (1982).

For a sinusoidal flow with two constant depth layers and with a constant mixing rate ω/Λ, the moment equations (19.4) can be written

$$\frac{d}{dt} \overline{c}^{(p)} = - \frac{pU}{2[1 + s^2]^{1/2}} \sin\omega t \, c'^{(p-1)} \tag{19.43a}$$

$$\left(\frac{d}{dt} + \Lambda \right) c'^{(p)} = \frac{pU}{2[1 + s^2]^{1/2}} \sin\omega t \left[sc'^{(p-1)} - \frac{1}{2} \overline{c}^{(p-1)} \right] \tag{19.43b}$$

with

$$\overline{c}^{(0)} = \overline{q} \quad \text{and} \quad 0 = c'^{(0)} = \overline{c}^{(1)} = \dots \text{ at } t = t_0 \tag{19.43c,d}$$

The first few terms have simple solutions

$$\bar{c}^{(0)} = \bar{q}, \quad c'^{(0)} = 0, \quad \bar{c}^{(1)} = 0 \tag{19.44a,b,c}$$

Rather than seeking the full solution, we shall restrict our attention to flows with ω/Λ small and to times long enough after discharge for there to be thorough mixing across the flow.

For example, the exact result

$$c'^{(1)} = -\frac{\bar{q}U}{2(1 + s^2)^{1/2}} \left\{ \frac{\Lambda\sin\omega t - \omega\cos\omega t}{\Lambda^2 + \omega^2} - \exp[-\Lambda(t-t_0)]\frac{(\Lambda\sin\omega t_0 - \omega\cos\omega t_0)}{\Lambda^2 + \omega^2} \right\} \tag{19.45a}$$

will be approximated by

$$c^{(1)} = -\frac{\bar{q}U}{2(1 + s^2)^{1/2}\Lambda} \{sin\omega t + \dots\} \tag{19.45b}$$

The corresponding approximate solutions for the next few terms are

$$\bar{c}^{(2)} = \frac{\bar{q}U^2}{4(1 + s^2)\Lambda\omega} \left\{ \omega(t - t_0) - \frac{1}{2}\sin2\omega t + \frac{1}{2}\sin2\omega t_0 + \dots \right\} \tag{19.46a}$$

$$c'^{(2)} = -\frac{\bar{q}U^2 s}{4(1 + s^2)\Lambda^2} \{1 - \cos2\omega t + \dots\} \tag{19.46b}$$

$$\bar{c}^{(3)} = \frac{\bar{q}U^3 s}{16(1 + s^2)^{3/2}\Lambda^2\omega} \{9\cos\omega t_0 - 9\cos\omega t - \cos3\omega t_0 + \cos3\omega t + \dots\} \tag{19.46c}$$

The asymptotic expression for the skewness is

$$\gamma(t, t_0) = \frac{s}{2}\left(\frac{\omega}{\Lambda}\right)^{1/2} \frac{\{9\cos\omega t_0 - 9\cos\omega t - \cos3\omega t_0 + \cos3\omega t + \dots\}}{\left\{\omega(t - t_0) - \frac{1}{2}\sin2\omega t + \frac{1}{2}\sin2\omega t_0 + \dots\right\}^{3/2}} \tag{19.47}$$

The direct proportionality between γ and s justifies the designation of s as the skewness measure.

At large times the skewness is of order $(\omega/\Lambda)^{1/2}$. So, as shown by Smith (1983), a symmetric Gaussian Taylor dispersion model can replicate the concentration distributions as computed by Allen (1982). However, for small times after discharge we have the non-uniform asymptotes

$$\gamma \sim \frac{3s}{[2\Lambda(t - t_0)]^{1/2}} \quad \text{for } t_0 \neq 0 \tag{19.48a}$$

$$\gamma \sim \left(\frac{3}{2}\right)^{5/2}\frac{s}{[\Lambda t]^{1/2}} \quad \text{for } t_0 = 0 \tag{19.48b}$$

[Strictly these formulae are restricted to large values of $\Lambda(t - t_0)$]. Appropriately, Allen (1982) focuses her attention on the case $t_0 = 0$, for which the skewness is more marked.

In her random walk simulation of contaminant dispersion in the Mersey estuary Allen (1982) used the Bowden and Fairbairn (1952) velocity profile (Equation 19.30) with

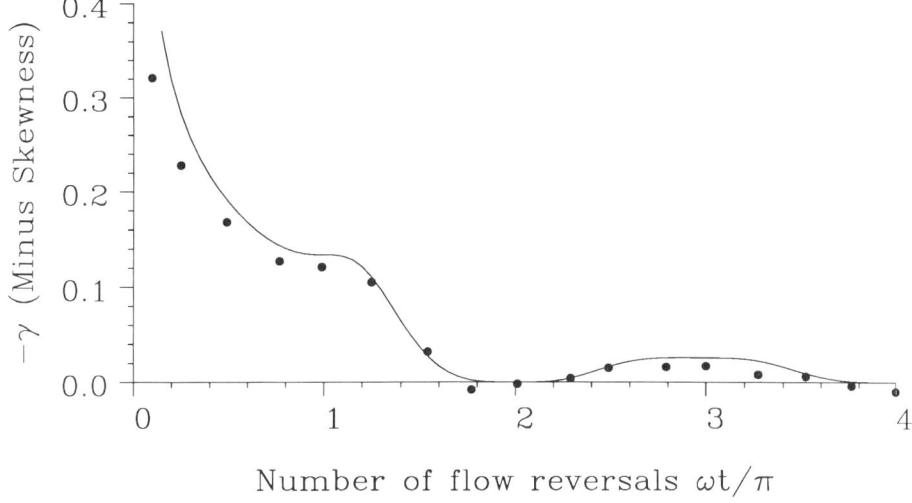

Figure 19.2 Comparison between the skewness of the longitudinal concentration distribution (●) as computed by Allen (1982; Figure 7b) and (—) as given by the two-layer approximation (Equation 19.47)

$$U = 5.6u_*, \quad \omega = 0.0703u_*h \qquad (19.49a,b)$$

where u_* is the friction velocity. Her random walk specification of the vertical mixing process is equivalent to a constant eddy diffusivity

$$\bar{\kappa} = 0.055hu_* \qquad (19.50)$$

With these parameters (Equations 19.49 and 19.50) the ratio ω/Λ has the small value

$$\omega/\Lambda = \omega h^2/9\bar{\kappa} = 0.142 \qquad (19.51)$$

Figure 19.2 compares $-\gamma$ as computed by Allen (1982; Figure 7b) with the approximation Equation 19.47. The high level of agreement (despite the different diffusivity models) can be attributed to the robustness of the definition (Equation 19.31b) for K_0.

19.9 CONCLUDING REMARKS

Conventionally the parameters s, u', μ in a two-layer model would be determined by recursive estimation from field observations of the concentration (Young, 1984). Instead, this chapter relates these parameters to the hydrodynamic velocity and eddy diffusivity profiles $u(z, t)$ and $\kappa(z, t)$.

The physical arguments underlying the two-layer model suggest that the skewness measure $s(t)$ should be almost constant, that the velocity jump $u'(t)$ should be in close synchronization with the actual velocity differences $u(z, t) - \bar{u}(t)$, and that the e-folding rate $\mu(t)$ should synchronize with the vertical mixing rate $\kappa(z, t)$. The outcome of this chapter strongly supports those suggestions. Indeed, the illustrative example studied in Section 19.7 reveals the robustness of the two-layer specification to changed flow conditions: the same splitting of the water depth into two layers was appropriate for all estuary depths and at all stages of the tide.

REFERENCES

Allen, C.B. (1982) Numerical simulation of contaminant dispersion in estuary flows. *Proc. R. Soc. London A* **381**, 179–194.

Aris, R. (1956) On the dispersion of solute in a fluid flowing through a tube. *Proc. R. Soc. London A* **235**, 67–77.

Beer, T. and Young, P.C. (1983) Longitudinal dispersion in natural streams. *J. Environ. Engin.* **109**, 1049–1067.

Bowden, K.F. and Fairbairn, L.A. (1952) A determination of the frictional forces in a tidal current. *Proc. R. Soc. London A* **214**, 31–392.

Chatwin, P.C. (1970) The approach to normality of the concentration distribution of a solute in a solvent flowing along a straight pipe. *J. Fluid Mech.* **43**, 321–352.

Chatwin, P.C. and Sullivan, P.J. (1984) The effect of the interaction between mean and fluctuating velocity components on turbulent dispersion in unsteady turbulent boundary layers. In: *Unsteady Turbulent Boundary Layers and Friction — FED.* Vol. 12. (Eds D.G. Wiggert and C.S. Martin). ASME, New York, 45–49.

Holley, E.R., Harleman, D.R.F. and Fischer, H.B. (1970) Dispersion in homogeneous estuary flow. *J. Hydr. Div. ASCE* **96**, 703–724.

Smith, R. (1983) The contraction of contaminant distributions in reversing flows. *J. Fluid Mech.* **129**, 137–151.

Smith, R. (1985) When and where to put a discharge in an oscillatory flow. *J. Fluid Mech.* **153**, 479–499.

Smith, R. (1987) A two-equation model for contaminant dispersion in natural streams. *J. Fluid Mech.* **178**, 257–277.

Taylor, G.I. (1953) Dispersion of soluble matter in solvent flowing slowly through a tube. *Proc. R. Soc. London A* **219**, 186–203.

Thacker, W.C. (1976) A solvable model of shear dispersion. *J. Phys. Oceanogr.* **6**, 66–75.

Young, P.C. (1984) *Recursive Estimation and Time Series Analysis: an Introduction.* Springer Verlag, Berlin.

20 Turbulence in the River Severn: a Dynamic Systems Approach

LEONARD A. SMITH

Mathematical Institute, University of Oxford, Oxford, UK

20.1 INTRODUCTION

It is well known that the analysis of complex geophysical time series is a difficult task and that recent developments in non-linear dynamic system theory have suggested a variety of new approaches including prediction techniques (Casdagli, 1989; Farmer and Sidorowich, 1988). In addition to their obvious applications, these methods may be used to test implicit assumptions made in alternative techniques (e.g. that the series is a realization of a linear stochastic process) through the method of surrogate data (Smith, 1992; Theiler *et al.*, 1992). The comparison with surrogate series provides a quantitative measure of the quality of different models of the data and, in turn, an indication of whether the underlying dynamics is linear or non-linear, deterministic or stochastic.

This chapter illustrates non-linear prediction by way of two-dimensional interpolation in a toy model of particulate transport, and then applies it to the prediction of observational data. The particulate transport model used, one of the first to show the effects of chaotic advection, is used for spatial and temporal predictions. A preliminary analysis of velocity data collected from the River Severn concludes the chapter. The results of this analysis reject the hypothesis that this signal is linear red noise, yet no evidence of low dimensional determinism is obtained. Several suggestions for the design of future data collection are proposed.

20.2 STOMMEL FLOW

In this section we introduce non-linear prediction with a simple two-dimensional time-dependent flow which can exhibit chaotic advection (Smith, 1984; 1987; Smith and Spiegel, 1985). The flow was previously considered as a model for the motion of particles suspended in a time-dependent flow, a situation of interest in many areas of geophysics (Huppert, 1984; Maxey and Corrsin, 1985). By introducing periodic time dependence, we generalize the laminar flow originally considered by Stommel (1949) to allow chaotic advection. In the chaotic regime, particles may take much longer to traverse a series of cells than the retention time indicated by dimensional calculations. Related flows have been investigated experimentally by Tooby, Wick and Isaacs (1977), Gollub and Solomon (1987) and Chaiken *et al.* (1987).

Mixing and Transport in the Environment. Edited by K.J. Beven, P.C. Chatwin and J.H. Millbank
© 1994 John Wiley & Sons Ltd

20.2.1 Steady flow

Intrigued by the observation that the yield of plankton tows taken along the direction of the wind were more variable than those taken perpendicular to the wind, Stommel (1949) considered the trajectories of negatively buoyant bodies immersed in steady fluid rolls. If the fluid motion is described by a stream function $\Psi(x,y,t)$, then, when particle inertia is negligible, the particle trajectories also allow a stream function. That is, the trajectories are solutions of

$$v_x = \frac{d\psi(x,y,t)}{dy} \tag{20.1}$$

$$v_y = -\frac{d\psi(x,y,t)}{dx} \tag{20.2}$$

where the particle stream function is

$$\psi(x,y,t) = \Psi(x,y,t) + v_s x \tag{20.3}$$

and v_s is the Stokes velocity (the terminal velocity of a particle in free fall through a quiescent, viscous fluid). Stommel considered a two-dimensional, vertical cross-section and adopted the stream function

$$\Psi(x,y) = A\sin x \sin y \tag{20.4}$$

Streamlines of this flow are shown in Figure 20.1. The particle stream function is then

$$\psi(x,y) = A\sin x \sin y + v_s x \tag{20.5}$$

where x represents the horizontal direction and y the vertical direction. Particle stream lines

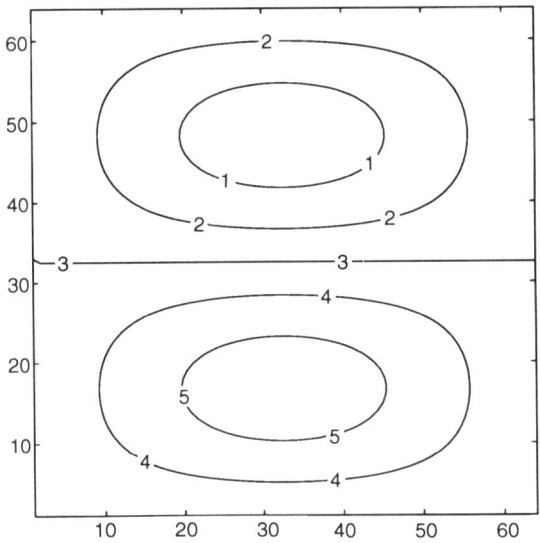

CONTOUR KEY	
1	-0.800
2	-0.400
3	0.000
4	0.400
5	0.800

Figure 20.1 Fluid streamlines of the steady flow. Here x is the horizontal axis and y is the vertical axis; the figure shows a single cell (i.e. $0 \le x \le \pi$, $0 \le y \le 2\pi$)

for $v_s = 0.25$ are shown in Figure 20.2. Stommel (1949) concluded that the flow divided into two classes: one which swept through vertically stacked cells and another, trapped within the closed contours, which formed a region of retention in each cell. This model may be interpreted in a downstream guise (where x is the cross-stream direction and y the downstream direction) with the Langmuir cells corresponding to dead zones due to boundary conditions not explicitly modeled; v_s then corresponds to a superimposed downstream velocity. In the steady flow case, downstream motion is trivial. Dead zones really are dead as there is no transport across the stream lines, and the streamlines themselves have the simple structure shown in the figures. We could consider more spatially complicated flows by superimposing additional structure periodic with half the wavelength of this flow. Repeating the process at still smaller length scales would yield a flow similar to the β model (Frisch, Sulem and Nelkin, 1978). As shown in the following, the particle trajectories from Equation 20.5 are already fairly intricate when the flow is time dependent.

20.2.2 Time dependent flow

When ψ is independent of time, topological constraints prevent trajectories from displaying chaos. A time-dependent stream function, however, corresponds to a non-autonomous Hamiltonian system; in this instance the system has a three-dimensional phase space and admits qualitatively different trajectories (see, for example, Mackay and Meiss, 1987). To investigate this case, consider

$$A(t) = A_0[1 + \epsilon \sin(\omega t)] \tag{20.6}$$

where ϵ quantifies the strength of the time dependent element of the flow. For $\epsilon \neq 0$, this has the immediate effect of detaching the dead zone from the wall; for small ϵ an isolated

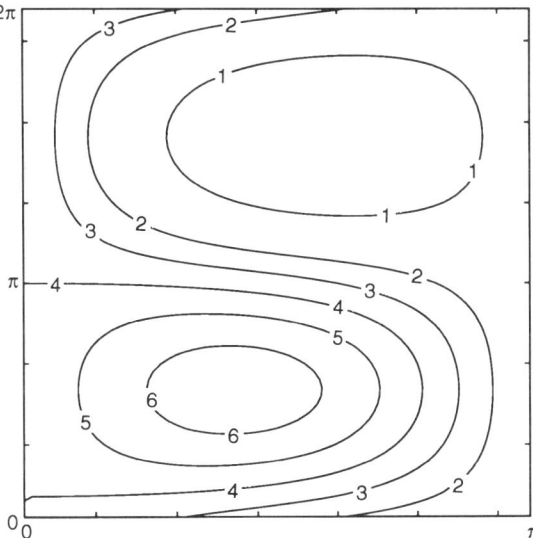

CONTOUR KEY	
1	-0.999
2	-0.500
3	-0.250
4	0.000
5	0.250
6	0.500
7	0.999

Figure 20.2 Particle streamlines in the steady flow with $v_s = 0.25$, $A = 1$. Again the figure shows a single cell which may either be considered as one of a series of vertically stacked cells, or to have periodic boundary conditions identifying the bottom with the top of the cell

dead zone(s) persists, but the boundaries can be very complicated; whether or not a particle will be retained for another period of the flow depends sensitively on its location. We will consider the case $A_0 = 1.0$, $v_s = 0.25$, $\omega = 2\pi/4.5$ and apply periodic boundary conditions in y, so that particles passing through the bottom of the cell are reintroduced at the top.

This complexity is reflected in Figure 20.3 in the total displacement, δ, of a particle from its initial position after one period [$\delta = |x(2\pi/\omega) - x(0)|$]. The concentric contours 1, 2 and 3 within the dead zone reflect stable, roughly circular motion, whereas the oblong contour 5 represents particles which fall out of the cell. Similar contours for longer evolution times maintain the relatively simple structure within the dead zone, but reveal complex structures near its boundary as some particles fall repeatedly through cells while their near neighbours become entrained (temporarily) near the dead zones of other cells. Examining the variation in residence time of a particle near the dead zone boundary reveals the sensitive dependence of residence time on initial position.

In the next section, we shall consider the question of interpolation in this field, specifically: given the initial position of an ensemble of particles and their final position after one period of the fluid forcing, how can we approximate the final position of some other initial condition (if the underlying equations are unknown)? First, we consider the results of a long run shown in Figure 20.4. Here we have plotted the location of a single initial condition ($v_s = 0.25$) once per cycle of the background forcing and applied periodic boundary conditions identifying the top and bottom of the cell; this stroboscopic graph is equivalent to a Poincaré section. The figure sketches the region accessible to migratory particles. Note that within each cell the dead zone is divided into several disconnected regions; this is typical of Hamiltonian systems of this form. If a line of particles is introduced (Smith and Spiegel, 1985), it is quickly contorted into a complex shape that appears to be self-similar; as experimental techniques of flow visualization typically rely on tracers (see, for example, Corrsin, 1950), and as material lines

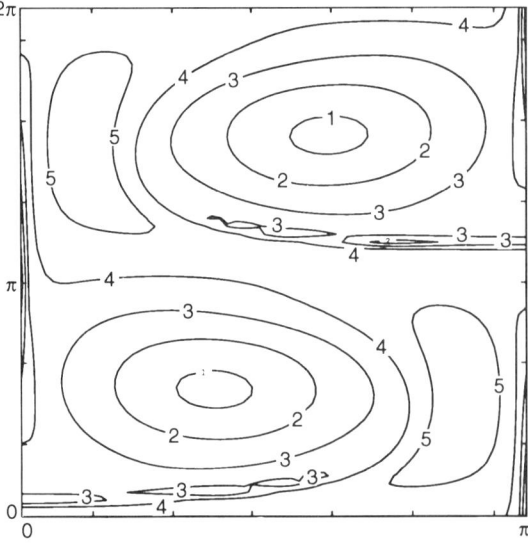

CONTOUR KEY	
1	0.065
2	0.188
3	0.311
4	0.435
5	0.558

Figure 20.3 Contours of the displacement after one period as a function of initial position for the case $A_0 = 1.0$, $v_s = 0.25$, $\omega = 2\pi/4.5$. The fine structure already visible near the centre and bottom of the cell (contours 2,3,4) becomes much more intricate as the number of periods increases

in a turbulent flow are expected to develop into fractal structures, it is interesting to note that such structures can occur in a periodic laminar flow, leading to the formation of self-similar distributions. There is nothing turbulent here; the flow is smooth but chaotic. Pollutant dispersal may be more widespread and more concentrated than expected from laminar flows without requiring turbulent mixing. Additional discussion of this and related flows is given in Smith and Spiegel (1985), Pasmanter (1987), Smith (1987), Broomhead and Ryrie (1988), Crisanti *et al.* (1990; in press), Yu, Grebogi and Ott (1990) and Huppert (1991).

20.2.3 Interpolation of vertical displacement in two dimensions

Consider the problem of interpolating a function $s(x)$ which determined the contours shown in Figure 20.3. We shall initially consider a global predictor (or map), $F(x):R^2 \rightarrow R^1$, which estimates s for any x. First, choose n_c base points or centers in the two-dimensional space

$$c_j, \quad j = 1,2,\ldots,n_c; \quad c_j \in R^2 \qquad (20.7)$$

We will consider $F(x)$ of the form

$$F(x) = \sum_{j=1}^{n_c} \lambda_j \phi(\|x - c_j\|) \qquad (20.8)$$

where $\phi(r)$ are radial basis functions (Powell, 1985). We will consider $\phi(r) = r^3$ and $\phi(r) = e^{-r^2/c^2}$ where the constant is based on a multiple of the average distance between data

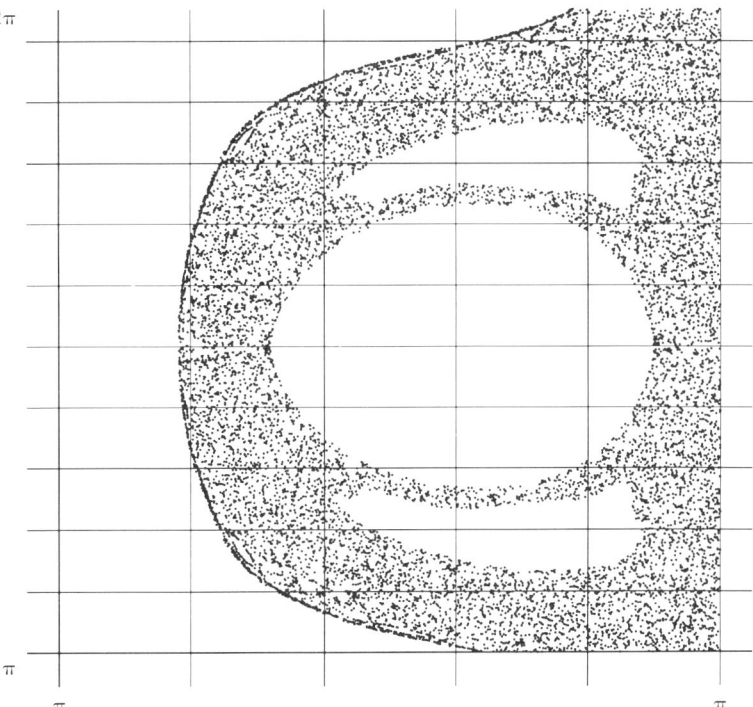

Figure 20.4 Stroboscopic view of a single particle falling for 2^{12} periods with periodic boundary conditions. Only the upper half of the cell is shown

points, d_{nn}. The λ_j are constants which are determined by observations

$$F(x_i) \approx s_i, \quad i = 1,2,\ldots,n_L \qquad (20.9)$$

where the x_i are the initial conditions of the n observations and the s_i are their observed displacements. We shall call the data used in determining the λ_j the *learning set*.

Determining the λ_j corresponds to the solution of the (linear) problem

$$b = A\lambda \qquad (20.10)$$

where λ is a vector of length n_c whose jth component is λ_j and A and b are given by

$$A_{ij} = \omega_i \phi(\|x_i - c_j\|) \qquad (20.11)$$

and

$$b_i = \omega_i s_i \qquad (20.12)$$

where $i = 1,\ldots n_L$ and $j = 1,\ldots,n_c$. Traditionally, the weights ω_i reflect the confidence associated with the ith observation, we shall restrict attention to the case where all ω_i are equal, assuming that the errors are independent with equal variance (but note the discussion in Section 3.2 of Smith, 1992). When there are fewer centres than data points, A is not square and we are left with a standard least-squares problem of finding a λ which minimizes $\chi^2 = \|b - A\lambda\|^2$. The solution which also minimizes $\|\lambda\|^2$ corresponds to

$$\lambda = A^+b \qquad (20.13)$$

where A^+ is the Moore−Penrose pseudo-inverse of A. Efficient methods to calculate A^+ are given in Press *et al.* (1987). Additional discussion of the details of the construction of this type of predictor is provided by Broomhead and Lowe (1988), Casdagli (1989), Farmer and Sidorowich (1988) and Smith (1992).

A function F constructed in this way will be called a radial basis function (RBF) predictor. We shall also consider local linear (and quadratic) interpolation. In all instances the data in the learning set used to construct the predictor are kept distinct from the test data set on which it is evaluated. Thus the time series results represent out of sample prediction. Out of sample statistics are crucial if we are to establish that an RBF predictor is robust in the presence of noise; RBF predictors can minimize in-sample error by overfitting the noise in the learning set.

20.2.3.1 *Global and local prediction*

We now contrast the results of constructing one global predictor for the entire cell with those from constructing a local predictor for each point of interest based on that point's near neighbours. This simple two-dimensional example illustrates qualitative features of global reconstructions less easily visualized in higher dimensional constructions. Examining contours of the predictor error as a function of location in the cell shows that the largest prediction errors occur near the corners where the gradient is greatest; when the data density is low, global predictors may maintain a smoothness over the cell lost in local predictors. In this instance, using a grid of 32 × 32 sample points as the learning set, the 32 neighbour local RBF predictor does a superior job of resolving the details of the flow. When the data density is very high, local linear predictors should be most accurate; at this data density, however, local quadratics were found to give the smallest mean absolute error.

The predictions of one local RBF with $\phi(r) = r^3$, one local linear, and one global predictor

fraction

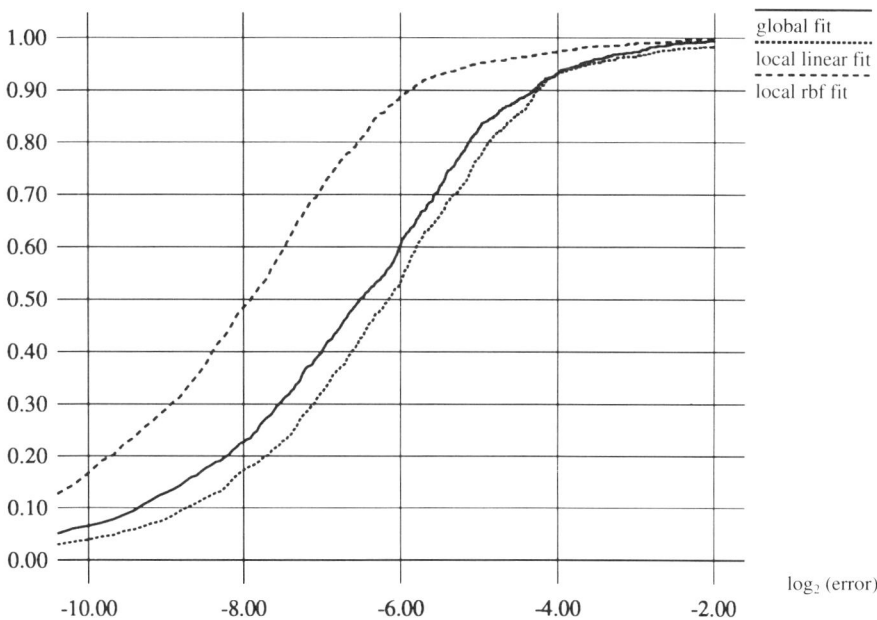

Figure 20.5 Comparison of prediction error profile for (a) global and two 32 neighbour local predictors. The horizontal axis is the $\log_2$ of the error balance. The local predictors are (b) local linear and (c) RBF with $\phi(r) = r^3$ with $n_c = 16$

are summarized in the predictor error profiles of Figure 20.5. Each curve shows the fraction of the out of sample points which can be predicted to within a given (absolute) accuracy; for long test sets this should converge to the cumulative probability distribution of the prediction errors. The local RBF predictor is clearly the best; for example, 50% of its predictions have an error of less than $2^{-8} \approx 0.004$, almost a factor of four less than the corresponding level for the local linear predictor. Evaluating a predictor with a single measure of an 'average' error in the prediction may be misleading when a few per cent of the predictions have very large errors. In extreme cases this may result in a predictor which is more accurate 90% of the time having a greater average absolute error. As long as the prediction error profiles are well separated there is no confusion, but if they cross (as in Figure 20.8), it is important to ascertain which properties of the predictions are considered the more important when evaluating predictors; in this instance, different definitions of the error can result in different 'optimal' predictors.

20.3 TIME SERIES APPLICATION

20.3.1 Residence times: transition to a Lagrangian frame

In the time-dependent Stommel flow, the velocity components at a fixed position each show a regular periodic oscillation which is straightforward to predict. We therefore adopt a Lagrangian reference frame, moving with a particle, to show the practical problems involved with the prediction of chaotic systems, and consider the series of residence times in each cell (defined as the time interval between passages through the lower periodic boundary of the cell). At first glance, this problem appears far removed from the interpolation problem presented earlier. On reflection, however, they are seen to be very similar; Casdagli (1989) and Farmer and Sidorowich (1987; 1988) were the first to make this application to chaotic time series.

Figures 20.6 and 20.7 are derived from two segments of the trajectory which generated Figure 20.4; here the residence time in consecutive cells is given. This spiky appearance, with regions of activity separated by quiescent periods, is common to many observed series in geophysics and is difficult to predict. This series is extremely varied (more so than can be shown in these figures); there are stretches of time during which the particle will fall through dozens of cells without being re-entrained; alternatively, there are cells in which a particle becomes trapped for a large fraction of the total integration time. Although there can be no attractors in this Hamiltonian system, the probability density in these 'reef' regions where the particle is temporarily trapped can become fairly large (Meiss, 1986) and may appear as a strange accumulator (Smith and Spiegel, 1987).

20.3.2 Method of delays

The first step in applying this method of prediction is to *reconstruct* the time series into a geometrical framework. [An introduction to this method is given by Broomhead and Jones (1989).] In many instances, $s(t)$ would be a physical variable measured at regular intervals

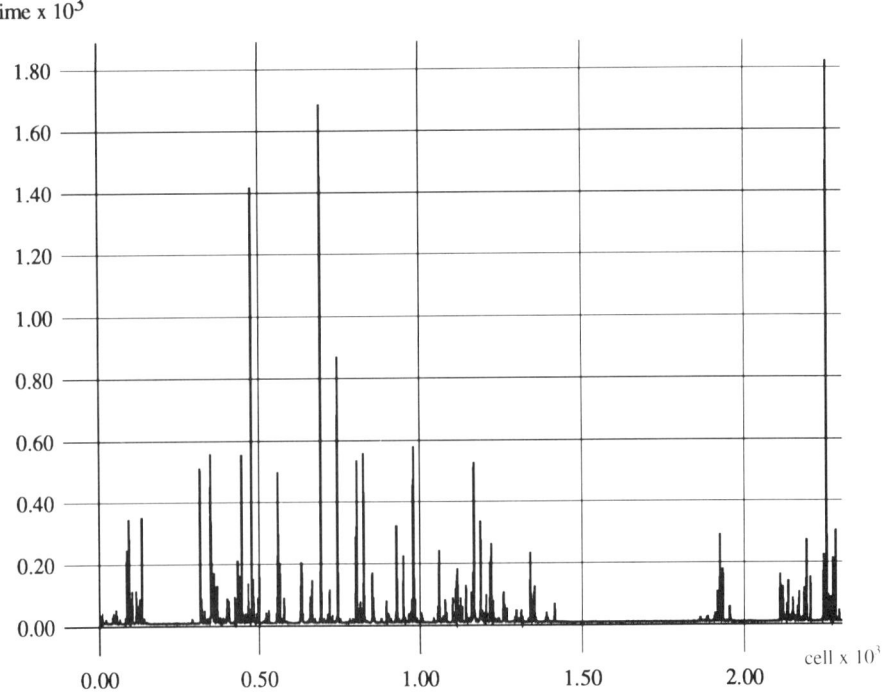

Figure 20.6 Segment of the series of time a particle remains in a given cell; the data corresponds to the same trajectory as the stroboscopic section shown in Figure 20.4. The x axis corresponds to the ith cell; the residence times in a few thousand cells are shown

(τ_s, the sampling time)

$$s_i = s(i\tau_s) \quad i = 1,2,\ldots,n_s \tag{20.14}$$

In our case a 'sampling time' is exactly what we are trying to predict, and the s_i are taken as consecutive observations.

A trajectory, $x(t)$, of this system is reconstructed in M dimensions from a time series of a single observable, $s(t)$, by the method of delays to yield a series of vectors

$$x_i = (s_i, s_{i-j}, \ldots, s_{i-j(M-1)}) \tag{20.15}$$

where j (or $j\tau_s$) is called the delay time, τ_d. (As noted by a referee, the phrase 'delay time' may be misleading in this case, as the subscript i relates to the time spent in the ith cell.) For a deterministic system with phase space dimension M_s and a generic observable, this reconstruction preserves many of the characteristics of the original system for sufficiently large M (see, for example, Casdagli, 1992; Packard *et al.*, 1980; Sauer, Yorke and Casdagli, 1991; Takens, 1981). As shown in the following, multivariate series can also be considered, often with significantly shorter time series in terms of the total duration of the experiment. This is easily understood as multivariate probes can distinguish well separated states in phase space which appear similar to univariate probes due to projection effects. For example, combining the phase of the background flow when the particle entered the cell with the residence time series could improve the predictions *significantly* more than doubling the length of the residence time series. It is the information content, not the length of the data set, which is more important.

Consider the problem of predicting a fixed distance $\tau_p = k$ steps into the future. Once we have the set of observed initial conditions x_i and later observations s_{i+k}, the same

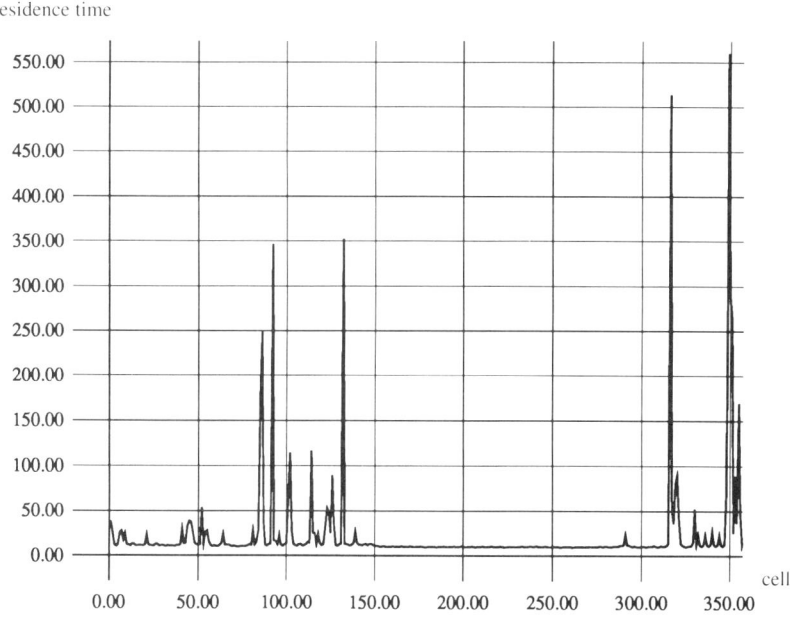

Figure 20.7 Initial segment of the series in the previous figure; only the first 350 residence times are shown here. Note the change in the scale of the vertical axis

fraction

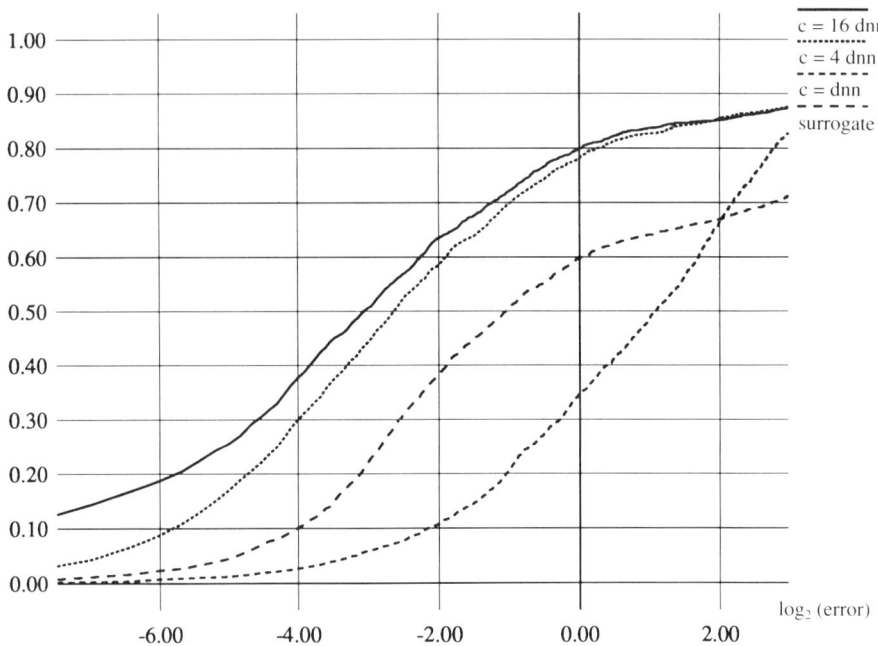

Figure 20.8 Cumulative predictor error used to optimize basis function parameters and show the significance with respect to surrogate data for the prediction of particle residence times. RBF predictors using $\phi(r) = e^{-r^2/c^2}$ for $c = 1$, 4 and 16 times each *local* average nearest neighbour distance. Local linear prediction is of similar quality as the best RBF predictor

machinery set up to solve the two-dimensional example may be used to predict the value of s, k steps ahead in this M-dimensional case. The most immediate difficulties to arise are those of data density in higher dimensional spaces and, of course, visualization of the result.

We will use a learning set consisting of the first 2000 points of a 6000 point series and evaluate the predictors by making one step (i.e. cell) ahead predictions on each of the remaining points. Figure 20.8 provides the prediction error profiles for this example. Three of the curves correspond to different choices for the constant c in the basis function $\phi(r) = e^{-r^2/c^2}$, illustrating how the profile may be used to optimize free parameters. In this instance we knew there was an underlying deterministic system; how would we evaluate the quality of this result if we did not?

20.3.3 Null hypothesis testing

The significance of a result is determined through the consideration of surrogate data and surrogate predictors; other algorithms developed from non-linear dynamical systems theory can also be evaluated in this manner (Smith, 1992; Theiler *et al.*, 1992). In brief, the procedure is to construct a stochastic data series with statistical characteristics similar to the original data and then to determine whether a given algorithm can distinguish the two series. An ensemble of surrogates may be analysed and the probability of a surrogate series producing the observed result may be estimated. Alternatively, surrogate predictors quantify the errors that should be expected by even the most naïve forecasts, the simplest forecast being either the persistence of the last observed value, or a random choice from the observed distribution.

20.3.4 Construction of surrogate data

The method of surrogate data can be used to quantify the significance of a claim for non-linearity, chaos, or even periodicity. There are a variety of methods for constructing surrogate series; the appropriate method in a specific example will depend on the statistic evaluated and null hypothesis to be tested. In the following we will consider examples which reproduce the distributions observed in, or the autocorrelation function of, the data set. In general, any method of simulation (including, of course, bootstrap and parametric bootstrap methods well known in statistics; Tong, 1990) may be used; however, the correspondence between an algorithm for generating surrogates and a well posed null hypothesis may be obscure.

The basic idea is to take surrogate series from a process which does not, for example, reproduce the sensitivity to initial value found in chaotic series; discovering whether a given algorithm can detect this difference (by distinguishing the chaotic series within an ensemble of surrogates) is the key point. In short, we want the surrogates to reproduce the statistics of the series, but not the physics of the system.

20.3.4.1 Shuffling

A typical method for generating surrogates for iterated systems such as the retention times is to draw from the observed distribution at random; the resulting surrogate series are independent and identically distributed (IID) random variables with the same distribution as the observations. (This is similar to shuffling the data, but allows the construction of arbitrarily long series.) The true retention time signal is easily distinguished from surrogate signals of this type. This should be the case whenever there is some memory in the system — that is, when the expected value of the next observation is conditioned on the current value. A simple method to simulate this conditional probability distribution is given in the following.

20.3.4.2 Malicious shuffling

A method of shuffling which retains the general association between consecutive observations provides more realistic surrogates. For the residence time series, short residence times often follow very short residence times, whereas long residence times tend to follow those of intermediate length. We may produce a surrogate series with this quality from a series of N observations with the following algorithm:

(1) Sort the first $(N - 1)$ members of the series into increasing order, recording the value which immediately followed each member in the unsorted series (i.e. its image).
(2) Divide the sorted list into K subgroups.
(3) Pick a member from the entire series at random; this is the first data point of the surrogate series.
(4) Determine which of the K subgroups the *image* of this value falls in, and pick an element from this subgroup at random as the next surrogate data point.
(5) Repeat step 4 until a series of the desired length is achieved.

These surrogates cannot only preserve the distribution of the original data set, but can also approximate the (one-step) conditional probability distribution [i.e. $P(s_i = c|s_{i-1} = b)$] depending on the value chosen for K and are, in essence, Markov chains. For $K = 1$, the surrogates are IID, alternatively, for $K \approx N$ segments of the original are reproduced. [Assuming that the observations are distinct (i.e. $s_i \neq s_j$ for $i \neq j$) thereby avoiding an ambiguity often introduced by quantization effects. In this case, the final image used must be grouped with an earlier observation, hence $K < N - 1$.] The choice of a particular value for K depends on the structure of the distribution. Although surrogate generators will provide the advertised results in the long run, it should always be verified that, for the desired length of surrogate series, the *observed* distributions of surrogate data are not sensitive to either the choice of K or that of the location of the boundaries between subgroups.

This variant of shuffling was originally envisaged for series which displayed distinct 'modes' of behaviour; modes which, in turn, were reflected by the magnitude of the observation. The subgroups are (individually) IID and there is a fixed probability for remaining in a given subgroup. As in the following example, K *was taken to be much less than* N. This appears more natural for iterated systems than for those which evolve smoothly in time: the one-step conditional probability distribution is often blind to 'trends' in stationary series. [Exceptions exist in processes whose conditional probability distributions change under time-reversal (i.e. $P(s_i = c|s_{i-1} = b) \neq P(s_{i-1} = b|s_i = c))$, like a saw-tooth series.] It should also be noted that it is straightforward to extend this method of shuffling to maintain correlations over periods longer than one step [e.g. $P(s_i = c|s_{i-1} = b, s_{i-3} = a)$].

Not surprisingly, the expected prediction error of surrogate series with $K = 5$ is lower than that for the IID surrogates produced by straight shuffling (equivalent to $K = 1$); none the less, the observed series can be distinguished from surrogates of this type, as illustrated in Figure 20.8.

20.3.4.3 Fourier transforms

Surrogates which test whether 'good' predictions result from the autocorrelation function alone may be generated with Fourier transforms (Osborne *et al.*, 1986; Smith, 1992; Theiler *et al.*, 1992). A surrogate is generated through the inverse Fourier transform of the observed Fourier amplitudes and random values for the phase of each frequency component. As noted by Theiler *et al.* (1992), this is equivalent to testing whether the signal is distinguishable from a linear stochastic noise. This method is used for the Severn data in the next section.

20.4 DYNAMIC RECONSTRUCTIONS OF THE RIVER SEVERN

We now turn to the analysis of velocity time series taken on the River Severn. The data consists of simultaneous, three-component velocity measurements taken in 1989 at the Leighton New Upstream Pool Site; details of the collection and previous analysis of this and related data sets are given elsewhere in this volume (Heslop *et al.*, 1994) and in Beven and Carling (1992) and Heslop and Allen (1990). It should be noted that, due to noise contamination, the analogue signal was redigitized and low-pass filtered (Holland, 1991); it is possible this would have a negative impact on the current analysis. Renormalization would not affect reconstructions of the individual components, but would affect the analysis of the total velocity series.

20.4.1 'Predicting' the total velocity from observed components

As stressed earlier, the non-linear models applied here are constructed completely from the data; no underlying model of the system is required (although additional information can be used). This makes the method sensitive to variations in the density of data points in the reconstruction space; in particular, when the system is in a region of reconstruction space not explored during the learning set, these predictions correspond to extrapolations (rather than interpolations) and their uncertainty is high even when the other assumptions of the technique are satisfied. To stress the importance of this effect, and the sensitivity to details of the reconstruction, we will pose a simple problem: Given the observations v_x, v_y, v_z, what is the total velocity? This test, and its intermittent failure, also serves to stress that no underlying model of the system is assumed.

Figure 20.9 shows the results of a local linear 'prediction' of the current total velocity given its three component parts (i.e. $\tau_p = 0$). Of course, this computation can be performed exactly as we know the underlying relationship; this information is not available to the non-

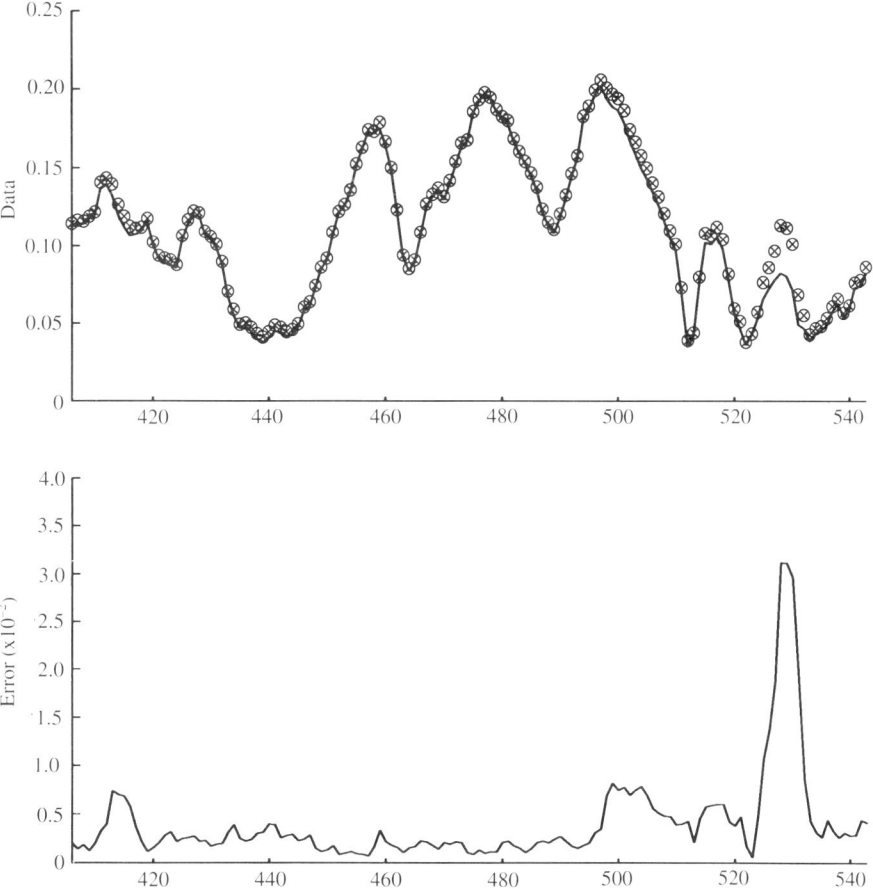

Figure 20.9 Observed (solid) and 'predicted' (symbol) values of the total velocity from a local linear predictor using 32 nearest neighbours (upper). Simultaneous time series of the prediction error (lower)

linear model; regions of poor fit (e.g. near $t = 525$) correspond to areas of relatively low data density.

Local non-linear predictors cope better with this difficulty, in part because they do not assume that the local behaviour is linear and hence can more accurately interpolate low data regions where it is indeed not linear. Non-linear predictors, in turn, can over fit the noise in the data and are sensitive to the additional parameters in their definition (for example, the constant c in $\phi(r) = e^{-r^2/c^2}$). What is needed is a robust method to determine how much of the local structure should be reproduced by the predictor; some progress in this area has been achieved in collaboration with A. Mees (manuscript in preparation). The results for the total velocity 'prediction' are summarized in Figure 20.10, where the optimum RBF predictor reduces the error with which 50% of the predictions are made by more than a factor of 32 over the local linear case illustrated in Figure 20.9.

20.4.2 Predicting short-term fluctuations in velocity

Finally, we consider the 'standard problem' of time series prediction. Given a single, univariate time series of the total velocity, $v(t)$, how well can we predict a future value $v(t + \tau_p)$? For the Severn data with $\tau_p = 2$ and $\tau_p = 7$, predictor error profiles indicate that the best predictors have a delay time $\tau_d = 1$ and either $M = 2$ or $M = 3$. This has implications for future data collection. The short delay time indicates that a shorter sampling rate would be useful, whereas the low embedding dimension also points to data density problems. In addition to requesting a longer, more densely sampled data stream, comparison with Fourier transform surrogate series indicates that, for short times, the observed series ($\tau_p = 2\tau_s$) may be slightly more predictable than the surrogates, but for longer time scales ($\tau_p = 7\tau_s$) the observed series can be distinguished as significantly *less* predictable than the linear stochastic surrogates.

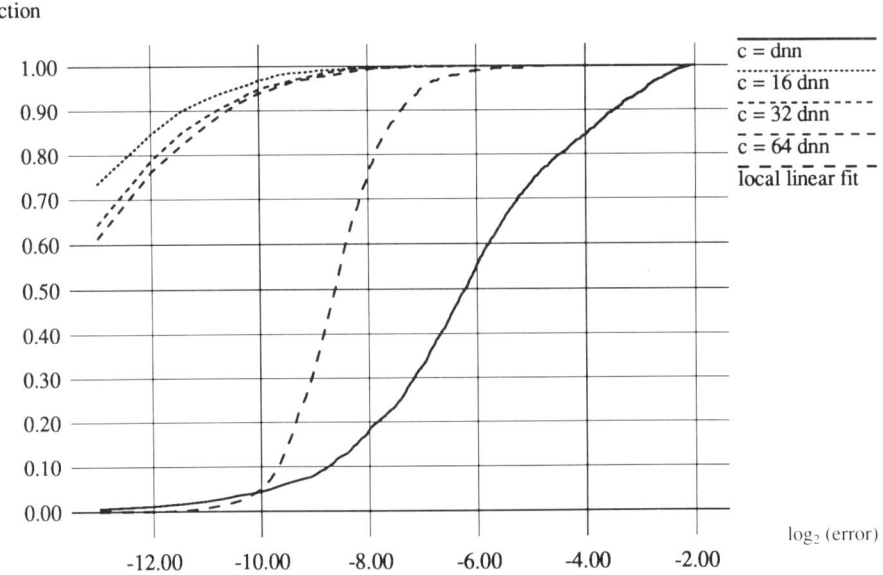

Figure 20.10 Prediction error profile for local linear and RBF predictors using $\phi(r) = e^{-r^2/c^2}$ for $c = 1, 16, 32$ and 64 times the *local* average nearest neighbour distance, d_{nn}

This result indicates that the series contains significant non-linearities which cannot be captured by linear models. In addition, it implies that the data density is too low to fill out the reconstruction space — that is, the data record is too short to allow sufficient recurrence in most regions of the reconstruction space even though the statistics of the distribution of data values may appear stationary. Examination of the series shows that many distinct types of behaviour are indeed observed.

Predictions from a three-dimensional reconstruction based on the three simultaneous velocity components resulted in an improvement in the case $\tau_p = 7\tau_s$. The average absolute errors and predictor profiles of this predictor are also better than those of a persistence predictor. This interesting result indicates that the orientation of the velocity vector is an important diagnostic of short time fluctuations.

20.5 DISCUSSION

We should also address the question of whether these techniques are worth applying. The most robust answer is yes. From the point of view of simplicity, low dimensional chaotic systems are intermediate between periodic (or static) systems and rich turbulent systems; without performing tests such as these, they cannot be dismissed. From a more pragmatic point of view, these techniques can be useful in the event that the dynamics are simple but *not* deterministic. For example, reasonable predictions are obtained when these methods are applied to data originating from non-linear stochastic models (and also for linear ARMA type models). If the system is equivalent to a linear ARMA model, the parameters of which are well estimated, then the techniques discussed here can add little more. If, on the other hand, the observations are non-linear and not easily transformed to linearity (Theiler *et al.*, 1992), then deterministic predictors which are robust in the presence of noise should yield good predictions (in the sense of expected values). The data requirements needed to apply these techniques are currently under investigation; ultimately, it will most likely be their relative speed of convergence (as the amount of data in the learning set increases) which will determine their usefulness.

We have illustrated a non-linear prediction technique and presented the initial results of an analysis of data from the River Severn. The analysis shows (1) that the velocity series contains important non-linearities which cannot be described within a linear model and (2) that the three-dimensional orientation of the velocity vector is relevant to the short-term evolution. The most serious limitations encountered appear to arise from a lack of recurrence (low data densities), implying the need for longer data series. There is little doubt that series such as this, and those of geophysical systems in general, will provide a tough proving ground for these techniques.

ACKNOWLEDGEMENTS

I thank Keith Beven for introducing me to this problem and for his patience, and Alistair Mees for discussion on a wide range of issues. I also thank Mark Muldoon and Catherine Lanone for a critical reading of the manuscript. I met Cath Allen only once, while visiting Lancaster, and am still struggling to solve several of the many questions she asked after my seminar. I would like to express my admiration of her courage and that of other scientists strong enough to battle both cancer and ignorance simultaneously.

REFERENCES

Beven, K. and Carling, P. (1992) Velocities, roughness and dispersion in the lowland River Severn. In: *Lowland Floodplain Rivers: Geomorphological Perspectives* (Eds P. Carling and G.E. Petts). Wiley, Chichester, 71–93.

Broomhead, D.S. and Jones, R. (1989) Time-series analysis. *Proc. R. Soc. London* **423**, 103–121.

Broomhead, D.S. and Lowe, D. (1988) Multivariable functional interpolation and adaptive networks. *J. Complex Syst* **2**, 321–355.

Broomhead, D.S. and Ryrie, S.C. (1988) Particle paths in wavy vortices. *Nonlinearity* **1**, 409–434.

Casdagli, M. (1989) Nonlinear prediction of chaotic time series. *Physica D* **35**, 335–356.

Casdagli, M. (1992) Chaos and deterministic *versus* stochastic non-linear modeling. *J. R. Statist. Soc.* **B 54**, 303–328.

Chaiken, J., Chu, C.K., Tabor, M. and Tan, Q.M. (1987) Lagrangian turbulence and spatial complexity in stokes flow. *Phys. Fluids* **30**, 687–694.

Corrsin, S. (1950) Turbulent flow. *Am. Sci.* **38**, 300–325.

Crisanti, S., Falcioni, Provenzale, A. and Vulpiani (1990) Passive advection of particles denser than the surrounding fluid. *Phys. Lett. A.* **150**, 79.

Crisanti, S., Falcioni, Provenzale, A. and Vulpiani Dynamics of passively advected impurities in simple two-dimensional flow models. *Phys. Fluids A*, in press.

Farmer, J.D. and Sidorowich, J. (1987) Predicting chaotic time series. *Phys. Rev. Lett.* **59**, 8.

Farmer, J.D. and Sidorowich, J. (1988) Exploiting chaos to predict the future and reduce noise. In: *Evolution, Learning and Cognition* (Ed. Y.C. Lee). World Scientific, 277.

Frisch, U., Sulem, P.-L. and Nelkin, M. (1978) A simple dynamical model of intermittent fully developed turbulence. *J. fluid Mech.* **87**, 719–736.

Gollub, J.P. and Solomon, T.H. (1987) Complex particle trajectories and transport in stationary and periodic convective flows. In: *Chaos and Related Nonlinear Phenomena: Where do we go from here?* (Ed. I. Procaccia.) Plenum Press.

Heslop, S.E. and Allen, C.M. (1990) Turbulence and dispersion in larger UK rivers. In: *Proceedings 23rd Congress of the International Association of Hydraulic Research, Ottawa, Canada*, D75–D82.

Heslop, S.E., Holland, M.J. and Allen, C.M. (1994) Turbulence measurements in the River Severn. In: *Mixing and Transport in the Environment* (Eds. K. Beven, P.C. Chatwin and J.H. Millbank). Wiley, Chichester, Ch. 3.

Holland, M.J. (1991) A detailed analysis of turbulence on the River Severn. Preprint. Dept. of Environmental Sciences, University of Lancaster.

Huppert, H.E. (1984) Lectures. In: *Lectures on Geological Fluid Dynamics* (Ed. W. Malcus). *Woods Hole Oceanogr. Inst. Tech. Rep. WHOI-84-44*, WHOI, Woods Hole, 1–71.

Huppert, H.E. (1991) Buoyancy-driven motions in particle-laden fluids. In: *Of Fluid Mechanics and Related Matters* (Eds R. Salmon and D. Betts). Scripps Institution of Oceanography Ref. Ser., San Diego, 141–159.

Mackay, R.S. and Meiss, J.D. (Eds) (1987) *Hamiltonian Dynamical Systems*. Adam Hilger, Bristol.

Maxey, M.R. and Corrsin, S. (1985) Gravitational settling of aerosol particles in randomly oriented cellular flow fields. *J. Atmos. Sci* **43**, 1112–1134.

Meiss, J.D. (1986) Class renormalization: Islands around islands. *Phys. Rev. A* **34**, 2375–2383.

Osborne, A.R., Kirwan, A.D., Provenzale, A. and Bergamasco, L. (1986) A search for chaotic behavior in large and mesoscale motions in the Pacific Ocean. *Physica D* **23**, 75–83.

Packard, N.H., Cruchfield, J:P., Farmer, J.D. and Shaw, R.S. (1980) Geometry from a time series. *Phys. Rev. Lett.* **45**, 712.

Pasmanter, R.A. (1987) Deterministic diffusion. In: *Proceedings of the International Symposium on Physical Processes in Estuaries* (Ed. W. Van Leussen). Springer Verlag, New York.

Powell, M.J.D. (1985) Radial basis functions for multivariate interpolation: a review. In: *IMA Conference on Algorithms for the Approximation of Functions and Data*. RMCS, Shrivenham.

Press, W.H., Flannery, B.P., Teukolsky, S.A. and Vetterling, W.T. (1987) *Numerical Recipies*. Cambridge University Press, Cambridge.

Sauer, T., Yorke, J.A. and Casdagli, M. (1991) Embedology. *J. Stat. Phys.* **65**, 579–616.

Smith, L.A. (1984) Particulate dispersal in a time-dependent flow. In: *Lectures on Geological Fluid Dynamics* (Ed. W. Malcus). Woods Hole Oceanogr. Inst. Tech. Rep. WHOI-84-44, WHOI, Woods Hole, 243–265.

Smith, L.A. (1987) Lacunarity and chaos in nature. *PhD thesis*, Columbia University, New York.

Smith, L.A. (1992) Identification and prediction of low-dimensional dynamics. *Physica D* **58**, 50−76.

Smith, L.A. and Spiegel, E.A. (1985) Pattern formation by particles settling in viscous flows. In: *Macroscopic Modeling of Turbulent Flows*. Vol. 230. *Lecture Notes in Physics*. (Eds. U. Frisch *et al.*). Springer-Verlag, New York.

Smith, L.A. and Spiegel, E.A. (1987) Strange accumulators. In: *Chaotic Phenomena in Astrophysics*. (Eds J.R. Buchler and H. Eichhorn). *Ann. NY Acad. Sci.* **497**, 61−65.

Stommel, H. (1949) Trajectories of small bodies sinking slowly through convection cells. *J. Mar. Res.* **8**, 24−29.

Takens, F. (1981) Detecting strange attractors in fluid turbulence. In: *Dynamical Systems and Turbulence*. Vol. 898. (Eds D. Rand and L.S. Young). Springer-Verlag, New York, 366.

Theiler, J., Eubank, S., Longtin, A., Galdrikan, B. and Farmer, J.D. (1992) Testing for nonlinearity in time series: The method of surrogate data. *Physica D* **58**, 77.

Tong, H. (1990) *Non-Linear Time Series Analysis*. Oxford University Press, Oxford.

Tooby, P.F., Wick, G.L. and Isaacs, J.D. (1977) The motion of a small sphere in a rotating velocity field: A possible mechanism for suspending particles in turbulence. *J. Geophys. Res.* **82**, 2096−2100.

Yu, L., Grebogi, C. and Ott, E. (1990) Fractal structure in physical space in the dispersal of particles in fluid. In: *Non-linear Structure in Physical Systems* (Eds L. Lam and H.C. Morris). Springer-Verlag, New York, 223−231.

21 Continuous Releases of Dense Fluid from an Elevated Point Source in a Cross-flow

P.F. LINDEN and J.E. SIMPSON

Department of Applied Mathematics and Theoretical Physics, University of Cambridge, Cambridge, UK

21.1 INTRODUCTION

In this chapter, a study of the release of dense material from an elevated source in an ambient wind is reported. This study has applications to the accidental release of dense, potentially hazardous gases into the atmosphere, to the motion of cool air descending under a thunderstorm, or to the discharge of buoyant material into a flowing stream. The aims of the work are to establish the spread of this material along the ground under calm conditions and under the influence of wind, and to determine concentrations as functions of distance from the source. Three-dimensional releases of negatively buoyant fluid from constant flux sources are considered in a stationary environment and in an ambient shear flow.

The motion of dense fluid released from a source is determined by the initial momentum and density of the fluid and the elevation of the source. From an elevated source the dense fluid will fall towards the ground in the form of a forced, negatively buoyant plume. The trajectory of the plume and the mixing along the path will be influenced by the velocity of the fluid through which it is falling. On impact with the ground the dense fluid then spreads out horizontally. This horizontal spread is also influenced by the motion of the ambient fluid, which tends to accelerate the dense fluid downwind.

We describe laboratory experiments which investigate both the descent and horizontal spreading phases. A wide range of source conditions is examined to cover the possible scenarios likely to be encountered in practice.

Many of the resulting properties of the flows in which we are interested can be deduced from dimensional analysis, and we describe how this approach was used through suitably chosen experiments to measure the variations in velocity, depth and density of the dense cloud. As these results were obtained in non-dimensional form, they could therefore be applied to flows at full scale.

Earlier experimental work (Simpson and Britter, 1980) has shown the great sensitivity of dense clouds spreading along the ground to any opposing or following flows in the environment. Both the rate of advance and the profiles of the fronts are affected by the ambient flow. The upwind front becomes thinner and wedge-shaped, whereas the downstream front is seen to be thicker and more diffuse.

This work follows on from the study of Linden and Simpson (1990) of two-dimensional

Mixing and Transport in the Environment. Edited by K.J. Beven, P.C. Chatwin and J.H. Millbank
©1994 John Wiley & Sons Ltd

releases of negatively buoyant fluid from an elevated source in calm surroundings. They found that most of the mixing occurs before the jet hits the ground, but then the dense cloud subsequently spreads out as a gravity current. These results are extended here to axisymmetrical releases and we find that the cloud on the ground contains a set of axisymmetrical 'rings' of dense fluid, between which there is almost unmixed environmental fluid. As a result, very high concentration fluctuations are observed compared with releases in two-dimensional flows.

21.2 EXPERIMENTS

The dense fluid was released into a water flume in which a continuous flow could be set up. The flow channel had dimensions of 7 m (length) and 78 cm (width), and could be filled with water to a depth of up to 25 cm. Free-stream velocities of up to 5 cm s^{-1} were used and in most of our experiments the adjustable overflow gate was set to a height just below the maximum operating depth of 25 cm. The flow through the tank was then controlled by raising and lowering the adjustable outflow gate. The return flow was carried through the catch tank to the metering weir and then to the sump.

In most of the experiments the ambient free-stream velocity was measured by releasing a patch of dye along the centreline of the flume at about 10 cm above the floor and measuring its progress by photography or video recordings.

As mentioned earlier, a metering weir was available in which a micrometer gauge could be used to measure the flow depth over a discharge head at a sharp-edged weir. This had the advantage of ensuring that repeatable experimental conditions could be obtained. From Batchelor (1967), the volume flow per unit width

$$Q = 1.05(2/3)^{3/2}g^{1/2}h^{3/2},$$

$$= 17.85h^{3/2} \text{ cm}^2 \text{ s}^{-1},$$

where h (cm) is the height of the free surface at the weir relative to the height far upstream. The width of the weir channel is 61.0 cm and the width of the tank is 78 cm, so that the velocity in tank per centimetre width is $13.96\,h^{3/2}$. Consequently, for a full depth of 25 cm, the velocity $V = 0.558h^{3/2}$ cm s^{-1}. Figure 21.1 shows a plot of V, calculated from this equation, together with a series of values measured at a height of about 10 cm from the dye patches as described. The measured values are always significantly greater than the prediction. To account for the discrepancy between the two estimates of the velocity it is necessary to consider the velocity variations over the cross-section of the channel.

Detailed measurements of velocity profiles with height of flow in this tank have already been made by Sullivan (1968), and two of his results are shown in Figure 21.2, chosen at Reynholds numbers close to those used in our experiments. For example, the depth, H, was usually 25 cm, and with a value for the viscosity $\nu = 0.01$ cm^2 s^{-1}, $Re = VH/\nu$ has the value 5000 for $V = 2$ cm s^{-1} and $Re = 12\,500$ for $V = 5$ cm s^{-1}.

These results enabled us to estimate the ratio between the velocity measured at the height of 10 cm with the 'bulk velocity' as measured by the micrometer needle. In the two examples of Sullivan shown here the ratio is 1.2. In addition, a similar factor also needs to be applied to account for the reduced velocity in the boundary layers at the side of the tank, giving a total ratio of these velocities of 1.4. The value of the corresponding ratio measured from our results in Figure 21.1 is 1.42, close to that deduced from Sullivan's figures.

In connection with a series of wind tunnel experiments on the dispersal of dense gases,

Hall, Hollis and Ishaq (1982) discuss the most appropriate reference velocity to be used in dense gas release experiments. They used an artificial surface roughness which was generated by an array of pegs mounted on the floor, and measured a logarithmic profile and established the roughness length $z_0 = 0.4$ mm. Hall, Hollis and Ishaq (1982) used the friction velocity u_* as the velocity measure which is connected via the roughness length z_0. On the other hand, Simpson and Britter (1980) in their tank experiments, where the velocity was measured well above the boundary layer, used a reference velocity at a height just above the dense cloud. In the Thorney Island trials, the wind speeds were also measured at heights slightly higher than the depth of the dense gas cloud and compared well with the results of Simpson and Britter (1980). These two approaches are equivalent provided that the profiles of the ambient wind are logarithmic and z_0 is scaled appropriately.

As a result of these considerations, and to allow for extrapolation to atmospheric flows, we use the mean velocity measured at 10 cm above the floor, either determined from direct measurements or from the weir flow using Figure 21.1.

A dense salt solution was released from a circular pipe with a diameter in the range 3−8 mm, at a length H above the base of the tank. The salt solution was pumped from a reservoir

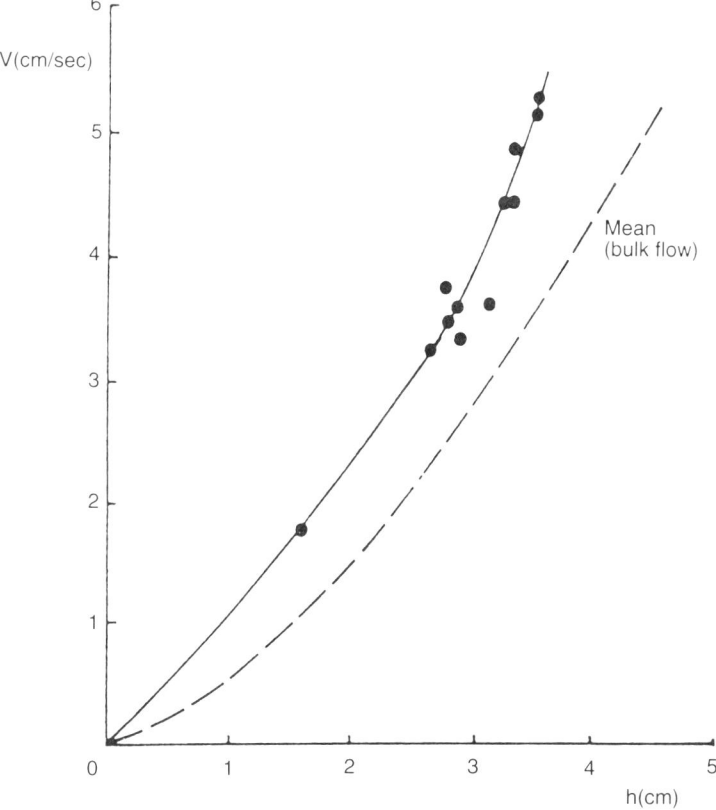

Figure 21.1 Velocity measured by dye patch at about 10 cm compared with bulk velocity over sharp-edged weir. The broken line is that estimated from the weir meter and the solid line is a curve drawn by hand through the data points

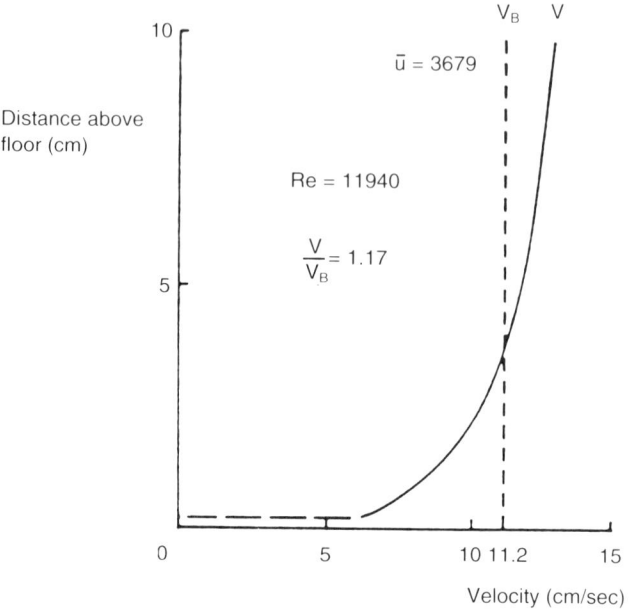

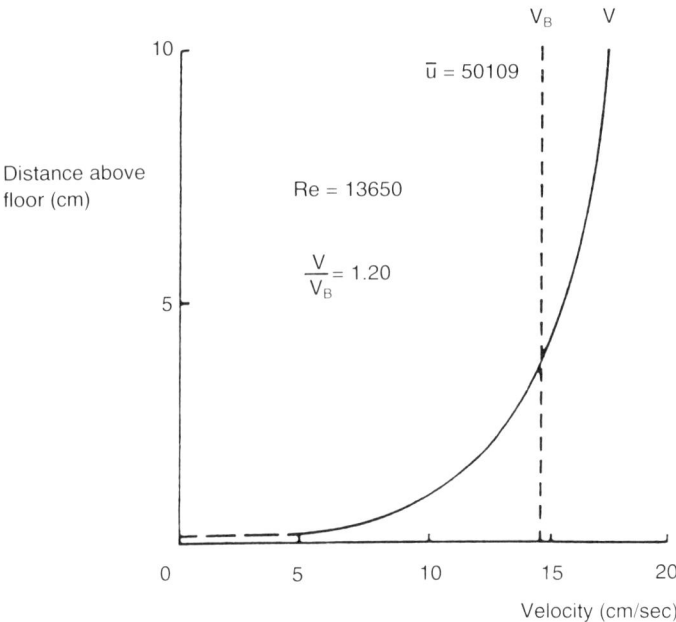

Figure 21.2 Two examples of velocity profiles, after Sullivan (1968)

to which a small amount of dye had been added for flow visualization and the volume flux was monitored using a flow meter. The density differences $\Delta\rho/\rho$ between the salt solution (density $\rho + \Delta\rho$) and the ambient fluid (density ρ) varied from 0.01 to 0.1. Two source heights ($H = 0.12$ and 0.24 m) were used and buoyancy fluxes were in the range 6×10^{-6} to $4.5 \times 10^{-5}\,\mathrm{m^4\,s^{-3}}$. Care was taken to ensure that the flow-rate remained constant and had no significant fluctuations during the course of an experiment.

Measurements were made from video recordings of the flow which were subsequently analysed using an image processing system DigImage. Concentration measurements were made using a conductivity probe, calibrated against density.

21.3 ELEVATED CONTINUOUS RELEASE INTO CALM SURROUNDINGS

The discharges of dense fluid are initially negatively buoyant jets with both momentum and buoyancy. The initial flow may be driven mostly by the momentum of the fluid leaving the orifice, but eventually all buoyant jets act as plumes (Morton, 1959). Analysis has often been based on the assumption that the discharge behaves almost entirely as a plume, but in many instances the initial momentum of the flow can be important. In this work we have investigated its effects on the speed and the dilution of the fluid advancing on the ground.

The primary variables governing the behaviour of axisymmetrical buoyant jets are: volume flux Q (dimensions L^3T^{-1}); momentum flux M (dimensions L^4T^{-2}); and buoyancy flux B (dimensions L^4T^{-3}).

Provided the Reynolds number, Re, is large enough, $Re > 4000$, for a round jet with the Reynolds number defined as $Re = M^{1/2}/\nu$, all the properties of the buoyant jet are determined by Q, M, B and the distance z from the source.

Dimensional analysis shows that the only characteristic length scale for a buoyant jet from a point source, involving both M and B, is given by the jet length $L_m = (M^{3/4})B^{-1/2}$. The parameter which controls whether a buoyant jet is jet-like or plume-like near the ground is the ratio of the height of release, H, to L_m. For $H/L_m \ll 1$ the flow is like a jet and the initial momentum effects are still important, and for $H/L_m \gg 1$ the flow is like a plume and is determined primarily by the buoyancy flux.

21.3.1 Speed and height of spread

Dimensional analysis shows that the distance x travelled along the ground in time t measured from the time of impact should be given by

$$x = C_1 (H/L_m) B^{1/4} t^{3/4}, \tag{21.1}$$

where B is the buoyancy flux. Figure 21.3 shows the diameter of the spread plotted against $t^{3/4}$. It can be seen that the graph soon attains the form of a straight line supporting the dimensional relationship (21.1) enabling the function C_1 to be measured. For this example, in which the value of $H/L_m = 9.25$, $C_1 = 0.63$.

Values of C_1 are found to increase for smaller values of H/L_m, and the graph in Figure 21.4a shows how the value of C_1 varies with the value of H/L_m. For momentum-dominated flows, $H/L_m \ll 1$, C_1 approaches a value of unity; it is about equal to 0.7 when $H/L_m = 1$ and approaches a value of 0.65 for large values of H/L_m. Figure 21.4b shows a log/log plot of these results.

The height, h, of the spreading fluid was measured using video and still photography.

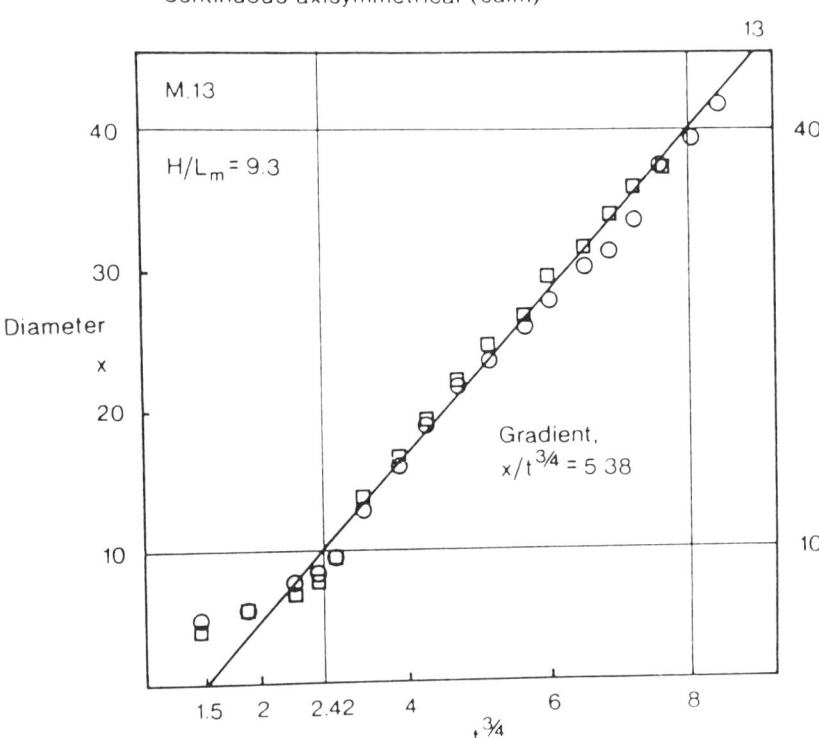

Figure 21.3 Diameter of spread of the dense cloud along the ground plotted against $t^{3/4}$

Dimensional analysis gives an expression for this height of

$$h = C_2(H/L_m)H^3B^{-1/2}t^{3/2} \qquad (21.2)$$

and the graph in Figure 21.5 shows experimentally how this height changed in one run, with $H/L_m \gg 1$, after the beginning of the spread. The value of $ht^{3/2}$ was thus obtained, and, using the appropriate value of buoyancy flux B and height of release H, it was found that $C_2(H/L_m) = hB^{1/2}t^{3/2}/H^3$ had the value of 0.17. There was no significant variation of C_2 with the parameter H/L_m.

21.3.2 Dilution measurements

A study was made of the density changes during the horizontal spread from the circle of impact, using a conductivity probe taking readings at a frequency of 10 Hz.

   For the purposes of the density measurements the flow can be divided into three regions, as shown in Figure 21.6. Section A is the descent from the source elevation to the ground, during which there is considerable mixing, the processes of which have been much studied in the past. Section C is also a familiar regime as it appears to behave like a gravity current, in which there is known to be very little internal mixing behind the head. As a result, the density in section C is nearly constant during any particular run.

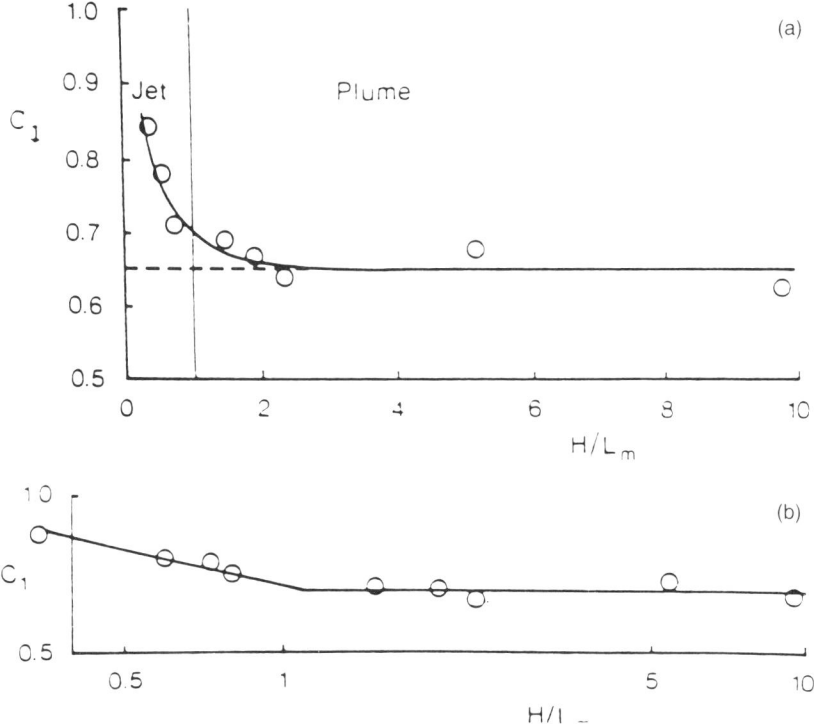

Figure 21.4 (a) Graph showing how the value of $C_1 = x/B^{1/4}t^{3/4}$ varies with the ratio of H to the 'jet length', $L_m = M^{3/4}/B^{1/2}$. (b) Log–log plot of C_1 with H/L_m for the data shown in (a).

Section B corresponds to the entraining hydraulic jump seen in two-dimensional flow and described in Linden and Simpson (1990). This was found to be a zone of intense mixing, with some special characteristics which we describe here.

A major new result is the observation that a 'zone of rings' is generated close to the impact circle. This zone consists of rings which pass at about 1 Hz and gradually become less clear with distance from the centre. Figure 21.7 shows examples of enhanced digitized records from the video showing plan views of the flow. The 'rings' can be seen as dark circles indicating the presence of deeper regions of dyed fluid. The squares shown measured 10 cm and a salinity probe was used at 10 cm from the centre point to produce a record, part of which appears in figure 21.8. This record shows a series of oscillations which were in phase with the waves seen in the plan view. It is notable that the waves in the salinity trace nearly all have the same minimum values, close to that of the external fluid.

Figure 21.9 shows the results of a series of measurements of the dimensionless density of the spreading cloud at increasing distances from the centre of impact. This distance is scaled in terms of the height, H, of release, and measures from 0 to 3. It appears that the RMS value of the oscillating density readings decrease to a steady value, $C_3 = g'H^{5/3}/B^{2/3}$, the non-dimensional value of the density, of about 2.5 for distances more than 1.5 times the height H. The lower graph shows how the variability decreases through this range, as the standard deviation of these results decreases to an almost steady low value of 0.2 for $x/H > 2$.

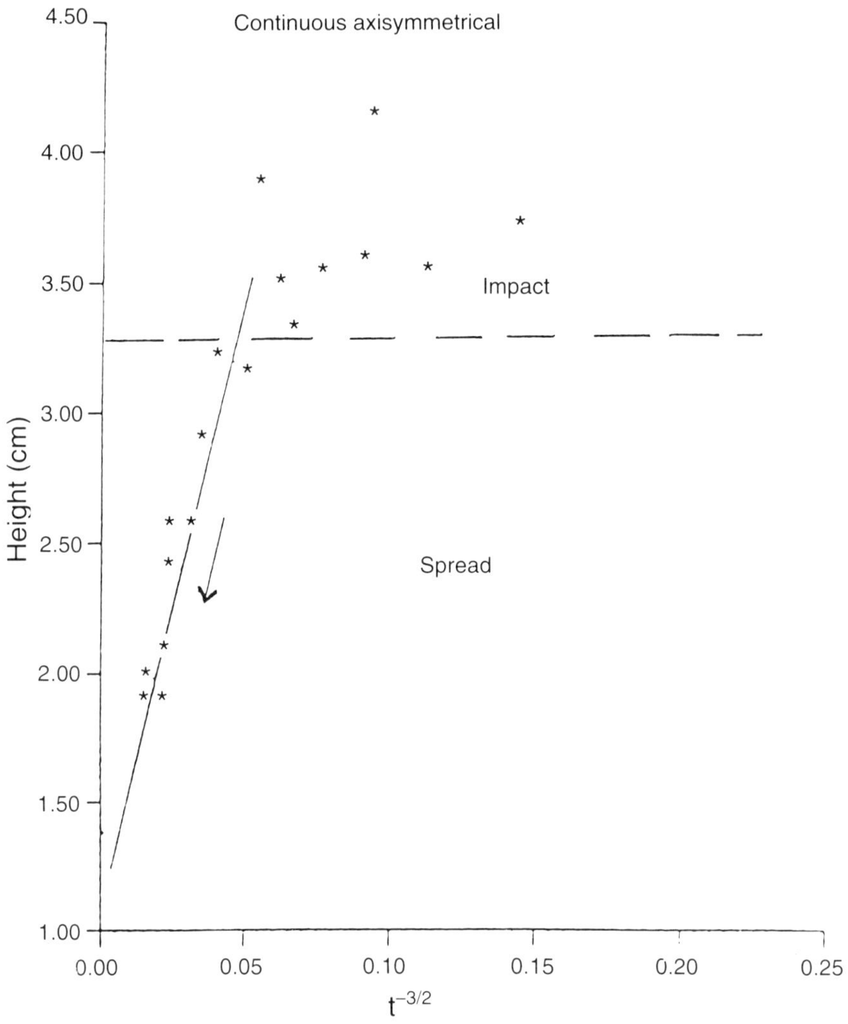

Continuous axisymmetrical

Height (cm)

$t^{-3/2}$

Impact

Spread

Figure 21.5 Height of the flow after impact $h = C_2 H^3 B^{-1/2} t^{-3/4}$. The points in the impact zone refer to those measured directly under the source, soon after the dense fluid has reached the ground.

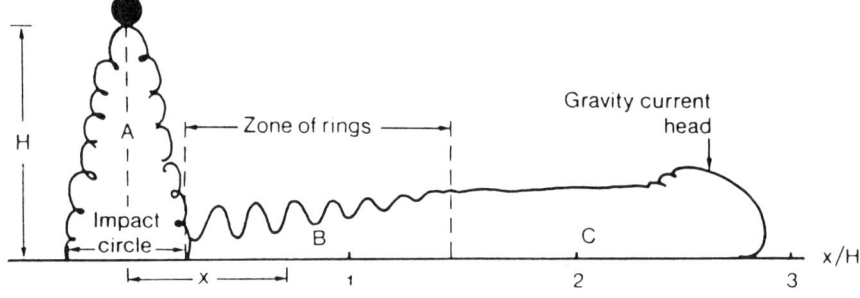

Figure 21.6 Flow from elevated release divided into three regimes

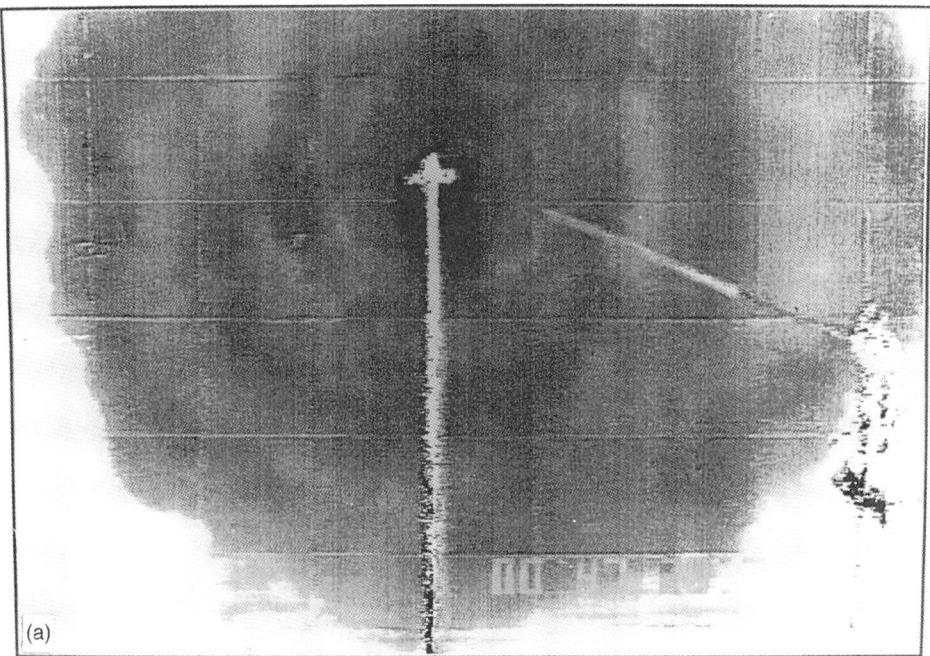

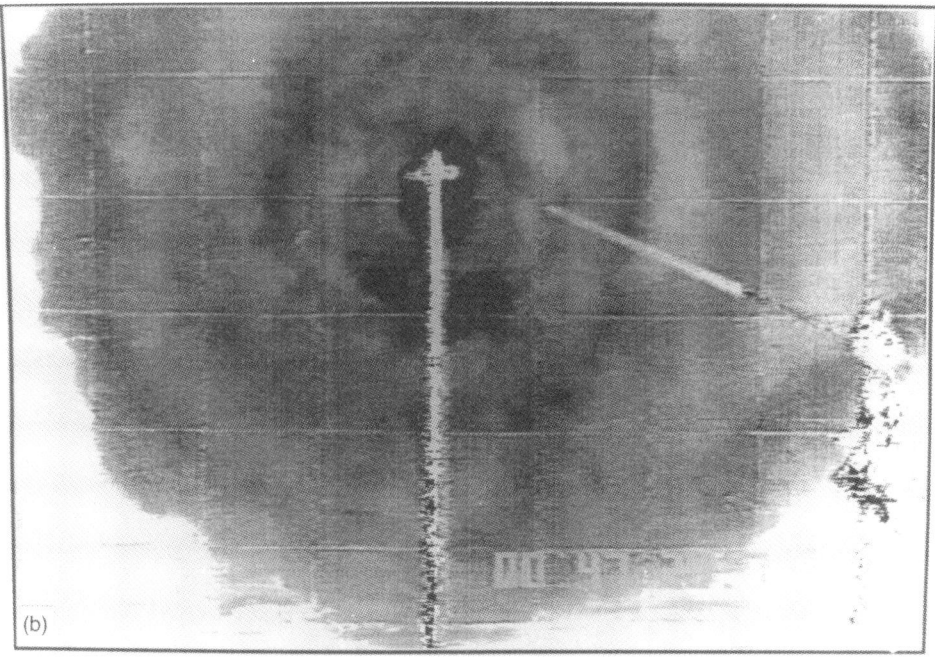

Figure 21.7 (a) Experiment Y47: plan view of the waves in mixing zone at $t = 24.4$ s. (b) Experiment Y47: plan view of the waves in mixing zone at $t = 24.9$ s

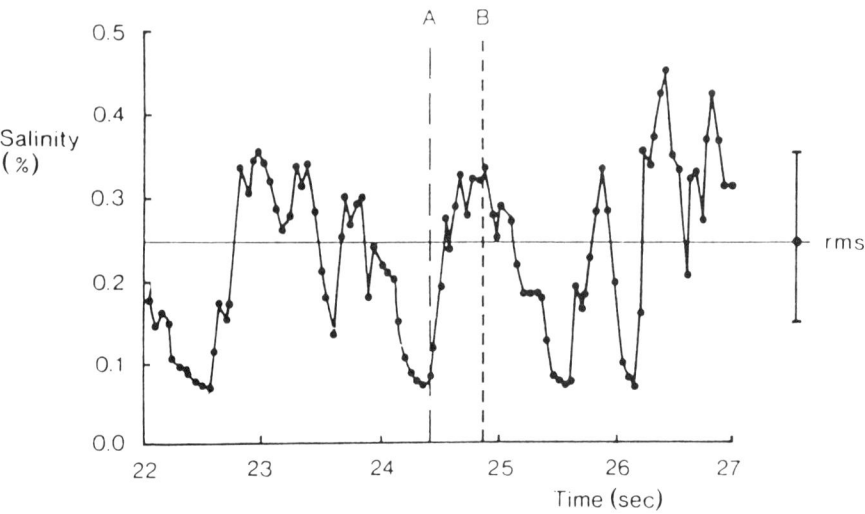

Figure 21.8 Salinity trace to illustrate the flow shown in Figure 21.6. A and B mark salinity at recording times 24.4 and 24.9 s, corresponding to the views shown in Figure 21.7

Now we have established the general nature of the flow in region B we are in a position to investigate the results of the final mixing as appearing in the gravity current regime, and its dependence on the initial momentum. These results appear in Figure 21.10 and show that the final density, g', measured far downstream, non-dimensionalized by $H^{5/3}/B^{2/3}$ is strongly dependent on the initial excess momentum, characterised by H/L_m. For a large initial momentum the values of C_3 are small and the values increase as we approach the pure plume state, reaching a final value of about four. The graph in Figure 21.11 is a log−log plot of the data shown in Figure 21.10. This shows, as before, how in the plume section, $H/L_m \geq 2$, the values approach the straight line $C_p = 4.2$. It also gives an indication of the behaviour in the uet section where the density values approach the line $C_j = 2.2H/L_m$. The lines cross at $H/L_m = 1.7$.

These results for descents in calm surrounding are important, as we shall see in the next section that the ambient flow has only a minor effect on the measured density.

21.4 CONTINUOUS FLUX IN CROSS-FLOW

21.4.1 Spread of the cloud

Figure 21.12 shows the appearance of the descent and spread of the dense fluid in a moderate ambient flow. As can be seen in Figure 21.12a, the side elevation, the thickness of the flow is less towards the upstream front, where the head is flattened. The downwind front is deeper and more diffuse. The plan form is shown in Figure 21.12b. Apart from the forms of the fronts, the chief difference seen between these results and those in calm surroundings is the fact that it was often possible to bring the upstream boundary of the spreading flow to rest. (In an unbounded domain, the upstream front will always be arrested by the external flow.

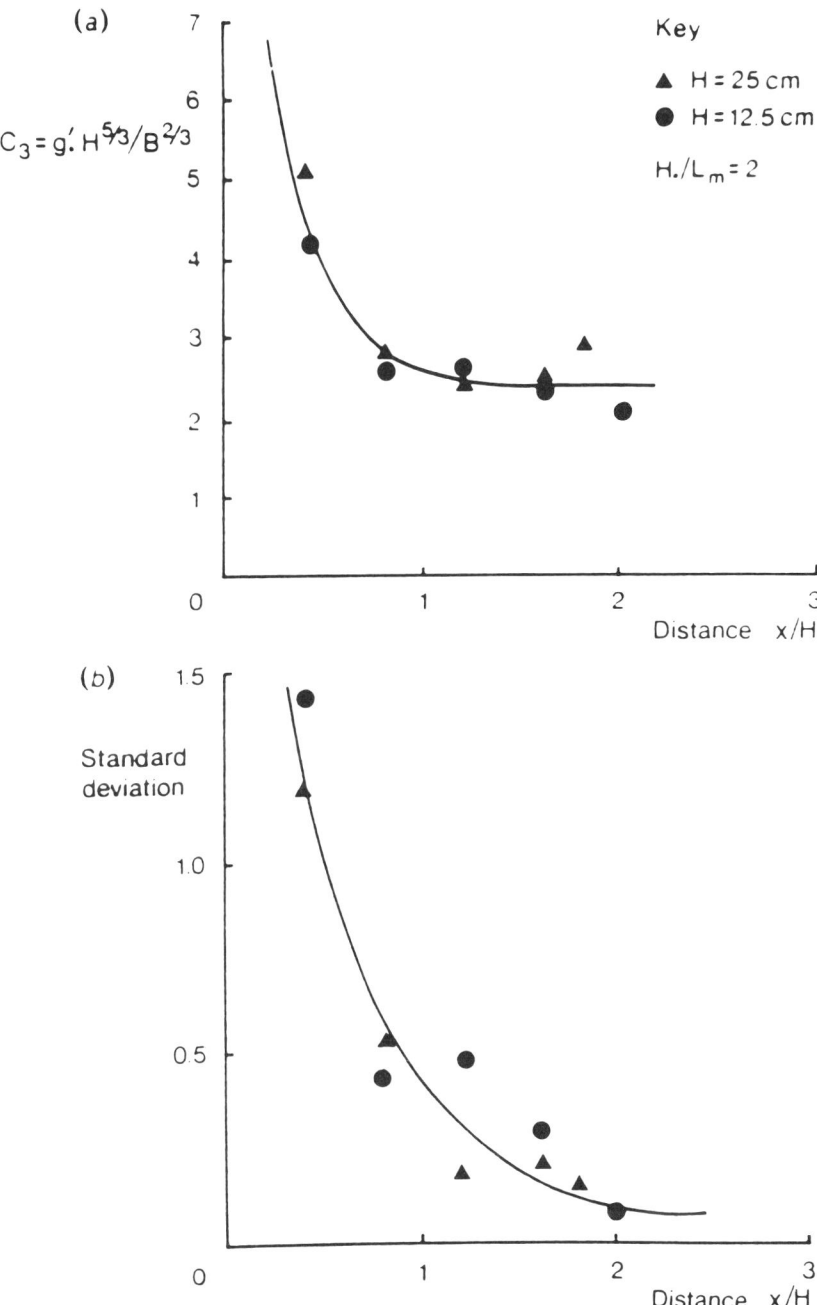

Figure 21.9 Dimensionless density measured at distance from impact circle. (a) RMS values; (b) standard deviation

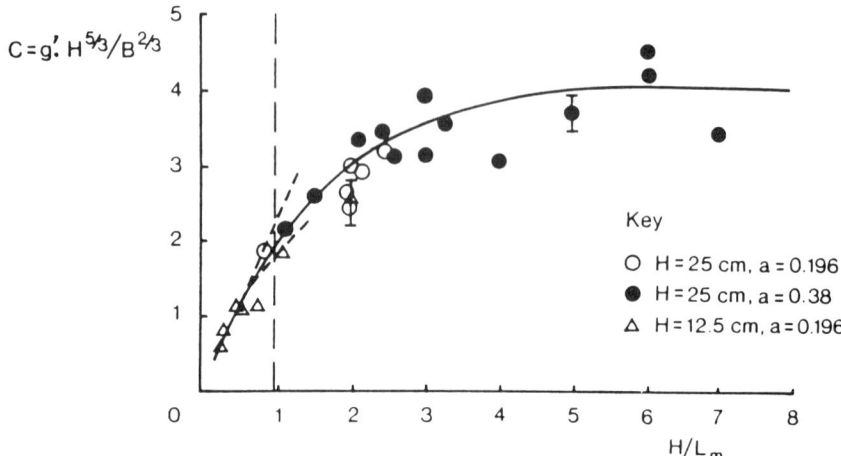

Figure 21.10 Relationship between the dimensionless density C_3 measured far downstream and H/L_m

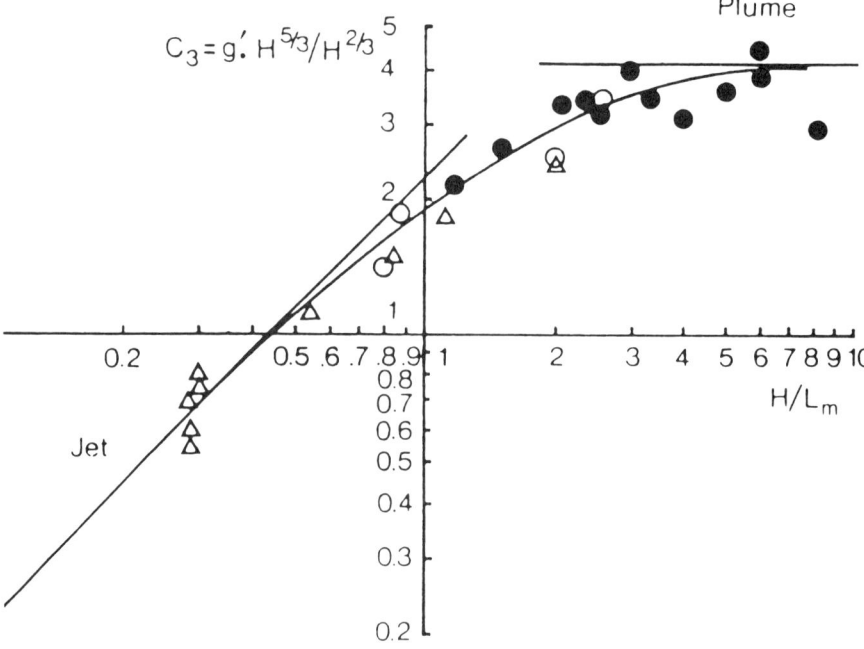

Figure 21.11 As Figure 21.10, but a log–log plot

In the confined geometry of the laboratory tank, some fronts can propagate continuously upstream once the dense fluid fills the full width of the flume.) This feature is shown in Figure 21.13, which shows the position of the front at intervals of two seconds.

The measurements from each experiment were plotted to show the change in position of the upstream and downstream boundaries with time. Figure 21.14 is an example of one of

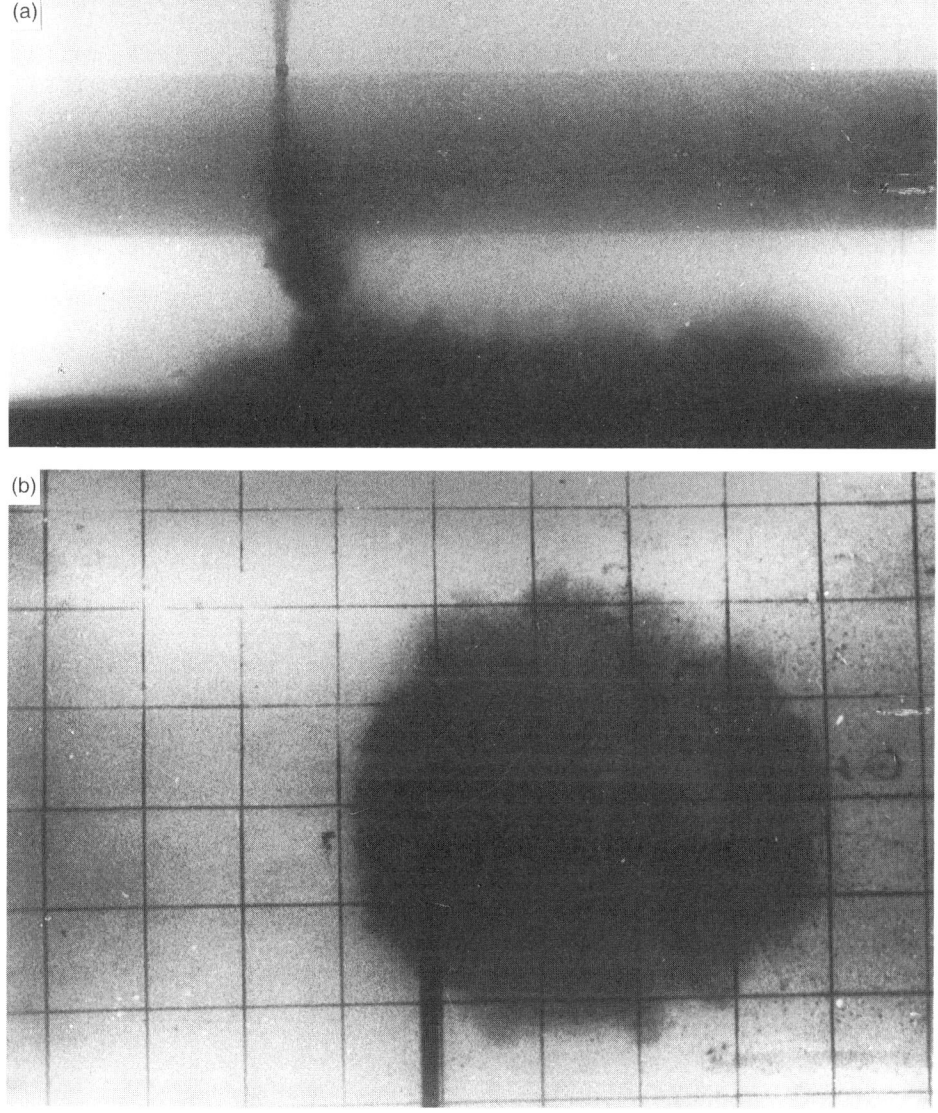

Figure 21.12 (a) Elevation and (b) plan views of descent and spread of constant flux into cross-flow, from left to right

these plots, which also includes a line showing the strength of the ambient flow, and also a (broken) line giving the advance of the centre of the spreading circle. From such graphs the spread of the fronts was deduced by drawing tangents to the curves and measuring the gradient of the tangents at intervals of about one second.

Dimensional analysis suggests that in a continuous release into a cross-flow the distance upstream where the front comes to rest is proportional to B/V^3. Using data derived from graphs of the form shown in Figure 21.14 it is possible to deduce the values of the speed at different distances, and these results are shown replotted in Figure 21.15 with both the

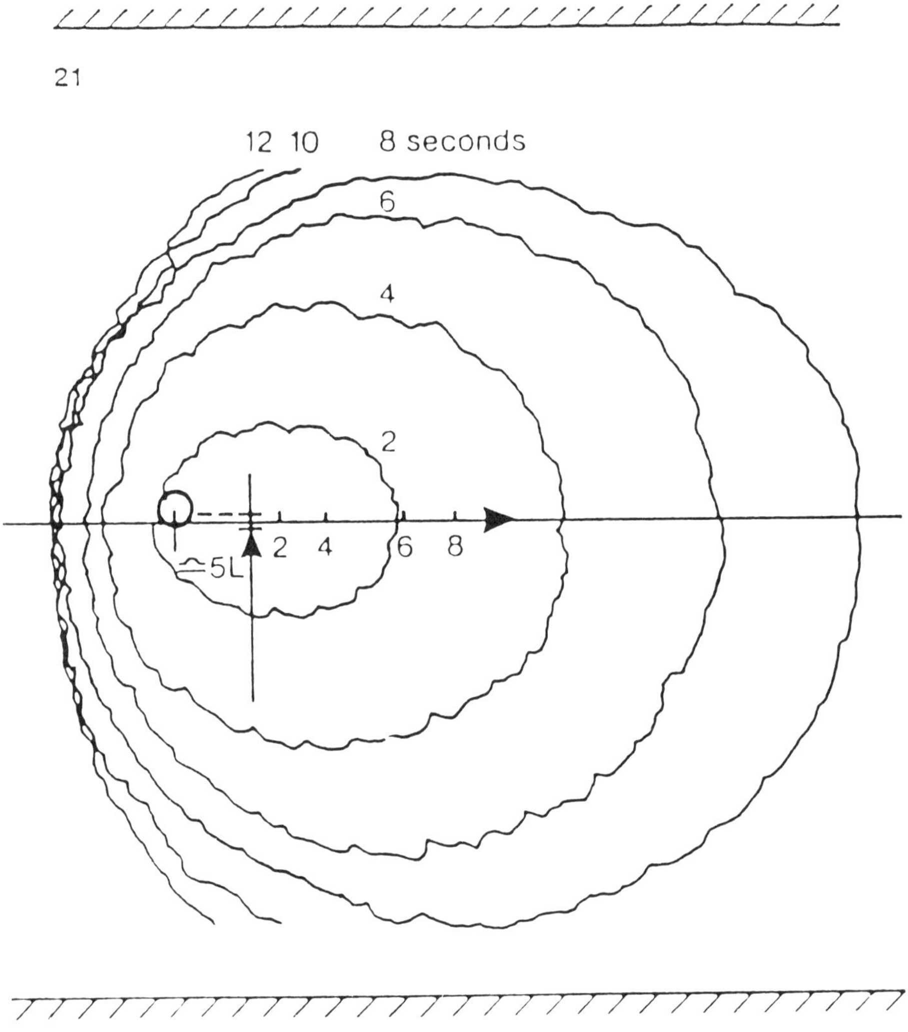

Figure 21.13 Plan form of the ground spread. The small circle indicates the position of the source and the marks on the centreline indicate the mid-point of the upstream and downstrean fronts at subsequent times (seconds)

speed and the distance suitably non-dimensionalized. These data show that the velocity of the downstream front approaches that of the cross-flow when x/BV^{-3} is about equal to 1.0. The upstream front comes to rest when x/BV^{-3} has about the same value. A rather large spread in the values shown in Figure 21.15 of x_*/BV^{-3}, where x_* is the distance of the upstream point of rest, can be partly explained by the failure to allow for the excess momentum present at the point of release. The magnitude of this effect was examined by plotting the non-dimensional distance $X = x_*/BV^{-3}$ against H/L_m, the ratio of the depth of descent to the 'jet length'. The results, plotted in Figure 21.16, show a clear steady variation of X with the jet length ratio. It is shown that, as the pure plume regime is approached, the distance, x_*, of the upstream foremost point approaches a value of $0.85BV^{-3}$.

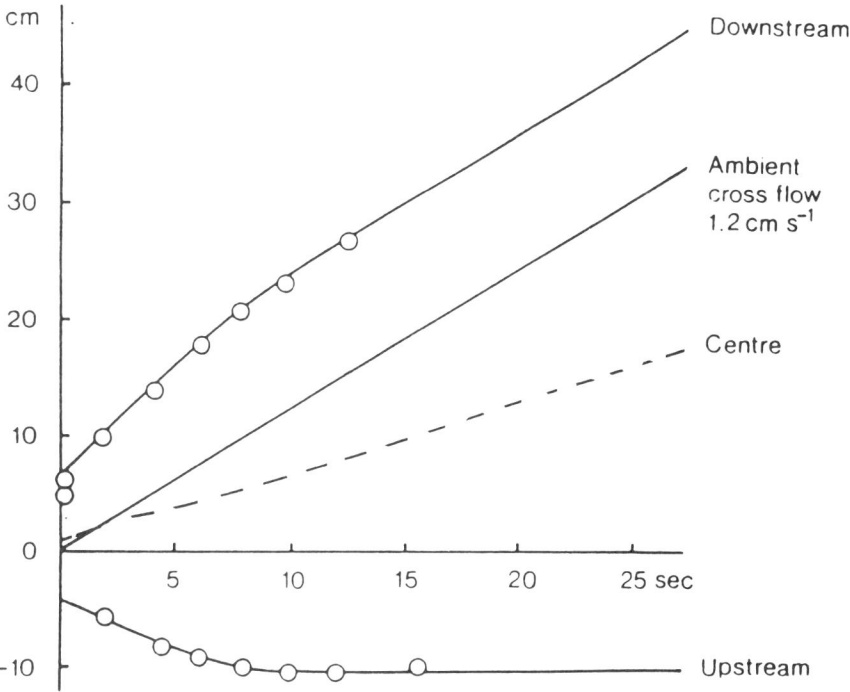

Figure 21.14 Upstream and downstream boundaries in ground spread

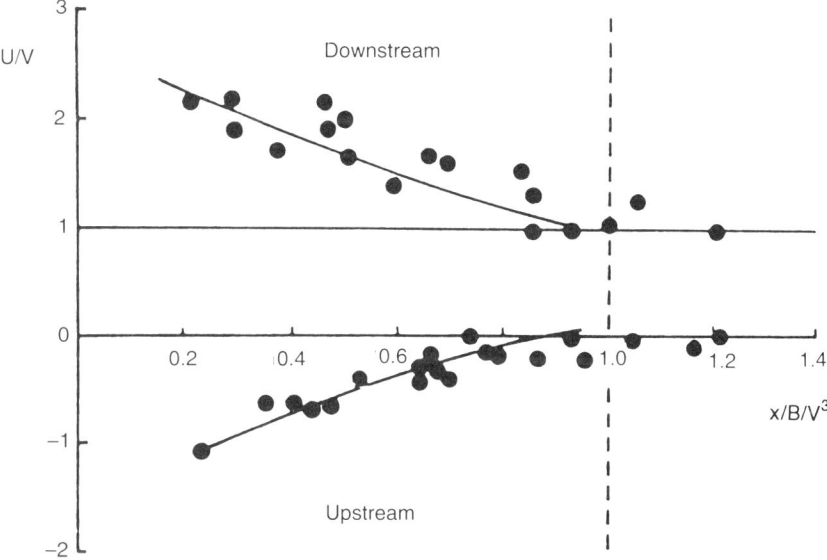

Figure 21.15 Velocities of upstream and downstream points

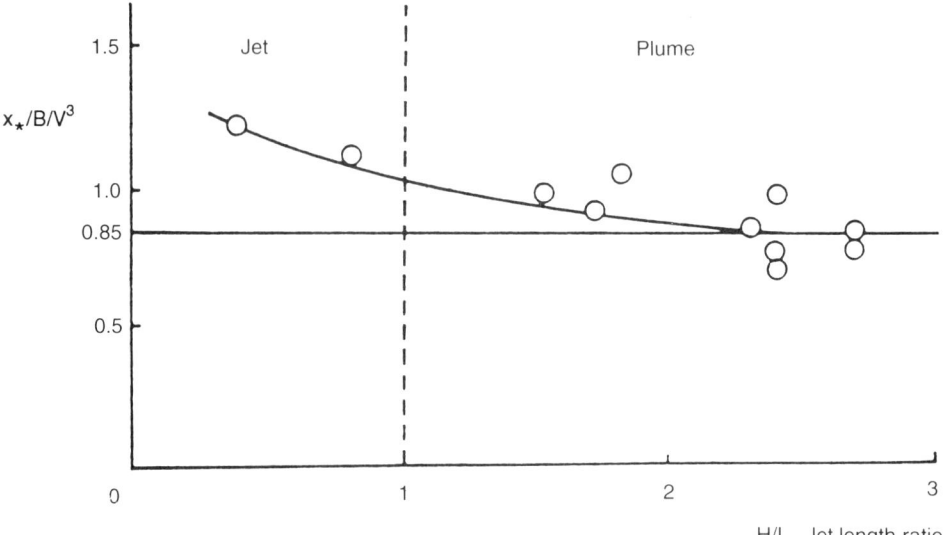

Figure 21.16 Dependence of the position of the maximum upstream point, x_*, of the dense cloud on the jet length ratio, L_m

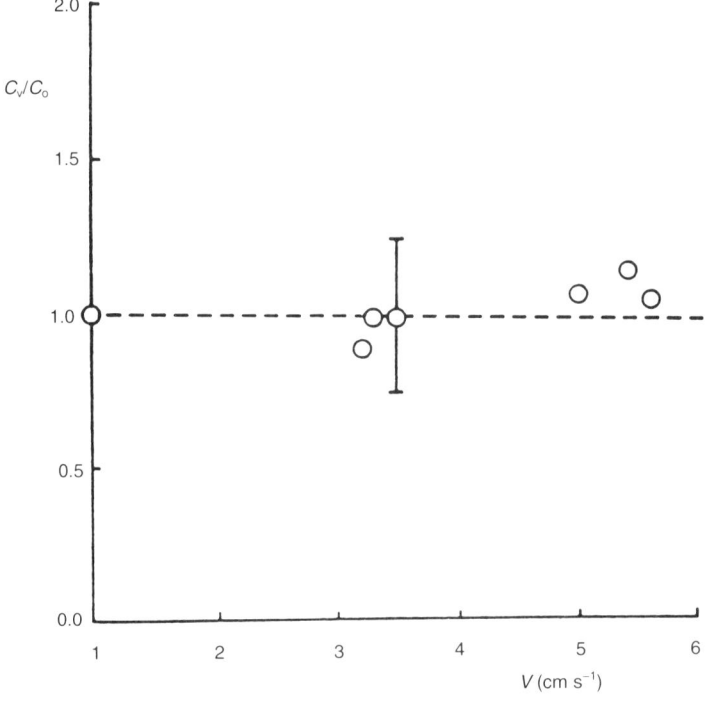

Figure 21.17 Effect of cross-flow on dilution. $C_V = g'/(H^{5/3})/(B^{2/3})$, measured for different values of V, the speed of the cross-flow. C_0 is the value of C_V in calm conditions

21.4.2 Dilution measurements

The mixing during the descent regime in a steady cross-flow cannot be affected in any way by a steady sideways advection without velocity shear. As most of the main mixing occurs during the descent, we would not expect the final dilution to be appreciatly different in the presence of a cross-wind. As stated in Section 21.3.2, the final density acquired in the gravity current regime after descent in calm surroundings is dependent only on the initial buoyancy flux, B, and the height of descent, H, and is given by $g' = C_3(H^{5/3})/(B^{2/3})$. We showed (see Figure 21.11) that the value of C_3 is strongly dependent on the value of H/L_m, and rises to a value of about four for a pure plume with negligible initial excess momentum.

To check the validity of the suggestion that the dilution does not depend on the value of the ambient flow a series of experiments was carried out. In each set of experiments runs were made with identical B and H and the value of $C_3 = g'/[(H^{5/3})/(B^{2/3})]$ was measured for three different strengths of the ambient flow, V. Denoting the determined value of C_3 in calm conditions by C_0 and that in cross-wind, V, by C_v, the value of C_v/C_0 was plotted against the value of V. The results in Figure 21.17 show that the measured difference of C_v/C_0 in every instance was less than that of the probably experimental error.

21.5 SUMMARY AND CONCLUSIONS

A series of laboratory experiments on the flow of a negatively buoyant jet from an elevated point source in a cross-flow have been carried out. As the descent of the jet is unaffected by a uniform air flow we have concentrated on the effects of the cross-flow on the spread of dense material along the ground.

The head of the upstream front on the ground becomes smaller and flatter than that at the downstream boundary front. the upstream front comes to rest where the non-dimensional distance, x/BV^{-3}, reaches a critcal distance, the value of which varies with the jet length ratio, H/L_m, from a value of 1.5 for a pure jet to about 0.85 for a pure plume. At about the same non-dimensional distance the downstream front reaches the speed of the ambient flow. As a result of the cross-flow the shape of the cloud is no longer circular, in contrast with what was found for constant volume releases. The downsteam front is accelerated until it acquires the same downstream speed as the ambient flow, whereas the upstream front remains fixed.

The final dilution is little affected by cross-wind and the final density reached is a value given by $g' = C_3 H^{5/3}/B^{2/3}$. The value of C_3 is strongly dependent on the jet length ratio, attaining a maximum value from an initially pure jet of about four. This value is attained asymptotically at large distances downstream. Nearer the source the ring structure also observed in calm conditions was found, but the shape of the rings was no longer circular. On the upstream side the rings were concentrated between the impact position and the foremost upsteam point, and as a result the fluctuations in concentration have larger spatial gradients there. On the downstream side the rings are stretched out by the mean flow.

ACKNOWLEDGEMENT

This work was supported by a grant from the Health and Safety Executive.

REFERENCES

Batchelor, G.K. (1967) *An Introduction to Fluid Dynamics*. Cambridge University Press, Cambridge.

Hall, D.J., Hollis, E.J. and Ishaq, M. (1982) A wind tunnel model of the Porton dense gas-spill field trials. *Rep. LR 394 (AP)*, Warren Spring Laboratory, Stevenage.

Linden, P.F. and Simpson, J.E. (1985) Buoyancy driven flows through an open door. *Air Infiltration Rev.* **6**, 4–5.

Linden, P.F. and Simpson, J.E. (1990) Continuous two-dimensional releases from an elevated source. *Loss Prev. Process Ind.* **3**, 82–88.

Morton, B.R. (1959) Forced plumes. *J. Fluid Mech* **5**, 151–163.

Simpson, J.E. (1987) *Gravity Currents in the Environment and the Laboratory*. Ellis Horwood, Chichester.

Simpson, J.E. and Britter, R.E. (1980) A laboratory model of an atmospheric mesofront. *Q. J. R. Meterol. Soc.* **106**, 485–500.

Sullivan, P.J. (1968) Dispersion in a turbulent shear flow. *PhD Dissertation*. University of Cambridge.

22 Particle Tracking Model of Sediment Transport

A. KELSEY, C.M. ALLEN and K.J. BEVEN

Centre for Research on Environmental Systems and Statistics, Institute of Environmental and Biological Sciences, Lancaster University, UK

and

P.A. CARLING

Institute of Freshwater Ecology, Windermere Laboratory, Ambleside, Cumbria, UK

22.1 SEDIMENT TRANSPORT

The total bedload transport of sediment in a river consists of the sum of the movements of individual particles set in motion by the flow. Owing to the number of particles in motion it is often possible to describe this transport in a continuum sense (e.g. Holly and Rahuel, 1990); however, the underlying system is discrete. The motion of the individual particles is influenced by the flow, the nature of the bed and interactions with other particles, whereas movements of individual particles will in time modify the flow, the bed and the behaviour of other particles. The system can be described by considering flow, transport and boundary components, together with feedback mechanisms occurring between them (see, for example, Leeder, 1983; Figure 1).

Both stochastic and physical models of the dynamics of particle motions have been used in the past. Einstein (1937) produced a model of bedload transport based on the exponential distributions of the distance moved and the time spent resting by particles, acknowledging the stochastic nature of sediment transport. This approach to sediment transport has also been used by Sayre and Hubbel (1965), Shen and Todorovic (1971), Stelczer (1981) and Hassan, Church and Schick (1991). Although these applications have shown the possibility of modelling sediment distribution using this technique, the number of particles necessary to identify the appropriate distribution is thought to be more than 10^3 (see Hassan, Church and Schick, 1991), while the downstream distribution of movements will in general be affected by large-scale structures in rivers, e.g. pool−riffle structures.

A physical approach to calculating particle movements has been used by Sekine and Kikkawa (1988), Wiberg and Smith (1985) and van Rijn (1984). In all these models the particle trajectory due to the mean flow is calculated. The queueing model of Naden (1987b) also uses a physical approach, but the particle trajectories used to describe the motion are those observed by Abbott and Francis (1977). In the model of Naden (1987b) the probability of initial motion of a particle due to turbulence is calculated and the particle motion is then calculated from the observed saltation characteristics, so incorporating the effects of turbulence on the saltation path. The models in which the particle tracks are calculated directly ignore this influence, though a

Mixing and Transport in the Environment. Edited by K.J. Beven, P.C. Chatwin and J.H. Millbank
© 1994 John Wiley & Sons Ltd

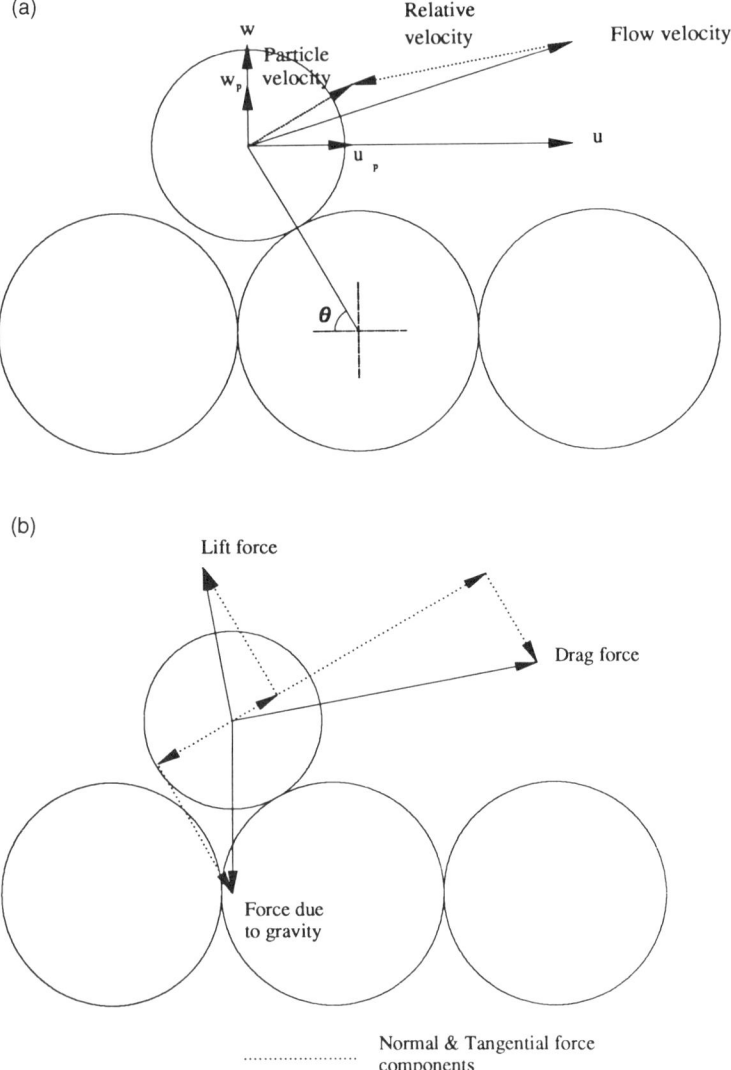

Figure 22.1 Rolling motion of particle. (a) Flow and particle velocities determining relative velocity. (b) Forces due to flow and gravity. The normal component of force determines whether the particle remains in contact with the bed, the tangential component of force affects the particle motion

stochastic element is introduced by including a random element in the impact process, allowing the calculation of a range of saltation trajectories (Sekine and Kikkawa, 1988; Wiberg and Smith, 1985).

Inclusion of the effects of turbulence in the fluvial sediment transport of bedload has also been modelled by a range of approaches. Bagnold (1973) splits particle motion into saltation, which he assumes to be purely ballistic with particle trajectories unaffected by turbulence, and suspended, where the turbulent fluctuations of the flow are capable of supporting particles completely. From laboratory observations Abbott and Francis (1977) suggest that although saltations should cover the purely ballistic trajectories there is a region between this and the fully suspended region where saltation trajectories are modified by turbulence, causing upward acceleration of the particle after it has left contact with the bed. They suggested that this region should also be described as suspended, and observed that modification of saltation trajectories could occur even at low transport stages, defined as u_*/u_{*cr} — that is, the ratio of mean bed shear velocity to the critical shear velocity required to initiate particle motion. Observations in the field (e.g. Drake *et al.*, 1988) also show the importance of turbulence in the transport process, with the high velocity turbulent fluctuations allowing a range of particles to be entrained into the flow and transported at low transport stages. In modelling the response of particles to turbulence in the aeolian environment, this region has been called the region of 'modified saltation' (Anderson, 1987; Hunt and Nalpanis, 1985).

The presence of trajectories modified by turbulence at even low transport stages and the importance of turbulence in the entrainment of particles indicates that, if possible, a particle based model of fluvial sediment transport should include the effects of turbulence. Here there effects are included by the use of a particle tracking approach to modelling particle and flow behaviour.

22.1.1 Particle tracking

In particle tracking models of dispersion the dispersion due to turbulence of a cloud of pollutant or tracer is modelled by tracking a large number of particles for a period of time or over a distance. The particles represent the pollutant or tracer; their positions are tracked from a source in a series of steps, taking account of the local mean and turbulent flow conditions at each step. Thus an individual path is calculated for each particle based on the local flow conditions experienced by that particle. This approach to modelling dispersion is therefore explicitly Lagrangian, with the position of each particle being dependent on its history. Particle tracking models have been developed to model dispersion in the atmosphere, rivers and estuaries. They also offer advantages in modelling the dispersion of materials that interact with the environment.

Most of the work on particle tracking models of dispersion in open channel flow has been for passive pollutants (e.g. Allen, 1982). In these models the behaviour of the particles is identical to that of the fluid; parameters describing the flow can be used directly in calculations describing the dispersion of a tracer. In sediment transport of bedload the particles are non-passive, lagging behind rather than following the flow completely. Particle tracking models have been developed for heavy, non-passive tracers; for example, Zhuang, Wilson and Lozowski (1989) modelled the data of Snyder and Lumley (1971), and Yvergniaux and Chollet (1989) modelled the behaviour of a single particle moving in suspension in an open channel flow observed by Sumer and Deigaard (1981). Anderson (1987) and Hunt and Nalpanis (1985) modelled the response of saltating particles to turbulent fluctuations in the aeolian environment. In this chapter a particle tracking approach to sediment transport in fluvial environments is developed and applied to the calculation of the movement of single particles observed in laboratory experiments. An extension to multiple particles of different sizes with stochastic boundary conditions is also briefly described.

22.2 FLOW MODELLING

As the model is two-dimensional, only the streamwise and vertical components of flow are calculated and the velocities are non-dimensionalized with respect to the mean bed shear velocity, u_*. The horizontal and vertical components of flow velocity, u, w, are modelled by the mean flow components, U, W, to which are added random fluctuating components of velocity, u', w', to represent turbulence.

$$u = U + u'$$

$$w = W + w'$$

At the start of each iteration the flow conditions are set according to the position of the particle centre and the flow state and these are then kept constant for the duration of the iteration.

22.2.1 Mean components of velocity

The mean streamwise component of flow, U, at height z is calculated from a log law profile for turbulent flow over a rough boundary

$$U = \frac{1}{\kappa} \log\left(\frac{30z}{k_s}\right)$$

The value of the roughness length, k_s is set from the experimental data used. The zero height is set at 0.2 characteristic diameters, d_{char}, below the tops of the bed particles and the velocity profile is assumed to apply down to the zero velocity height. The velocity profile normally diverges from the logarithmic profile close to the bed due to viscous effects and the presence of the elements making up the bed roughness, but these effects are ignored here. The mean vertical component of velocity, W, is zero from continuity considerations.

22.2.2 Fluctuating components of velocity

The fluctuating components of velocity are calculated from Gaussian distributions and the magnitudes of the fluctuations are set using the values of the standard deviations from the data of McQuivey (1973). The variation with depth has been calcualted (after Naden, 1987a) as

$$\frac{\sigma_u}{U} = 0.16\left(\frac{z}{d_{char}}\right)^{-0.65}$$

$$\sigma_w = 0.77\sigma_u$$

where σ_u and σ_w are the standard deviations of the velocity fluctuations u', w' at height z and d_{char} is the characteristic diameter of the bed particles.

22.2.3 Correlation of turbulent fluctuations

From the observations, the distributions of horizontal and vertical velocity fluctuations are almost Gaussian, whereas the distribution of the product of these fluctuations, $u'w'$, is observed to be skewed and have high kurtosis (Heathershaw, 1979). Assuming a Gaussian joint probability and knowledge of a value for the correlation coefficient, r, between u' and w', the values for the conditional mean, $\hat{W}$, and standard deviation, $\hat{\sigma}_w$, of the vertical velocity at an iteration can be calculated, given the value of the horizontal velocity fluctuation at that iteration

$$\hat{W} = r\frac{\sigma_w}{\sigma_u}u'$$

$$\hat{\sigma}_w = (1 - r^2)^{1/2}\sigma_w$$

Use of these expressions allows the calculation of turbulence with the same mean correlation coefficient as a set of data. The correlation between the streamwise and vertical velocity fluctuations was initially calculated from a correlation coefficient from Heathershaw (1979), giving a value of -0.18. This is a measurement for turbulent flow over the sea bed in a tidal current and may not be applicable for use in rivers. The correlation coefficient represents a mean value for a trace. However, in turbulent flow the mean correlation coefficient is the result of high correlation and activity occurring for short periods of time, separated by periods of relative inactivity. The periods of high activity associated with burst and sweep processes are characterized by high correlations between the velocity components and large contributions to the turbulent shear stress. Observations in the fluvial environment (Drake *et al.*, 1988) and in the marine environment (Williams, 1990) indicate that these structures are important in sediment transport processes, especially close to the threshold for sediment movement.

An attempt has been made to include the effects of these correlated signals by breaking the flow modelling into periods of flow with high negative correlation corresponding to bursts and sweeps, positively correlated flows corresponding to up-accelerations and down-decelerations, and a period of flow with low correlation corresponding to the quiet period.

22.3 PARTICLE PROCESSES

The calculation of the track of sediment particle motion depends on the representation of a number of processes. At the beginning of each iteration the appropriate process is selected, based on the present particle condition. The processes used to describe particle behaviour are entrainment, transport and impact. The transport process consists of two components, contact and non-contact motion, depending on the position of the particle with respect to the bed. The calculations performed are all non-dimensional, the parameters used to non-dimensionalize being the fluid density, ρ, the flow depth, h, and the mean bed shear velocity, u_*.

22.3.1 Entrainment process

The initial motion criterion used is based on the shear stress required to initiate particle motion compared with the instantaneous shear stress due to the turbulent flow. A range of values of shear stress is used, based on the variation observed by Fenton and Abbott (1977), and assuming that the Shields stress, θ_{cr}, for the initial motion of a coplanar particle is 0.06. A variation in shear stress for the initial motion is expected due to the position of the particle within the bed.

The instantaneous shear stress due to turbulent fluctuations of the flow is calculated from the assumption that the drag force per unit area near the bed is equivalent to the shear stress acting at the bed. Here the appropriate velocity for the calculations is assumed to be that at the particle centre.

The instantaneous shear stress acting at the bed, τ_0, can be calculated as

$$|\tau_0| = \frac{C_D}{2} u^2$$

$$= \frac{C_D}{2} U^2 \left[\left(1 + \frac{u'}{U} \right)^2 + \left(\frac{w'}{U} \right)^2 \right]$$

where C_D is the coefficient of drag of the particle.

From this the mean shear stress, $\bar{\tau}_0$, can be calculated as

$$\bar{\tau}_0 = \frac{C_D}{2} U^2 \overline{\left[\left(1 + \frac{u'}{U} \right)^2 + \left(\frac{w'}{U} \right)^2 \right]}$$

$$= \frac{C_D}{2} U^2 \left[1 + \left(1 + \frac{\sigma_u}{U} \right)^2 + \left(\frac{\sigma_w}{U} \right)^2 \right]$$

Thus, relating these expressions

$$|\tau_0| = \bar{\tau}_0 \frac{\left[\left(1 + \frac{u'}{U} \right)^2 + \left(\frac{w'}{U} \right)^2 \right]}{\left[1 + \left(1 + \frac{\sigma_u}{U} \right)^2 + \left(\frac{\sigma_w}{U} \right)^2 \right]}$$

To calculate whether the instantaneous shear stress is sufficient to initiate the motion of a particle, the above expression must be divided by the shear stress for the initiation of particle motion, τ_{0cr}, giving an equation

$$\frac{|\tau_0|}{\tau_{0cr}} = \frac{\bar{\tau}_0}{\tau_{0cr}} \frac{\left[\left(1 + \frac{u'}{U} \right)^2 + \left(\frac{w'}{U} \right)^2 \right]}{\left[1 + \left(1 + \frac{\sigma_u}{U} \right)^2 + \left(\frac{\sigma_w}{U} \right)^2 \right]}$$

If this quantity is greater than unity, then particle motion is initiated.

22.3.2 Particle transport

Transport of particles by the flow occurs in two modes: a contact mode, where the particle rolls along in contact with the rough bed, and a non-contact mode. In both modes fluid forces due to the relative motion of the particle and flow are calculated and are used to calculate the particle motion. The equations of particle motion are solved numerically for the duration of an iteration to give the particle position and velocity. The particle state at the end of one iteration is used as the initial condition for the next iteration, flow conditions being set for the new particle position.

The transport equations contain coefficients of drag, C_D, lift, C_L, and added mass, C_A. The value of the coefficient of drag is calculated from the curve for an isolated sphere in steady motion, using a fit by Morsi and Alexander (1972), allowing the calculation of the coefficient of drag for a range of particle Reynolds number. This ignores any effects due to the particle motion not being steady state, or any variation due to the proximity of a boundary, though measurements by Coleman (1972) and Bagnold (1974) showed any variation in the coefficient of drag due to the presence of a boundary to be small. There is less information on the coefficient of lift, in particular about the variations of the coefficient of lift with the particle Reynolds number. In part this is due to the mechanisms generating forces normal to the direction of a flow. These are fluid shear and particle rotation, which can act at the same time for any particle Reynolds number. The presence of different mechanisms generating the lift force means that the measurement of one contribution to the lift force will often constrain the system in such a way that other contributions cannot be measured. The situation is further confused by the variation in lift observed approaching surfaces (Bagnold, 1974; Sumer, 1984). The range of values of the coefficient of lift obtained experimentally is partly due to the measurements being of different contributions to the lift force, and partly due to the use of different definitions for the coefficient of lift. The contributions due to different mechanisms acting to generate lift on a particle close to a surface in turbulent flow are hard to determine, as is the variation in lift moving away from the surface. This makes parameterization of these quantities difficult. Thus the effect of uncertainty in the estimated values of these parameters might usefully be examined. In the presence of a fluid the effective mass of an accelerating particle is increased by an amount called the added mass. This extra inertia is due to the fluid accelerated with the particle; for an isolated sphere this added mass is equal to the mass of fluid that would occupy half the volume of the sphere (Milne-Thompson, 1968), giving an added mass coefficient, C_A, of 0.5.

22.3.2.1 Contact mode

The calculation of particle movement while in contact with the bed is similar to the description of Francis (1973) and Sekine and Kikkawa (1988), though here solved for flow conditions on an iteration by iteration basis. The equations describe the motion of a sphere rolling over another sphere without slippage. The relative velocities and forces acting are as shown in Figure 22.1. The tangential components of the forces acting determine the rolling motion of the particle and the normal components determine whether the particle remains in contact with the bed. The equations of rolling motion are solved for each iteration after the initial motion and also after impact until the particle either loses contact with the bed or stops rolling. If the particle loses contact with the bed, the particle position and velocity are used as the initial condition for non-contact motion at the next iteration.

The angular velocity, ω, of the particle due to tangential components of force (gravity, drag and lift) is calculated. The ordinary differential equation for the particle angular velocity is solved numerically, giving values for both the angular velocity and the contact angle θ.

$$\frac{d\omega}{dt} = \frac{10}{7\frac{\pi d^3}{6}(\rho_s + C_A)(d + d_{char})} \left(-\frac{\pi d^3}{6}(\rho_s - 1)g\cos\theta \right.$$

$$+ \left\{ \frac{C_D}{2}\frac{\pi d^2}{4}(u - u_p)[(u - u_p)^2 + (w - w_p)^2]^{1/2} \right.$$

$$\left. - \frac{C_L}{2}\frac{\pi d^2}{4}(w - w_p)[(u - u_p)^2 + (w - w_p)^2]^{1/2} \right\}\sin\theta$$

$$+ \left\{ \frac{C_D}{2}\frac{\pi d^2}{4}(w - w_p)[(u - u_p)^2 + (w - w_p)^2]^{1/2} \right.$$

$$\left.\left. + \frac{C_L}{2}\frac{\pi d^2}{4}(u - u_p)[(u - u_p)^2 + (w - w_p)^2]^{1/2} \right\}\cos\theta \right)$$

where d is the diameter of the moving particle, ρ_s is the density of the sediment, g is the acceleration due to gravity and u_p and w_p are the horizontal and vertical components of particle velocity, respectively.

22.3.2.2 Non-contact mode

The non-contact mode solution is for high Reynolds number flow — that is, a flow where viscous forces are not important. It is a solution for inviscid, irrotational flow similar to that of Wiberg and Smith (1985), though with the pressure term removed as the magnitude of this component is small. The relative velocities and horizontal and vertical components of the forces acting are shown in Figure 22.2. The equations of motion of a particle are solved for horizontal and vertical motion due to forces of drag, lift and gravity.

Ordinary differential equations for the velocity components of a single particle can be written as

$$\frac{\pi d^3}{6}(\rho_s + C_A)\frac{du_p}{dt} = \frac{C_D}{2}\frac{\pi d^2}{4}(u - u_p)[(u - u_p)^2 + (w - w_p)^2]^{1/2}$$

$$- \frac{C_L}{2}\frac{\pi d^2}{4}(w - w_p)[(u - u_p)^2 + (w - w_p)^2]^{1/2}$$

$$\frac{\pi d^3}{6}(\rho_s + C_A)\frac{dW_p}{dt} = \frac{C_D}{2}\frac{\pi d^2}{4}(w - w_p)[(u - u_p)^2 + (w - w_p)^2]^{1/2}$$

$$+ \frac{C_L}{2}\frac{\pi d^2}{4}(u - u_p)[(u - u_p)^2 + (w - w_p)^2]^{1/2}$$

$$- \frac{\pi d^3}{6}(\rho_s - 1)g$$

22.3.3 Impact process

The impact process is modelled by conserving some fraction of the normal and tangential components of momentum at impact, as shown in Figure 22.3. Particles can impact during any iteration when a particle is in motion. The vertical particle position at the end of an iteration is checked to see whether the particle could have made contact with the bed. If contact could have occurred then an impact height is chosen from a uniform random distribution, between

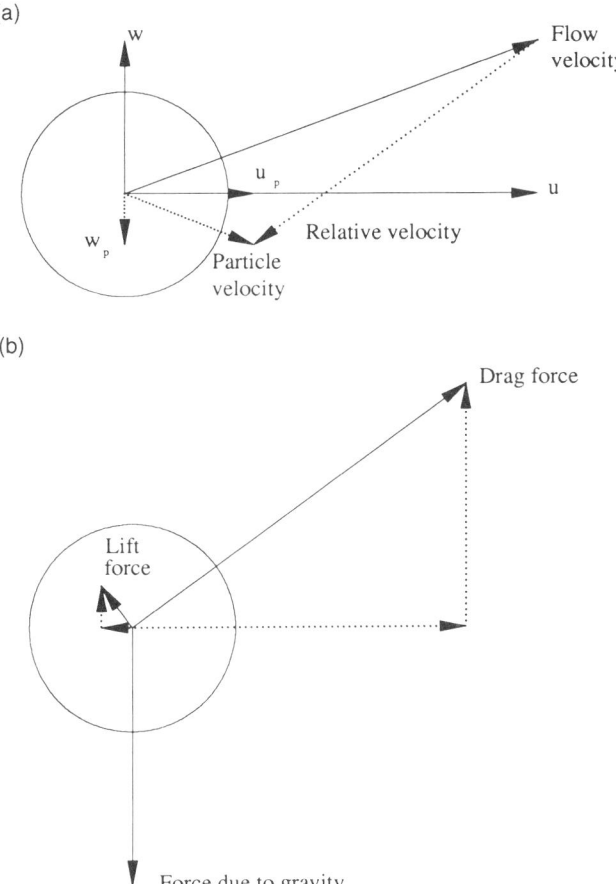

Figure 22.2 Non-contact motion of particle. (a) Flow and particle velocities determining relative velocity. (b) Forces due to flow and due to gravity. The horizontal and vertical components of each of these forces are shown indicating contributions to the horizontal and vertical forces acting on the particle

the zero velocity height and the maximum height at which the particle could have come into contact with the bed, this height specifying a point and hence a contact angle on the bed particle. The non-contact motion equations are solved for this height to give the streamwise position of the particle and its velocity, and the normal and tangential components of velocity are then calculated for the contact angle with a sphere of characteristic diameter.

The fraction of momentum conserved on impact is hard to determine and few data exist for the fluvial environment. The work of Gordon, Carmichael and Isackson (1972) can be interpreted as showing that the normal component of momentum is lost whereas the tangential momentum is conserved. Abbott and Francis (1977) found that trajectories appeared to be independent of the impact preceding them, although Naden (1987b) points out that as most of their observed data included a period of rolling between non-contact motions, this was only to be expected. In models of fluvial sediment transport including an impact process, Sekine and Kikkawa (1988) set a coefficient of restitution empirically, whereas Wiberg and Smith (1985) used the coefficients of restitution and friction as a single parameter to fit their

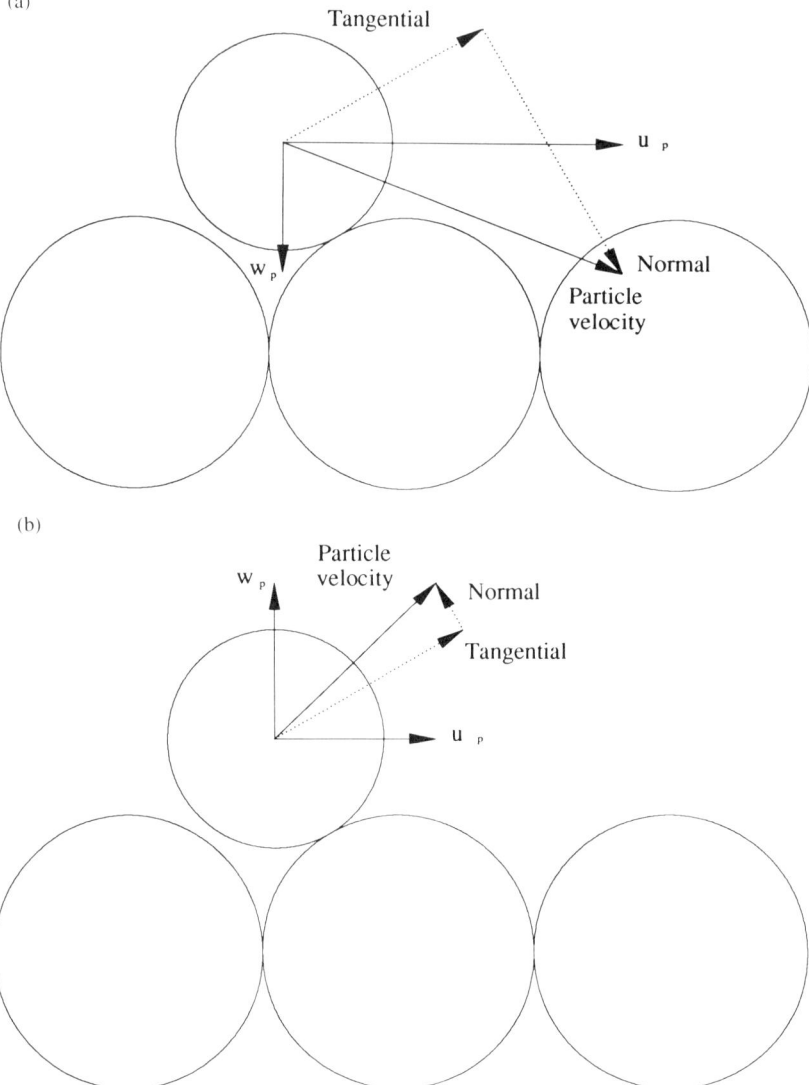

Figure 22.3 Effect of impact on particle velocities. (a) Particle velocity and components immediately before impact. (b) Particle velocity and components immediately after impact, with fractions of the normal and tangential components of velocity conserved

model to the available data. The fractions of tangential and normal momentum conserved at impact seem to be ideal parameters to examine in Monte Carlo simulations of particle trajectories, so that the effects of variations in their values can be assessed.

22.4   TWO-PHASE FLOW

In a particle tracking model of turbulent dispersion, a passive tracer, with similar physical properties to the fluid, is represented by a cloud of particles. The particles are moved according to the local flow conditions in a series of steps giving a Lagrangian model of the dispersion. The data necessary to calculate the dispersion are the magnitude of the turbulent fluctuations, the distribution of the fluctuations, and a description of the time and length scales of the turbulence. The fluctuation distributions are usually assumed to be Gaussian; the magnitudes of the fluctuations are then characterized by the standard deviation of the fluctuations. The time and length scales used are related to the Lagrangian integral time and length scales of the flow, these being the time and distance over which the velocity fluctuations remain correlated.

Sediment is a non-passive tracer and its behaviour differs from that of the fluid. When the fluid velocity experienced by a particle changes, a non-passive particle responds to this change over time. During this time the particle has a velocity relative to the flow and forces due to this velocity will be acting. The particle responses lag behind the behaviour of the fluid and the fluid leaves particles behind due to slippage of the particles: the 'crossing trajectories' effect of Csanady (1963). The time and length scales must then be modified to include the particle response. Approaches to this problem are either to use a modified time-scale (Csanady, 1963; also used by Hunt and Nalpanis, 1985; Anderson, 1987; Sawford and Guest, 1991), or to track the fluid and sediment particles until they diverge to a distance such that the correlation falls to zero (as used by Shuen et al., 1986; Zhuang, Wilson and Lozowski, 1989; Yverngniaux and Chollet, 1989). The latter approach is used here.

The time taken by a particle to respond to fluctuations in a flow can be characterized by a particle response time, t_r, which Hinze (1972) takes to be time for the particle relative velocity to fall to 50% of its initial value. At low particle Reynolds numbers, i.e. Stokes flow

$$t_r \propto d^2$$

whereas at high particle Reynolds numbers

$$t_r \propto \frac{d}{(u-u_p)^2 + (w-w_p)^2}$$

The high Reynolds number form of the particle response time is appropriate for the particles being modelled. The presence of the relative velocity term in the high Reynolds number response time makes the prediction of response times in advance difficult. Comparison of this response time with the time-scale of turbulent fluctuations indicates whether a size of particle will respond to the fluctuations of the flow.

The particle trajectories are calculated assuming that the fluctuating components of velocity exist on average for one eddy cycle (Sullivan, 1971). A time interval less than the time-scale of the eddy is used to calculate fluid and particle movements, allowing a constant time interval at each iteration while the eddy scale varies. When a particle enters an eddy the velocity of fluid coincident with the particle centre is calculated and the motion of the fluid and particle during the iteration are then calculated. At the end of each iteration the separation of the fluid and particle and the elapsed time in the eddy are compared with the eddy length and time-scales, respectively. If the lengths and times are less than these scales the flow conditions are retained; if not, a new set of velocities is calculated. As the scales with which comparisons are made are those for the eddy, then when they are exceeded no correlation remains with the previous values and the new conditions should be selected from an appropriate random distribution.

There are few Lagrangian data available from measurements in open channel flow. McQuivey and Keefer (1971) report some for surface dispersion and Sullivan (1971) some for particle dispersion through depth. Therefore the available Eulerian data may have to be used in a pseudo-Lagrangian way. Sullivan (1971) and Allen (1982) used a time scale, t_L, based on the assumption that a particle moved with velocity σ_w, the standard deviation of the vertical velocity fluctuations, over a distance L_E, the Eulerian integral length scale, which gave a pseudo-Lagrangian time-scale

$$t_L = \frac{L_E}{\sigma_w}$$

assuming that the fluctuating component of velocity exists on average for one eddy cycle.

The expression for the streamwise Eulerian integral length scale has been calculated from the data of McQuivey (1973) for flow over rough beds in a flume.

$$L_{Ex} = 2.676 \exp\left(-5.020 \frac{\sigma_u}{U}\right) \quad r = -0.744; \, n = 60$$

Published data giving information about integral length scales in open channel flow are very limited. The data of McQuivey (1973) only includes values for the streamwise Eulerian length scale; no values for the vertical length scale are given. The vertical Eulerian length scales were calculated as a fraction of the streamwise length scales. Owing to the limited amount of data of this type available for open channel flow, the data used were from turbulence measurements made on the River Severn (Heslop et al., 1994), with L_{Ez}, being set to 50% of the calculated horizontal Eulerian integral length scale, L_{Ex}.

As the model is to be run many times for the same conditions while calculating the effect of varying parameters, the suitability of the time interval and the influence of turbulence on particle movement can be checked by outputting the time-scales, the time spent in eddies and the particle responses to the flow at each iteration. The time-scales and time spent in eddies show whether the time interval used is always less than the eddy time-scale, whereas the particle responses indicate whether the inclusion of turbulence in the model is worthwhile.

22.5   USE OF MODELS

The paths of single particles were calculated using the model to produce examples of single particle trajectories and to find an appropriate time interval to use in the calculations. The non-dimensional time interval was set to a value of 0.01; the suitability of this value was checked by comparison with calculated time-scales for the different transport stages.

To test the effects of uncertainty on the lift and momentum parameters the single particle model was incorporated into a parallel harness, enabling Monte Carlo simulations to be run on the Meiko Computing Surface at Lancaster University. In the Monte Carlo simulations calculations were performed for the same sets of conditions with parameters randomly selected from ranges of values.

The lift parameters varied were the coefficient of lift and a value, n, used as an exponent to alter the variation of the coefficient of lift away from the bed, as in the expression

$$C_L = C_{L0}\left(\frac{0.5}{z/d}\right)^n$$

where C_{L0} is the coefficient of lift at a reference height at the stream bed. The momentum parameters varied were the coefficients of friction and restitution. The ranges over which the parameters were varied are shown in Table 22.1.

Table 22.1 Ranges of parameters used in uncertainty calculations

Coefficient of lift	0.2–0.6
n	1.0–5.0
Coefficient of friction	0.0–1.0
Coefficient of restitution	0.0–1.0

The model was run with 5000 sets of these varying parameters for each set of conditions used, each parameter set being used for 10 single particle runs. The values of the variable parameters were selected at random from the ranges shown in Table 22.1.

22.5.1 Data necessary to set model conditions

The data necessary to run the model are flow depth, h, bed roughness length, k_s, and an indication of the size of the particle relative to this quantity, bed shear velocity and particle density. These values are input as a flow Reynolds number, Re_{*h}, roughness Reynolds number, Re_{*ks}, non-dimensional specific particle weight, $g(\rho_s - \rho)h/\rho u_*^2$, and particle relative density, ρ_s/ρ. The other information required is the maximum number of iterations to perform and the time interval used for these iterations, also input in non-dimensional form.

22.5.2 Data used

The model has been compared with the data of Abbott and Francis (1977) for the movement of single particles over a fixed, flat rough surface in a flume. These data provide information on the saltation characteristics, both saltating and partially suspended (by their definition), and data on particle velocity and the percentage of time spent in different modes of motion: rolling, saltating and suspended. These are plotted for a range of values of transport stage (u_*/u_{*cr}) for a known depth over a flat rough bed (Figure 12 of Abbott and Francis, 1977). The conditions for the different transport stages were set with a critical shear velocity, u_{*cr}, calculated for a Shields stress of 0.06; calculations were performed for transport stages of 1.5, 2, 2.5 and 3.

22.6 SUITABILITY OF TIME INTERVAL

The suitability of the constant time interval used at each iteration was checked by comparing it with the calculated vertical and horizontal time-scales and the duration of the eddies. The

Table 22.2 Vertical Eulerian and pseudo-Lagrangian time-scales

Transport stage	Vertical Eulerian time-scale		Vertical pseudo-Lagrangian time-scale	
(u_*/u_{*cr})	Minimum	Mean	Minimum	Mean
1.5	0.0135	0.127	0.0265	0.212
2.0	0.0135	0.027	0.0277	0.288
2.5	0.0137	0.028	0.0237	0.290
3.0	0.0135	0.301	0.0242	0.514

calculated Eulerian and pseudo-Lagrangian time-scales in the vertical direction never dropped below the constant value of time interval used (Table 22.2). The values used were therefore reasonable, if computationally expensive.

22.6.1 Qualitative model behaviour

The qualitative prediction of particle behaviour made by the model is similar to laboratory and field observations. The calculated particle tracks show the effect of moving from one eddy to another, as seen in Figure 22.4 (see also Figures 13 and 16 of Abbott and Francis, 1977). The entrainment and deposition behaviour of the particles show the possibility of particles remaining immobile for some time before a sufficiently large turbulent fluctuation occurs to initiate particle motion, travelling for a variable number of hops before coming to rest and then either repeating these processes or coming to rest permanently.

22.6.2 Presentation of results

The variables output for each set of parameters are those shown in Figure 12 of Abbott and Francis (1977) — that is, saltation and suspension lengths and the mean maximum heights, mean particle velocities and the percentage of time that a particle spends rolling, at rest and in suspension. The mean particle velocity for a track was calculated as the distance travelled by a particle divided by the time spent travelling. The trajectory lengths and heights were calculated from the particle tracks, the distance between contacts with the bed, and the highest point between these contants. The split between saltating and partially suspended trajectories was calculated by checking for the occurrence of any upward accelerations in the particle trajectory between contacts with the bed.

To compare the observed and calculated results the normalized square of the residual was calculated, then subtracted from one — that is

$$\text{fit} = 1 - \left(\frac{y - y_c)}{y} \right)^2$$

where y is the observed value of a quantity and y_c is the mean of the calculated values for a given set of conditions for the same quantity. This expression gives a value of unity for a perfect fit and can be used to rank values in order of closeness to the observed value.

The results for the 5000 sets of parameters calculated for each transport stage were sorted using the sum of the values for the fit of the mean particle velocity, saltation and suspension length, divided by three. The number of variables was limited to three so that poor results for one variable were not averaged out by the other variables used. The lengths and heights

Table 22.3 Correlations of mean maximum saltation heights with saltation lengths

Transport stage $(u_*/u_{*\text{cr}})$	Regression slope	Regression constant	Correlation coefficient
1.5	0.786	0.144	0.869
2.0	0.719	0.155	0.917
2.5	0.596	0.159	0.966
3.0	—	—	—

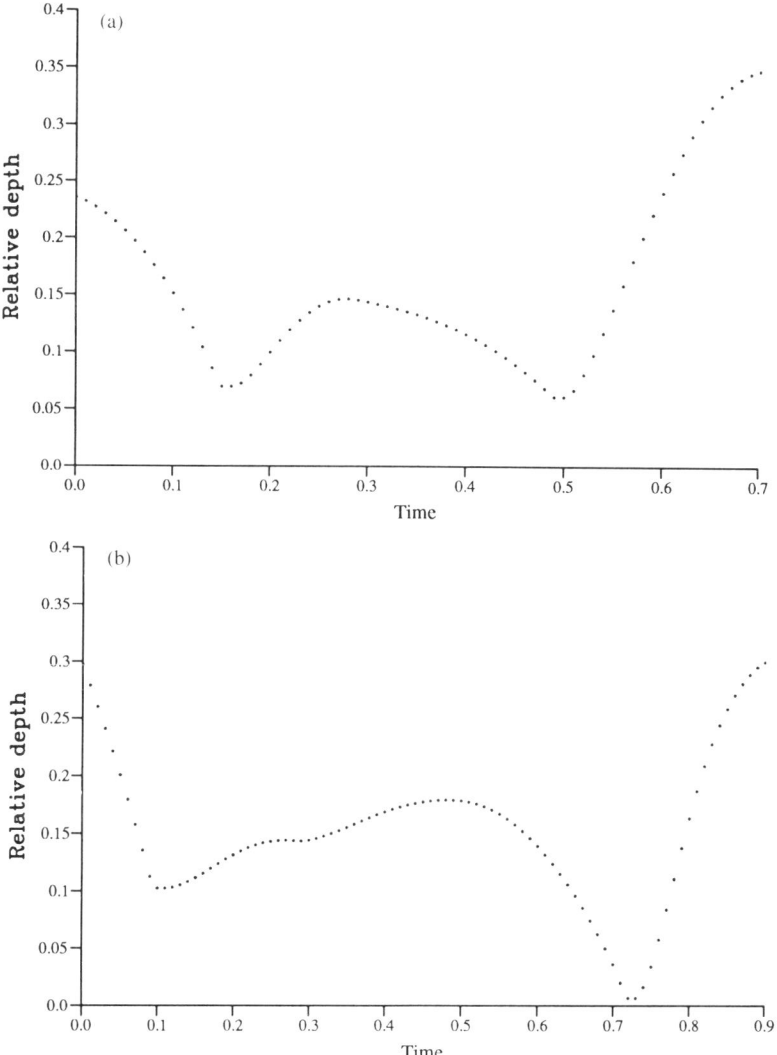

Figure 22.4 Saltations modified by the effects of turbulence. Calculated particle heights at each iteration are shown plotted against time for transport stages of (a) 2.0 and (b) 2.5. Units are non-dimensionalized with respect to flow depth and mean bed shear velocity

Table 22.4 Correlations of mean maximum trajectory heights with suspended trajectory lengths

Transport stage (u_*/u_{*cr})	Regression slope	Regression constant	Correlation coefficient
1.5	1.209	0.030	0.978
2.0	0.970	0.034	0.985
2.5	0.868	0.037	0.989
3.0	0.750	0.078	0.988

of trajectories are positively correlated over the whole parameter set, for each transport stage, so using these variables in the overall fit would give no useful increase in information (see Tables 22.3 and 22.4).

22.6.3 Effects of uncertainty in parameters

The parameters varied to check the effect of uncertainty on their values were those affecting the lift force acting on a particle and those affecting momentum conservation on impact.

The effect of uncertainty on the coefficient of lift and the variation in lift away from the bed is shown in Figure 22.5. The region of best fit forms a diagnonal with a positive slope. In this region increases in the coefficient of lift and hence lift force are matched by decreases in the persistence of lift force away from the bed. In the region above the diagonal particle motions are too short, low and slow, whereas below the diagonal the particle motions are too long, high and fast. The observed behaviour can be simulated by a range of coefficients of lift and different curves for the variation in lift away from the bed.

The variation of the coefficients of friction and restitution does not give rise to any structure across the range of parameter values used. From their observations Abbott and Francis (1977) found the initial particle velocity at the start of a trajectory to be independent of any prior rolling and there was little difference between the maximum trajectory heights from rest and those after an impact. They therefore asserted that momentum was not conserved on impact.

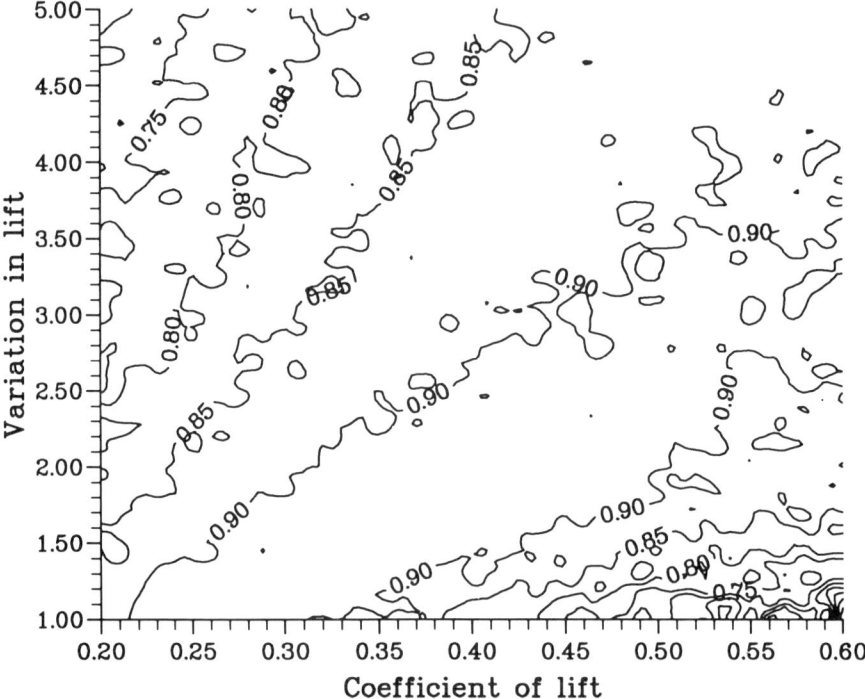

Figure 22.5 Effect of uncertainty in the values of the coefficient of lift on calculated fits, transport stage 3. The values of the coefficients of friction and restitution are also varying, selected from uniform random distributions

This work was reviewed by Naden (1985), who points out that the existence of the rolling and saltating motion does not eliminate the possibility of momentum being conserved and that the view of Abbott and Francis (1977) was based on rather sparse data. The model used here allows momentum to be conserved on impact but assumes that a period of rolling, however short, occurs after impact. Momentum conservation may therefore be important in the continuation or otherwise of particle motion, but the period of rolling appears to decouple the impact at the end of one saltation from the trajectory of the next saltation.

22.6.4 Model output

The fit of the calculated values from the model to the observed data as expressed by the sum of fits described earlier shows a trend of better fits with increasing transport stage, the better fits occurring at lower values of the coefficient of lift and higher values of n — that is, lower values of the coefficient of lift away from the wall (see Figure 22.6). The plots in Figure 22.6 only show a variation in the goodness of fit with the lift parameters, as only with these

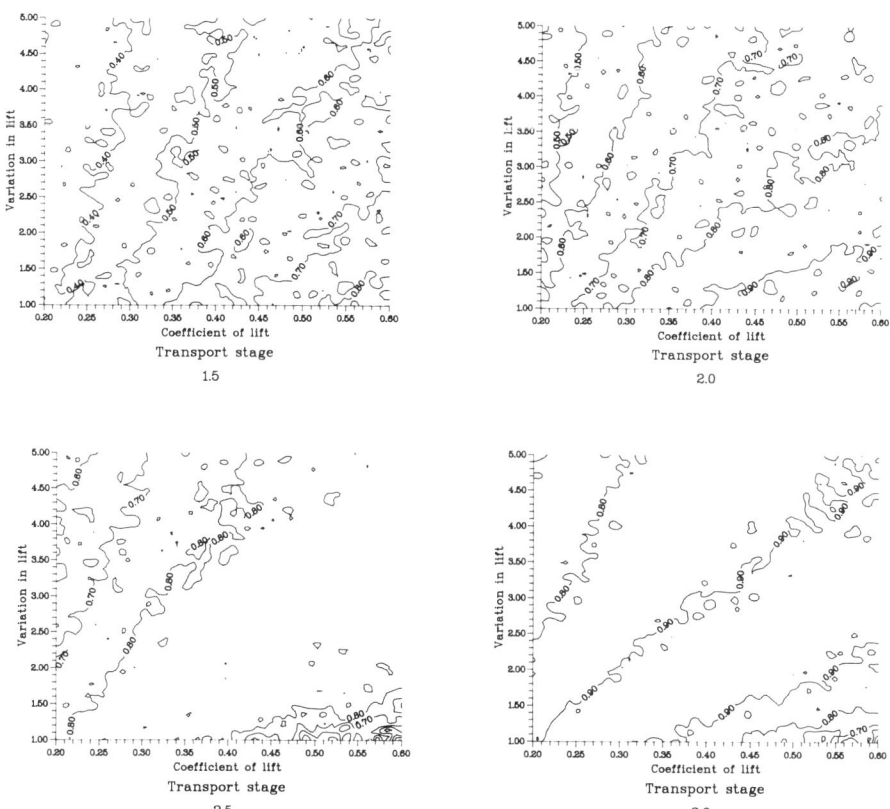

Figure 22.6 Variation in calculated fit of model to observed behaviour due to uncertainty in lift parameters used in model. The graphs show the calculated fits at transport stages 1.5—3 for the range of values of variation in lift and coefficient of lift shown in Table 22.1. The values of the coefficients of friction and restitution are also varying, selected from uniform random distributions

is there any structured variation. The coefficients of friction and restitution were also varied in these calculations, but caused no structure in the goodness of fit.

When the components of fit are examined, it is found that the predicted particle velocity is always lower than the observed, whereas the length scales show a range of fits from under- to overestimation. The regions of best fit for each parameter do not overlay each other completely, which accounts for some of the lack of overall fit for the model, but the main limitation on the accuracy of the model fit appears to be due to the poor fit to the mean particle velocity. Examining the interaction of this variable with the parameter sets and other calculated variables gives an indication of why the calculated mean particle velocity is always low.

The variation of particle velocity with the lift parameters shows an increasing accuracy of fit with increasing coefficient of lift, and with these higher values of coefficient of lift persisting away from the bed. The calculated particle velocity shows a positive correlation with the saltation and suspended trajectory length scales, for the best 500 fits, though the correlations are poor (see Tables 22.5 and 22.6). The other comparisons that may be made are of the time a particle spends in the different modes of transport.

The time spent rolling and hence the time spent near to the bed is overestimated at all transport stages. The distance travelled while rolling will be less than the distance that would be travelled if a particle was saltating or partially suspended; this therefore reduces the mean particle velocity. The variation of time spent rolling with particle velocity shows a trend of increasing particle velocity with decreasing time spent rolling at all the calculated transport stages, though the scatter of points below this line shows that other factors also affect the mean particle velocity. The time spent in modified saltation shows the reverse trend of increasing particle velocity with increasing time spent in modified saltation, though again the scatter of points shows that other factors are affecting the mean particle velocity.

The overestimation of time spent rolling at all transport stages could be due to the near-bed velcoity modelling, or the rolling model, or some combination of both. The improved fit of the model with increasing stage can at least partially be explained by the reduced time

Table 22.5 Correlations of saltation lengths with mean particle velocities for best 500 fits

Transport stage $(u_*/u_{*\mathrm{cr}})$	Regression slope	Regression constant	Correlation coefficient
1.5	0.228	0.451	0.064
2.0	1.112	0.121	0.314
2.5	1.454	0.261	0.368
3.0	—	—	—

Table 22.6 Correlations of suspended trajectory lengths with mean particle velocities for best 500 fits

Transport stage $(u_*/u_{*\mathrm{cr}})$	Regression slope	Regression constant	Correlation coefficient
1.5	0.685	0.085	0.253
2.0	1.013	0.096	0.315
2.5	1.095	0.252	0.344
3.0	1.846	−0.067	0.575

spent in contact with and close to the bed compared with the lower transport stages. The alteration in the fit with the variation in the uncertain parameters can also be seen to be due to decreased time in contact with the bed. Here this is caused by the lift force being large enough to reduce the time a particle spends in contact with the bed.

The rolling model used is very simple: sphere rolling over sphere with no slippage. However, the inclusion of a percentage slip factor in the calculation of particle rolling had little effect on the overall results, though the conditions under which particles lose contact with the bed could also affect the time spent close to the bed.

The near-bed velocity has been modelled using a logarithmic flow profile to calculate the mean streamwise velocities. This profile breaks down close to the boundaries, but to allow the calculation of forces on particles set into the surface of the bed the profile is here assumed to apply down to the zero velocity height. The use of a flow model which correlates the velocity fluctuations to give a degree of structure to the turbulence improves the fit of the model, more so at low transport stages than at high transport stages. At higher transport stages fluctuations in shear stress about the mean value need not be as large for the flow to become competent to cause sediment transport; fluctuations are therefore not as important in starting the transport process. The structure in the turbulence is only represented by different values of the correlation coefficient, assuming constant values for the standard deviations of the velocity fluctuations. Any systematic differences in these values during the high shear stress events is therefore ignored. The distributions used for horizontal and vertical fluctuations are still Gaussian, so the shear stress contributions due to a few large events or lots of small events will not be modelled adequately. As the model is failing due to particles spending too much time near the bed, the inclusion of such effects, causing the particle to be displaced form the region, might well improve the model further.

The low particle velocities are likely to be a result of a combination of the effects of these processes. The equations used to describe non-contact particle motion are simplified with terms that only made small contributions to the force acting on a particle not being included. They give reasonable predictions of the particle motion away from the bed, as can be seen from the predictions at higher transport stages. The accuracy of the test of the model may be examined by consideration of the results for a transport stage of 3. At this stage the observed time spent rolling was 4%, so the distance travelled in this mode of motion would also be expected to be small compared with the distances travelled in saltation and partially suspended saltation. If the mean particle velocity is modified by using only the observed time spent rolling and the calculated time spent in the other modes of motion this gives better fits, as shown by comparing Figure 22.7 with Figure 22.5. A similar calculation may be performed for the lower transport stages which does improve the calculated fit to the mean particle velocity. At lower transport stages the time spent rolling increases and so the distance travelled in this mode of transport becomes more significant; the fit remains worse.

22.7 DISCUSSION

The present model predicts the behaviour of single particles across a range of transport stages, though the quantitative results are not accurate at low transport stages and improve as the transport stage increases. The fit to the saltation characteristics is much better than to the mean particle velocities. The variation of goodness of fit with the coefficient of lift in these calculations shows that particle behaviour can be effectively reproduced by a range of values. Calculation of the behaviour of single particles allows comparison with available data and

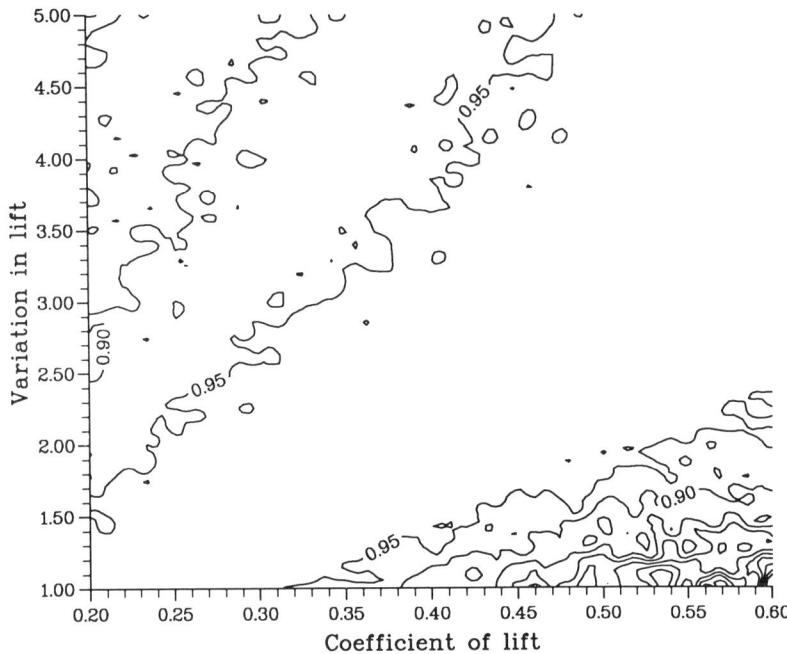

Figure 22.7 Effect of uncertainty in the values of coefficient of lift with the fit modified to account for the excess time spent rolling; transport stage 3. Observed time spent rolling at this transport stage is 4%; fit is modified assuming that the distance travelled while rolling is small compared with that saltating, and that the mean particle velocity can be altered by scaling the time in motion. The values of the coefficients of friction and restitution are also varying, selected from uniform random distributions

the results obtained may then be used in other calculations. The single particle models of van Rijn (1984) and Wiberg and Smith (1989) have been used to calculate total bedload transport, based on the profiles of sediment distribution through depth obtained from the models. Another use of particle models of sediment transport is in calculating the development of bedforms, as in Naden (1987b).

The ability to run a model many times in parallel can be used to perform multiple calculations across a range of parameters, investigating the effect of uncertainty on these parameters, as has been described here. Alternatively, the parameters may be fixed and multiple calculations performed for a single set of conditions. These multiple calculations allow the production of distributions in space and time for the motion of a single particle, as shown in Figure 22.8. These calculations may be repeated for a range of combinations of particle size, bed roughness and transport stage. In this way the distributions for a range of particles representative of a sediment distribution over a related range of bed roughnesses may be calculated. The calculation of sediment transport by direct calculation of the behaviour of a sufficient number of particles to represent a system would require large amounts of computer time, which would not be available with our present facilities. Pre-calculation of the distribution functions for a range of particles, and for a range of conditions allows sediment movement to be calculated from these distributions, removing the restriction on the number of particles whose behaviour can be calculated.

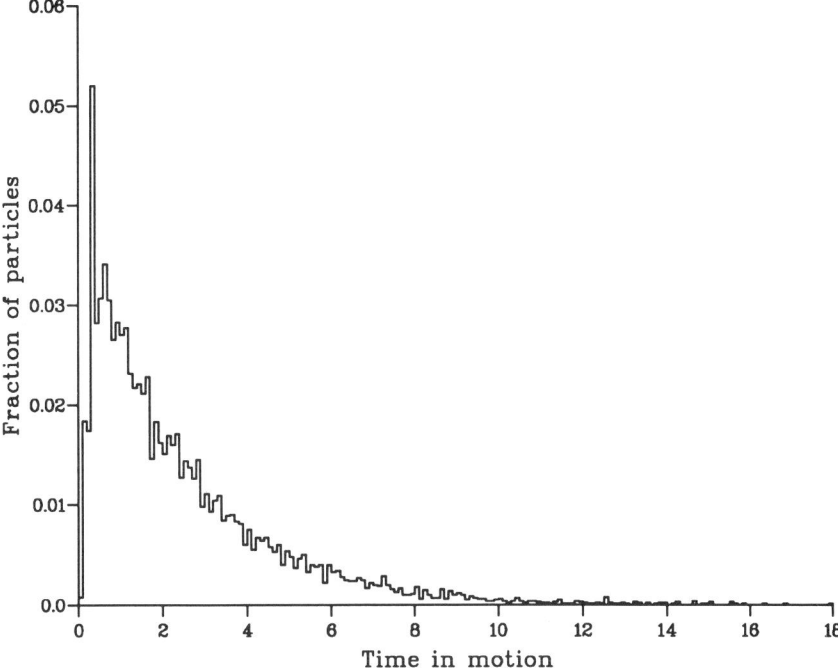

Figure 22.8 Calculated distribution of time particles spend in motion. The distributions are generated from calculations of the movement of 10 000 particles at transport stage 2. Each particle track is calculated for a particle initially at rest; the calculation is continued until the particle stops moving, giving values for single particle motions. The distribution only includes those particles that move from the initial rest position

The calculation of distribution curves has similarities with the stochastic approaches to calculating sediment transport. In this approach the distributions are calculated from the dispersion of sediment tracers in flumes or rivers (Kirkby, 1991), but there are problems in the retrieval of the tracer and in identifying the effects of the interactions of bedforms with the sediment transport process and hence in the number of particles necesary to identify distributions (see, for example, Hassan, Church and Schick, 1991). In comparison the modelled system is a simplification of the particle transport processes in its representation of the bed as a flat rough surface with no bedforms present. The distributions produced are not distorted by any interactions with bedforms and the final position of all particles is calculated. The assumptions made make the initial calculation of sediment distributions simple, but mean that the effects of any interactions with the bed when routing sediment may be superimposed later. This distribution function approach to modelling the total bedload transport is being explored in further studies of the particle tracking model.

ACKNOWLEDGEMENT

A.K. was in receipt of a NERC research studentship (GT4/89/AAPS/30) during this work.

REFERENCES

Abbott, J.E. and Francis, J.R.D. (1977) Saltation and suspension of solid grains in a water stream. *Phil. Trans.R. Soc. London* **284A**, 225−254.

Allen, C.M. (1982) Numerical simulation of contaminant dispersion in estuary flow. *Proc. R. Soc. London* **A381**, 179−194.

Anderson, R.S. (1987) Eolian sediment transport as a stochastic process: the effects of a fluctuating wind on particle trajectories. *J. Geol.* **95**, 497−512.

Anderson, R.S. (1989) Saltation of sand: a qualitative review with biological analogy. *Proc. R. Soc. Edinburgh* **96B**, 149−165.

Bagnold, R.A. (1973) The nature of saltation and of 'Bed load' transport in water. *Proc. R. Soc. London* **A332**, 473−504.

Bagnold, R.A. (1974) Fluid forces on a body in shear flow; experimental use of 'stationary flow'. *Proc. R. Soc. London* **A340**, 147−171.

Coleman, N.L. (1972) The drag coefficient of a stationary sphere on a boundary of similar spheres. *La Houille blanche* **27**, 17−21.

Csanady, G.T. (1963) Turbulent diffusion of heavy particles in the atmosphere. *J. Atmos. Sci.* **20**, 201−208.

Drake, T.G., Shreve, R.L., Dietrich, W.E., Whiting, P.J. and Leopold, L.B. (1988) Bedload transport of fine gravel observed by motion-picture photograph. *J. Fluid Mech.* **192**, 193−217.

Einstein, H.A. (1937) Bed load transport as a probability problem. *DSc Thesis.* Federal Institute of Technology, Zurich. Reprinted in: *Sedimentation — Symposium in Honour of Professor H.A. Einstein* (Ed. H.W. Chen), Fort Collins Co., H.W. Shen, C1−C105.

Fenton, J.D. and Abbott, J.E. (1977) Initial movement of grains on a stream bed: the effect of relative protrusion. *Proc. R. Soc. London* **A352**, 523−537.

Francis, J.R.D. (1973) Experiments on the motion of solitary grains along the bed of a water stream. *Proc. R. Soc. London* **A322**, 443−471.

Gordon, R., Carmichael, J.B. and Isackson, F.J. (1972) Saltation of plastic balls in a 'one-dimensional flume'. *Wat. Resourc. Res.* **8**, 444−459.

Hassan, M.A., Church, M. and Schick, A.P (1991) Distance of movement of coarse particles in gravel bed streams. *Wat. Resour. Res.* **27**, 503−511.

Heathershaw, A.D. (1979) The turbulent structure of the bottom boundary layer in a tidal current. *Geophys. J. Roy. Astron. Soc.* **58**, 395−430.

Heslop, S., Holland, M.J. and Allen, C.M. (1994) Turbulence measurements in the River Severn. In: *Mixing and Transport in the Environment* (Eds. K.J. Beven, P.C. Chatwin and J.H. Millbank). Wiley, Chichester, Ch. 3.

Hinze, J.O. (1972) Turbulent fluid and particle interaction. *Prog. Heat Mass Transfer* **6**, 433−452.

Holly, F.M. Jr and Rahuel, J.L. (1990) New numerical/physical framework for mobile bed modelling. Part I: numerical and physical principles. *J. Hydr. Res.* **28**, 401−416.

Hunt, J.C.R. and Nalpanis, P. (1985) Saltating and suspended particles over flat and sloping surfaces. I. Modelling concepts. In: *Proceedings of the International Workshop on the PHysics of Blown Sand* (Eds O.E. Barndorff-Nielsen *et al.*). Vol. 1. University of Aarhus, 9−36.

Kirkby, M.J. (1991) Sediment travel distance as an experimental and model variable in particulate movement. *Catena Suppl.* **19**, 111−128.

Leeder, M.R. (1983) On the interactions between turbulent flow, sediment transport and bedform mechanics in channelized flows. In: *Modern and Ancient Fluvial System* (Eds J.D. Collinson and J. Lewin). *Assoc. Sedimentol. spec. Publi.* **6**, 5−18.

McQuivey, R.S. (1973) Summary of turbulence data from rivers, conveyance channels and laboratory flunes. *turbulence in Water.* US *Geol. Sur. Prof. Pap.* 802-B.

McQuivey, R.S. and Keefer, T.N. (1971) Turbulent diffusion and dispersion in open channel flow. In: *Stochastic Hydraulics* (Ed. C.-L. Chiu). University of Pittsburgh, 231−250.

Milne-Thomson, L.M. (1968) *Theoretical Hydrodynamics.* 5th edn. Macmillan, London.

Morsi, S.A. and Alexander, A.J. (1972) An investigation of particle trajectories in two phase flow systems. *J. Fluid Mech.* **55**, 193−298.

Naden, P.S. (1985) Gravel bedforms — the development of a sediment transport model. *PhD Thesis.* University of Leeds.

Naden, P.S. (1987a) An erosion criterion for gravel bed rivers. *Earth Surf. Processes Landforms* **12**, 83–93.

Naden, P.S. (1987b) Modelling gravel-bed topograhpy from sediment transport. *Earth Surf. Processes Landforms* **12**, 353–367.

Sawford, B.L. and Guest, F.M. (1991) Lagrangian statistical simulation of the turbulent motion of heavy particles. *Boundary Layer Meterorol.* **54**, 147–166.

Sayre, W.W. and Hubbell, W.W. (1965) *Transport and Dispersion of Labelled Bed Material North Loup River, Nebraska. US Geol. Sur. Prof. Pap. 433-C.*

Sekine, M. and Kikkawa, H. (1988) A fundamental study on the sediment transport in an open channel flow. *Mem. School Sci. Engin. Wasdea Univ. No. 52.*

Shen, H.W. and Todorovic, P. (1971) A general stochastic model for the transport of sediment bed material. In: *Stochastic Hydraulics* (Ed. C.-L. Chiu). University of Pittsburgh, 489–503.

Shuen, J.-S., Solomon, A.S.P. and Faeth, G.M. (1986) Drop-turbulence interactions in a diffusion flame. *AIAA.* **24**, 101–108.

Snyder, W.H. and Lumley, J.L. (1971) Some measurements of particle velocity autocorrelation functions in a turbulent flow. *J. Fluid Mech.* **48**, 41–71.

Stelczer, K. (1981) *Bed-load Transport, Theory and Practice.* Water Resources Publications, Littleton Co.

Sumer, B.M. (1984) Lift forces on moving particles near boundaries. *J. Hydr. Engin.* **110**, 1272–1278.

Sumer, B.M. and Deigaard, R. (1981) Particle motions near the bottom in turbulent flow in an open channel. Part 2. *J. Fluid Mech.* **109**, 311–338.

Sullivan, P.J. (1971) Longitudinal dispersion within a two-dimensional shear flow. *J. Fluid Mech.* **49**, 551–576.

van Rijn, L.C. (1984) Sediment transport, Part I: bed load transport. *J. Hydr. Engin.* **110**, 1431–1456.

Wiberg, P.L. and Smith, J.D. (1985) A theoretical model for saltating grains in water. *J. Geophs. Res.* **90**, 7341–7354.

Wiberg, P.L. and Smith, J.D. (1989) Model for calculating bed load transport of sediment. *J. Hydr. Engin.* **115**, 101–123.

Williams, J.J. (1990). Video observations of marine gravel transport. *GeoMar. Lett.* **10**, 157–164.

Yvergniaux, P. and Chollet, J.P. (1989) Particle trajectories modelling based on a Lagrangian memory effect. In: *IAHR XXIII Congress: Hydraulics and the Environment, A: Turbulence in Hydraulics.* The National Research Council Canada, Ottowa, Canada, 301–313.

Zhuang, Y, Wilson, J.D. and Lozowski, E.P. (1989) A trajectory simulation model for heavy particle motion in turbulent flow. *J. Fluid Engin.* **111**, 492–494.

NOMENCLATURE

C_A Added mass coefficient

C_D Coefficient of drag

C_L Coefficient of lift

C_{L0} Coefficient of lift at reference height

d Particle diameter

d_{char} Characteristic particle diameter

g Acceleration due to gravity

h Flow depth

k_s Roughness length

L_{Ex} Horizontal Eulerian integral length scale

L_{Ez} Vertical Eulerian integral length scale

Re_{*h} Flow Reynolds number $= u_* h / v$

Re_{*ks} Particle Reynolds number $= u_* k_s / v$

t_L Eddy time scale

t_r Particle response time

U Mean streamwise component of velocity at height z

u Instantaneous streamwise component of velocity at height z

u' Fluctuating streamwise component of velocity at height z

u_* Mean bed shear velocity

u_{*cr} Critical shear velocity to initiate particle motion

u_p Streamwise particle velocity

W Mean vertical component of velocity at height z

$\hat{W}$ Conditional mean vertical component of velocity at height z at an iteration

w Instantaneous vertical component of velocity at height z

w' Fluctuating vertical component of velocity at height z

w_p Vertical particle velocity

y Observed value of quantity

y_c Calculated value of quantity

z Height above velocity zero

γ_s Specific weight of sediment $= [g(\rho_s - \rho)h]/\rho u_*^2$

θ Contact angle

θ_{cr} Critical Shields stress $= \rho u_{*cr}^2 / [g(\rho_s - \rho)d]$

ω Angular velocity of particle $= d\theta/dt$

ν Kinematic viscosity

ρ Density of fluid

ρ_s Density of sediment

σ_u Standard deviation of streamwise velocity fluctuations at height z

σ_w Standard deviation of vertical velocity fluctuations at height z

$\hat{\sigma}_w$ Correlated standard deviation of vertical velocity fluctuations at height z

τ_0 Shear stress

τ_{0cr} Critical shear stress to initiate particle motion

Index